Renewable Energy: Current Developments

Renewable Energy: Current Developments

Edited by David McCartney

SYRAWOOD
PUBLISHING HOUSE

New York

Published by Syrawood Publishing House,
750 Third Avenue, 9th Floor,
New York, NY 10017, USA
www.syrawoodpublishinghouse.com

Renewable Energy: Current Developments
Edited by David McCartney

International Standard Book Number: 978-1-68286-665-8 (Hardback)

Cataloging-in-Publication Data

Renewable energy : current developments / edited by David McCartney.
    p. cm.
Includes bibliographical references and index.
ISBN 978-1-68286-665-8
1. Renewable energy sources. 2. Power resources. I. McCartney, David.
TJ808 .R46 2019
333.794--dc23

TABLE OF CONTENTS

PREFACE

The energy derived from renewable sources such as sun, wind, rain, tides, etc. is known as renewable energy. These sources are preferred over conventional sources of petroleum and coal because they are naturally replenished in small timescales. They can be found in almost every geographical area unlike conventional resources. Some of the most widely used renewable energy technologies include wind power, hydropower, geothermal energy and solar power. They are primarily used for the purpose of electricity generation, apart from usage in the transportation industry. This book is a compilation of chapters, which discuss the most vital concepts and emerging trends in this field. It includes some of the most significant pieces of work being conducted across the world, on various topics related to renewable energy. Researchers and students actively engaged in this field will be assisted by this book.

Various studies have approached the subject by analyzing it with a single perspective, but the present book provides diverse methodologies and techniques to address this field. This book contains theories and applications needed for understanding the subject from different perspectives. The aim is to keep the readers informed about the progresses in the field; therefore, the contributions were carefully examined to compile novel researches by specialists from across the globe.

Indeed, the job of the editor is the most crucial and challenging in compiling all chapters into a single book. In the end, I would extend my sincere thanks to the chapter authors for their profound work. I am also thankful for the support provided by my family and colleagues during the compilation of this book.

Editor

Release Profile of Volatiles in Fluidised Bed Combustion of Biomass

Jaakko Saastamoinen*

VTT Technical Research Centre of Finland, Jyväskylä, Finland

Abstract

A simplified model for the release profile of volatiles in fluidised and circulating fluidised beds is presented. The location of the release of volatiles in the furnace depends on the interplay of flow of fuel particles affecting their location and heating of fuel particles affecting rates of drying and pyrolysis and consequently the mass of fuel particles. The coupled equations describing heating, de-volatilization including drying, elutriation and movement of particles are solved to find the release profile. Large heavy particles remain in the dense bed whereas light particles are entrained by the gas and elutriated from the bed to the riser. The situation is dynamic, since fuel particles lose weight due to release of moisture and volatiles and decrease in size due to shrinkage affecting their movement and escape from the lower dense bed to the riser. The model is based on the following the fate of a small batch of fuel particles undergoing de-volatilization and elutriation from the dense bed. The model predicts the location of the release of volatiles and the profile of the release of volatiles in the riser. Part of volatiles is released in the dense bed and the rest above the bed in the riser.

Keywords: Biomass fuel; De-volatilization

Introduction

There is an increasing interest for power production using biomass fuels. These fuels are considered as CO_2 neural, since CO_2 released during burning is again captured by growing plants. When applying CCS technology along with biomass combustion a net removal of CO_2 from atmosphere takes place [1]. Fluidised bed combustion has long been applied as a flexible means to burn biomass. Oxy-fuel circulating fluidized bed combustion technology has been under development [2]. The combustion is achieved in a mixture of oxygen and recirculated flue gases instead of air and the flue gases consist of carbon dioxide and water vapour allowing CO_2 capture. Especially the use of waste products from wood and food industry is beneficial, since no extra land area is required. Such fuels are plentiful available, but they are dispersed in wide area. There are two ways to utilize this biomass resource. It can be co-combusted in a large boiler plant using coal as the main fuel [3-6]. The other option is to burn the fuel in smaller plants producing electric energy. The efficiency for generation of electricity in smaller plants is lower, but in cold climate the waste heat can be utilized in the local heat production. The combined heat and power is widely applied in Finland using local wood and peat as auxiliary fuel. In hot climate the waste heat could be used for producing cooling by the absorption heat pump system. In fact, this technology is applied even in cold climate in Helsinki in Finland for district cooling of office buildings.

Fluidised bed combustion technology is advantageous for various fuels [3-8] since it is not always necessary to process the waste to more refined fuels but it can usually be burned directly. However, processing the fuel for example by pelletizing before combustion could be beneficial [9]. Drying of fuel with low temperature waste or solar heat is means to increase the availability of energy from biomass by increasing the heating value. Biomass fuels contain a large faction of volatile matter, which is even higher than the proximate value due to high heating rate in the furnace, which increases the part of fuel released as volatiles and reduces the amount of char. In this paper a simplified method to predict release profile of volatiles in fluidised bed (FB) and circulating fluidized bed (CFB) boilers is presented.

From boiler manufacturers' point of view it is important to know burning profiles for different fuels. This enables an optimal dimensioning and placement of heat exchange surfaces in a furnace. Volatile release from fuel is divided between the dense bed and the riser. The burning of the volatiles and heat release take place only after the volatiles are released from the particles. Combustion and gas concentration profiles affect also the temperature profile and emissions of CO, NO_x and SO_x in fluidized and circulating fluidized beds [10-12]. The knowledge of the profile of the release of volatiles gives means to optimize primary, secondary and tertiary air locations and rates. CO emissions can become high if much of the volatiles are released high above the dense bed leaving too short time for oxidation. This could be avoided for specific fuels by the design of the combustor (gas velocity) and by the processing of the biomass to optimal size or by pelletizing to suitable size and density. There are several studies [7,8,9,13-18] concerning combustion of high volatile fuel in fluidized or circulating fluidized beds. Circulation in CFB has been studied [19,20].

There is a wide literature concerning the computational methods for calculating the flow of particles in fluidized beds [21,22], but they require heavy computing and the accuracy is not so good due to the fact that flow phenomena are complex and biomass fuels are rather heterogeneous by size and shape [23]. A simplified method is developed here to predict volatile release profiles. The prediction is based on estimating the physical location and distribution of fuel particles and release of volatiles depending on particle size after feeding a fuel batch into the boiler. The final release profile is found by summing up the contributions from different times and from different size fractions.

Heating, Drying and De-volatilization of a Biomass Particle

The studies on de-volatilization times of coal particles in fluidized-beds have been reviewed [24]. Effects of moisture u, particle shape

---

***Corresponding author:** VTT Technical Research Centre of Finland, Box 1603, 40101 Jyväskylä, Finland, E-mail: jaakko.saastamoinen@vtt.fi (sphericity) φ_s, size d and ambient temperature on devolatilization and burning of wood particles have been studied [25,26]. Equations of the form

$$t_v = f(T)(d\varphi_s)^n(1+Cu) \qquad (1)$$

have been presented for the de-volatilization time of a wood particle. $f(T)$ is an experimental function and n and C are experimental constants. The knowledge of the total time of de-volatilization is not enough, when considering the location, where the volatiles are released. The de-volatilization as function of time can be presented by experimental correlation for example by suitable cumulative distribution. The Rosin-Rammler type

$$(v_0 - v)/v_0 = 1 - m/m_0 = X = 1 - \exp\{-[(t-t_d)/\tau_v]^n\} \qquad (2)$$

has been applied [27]. Pyrolysis does not start immediately after feeding to the boiler and here a time lag t_d due to drying and heating up (to about 473 K) is added to shift the zero point of the time scale. The parameters t_d, τ_v and n can be found from batch combustion or pyrolysis tests.

More fundamental calculations require the knowledge of the kinetics of pyrolysis. Several models for the kinetics of pyrolysis of biomass have been presented [28]. Kinetics data for different biomass

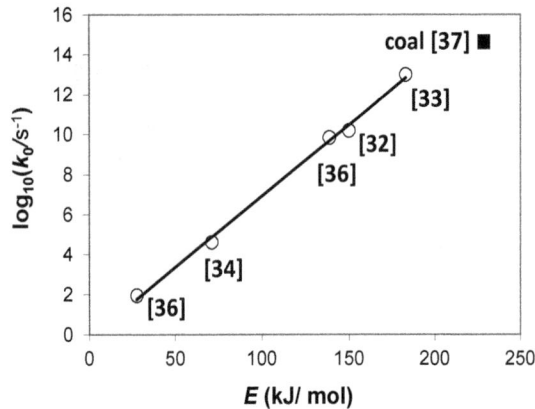

Figure 1: Autocorrelation between activation energy and frequency factor for biomass pyrolysis. Average value for some coals (Black Square for coal).

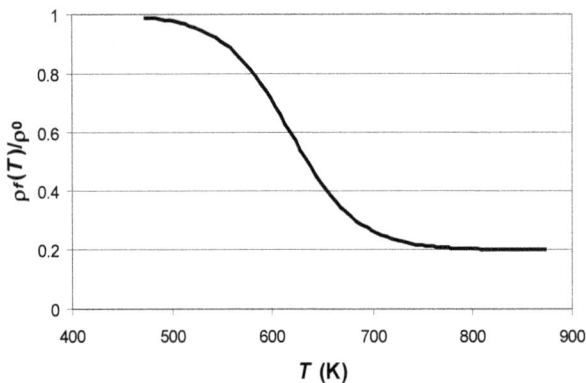

Figure 2: Temperature dependent final density of wood in pyrolysis in inert atmosphere.

has been presented [29]. In some models the biomass is divided into different compounds with different reaction kinetics. Some models developed for the kinetics of solid fuels can predict the composition of the gases. The single step Arrhenius-type reaction

$$-\frac{d\rho}{dt} = k(\rho - \rho_f) \qquad (3)$$

is generally applied to describe the rate of pyrolysis with a constant final density ρ_f and Arrhenius type rate coefficient $k = k_0 \exp(-E/RT_p)$. Even this model is frequently used; it is physically not valid, because ρ_f depends on some final temperature of the particle. The particle does not reach this temperature immediately after injection to the furnace, but pyrolysis takes place during the heat-up period to a great extent. During heat-up the particle cannot "predict" its future final temperature, which depends on the conditions of the ambient (which can even be changed with time). For coals a correlation has been presented [30,31] for the final density or yield of volatiles depending on final temperature. There is wide variation between the different values of pre-exponential coefficient and activation energy E measured by different researchers [32-36] with different heating rates and conditions (Figure 1). The higher activation energy for coal [37] shows that pyrolysis takes place at higher temperatures.

The distributed activation energy model accounts for reactions taking place on a broader temperature range for coals. Parameters for this model for wood have been reported [38]. The model of Kobayashi [39,40] for coal consists of two competing reactions. This model has the advantage that it can predict char yield depending on the heating rate.

A model, in which the driving force is the difference between the instantaneous density and the final density at that temperature, $\rho - \rho_f(T_p)$, instead of commonly applied difference between instantaneous density and the final density at some indefinite future temperature T_f, $\rho - \rho_f(T_p)$, better describes the real situation [41]. In this case the rate of pyrolysis is modelled by Eq. (1) but with temperature dependent final density. The temperature changes with time depending on the heating conditions. The temperature dependent final density [42] can also be approximated by a logistic function as

$$\rho_f(T)/\rho_0 = 1 - v_0/\{1 + \exp[-C(T_p - T_{ref})]\} \qquad (4)$$

shown in Figure 2 with values $C \approx 0.03/K$ and $T_{ref} \approx 616K$. The fraction released as volatiles is $v_0 \approx 0.75...0.8$ for wood at relative low heating rate. The amount of volatiles v_0 depends on the heat-up rate of the particle and with high heating rate v_0 is higher even 0.9-0.95 [43]. This heating rate depends on particle size and ambient temperature (gas, walls). The reaction rate coefficient for Eq. (3) with temperature dependent ρ_f has been reported [42] and its dependence on temperature is milder than that by the usual exponential Arrhenius rate equation, because the strong temperature dependency is now accounted for by the final density, which is not a constant but based on the prevailing instantaneous temperature.

Different models have been developed for de-volatilization of single particles (models with spatially uniform temperature and density, shrinking core models, models for particles with non-uniform temperature). In the model assuming uniform temperature and density the particle temperature is calculated by the equation

$$h_e S_p(T_e - T_p) = \rho_p c_p V_p \frac{dT_p}{dt} + l_p V_p \frac{d\rho_p}{dt} \qquad (4)$$

If the effect of heat of pyrolysis is assumed insignificant, Eq. (4) is becomes

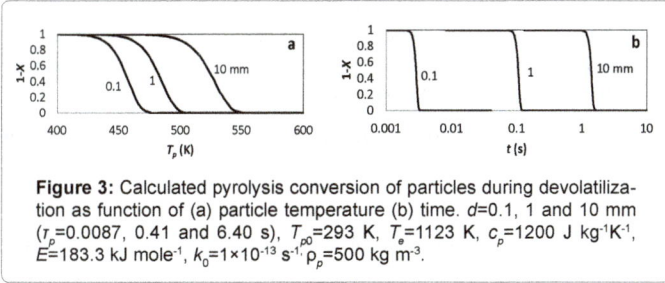

Figure 3: Calculated pyrolysis conversion of particles during devolatilization as function of (a) particle temperature (b) time. d=0.1, 1 and 10 mm (τ_p=0.0087, 0.41 and 6.40 s), T_{p0}=293 K, T_e=1123 K, c_p=1200 J kg^{-1}K^{-1}, E=183.3 kJ mole^{-1}, k_0=1×10^{-13} s^{-1}, ρ_p=500 kg m^{-3}.

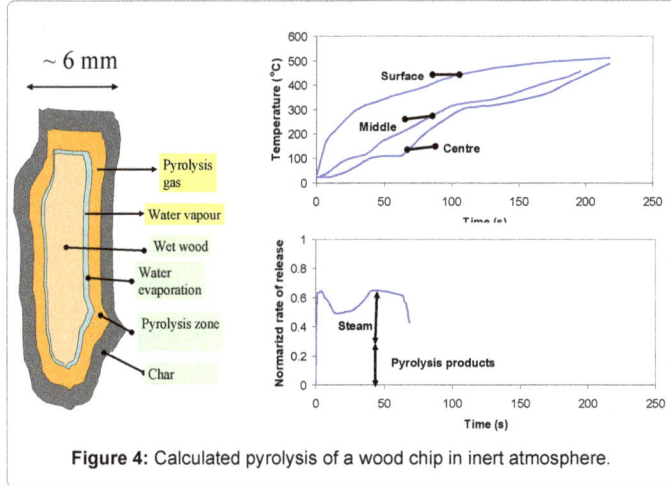

Figure 4: Calculated pyrolysis of a wood chip in inert atmosphere.

$$T_e - T_p = \tau_p \frac{dT_p}{dt} \tag{5}$$

If the time constant for particle heating $\tau_p = \rho_p c_p (d/2)/[h_e(1+\Gamma)]$ is assumed constant, the particle temperature T_p can be solved

$$T_p = T_f + (T_0 - T_f)\exp(-t/\tau_p) \tag{6}$$

In reality the time constant depends on temperature and particle density, which change with time. By eliminating time t Eqs. (1) and (5) lead to

$$\frac{d\rho}{\rho - \rho_f} = \frac{\tau_p}{T_e - T_p} e^{-E/RT} dT_p \tag{7}$$

This equation can be integrated [44]. The result in slightly different form is

$$1 - X = (\rho - \rho_f)(\rho_0 - \rho_f) = \exp[\tau_p k_0 f(T_p] \tag{8}$$

where

$$f(T) = E_1(1/\Theta_0) - E_1(1/\Theta_p) + [E_1(1/\Theta_p - 1/\Theta_e) - E_1(1/\Theta_0 - 1/\Theta_e)]\exp(-1/\Theta_e) \tag{9}$$

where $\Theta = RT/E$ and E_1 is exponential integral. This is illustrated in Figure 3.

In reality the particle density and specific heat capacity and heat transfer coefficient change with time. There may also be a flame around the particle increasing the effective temperature of the ambient [45-47]. Then a numerical method can be applied to calculate the particle temperature and mass as function of time [45]. In practice the temperature of the particle is also not uniform. The internal heat transfer resistance of the particle can approximately be taken into account by the coefficient φ_r [45,47] in the effective heat transfer coefficient $h_e = h(1 + 2B \phi_r)$. For complicated particle shapes it can

be found numerically [48]. The non-uniform temperature affects also on the average pyrolysis kinetics and a correction coefficient has been presented [47].

In reality there will be temperature differences even inside small pyrolysing particles. Then more accurate approach is to account for non-uniform temperature distribution. The energy equation for the particle including storage, conduction and convection of drying and pyrolysis products is

$$\rho_p c_p \frac{\partial T_p}{\partial t} = \frac{1}{r^\Gamma} \frac{\partial}{\partial r}\left(\lambda_p r^\Gamma \frac{\partial T_p}{\partial r} - \dot{m}'' c_g r^\Gamma T_p\right) + l_u \dot{\rho}_u + l_v \dot{\rho}_v \tag{10}$$

This equation is coupled with Eq. (3) for the local pyrolysis inside the particle. Numerical methods to solve the equations have been presented (see e.g. [41]).

Wood fuels usually have higher moisture content that coal. The drying delays devolatilization and the processes become overlapping [41] for large particles. This is illustrated in Figure 4.

The application of the detailed model, Eq. (10), is cumbersome as a sub-model. Drying and pyrolysis take place in narrow temperature ranges inside a biomass particle leading to steep moisture and density distributions even for relatively small particles [46]. Simplified shrinking core models for drying [49,50] and simultaneous drying and pyrolysis [51] have been presented.

Division of Combustion of a Particle in the Dense bed and above it

Drying may affect the release profile of volatiles inside the boiler in some cases. If the particle is completely dried before its weight reduces to the limit where its free settling velocity becomes below the gas velocity, moisture has no effect on the release profile, because the particle stays in the dense bed. A smaller particle may reach this limit or can even initially be below the limit. Then drying may have a great influence on the release profile. The heat up of the particle also affect the release profile of the volatiles if the particle mass is below the critical mass. For small particle heating up and drying take place before significant pyrolysis takes place and much of the volatiles can be released above the bed. For a large particle drying and pyrolysis can be overlapping [41]. Then more volatiles can be released in the bed, since moister particle is also heavier. However, particles may be completely dried before pyrolysis and before reaching the critical weight and then drying will have no effect on the release profile. Models can be used to estimate, if drying and pyrolysis are overlapping [41,50].

The mass of particle decreases during drying and devolatilization and its size reduces during char combustion. A model for division of the release profiles of volatiles and char combustion in the bed and above it [17] is further developed here. It was based on the free settling velocity of a single particle. If the free settling velocity is lower than the fluidization velocity, the particle remains in the dense bed. If it is higher, then the particle may escape the bed, but does not necessarily do so.

The movement of particles in the dense bed is rather chaotic. Above the bed the concentration of solids decreases [52,53]. Then it is possible to estimate the relation between diameter and density of particles that will be entrained with the flow, if the gas above the dense bed is assumed to be dilute enough. Gravity will make the particles with high density to go back to bed, but small light particles escape the bed. The propensity of the particles to burn in the bed or in the riser is estimated by using the equation of motion of a single particle. A border between

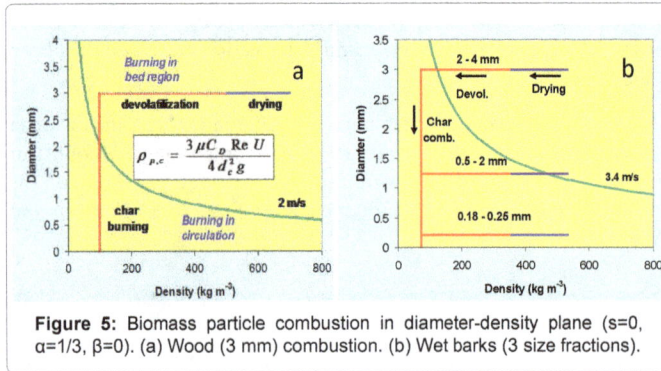

Figure 5: Biomass particle combustion in diameter-density plane (s=0, α=1/3, β=0). (a) Wood (3 mm) combustion. (b) Wet barks (3 size fractions).

these regions corresponds to situation, in which the free settling velocity of the particle is equal to the gas velocity (Figure 5). Thus a relation between the critical particle (volume average) density and size is obtained between the regions. The location of the border line depends on the gas velocity in the riser. So it is possible to estimate the proportions of drying, pyrolysis and char burning that take place in the dense bed at minimum, but in reality the proportions can be higher, since the particles may stay in dense bed for some time before the escape even the free settling velocity is lower that the fluidization velocity.

Burning Regions (Dense Bed and above Dense in Riser)

A particle first dries (blue line) and releases volatiles (red line) with decreasing density shown by the horizontal lines (Figure 5). It is possible to see from such a figure the part of de-volatilization that will take place in the bed region at minimum, since heavy particles cannot easily leave the bed. However, small and light particles do not necessary leave the bed, but their burning can take place also in the dense bed. Then some analysis about their residence time in the bed is required. The borderline is determined by the free settling velocity of the particles and it is in reality only a rough approximation for the border of the dense bed and dilute region, since the shape of the fuel is not spherical and the particles do not behave as single particles. The border is also not abrupt, but its shape and location depend on conditions. The de-volatilization and char burning may also be overlapping especially for small particles [45]. It is seen (Figure 5a) that most of the burning of a large heavy particle takes in the bed and only after the mass is decreased they can escape the bed.

For large particles (Figure 5a) drying and de-volatilization take place completely in the bed regions, since the burning (red) line crosses the (green) free settling velocity curve only in the char combustion stage. Much of char burning also takes place in the dense bed (because char mass ~ d^3). The char after crossing the critical intersection point (in the region "burning in circulation") can partly be burned in the dense bed and partly in the riser. Part of volatiles is released above the dense bed for lighter large particles and all volatiles from small particle can be released above the bed (Figure 5b).

After fuel is fed into a boiler drying de-volatilization and char combustion take place. Whether a particle escapes the bed or not depends on its free settling velocity, which depends on size and mass of the fuel particle, which change due to drying, pyrolysis and char oxidation. Heavy particles fall back into the dense bed while the lighter ones are entrained by the gas flow. However, even the free settling velocity of the particles is lower than the gas velocity statistically it takes some time that they reach the boundary region between the dense bed and the riser and eventually escape the dense bed. This can be

described by the elutriation rate constant or its inverse, the elutriation time constant. Most of the volatiles are released in the bed for particles residing long in the bed, but for smaller particles with shorter residence time, much of the de-volatilization can take place in the riser.

The particle velocity in an upward flow is described by the equation

$$\frac{dU_p}{dt} = -\frac{3\mu}{4\rho_p d^2} C_D \operatorname{Re}(U_p - U_g) - g \tag{11}$$

There is a great diversity of shapes of biomass particles. Coal particles are more regular in shape. There exists a vast literature on drag coefficient of particles of different shapes in different conditions [54-71]. In the present paper the drag coefficient is calculated from a correlation for a spherical particle [72] $C_D = 24[1 + 0.15(\varepsilon \operatorname{Re})^{0.687}]/(\varepsilon \operatorname{Re})$ when Re≤1000 and C_D=0.44 when Re>1000, but the calculation method can be applied to other correlations for C_D as well as distributed values of C_D. The particle does not move, if the free settling velocity is equal to the gas velocity. In this case dU_p/dt=0 and U_p=0. Equation (11) gives the explicit relation between critical density and critical particle size at different gas velocities

$$\rho_{p,c} = 3\mu C_D \operatorname{Re} U_g / (4gd_c^2) \tag{12}$$

Thus, it is possible to get information about the minimum fraction of fuel released from particles (as volatiles or char combustion products) in the dense bed. In reality the fraction can be larger, since the particles are not immediately escaping the dense bed when critical density and size are reached. If de-volatilization takes place with a constant size and decreasing density, the minimum fraction of volatiles released in bed is

$$X_{v,c} = (1 - \rho_{p,c} / \rho_{p,0}) / v_0 = [1 - 3\mu C_D \operatorname{Re} U_g / (4gd_0^2)]/v_0 \tag{13}$$

when $v_0\rho_{p0} > \rho_{p,c}$ and $d_0 > d_c$. When $v_0\rho_{p0} < \rho_{p,c}$ and $d_0 > d_c$ all volatiles are released in the bed ($X_{v,c} = 1$), but if $d_0 < d_c$, all volatiles can be released above the bed $X_{v,c}$=0. The minimum fraction of char released in the dense bed is

$$X_{C,c} = 1 - (d_c / d_0)^3 \tag{14}$$

when $v_0\rho_{p0} > \rho_{p,c}$ and $d_0 > d_c$. When $v_0\rho_{p0} < \rho_{p,c}$ and $d_0 > d_c$, $X_{C,c}$=0 char burning can take place even totally above the dense bed.

The above assumptions, de-volatilization with constant particle size and char combustion with constant density [17], are frequently applied in combustion modelling of solid fuels. In reality particles may shrink and oxidation takes place also inside the particle. Then the burning routes in reality are curved instead of the direct lines as shown in Figure 6 due to shrinkage during pyrolysis and internal combustion during char combustion. For some coals swelling also takes place affecting particle size and shape and its movement.

The volume change due to shrinkage during de-volatilization is approximated to depend linearly on degree of conversion

$$V / V_0 = (d / d_0)^3 = 1 - sX_v \tag{15}$$

where s is constant.

The mass of a particle is $m_p = \pi \rho_p d^3 / 6$. Thus during de-volatilization

$$m / m_0 = (\rho_p / \rho_{p0})(d / d_0)^3 \tag{16}$$

The particle mass decreases during de-volatilization

$$m = m_0(1 - v_0 X_v) \tag{17}$$

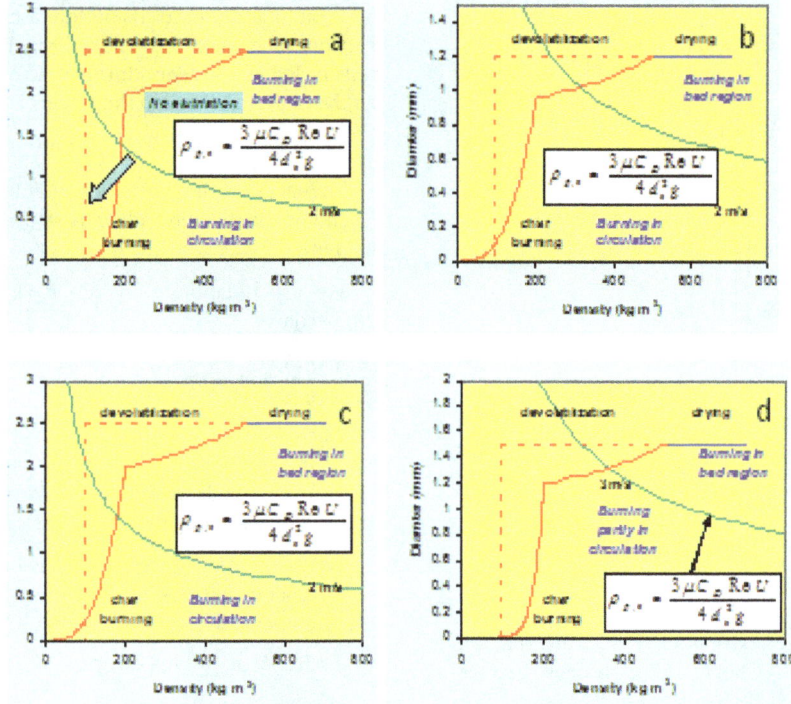

Figure 6: Burning of a wood particle in density-diameter plane. Solid red line shows the effect of shrinkage during devolatilization and effect of internal burning of char. Dashed red line is idealized case with no shrinkage and shrinking particle model during char burning. (a) Combustion regimes, when s=0.5 and α =0.32. Increasing elutriation from dense bed is shown by the arrow. (b) Fuel particle (1.2 mm) combustion. dc=1.02 mm, ρc=313.5 kg/m3, Xv,c=0.767 (0.651 without shrinkage) and XC,c=0 (0 with shrinkage and α=1/3), when s=0.5 and α =0.33. (c) Fuel particle (2.5 mm) combustion. dc=1.43 mm, ρc=179.4 kg/m3 and Xv,c=1 (1 without shrinkage) and XC,c=0.663 (0.443 with shrinkage and α=1/3), when s=0.5 and α =0.33. (d) Burning of a 1.5 mm wood particle in density-diameter plane, when α =0.32.

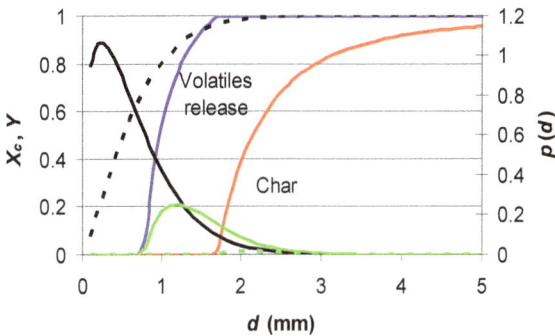

Figure 7: Effect of particle size on the (minimum) degrees (X_c) of release of volatiles and char oxidation in the dense bed. PSD of fuel ($p(d)$, solid black line) and cumulative PSD (Y, dashed black line) are also given.

Eqs. (15)-(17) give the relation for particle density during devolatilization

$$\rho_p / \rho_{p0} = (d_0 / d)^3 (1 - v_0 X_v) = (1 - v_0 X_v) / (1 - s X_v) \quad (18)$$

In the right hand size term the numerator accounts for pyrolysis and the denominator for shrinkage or swelling. Both the size and density can change during char oxidation [73,74]

$$d / d_0 = (1 - X_C)^\alpha \quad (19)$$

$$\rho / \rho_0 = (1 - X_C)^\beta \quad (20)$$

where $3\alpha + \beta = 1$. The part of volatiles released in the bed can be calculated by

$$\Delta m_{vb} / m_v = X_{v,c} = [1 - (\rho_{p,c} / \rho_{p,0})(d_c / d_0)^3] / v_0 \quad (21)$$

$X_{v,c}$ can be found by substituting $\rho_{p,c} = \rho_{p0}(1 - v_0 X_{vc}) / (1 - s X_{vc})$ and $d_c = d_0 (1 - s X_{v,c})^{1/3}$. The result is

$$X_{v,c} = (\rho_{p,0} - \rho_{p,c}) / (v_0 \rho_{p,0} - s \rho_{p,c}) \quad (22)$$

and $X_{C,c} = 0$ when $\rho_{p,c} > \rho_{p0}(1 - v_0) / (1 - s)$. When $\rho_{p,c} < \rho_{p0}(1 - v_0) / (1 - s)$, $X_{v,c} = 1$ and

$$X_{C,c} = 1 - (d_c / d_{v,f})^{1/\alpha} = 1 - [(1 - s)^{-1/3} d_c / d_0]^{1/\alpha} \quad (23)$$

The PSD of the coal can be described by Rosin-Rammler type distribution for a size interval $d_{min}...d_{max}$. Then the PSD and cumulative size are

$$p(d) = \frac{bnd^{n-1} \exp(-bd^n)}{\exp(-bd_{min}^n) - \exp(-bd_{max}^n)}, \quad Y(d) = \frac{\exp(-bd_{min}^n) - \exp(-bd^n)}{\exp(-bd_{min}^n) - \exp(-bd_{max}^n)} \quad (24)$$

The PSD becomes the normal Rosin-Rammler distribution, with $d_{min} = 0$ and $d_{max} = \infty$.

Figure 7 shows the effect of particle size on the (minimum) degree of release of volatiles and degree of char oxidation in the dense bed. The PSD is also shown with values n=1.32 and b=1.63 mm$^{-1.32}$, $d_{min} = 0$, $d_{max} = 20$ mm. The green lines describe the product $p(d_0)X_c$ (solid for volatiles, dashes for char).

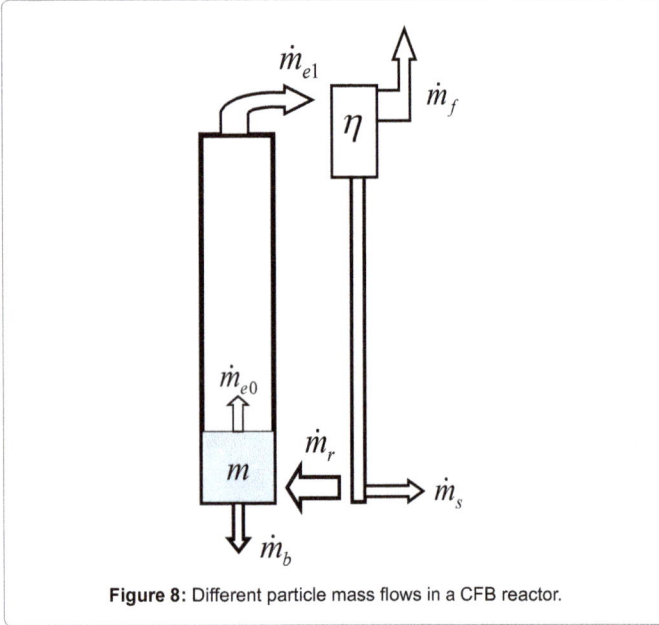

Figure 8: Different particle mass flows in a CFB reactor.

For a fuel with particle size distribution (PSD) the fraction of volatiles ($i=v$) released and char ($i=C$) oxidized in the dense bed are

$$\overline{X}_{i,c} = \int_0^{d_{max}} X_{v,c}(d_0)p(d_0)\mathrm{d}d_0 \tag{25}$$

Minimum 19.7 % of volatiles is released and 1.41 % of char is oxidized in the dense bed with the PSD given.

Mass Balance Model for Dense Bed

We consider the de-volatilization stage of particles. A fuel batch with particles of single size is fed in the CFB reactor. It may be considered as a differential batch in a continuous feed so that the batch itself does not affect the steady state environment in the reactor. The system under study is illustrated in Figure 8. The change of volatile mass in particles in the dense bed due to this batch is described by

$$\frac{\mathrm{d}m}{\mathrm{d}t} = \dot{m}_0 - \dot{m}_{e0} - \dot{m}_v - \dot{m}_b + \dot{m}_r \tag{26}$$

The term on the left hand side is the change of mass. The first term on the right hand side is the feed. The second term is the mass flow rate of volatiles in elutriated particles $\dot{m}_{e0} = K_e m$, the third term is due to mass loss by de-volatilization $\dot{m}_v = K_v m$, the fourth term is mass loss due to removal of bottom ash $\dot{m}_b = K_b m$. The last term accounts for the mass flow of volatile mass in particles returned into the bed. De-volatilization can take place completely in the bed, if the free terminal velocity U_t is higher than the gas velocity (i.e. the elutriation coefficient remains zero, $K_e=0$).

The particle residence time in the riser part of the loop t_{r1} depends on the initial density and particle size. Thus the mass flow rate in the riser changes due to de-volatilization and it is described by $\dot{m}_e / \dot{m}_{e0} = \exp(-K_v t)$. The mass reduction during the stay in the riser is

$$\dot{m}_{e1} / \dot{m}_{e0} = \exp(-K_v t_{r1}) \tag{27}$$

Devolatilization can practically be completed in the riser. After that a part may be escaped as fly ash though the cyclone

$$\dot{m}_f = K_e m_v (t - t_{r1}) \exp(-K_{ve} t_{r1})(1 - \eta) \tag{28}$$

Part of the material $\dot{m}_s = \kappa \dot{m}_{e1}$ can be removed also from the loop seal in addition to the bottom ash removal. It is assumed that no reactions take place in the flow down, but they could also be included by using longer residence time instead of t_{r1}. The mass flow rate of partly pyrolysed particle back to the reactor is

$$\dot{m}_r = K_e (\eta - \kappa) m_v (t - t_r) \exp(-K_{ve} t_{r1}) \tag{29}$$

The fact that the particles are retuned with an average time delay depending on the prevailing past bed mass is shown by the time difference $t - t_r$ in the bed mass.

We use notations $\tau_i = 1/K_i$, $K_t = K_v + K_e + K_b$, $\tau_t = 1/K_t$. Equation (26) becomes

$$\frac{\mathrm{d}m(t)}{\mathrm{d}t} = \dot{m}_0 - m(t)/\tau_t + m(t - t_r)/\tau_{e*} \tag{30}$$

where $\tau_{e*} = \tau_e \exp(t_{r1}/\tau_v)(\eta - \kappa)$. The time constant for elutriation depends on the particle size and density. Particle density decreases during de-volatilization. Particle size may also change due to shrinkage (or swelling for some fuels).

We follow the subsequent behaviour, when a batch $\dot{m}_{v,0} = m_0 \delta(t)$ with initial mass m_0 is fed in the boiler at time $t=0$. The bed mass can be solved numerically as function of time

$$m_{t+\Delta t} = [(1 - \tfrac{1}{2}\Delta t / \tau_t)m_t + \tfrac{1}{2}(\Delta t / \tau_{e*})(m_{t-t_r} + m_{t+\Delta t - t_r})] / (1 + \tfrac{1}{2}\Delta t / \tau_t) \tag{31}$$

The mass volatile matter fed in the batch of particles is divided in different parts with different fates. Volatile matter released as volatiles inside the bed ($i=v$), volatile matter in particles removed as bottom ash ($i=b$), volatile matter (inside particles) removed as fly ash ($i=f$) and volatile matter in particle removed in the loop seal ($i=s$) are

$$m_i = \int_0^\infty \dot{m}_i \mathrm{d}t \tag{32}$$

The rest of the volatiles are released above the bed in the circulation (riser)

$$m_e = m_0 - m_v - m_f - m_b - m_s \tag{33}$$

Analytical solutions are discussed in Appendix. These solutions are useful in rough evaluation of the effects of different time constants on the bed mass.

Elutriation of Particles from the Dense Bed

Elutriation is described by the equation [52]

$$-\frac{1}{A}\frac{\mathrm{d}m_i}{\mathrm{d}t} = K_* \left(\frac{m_i}{m_t}\right) \tag{34}$$

There is a great number of correlations for elutriation rate constant in the literature. Some of these correlations are summarized [75-81]. As an example some of the equations for the elutriation coefficient can be presented in the form

$$K_* = c_1[\rho_g(U_g - U_t)]\mathrm{Fr}^{c_2}\mathrm{Re}^{c_3}[(\rho_p - \rho_g)/\rho_g]^{c_4} \tag{35}$$

where c_i are constants presented in Table 1. Other more suitable correlations based on measurements or results from comprehensive flow models could also be used.

Particle Motion in the Riser

The residence time of the particles is obtained by solving the equation of motion, Eq. (11) for a single particle. Eq. (11) can also be presented as

Reference Number	c_1	c_2	c_3	c_4
[79]	2.07×10^{-4}	$Re^{-0.4}$	1.6	0.61
[80]	1.52×10^{-4}	0.5	0.725	1.15
[81]	4.6×10^{-2}	0.5	0.3	0.15

Table 1: Constants for equation (35).

$$\frac{dU_p}{dt} = -(U_p - U_g)/\tau_w - g \qquad (36)$$

using the time constant $\tau_w = 4\rho_p d_p^2 (3\mu C_D Re)$. If the time constant is assumed constant or average value is applied, the solution is

$$U_p = (U_g - \tau_w g)(1 - e^{-t/\tau_w}) + U_{p0} e^{-t/\tau_w} \qquad (37)$$

for large times $t \gg \tau_w$

$$U_p \approx U_g - U_t \qquad (38)$$

where the terminal velocity $U_t = g\tau_w$ can be calculated by iteration. The particle size decreases due to the shrinkage according to Eq. (15) giving $d = d_0 (1 - sX_v)^{1/3}$. Density decreases due to pyrolysis but it is also affected by the shrinkage as described by Eq. (18).

The terminal velocity will change during pyrolysis, since particle size and density are changing

$$U_t = [4\rho_{p0} d_0^2 g(1 - vX_v)]/[3\mu C_D Re(1 - sX_v)^{1/3}] \qquad (39)$$

$C_D Re$ also depends on the changing particle size (especially for large particle). The velocity of the particle can be solved numerically using the equation

$$U_{p,t+\Delta t} = [(1 - z_w)U_{p,t} + z_w(U_{g,t+\Delta t} + U_{g,t}) - g\Delta t]/[1 + z_w] \qquad (40)$$

where $z_w = (3\mu C_D Re \Delta t)/(8\rho_p d_p^2) = \Delta t/(2\tau_w)$ changes during pyrolysis and average values for variables (at t and at $t+\Delta t$) are applied.

However, if the time constant or relaxation time τ_w is small, the particle reaches its equilibrium velocity $U_p = U_g - U_t$ fast, which can be directly applied. In addition, the starting velocity of particles from the dense bed varies. Some particles are accelerated after leaving the bed and some particles have already excess velocity and are decelerated. This distribution of initial velocities is not well-known and it gives some dispersion for the volatile release due to particle-to-particle differences. In this context it should be noted that particles of same size and density have not in reality the same residence time in the riser, but there is a residence time distribution (RTD) (see e.g. [82-83]). The gas does not flow in plug-flow manner, but there is a velocity profile affecting also to the particle velocities in the riser. Part of the particles may flow downwards in near the wall region. The particles are also not of the same shape and oriented similarly, which gives some particle-to-particle differences in values of C_D. Thus, the release profile in the riser that is found here for one particle type is smoothened in reality due to this RDT. The analysis at different locations could be carried out with a measured or modelled RTD [84], but the weight and size reduction of the particles complicate the use of such methods.

The particle velocity $U_p = dx/dt$ gives the dependence between location of particle (distance from the dense bed) and time

$$x = \int U_p dt \qquad (41)$$

Solving the particle location and release of volatiles as function of time after it is removed from the bed by elutriation poses a difficulty in finding the steady state release profile of volatiles in the riser based

on the unsteady state response of a batch. The particles are reducing in mass and size and their velocity is changing. This means that x-values calculated with the above equation will change. It would be better to find the particle velocity and release of volatiles directly as function of co-ordinate x. Using relation $U_p = dx/dt$ equation (36) can be presented in the form

$$\frac{dU_p^2}{dx} = 2(U_g - U_p)/\tau_w - 2g \qquad (42)$$

which can be presented in the form for numerical calculations

$$U_{p,x+\Delta x} = \sqrt{(\tfrac{1}{2}\Delta x/\tau_w)^2 + U_{p,x}^2 + (U_{g,x} + U_{g,x+\Delta x} - U_{p,x})\Delta x/\tau_w - 2g\Delta x} - \tfrac{1}{2}\Delta x/\tau_w \qquad (43)$$

Then the corresponding time is found by

$$t = t_0 + \int_0^x \frac{1}{U_p} dx \qquad (44)$$

When the fine and coarse particles were mixed, the carry-over rate was increased significantly. A simple equation and calculation procedure has been presented [85]. The use of the effective gas density have been recommended [14] $\bar{\rho}_g = \varepsilon\rho_g + (1-\varepsilon)\rho_s$ for calculation motion of large biomass particles in flow of fine particles.

Discussion

The method is illustrated in the following for a pilot CFB reactor. Particles with size 1.5 mm are fed to the boiler. In this case the drying occurs completely in the bed (see Figure 6d). The gas velocity is assumed to be 3 m/s. The heat up time t_d and de-volatization time constant τ_v depend on particle size ($\sim d^n$, $n \approx 1.5$). In the illustration heating up

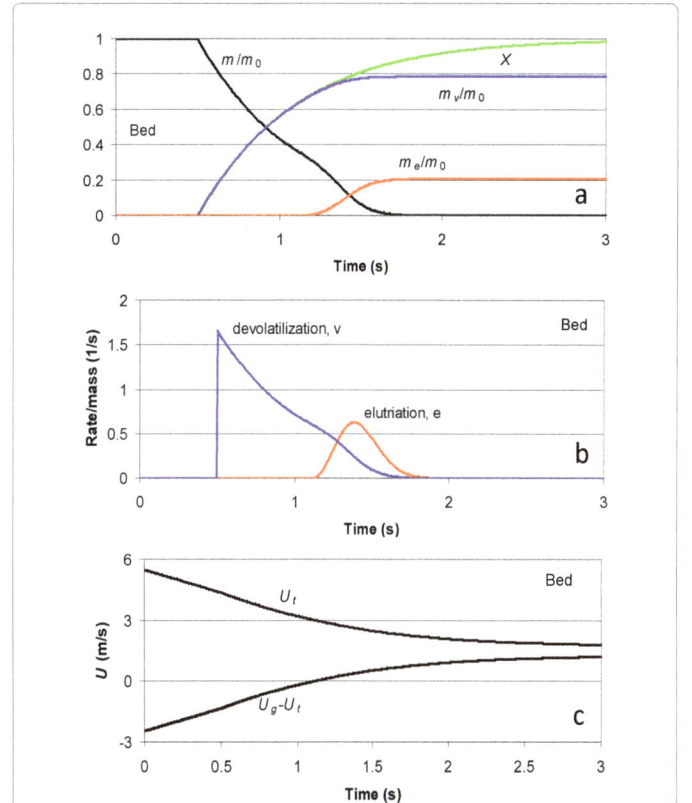

Figure 9: (a) The conversion, relative total mass in bed, mass of wood de-volatilized in bed and mass elutriated from the bed. (b) Rate of de-volatilization in the bed and rate of elutriation of volatile matter inside particles from the bed. (c) Free settling velocity and difference between gas and free settling velocity.

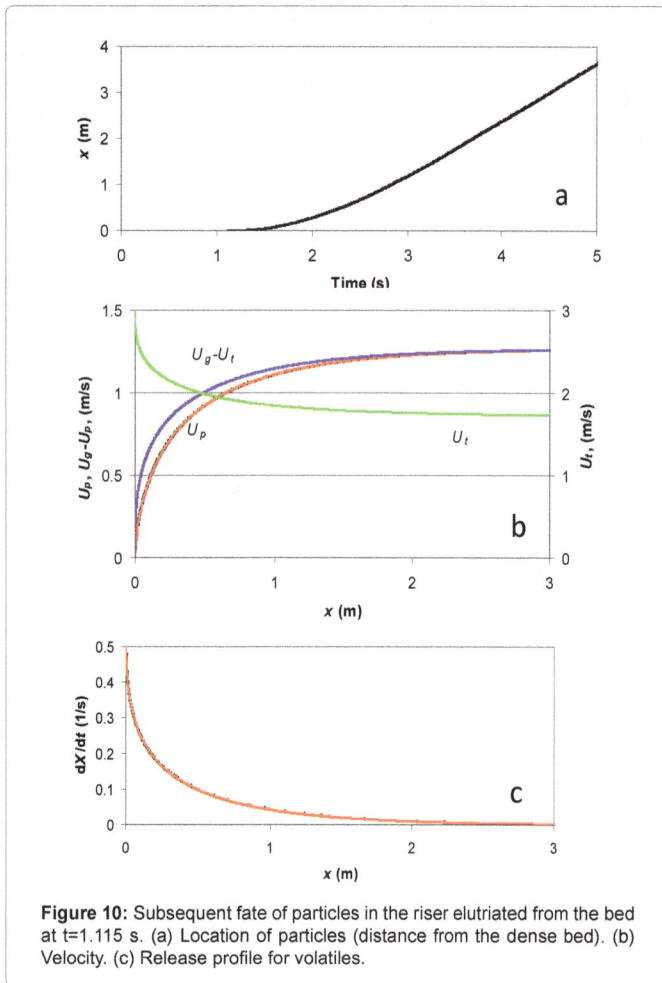

Figure 10: Subsequent fate of particles in the riser elutriated from the bed at t=1.115 s. (a) Location of particles (distance from the dense bed). (b) Velocity. (c) Release profile for volatiles.

taken into account. It is seen that about 80 % of volatiles is released in the bed in this case and the rest is released in the riser.

The subsequent distance above from the dense bed for these particles (elutriated at t=1.115 s) is shown as function of time in Figure 10a. If the devolatilization is considered to be competed at 3 s, the riser height < 2 m (not including the bottom bed) is enough. Velocities of particles are shown in Figure 10b. The particle velocities were calculated using two methods (Eq. (36) black line, Eq. (38) red line), which give practically the same line (the black line is under the red line). It takes some time before the particle velocity reaches the velocity U_g-U_t in this case with relative large particles. Release profile for volatiles is shown in Figure 10c.

The situation in the riser (in this batch study) changes with time, since the elutriated particles become lighter when pyrolysing in the bed. Figure 11 shows the situation for particles entrained at t=1.4 s. In this case with large particles the curves are rather similar as earlier, since the particles have released much weight already before elutriation. Small particles can be elutriated already even before pyrolysis starts so that the profiles in the riser will change with time due to the increasing conversion of the inlet particles to the riser.

Thus, the release profile in the riser changes (for the batch process) as can be seen by comparing Figures 10 and 11. The conversion as function of location in the riser is shown in Figure 12 at times t=1.115, 1.415, 1.715, 2.015 and 2.315 s. The inlet conversion to the riser increases with time. The average (steady state) conversion in the riser and its derivative are shown in Figure 12b. The profile is steep close

and drying time are given. Here we assume that t_d=0.5 s (which has no effect in this case on the release profile) and τ_v=0.5 s. These values depend on particle size and can be estimated from literature or from bench scale de-volatilization tests. A linear decrease of moisture during this initial period is assumed. Particle density is described simple by equation $\rho_p = \rho_{p0}[1+u_0(t_d-t)/t_d]$ during heat up and drying. The time lag t_d accounts for heat up and drying time before pyrolysis starts. The particle diameter and density are changing simultaneously with the flight due to de-volatilization as described by Eqs. (15) and (18). In the illustration de-volatilization is assumed here to obey Eq. (2) with n=1, but more comprehensive models to calculate pyrolysis and mass as function of time could also be used. No fragmentation is assumed during pyrolysis but it the effects could be included provided that the sufficient information is available.

The values of Tanaka et al. [81] (Table 1) are applied for elutriation. In this illustration it is assumed that the total bed mass is 30 kg/m². The release of volatiles in the bed is illustrated in Figure 9. The mass (volatiles in particles) starts to decrease at t=0.5 s after the particles have been dried and heated up. Elutriation starts at t=1.115 s, when the particles have lost weight enough so that $U_t<U_g$ and some of the particles are entrained by the gas. The rates of devolatilization and elutriation are also shown in Figure 9. Elutriation rate reaches a maximum (at t=1.4 s) when the particles have lost enough weight but still contain volatile matter. It should be noted here that m is the mass of volatiles. Char and ash are not included but their effect on particle mass and elutriation is

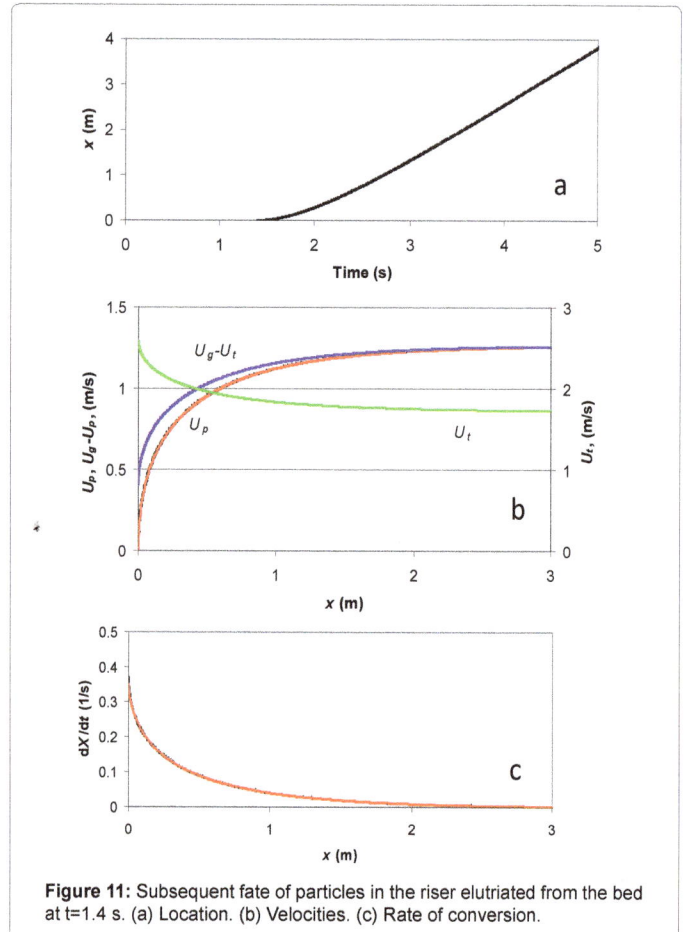

Figure 11: Subsequent fate of particles in the riser elutriated from the bed at t=1.4 s. (a) Location. (b) Velocities. (c) Rate of conversion.

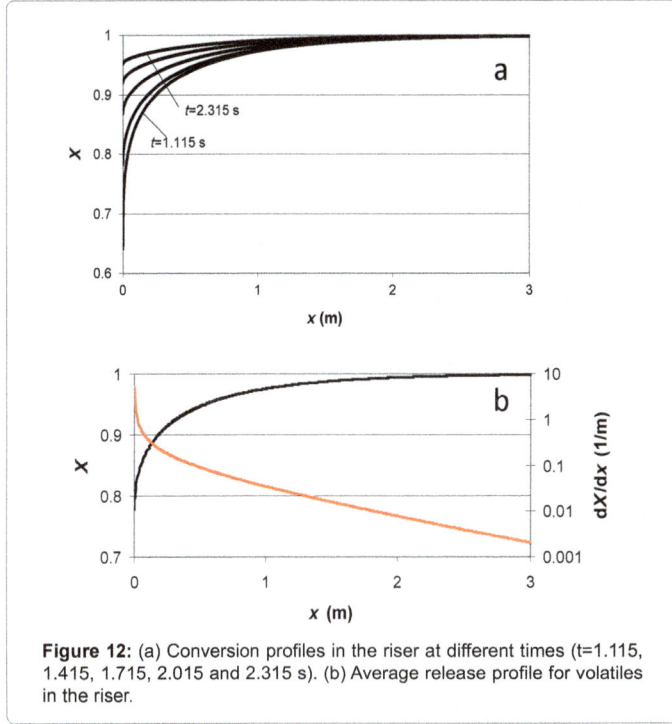

Figure 12: (a) Conversion profiles in the riser at different times (t=1.115, 1.415, 1.715, 2.015 and 2.315 s). (b) Average release profile for volatiles in the riser.

to bed because the velocity of particles close to bed is low, since their mass is still close to the critical one. It is practically the same as the conversion at 1.415 s at which the elutriation reaches maximum. This average conversion was calculated by

$$\overline{X}(x) = \left(\int_0^\infty \dot{m}_e(t) X(t,x) \mathrm{d}t \right) \bigg/ \left(\int_0^\infty \dot{m}_e(t) \mathrm{d}t \right) \quad (45)$$

Conclusions

Analytical formulas for the (minimum) amounts of volatiles released and char oxidized in the dense bed are presented. The effect of shrinkage of particles was included. The effect of fragmentation can be considered in further development.

A model for the release profile of volatiles in FB and CFB boilers has been presented. It gives the division of the release of volatiles in the dense bed and in the riser. It is based on the fact that large and heavy particles do not escape the dense bed before complete de-volatilization. Smaller particles also reside some time in the dense bed releasing volatiles and are entrained by the gas from the dense bed the bed to the riser. This elutriation depends on the terminal velocity of the particle. Particle size and density decrease also due to de-volatilization. Method to calculate the release profile of the volatiles in the riser is also presented. The propensity for elutriation increases with decreasing particle size and density. The analysis is carried out for all size fractions and the final profile is found by summing up each contribution. The model gives also the possibility to get information of the release distributions of fuel nitrogen and sulphur from particles during de-volatilization stage. The extension of the method to include char combustion stage is possible.

Acknowledgment

This work was performed with funding from the project ERA-NET Biomodelling—Advanced biomass combustion modelling for clean energy production within the framework of ERA-NET Bioenergy Joint Call on Clean Biomass Combustion.

Appendix: Analytical Solutions for Bed Mass

Here we consider simplified solutions with constant values for the coefficients K. In reality especially K_e is not constant but depends on the changing size and density of the particle. K_e is, however, constant for an inert particle. This inert case can be useful in evaluating elutriation time constant using cold reactor tests without devolatization or using tests at higher temperatures with inert material. By applying the Laplace transform

$$\overline{m}(s) = \int_0^\infty e^{-s} m(t) \mathrm{d}t \quad (A1)$$

Eq. (30) assuming constant time constants is transformed into

$$s\overline{m} - m_0 = \overline{\dot{m}}_{0v} - \overline{m}/\tau_t + \overline{m}e^{-t_r s}/\tau_{e*} \quad (A2)$$

which can be solved

$$\overline{m} = \frac{m_0 + \overline{\dot{m}}_0}{s + 1/\tau_t - e^{-t_r s}/\tau_{e*}} = (m_0 + \overline{\dot{m}}_0) \sum_{n=0}^\infty \frac{\exp(-n t_r s)}{\tau_{e*}^n (s + 1/\tau_t)^{n+1}} \quad (A3)$$

For a batch feed $\dot{m}_{0v} = \overline{\dot{m}}_{0v} = 0$. The inverse transform can be found using tabulated properties of Laplace transform. The solution for the mass in the bed for a batch is

$$m/m_0 = f(t) = \sum_{n=0}^\infty \frac{1}{n!} U(t - n t_r)[(t - n t_r)/\tau_{e*}]^n \exp[-(t - n t_r)/\tau_t] \, (A4)$$

where $U(t-a)$ is Heaviside's unit step function having the properties $U(t-a)=1$ when $t \geq a$ and $U(t-a)=0$ when $t < a$. The mass of the bed after long time is found by applying the limit property of Laplace transform. When $s \rightarrow 0$, $m(\infty) = \lim s\overline{m}$. It is zero except in the special case $\tau_t = \tau_{e*}$ the steady state mass in the bed becomes

$$m_\infty/m_0 = 1(\,1 + t_r/\tau_{e*}) \quad (A5)$$

The rest of the mass $(t_r/\tau_{e*})(\,1 + t_r/\tau_{e*})$ is in the circulation. This equation could be used to evaluate the elutriation time constant.

Next we consider the case $m_0 = 0$ and $\dot{m}_0 = 0$ when $t < 0$ and $\dot{m}_0$ =constant when $t \geq 0$. In this case $\overline{\dot{m}}_0 = \dot{m}_0/s$

$$\overline{m}/\dot{m}_0 = \sum_{n=0}^\infty \frac{\exp(-n t_r s)}{\tau_{e*}^n s(s + 1/\tau_t)^{n+1}} \quad (A6)$$

The steady state mass in the bed will be

$$m_\infty/\dot{m}_0 = 1(\,1/\tau_t - 1/\tau_{e*}) \quad (A7)$$

Nomenclature

A	cross-section area, m^2
Bi	Biot number, Bi=hR_p/λ_p, -
C_D	drag coefficient,-
c	specific heat capacity, J kg^{-1}K^{-1}
c_i	constant, -
d	particle diameter, m
E	activation energy, J mol^{-1}
Fr	Froude number, $\mathrm{Fr} = (U_g - U_t)^2/(gd)$,
g	specific gravity, ms^{-2}
h	heat transfer coefficient including convection and radiation, Wm^{-2}K^{-1}
k	reaction rate coefficient, s^{-1}
k_0	pre-exponential coefficient, s^{-1}
K	coefficient, s^{-1}
K_*	elutriation coefficient, $K_* = m_t K_e/A$, kg m^{-2}s^{-1}
l_u	heat of evaporation J kg^{-1}
l_v	heat of pyrolysis, J kg^{-1}

m mass, kg

$\dot{m}$ mass flow rate, kg s^{-1},

$\dot{m}''$ mass flux, kg s^{-1}m^{-2}

n exponent, -

p particle size distribution, PSD, m^{-1}

R universal gas constant, 8.314 J mol^{-1}K^{-1}

r radial co-ordinate, m

Re   Reynolds' number, $\mathrm{Re} = dU_t\rho_g / \mu$,

s shrinkage or swelling ratio, Laplace transform variable

T temperature, K

t time, s

t_c conversion time, s

U velocity, m/s

u moisture in fuel (dry basis)

V volume, m^3

v mass fraction of volatiles in fuel,-

X degree of conversion, -

$\bar{X}$ average conversion, -

x co-ordinate, distance from the dense bed, m

Y cumulative PSD, -

α exponent for change of diameter, -

β exponent for change of density, -

d Dirac delta function, s^{-1}

ε voidage, gas volume/total volume, -

η cyclone separation efficiency, -

Γ geometric shape factor, 0 for plate, 1 for cylinder, 2 for sphere, -

φ_r correction coefficient for internal thermal resistance, $\varphi_r = 1/(6+2\Gamma)$, -

φ_s sphericity, -

κ mass ratio, -

λ thermal conductivity, Wm^{-1}K^{-1}

μ viscosity, kg m^{-1}s^{-1}

ρ density, kg m^{-3}

$\dot{\rho}$ change of density, kgm^{-3}s^{-1}

τ time constant, s

Subscripts

0   initial, inlet

1   surface

b bottom ash

C char

c critical

d heat up and drying

e elutriation, effective

f fly ash, final

g gas

i index i

k convection

p particle

r back to reactor, radiation

t total

u water

v devolatilization, volatiles

w movement

References

1. Obersteiner M, Azar C, Kauppi P, Mollersten K, Moreira J, et al. (2001) Managing climate risk. Science 294: 786-787.

2. Pikkarainen T, Saastamoinen J, Saastamoinen H, Leino T, Tourunen A (2014) Development of 2nd generation oxyfuel CFB technology – small scale combustion experiments and model development under high oxygen concentrations. Energy Procedia.

3. Gayan P, Adanez J, de Diego LF, García-Labiano F, Cabanillas A, et al (2004) Circulating fluidised bed co-combustion of coal and biomass. Fuel 83: 277-286.

4. McIlveen-Wright DR, Huang Y, Rezvani S, Wang Y (2007) A technical and environmental analysis of co-combustion of coal and biomass in fluidised bed technologies. Fuel 86: 2032-2042.

5. Coda Zabetta E, Barisic V, Peltola K, Sarkki J, Jantti T (2013) Advanced technology to co-fire large shares of agricultural residues with biomass in utility CFBs. Fuel Process Technol 105: 2-10.

6. Silvennoinen J, Hedman M (2013) Co-firing of agricultural fuels in a full-scale fluidized bed boiler. Fuel Process Technol 105: 11-19.

7. Werther J, Saenger M, Hartge EU, Ogada T, Siagi Z (2000) Combustion of agricultural residues. Prog Energ Combust Sci 26: 1-27.

8. Scala F, Chirone R (2004) Fluidized bed combustion of alternative solid fuels. Experiment Therm Fluid Sci 28: 691-699.

9. Chirone R, Salatino P, Scala F, Solimene R, Urciuolo M (2008) Fluidized bed combustion of pelletized biomass and waste-derived fuels. Combust Flame 155: 21-36.

10. 10. Schütte K, Wittler W, Rotzoll G, Schügerl K (1989) Spatial distributions of O$_2$, CO$_2$ and SO$_2$ in a pilot-scale fluidized bed combustor operated with different coals. Fuel 68: 1499-1502.

11. Zhao J, Brereton C, Grace JR, Lim J, Legros R (1997) Gas concentration profiles and NO$_x$ formation in circulating fluidized bed combustion. Fuel 76: 853-860.

12. Chirone R, Miccio F, Scala F (2004) On the relevance of axial and transversal fuel segregation during the FB combustion of a biomass. Energ Fuels 18: 1108-1117.

13. Ogada T, Werther J (1996) Combustion characteristics of wet sludge in a fluidized bed. Release and combustion of volatiles. Fuel 75: 617-626.

14. Smolders K, Honsbein D, Bayens J (2001) Operating parameters for the circulating fluidized bed (CFB) combustion of biomass. Progr Thermochem Biomass Convers 1: 766-778.

15. Scala F, Salatino P (2002) Modelling fluidized bed combustion of high-volatile solid fuels. Chem Engg Sci 57: 1175-1196.

16. Miccio F, Scala F, Chirone R (2003) FB combustion of a biomass fuel: Comparison between pilot scale experiments and model simulations. 17th International Fluidized Bed Combustion conference, Morgantown, Monongalia.

17. Saastamoinen J, Häsä H, Pitsinki J, Tourunen A, Hämäläinen J (2005) A simplified method to predict heat release profiles in a circulating fluidized bed reactor. Circulating Fluidized Bed Technology VIII, International Academic Publishers, World Publishing Corporation, Beijing: 313-320.

18. Nevalainen H, Jegoroff M, Saastamoinen J, Tourunen A, Jäntti T, et al. (2007) Firing of coal and biomass and their mixtures in 50 kW and 12 MW circulating fluidized beds - Phenomenon study and comparison of scales. Fuel 86: 2043-2051.

19. Munts VA, Baskakov AP, Feforenko YN, Kozlova YG (1990) The circulation factor in furnaces with a circulating fluidised bed. Thermal Engineering 37: 176-180.

20. Huang Y, Turton R, Famouri P, Boyle EJ (2009) Prediction of solids circulation rate of cork particles in an ambient-pressure pilot-scale circulating fluidized bed. Ind Engg Chem Res 48: 133-141.

21. Farzaneh, M, Sasic S. Almstedt AE, Johnsson F, Pallarès D (2013) A study of fuel particle movement in fluidized beds. Ind Engg Chem Res 52: 5791-5805.

22. Nikku M, Myöhänen K. Ritvanen J, Hyppänen T (2014) Three-dimensional modelling of fuel flow with a holistic circulating fluidized bed furnace model. Chem Engg Sci 117: 352-363.

23. Zhong W, Jin B, Zhang Y, Wang X, Xiao R (2008) Fluidization of biomass particles in a gas-solid fluidized bed. Energ Fuel 22: 4170-4176.

24. Ross DP, Heidenreich CA, Zhang DK (2000) De-volatization times of coal particles in a fluidised-bed. Fuel 79: 873-883.

25. Huff ER (1985) Effect of size, shape, density, moisture and furnace wall temperature on burning times of wood pieces. Fund Thermochem Biomass Convers: 761-775.

26. de Diego LF, García-Labiano F, Abad A, Gayán P, Adánez J (2003) Effect of moisture content on devolatilization times of pine wood particles in a fluidized bed. Energ Fuel 17: 285-290.

27. Johansson A, Johnsson F, Niklasson F, Åmand LE (2007) Dynamics of furnace processes in a CFB boiler. Chem Engg Sci 62: 550-560.

28. Grønly M (1996) A Theoretical and Experimental Study of the Thermal Degradation of Biomass. PhD Thesis, The Norwegian University of Science and Technology.

29. Grammelis P, Kakaras E (2005) Biomass combustion modeling in fluidized beds. Energ Fuel 19: 292-297.

30. 30. Fu WB, Yu WD (1992) Application of the general devolatilization model of coal particles in a combustor with non-isothermal temperature distribution. Fuel 71: 793-795.

31. Zhang YP, Mou J, Fu WB (1990) Method for estimating final volatile yield of pulverized coal devolatilization. Fuel 69: 401-403.

32. Wagenaar BM, Prins W, van Swaaij WPM (1994) Pyrolysis of biomass in the rotating cone reactor: modelling and experimental justification. Chem Engg Sci 49: 5109-5126.

33. Liden AG, Berruti F, Scott DS (1988) A kinetic model for the production of liquids from the flash pyrolysis of biomass. Chem Engg Comm 65: 207-221.

34. Nunn TR, Howard JB, Longwell JP, Peters WA (1985) Product composition and kinetics in the rapid pyrolysis of sweet gum hardwood. Ind Engg Chem Process Design Develop 24: 836-844.

35. van den Aarsen FG, Beenackers AA, van Swaaij WP (1985) Wood pyrolysis and carbon dioxide char gasification kinetics in a fluidized bed. Fund Thermochem Biomass Convers 36: 691-715.

36. Lewellen PC., Peters WA, Howard, JB (1977) Cellulose pyrolysis kinetics and char formation mechanism. 16th Symposium (International) on Combustion, the Combustion Institute 16: 1471-1480.

37. 37. Solomon PR, Fletcher TH, Pugmire RJ (1993) Progress in coal pyrolysis. Fuel 72: 587-597.

38. de Diego LF, Garcia-Labiano F, Abad A, Gayán P, Adánez J (2002) Modeling of the devolatilization of nonspherical wet pine wood particles in fluidized beds. Ind Engg Chem Res 41: 3642-3650.

39. Kobayashi H, Howard JB, Sarofim AF (1976) Sixteenth Symposium (International) on Combustion, The Combustion Institute, Pittsburgh: 411-425.

40. Ubhayakar SK, Stickler DB, Rosenberg CWV, Gannon RE (1977) Sixteenth Symposium (International) on Combustion. The Combustion Institute, Pittsburgh: 427-436.

41. Saastamoinen J, Richard JR (1996) Simultaneous drying and pyrolysis of solid fuel particles. Combust Flame 106: 288-300.

42. Saastamoinen JJ, Hämäläinen JP, Kilpinen P (2000) Release of nitrogen compounds from wood particles during pyrolysis. Environ Combust Technol 1: 289-316.

43. Zanzi R, Sjöström K, Björnbom E (1996) Rapid high-temperature pyrolysis of biomass in a free-fall reactor. Fuel 75: 545-50.

44. Field MA, Gill DW, Morgan BB, Hawksley PGW (1967) Combustion of Pulverized Coal. The British Coal Utilization Research Association, Letherhead.

45. Saastamoinen J, Aho M, Linna V (1993) Simultaneous pyrolysis and char combustion. Fuel 72: 599-609.

46. Saastamoinen J, Aho M, Moilanen A, Sørensen LH, Clausen S, et al. (2010) Burnout of pulverized biomass particles in large scale boiler – single particle model approach. Biomass Bioenerg 34: 728-736.

47. Saastamoinen J (2000) Combustion of a pulverised wood particle - a modelling study. Nordic Seminar on Thermochemical Conversion of Biofuels. Trondheim, Norway: 28.

48. Saastamoinen J (2004) Heat exchange between two moving beds by fluid flow. Int J Heat Mass Trans 47: 1535-1547.

49. Saastamoinen JJ, Impola R (1995) Drying of solid fuel particles in hot gases. Drying Technol 13: 1305-1315.

50. Saastamoinen J (2011) Drying of biomass particles under control of heat transfer. 6th Baltic Heat Transfer Conference, Tampere, Finland: 8.

51. Saastamoinen JJ (2006) Simplified model for calculation of de-volatilization in fluidized beds. Fuel 85: 2388-2395.

52. Kunii D, Levenspiel O (1968) Fluidization Engineering. John Wiley & Son, New York.

53. Leckner B (1998) Fluidized bed combustion: Mixing and pollutant limitation. Prog Energ Combust Sci 24: 31-68.

54. Malte PC, Dorri PC (1981) The behaviour of fuel particles in wood-waste furnaces. Spring Meeting of the Western States Section of the Combustion Institute.

55. Cui H, Grace JR (2007) Fluidization of biomass particle: A review of experimental multiphase flow aspects. Chem Engg Sci 62: 45-55.

56. Dunnu G, Maier J, Schnell U, Scheffknecht G (2010) Drag coefficient of solid recovered fuels (SRF). Fuel 89: 4053-4057.

57. Fan L, Mao ZS, Yang C (2004) Experiment on settling of slender particles with large aspect ratio and correlation of the drag coefficient. Ind Eng Res 43: 7664-7670.

58. Hilton JE, Cleary PW (2011) The influence of particle shape on flow modes in pneumatic conveying. Chem Engg Sci 66: 231-240.

59. 59. Hilton JE, Mason LR, Cleary PW (2010) Dynamics of gas–solid fluidised beds with non-spherical particle geometry. Chem Engg Sci 65: 1584-1596.

60. Islam MA, Krol S, de Lasa HI (2010) Slip velocity in downer reactors: Drag coefficient and the influence of operational variables. Ind Engg Chem Res 49: 6735-6744.

61. Lau R, Hassan MS, Wong W, Chen T (2010) Revisit of the wall effect on the settling of cylindrical particles in the inertial regime. Ind Engg Chem Res 49: 8870-8876.

62. Papadikis K, Gu S, Bridgwater AV (2010) 3D simulation of the effects of sphericity on char entrainment in fluidised beds. Fuel Process Technol 91: 749-758.

63. Papadikis K, Gu S, Fivga A, Bridgwater AV (2010) Numerical comparison of the drag models of granular flows applied to the fast pyrolysis of biomass. Energ Fuel 24: 2133-2145.

64. Ren B, Zhong W, Jin B, Lu Y, Chen X, et al. (2010) Study on the drag of a cylinder-shaped particle in steady upward gas flow. Ind Engg Chem Res 50: 7593-7600.

65. Scully J, Frawley P (2011) Computational fluid dynamics analysis of the suspension of nonspherical particles in a stirred tank. Ind Engg Chem Res 50: 2331-2342.

66. Yin C, Rosendahl L, Kær SK, Sørensen H (2003) Modelling the motion of cylindrical particles in a nonuniform flow. Chem Engg Sci 58: 3489-3498.

67. Yin C, Rosendahl L, Kær SK, Sørensen, H (2011) Corrigendum to "Modelling the motion of cylindrical particles in a nonuniform flow". Chem Engg Sci 66: 17.

68. Zhou ZY, Pinson D, Zou RP, Yu AB (2011) Discrete particle simulation of gas fluidization of ellipsoidal particles. Chem Engg Sci 66: 6128-6145.

69. Nikolopoulos A, Papafotiou D, Nikolopoulos N, Grammelis P, Kakaras E (2010) An advanced EMMS scheme for the prediction of drag coefficient under a 1.2 MWth CFBC isothermal flow—Part I: Numerical formulation. Chem Engg Sci 65: 4080-4088.

70. Nikolopoulos A, Atsonios K., Nikolopoulos N, Grammelis P, Kakaras E (2010) An advanced EMMS scheme for the prediction of drag coefficient under a 1.2 MWth CFBC isothermal flow—Part II: Numerical implementation. Chem Engg Sci 65: 4089-4099.

71. Labowsky M (2011) An entrainment parameter for volatile particles. Chem Engg Sci 66: 810-812.

72. Gosman AD, Ioannides, E (1983) Aspects of computer simulation of liquid-fueled combustors. J Energ 7: 482-490.

73. Smith IW (1982) The combustion rate of coal chars: A review. Nineteenth Symposium (International) on Combustion, The Combustion Institute: 1045-1065.

74. Essenhigh RH (1988) An integration path for the carbon-oxygen reaction with internal reaction. Twenty-Second Symposium (International) on Combustion, The Combustion Institute: 89-96.

75. Wen CY, Chen LH (1982) Fluidized bed freeboard phenomena: Entrainment and elutriation. AIChE J 28: 117-128.

76. Chang YM, Chou CM, Su KT, Hung CY, We CH (2005) Elutriation characteristics of fine particle from bubbling fluidized bed incineration for sludge cake treatment. Waste Manag 25: 249-263.

77. Olofsson J (1980) Mathematical modelling of fluidised bed combustors. IEA Coal Research, London: 100.

78. Oka S (2003) Fluidized Bed Combustion, Dekker.

79. Kato K, Kanbara S, Tajima T, Shibasaki H, Quaswa K, et al. (1987) Effect of particle size on elutriation rate constant for a fluidized bed. J Chem Engg Japan 20: 498-504.

80. Wen CY, Hashinger RF (1960) Elutriation of solid particles from a dense fluidized bed. AIChE J 6: 220-226.

81. Tanaka I, Shinohara H, Hirosue H, Tanaka Y (1972) Elutriation of fines from fluidized bed. J Chem Engg Japan 5: 51-57.

82. Harris AT, Davidson JF, Thorpe RB (2003) Particle residence time distributions in circulating fluidised beds. Chem Engg Sci 58: 2181-2202.

83. Harris AT, Davidson JF, Thorpe RB (2003) The influence of the riser exit on the particle residence time distribution in a circulating fluidised bed riser. Chem Engg Sci 58: 3669-3680.

84. Hua L, Wang J, Li J (2014) CFD simulation of solids residence time distribution in a CFB riser. Chem Engg Sci 117: 64–282.

85. Geldart D, Cullinan J, Georhiades S, Gilvray D, Pope DJ (1979) The effect of fines on entrainment from gas fluidized beds. Trans Inst Chem Eng 57: 269-275.

2

Anaerobic Co-Digestion of Wastewater Activated Sludge and Rice Straw in Batch and Semi Continuous Modes

Nabil N Atta, Amro A El-Baz, Noha Said and Mahmoud M Abdel Daiem*

Department of Environmental Engineering, Zagazig University, Egypt

Abstract

Co-digestion of sewage sludge with rice straw may be attractive option from energetic, as well as, environmental viewpoints. In this study, co-digestion of wastewater activated sludge (WWAS) with grinded rice straw at different ratios (0.5, 1.0, 1.5 and 3.0%), straw to WWAS based on weight, was performed using batch reactors. Moreover, a semi-continuous model was developed for sludge digestion and co-digestion with rice straw. The results showed that the co-digestion of WWAS with rice straw improved the carbon to nitrogen ratio (C/N) and consequently increased biogas production compared to sludge digestion. Moreover, total solids, total volatile solids and chemical oxygen demand were reduced by digestion during the reaction time. Furthermore, the biogas yield increased by increasing mixing ratio in co-digester and reached four times at maxing ratio 3.0% compared to sludge digestion. The semi-continuous model showed that the co-digestion increased total biogas amount continuously and methane was the main component in biogas released from digester and co-digester.

Keywords: Co-digestion; Activated sludge; Rice straw; Biogas production

Introduction

Anaerobic digestion has been traditionally used as an effective, environmental sustainability and economical technology for the biological treatment of sewage sludge which enables energy production as heat, electricity and/or vehicle fuel, as well as stabilization of volume reduction of sludge [1,2] (Figure 1).

In Egypt, about 2.0×10^6 tons of sewage sludge are produced annually from 303 wastewater treatment plants (WWTPs) [3,4]. One of the most important WWTP in Egypt is Al Gabel Asfer that having a current sewage treatment capacity of 1.8×10^6 m³/day, which is expected to be doubled in 2020 [3]. The application of anaerobic digestion technology for sludge stabilization and power generation in Al-Gabel Al-Asfer WWTP has achieved good results and considerable experience of operation and maintenance has been gained. A large portion of the biogas produced is currently used for the operation of hot water boilers, which are used to heat the raw sewage sludge in the primary digesters. Dual fuel generators use the excess digested gas to generate electricity that representing about 37-68% of the power consumed by this WWTP [3].

In anaerobic digester, proper carbon to nitrogen (C/N) ratio is important for efficient digestion [2]. However, unbalanced C/N ratio in the sewage sludge inhibits the anaerobic digestion efficiency due to the formation of ammonia and volatile fatty acids, which, if accumulate too much in the digester, would inhibit the methanogen activity [5]. Typically, the C/N ratio of sewage sludge is ranged between 6-16 [6]. However, the optimal C/N ratio for anaerobic digestion should be in the range of 20-30 [7]. Recently, different organic waste materials with higher content of organic carbon have been mixed with sewage sludge in anaerobic digester (anaerobic co-digester) to improve C/N ratio [8]. Therefore, the combination is leading to an increase in biogas production [1,5,9-12].

Anaerobic co-digestion of sewage sludge with different organic waste material has been studied such as: source-sorted organic fraction of municipal solid waste [12,13], confectionery waste [14], municipal solid waste [9,15], food waste [11,16-18], sludge from pulp and paper

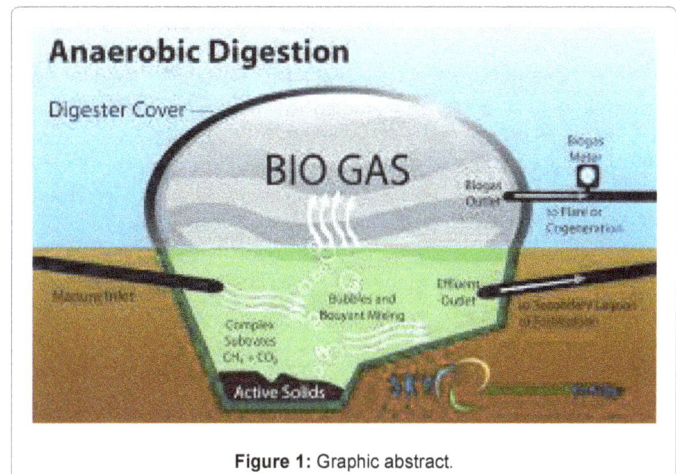

Figure 1: Graphic abstract.

industry and enzyme production [19], lixiviation of sugar beet pulp [20], grease trap sludge from a meat processing plant [1], macroalgae from the lagoons [21], sterilized solid slaughterhouse waste [22], corn straw [23], and rice straw [2,5,24,25].

In particular, rice straw is one of the most problematic agriculture wastes in Egypt, where it is generated in huge amounts (3.0 to 7.0 × 10^6 tons/year) over a limited harvesting period, representing the largest amount of unused agriculture residues by a wide margin [26-28]. A small amount of rice straw being used for livestock feedstuff of

*Corresponding author: Mahmoud M Abdel Daiem, Department of Environmental Engineering, Faculty of Engineering, Zagazig University, 44519, Zagazig, Egypt, E-mail: engdaim@hotmail.com

fertilizer and the majority of the waste is burned in open field, causing great impact on the greenhouse aspect as gas emissions as well as air pollution and consequently affects public health [29]. Moreover, the utilization of rice straw as a source of energy production is urgently needed in Egypt [26].

Komatsu et al. [5] investigated the feasibility of anaerobic co-digestion of sewage sludge and rice straw via mesophilic and thermophilic digesters using feeding ratio 1.0:0.5 based on the total solids (TS) of sewage sludge to rice straw and the results showed that the presence of rice straw increased methane production by 66-82% in case of mesophilic digester that was more effective than thermophilic one. Lei et al. [25] studied the influence of the effect of phosphate supplementation to anaerobic sludge in the presence of rice straw and the results showed that the phosphate seemed to have a little evident effect of the performance of anaerobic co-digestion, such as total volume of biogas or methane and accelerated the process. Kim et al. [2] investigated effective biohydrogen production process through anaerobic co-digestion of rice straw and sewage sludge via batch test. Based on their results, adding of rice straw showed that the C/N ratio of 25, pH range of 4.5-5.5, and untreated sludge were optimum for H_2 production. El-Bery et al. [24] studied the effect of thermal pre-treatment on inoculum sludge for continuous H_2 production from alkali hydrolyzed rice straw by using two mesophilic anaerobic baffled reactors with untreated and thermally treated sludge, ABR1 and ABR2, respectively; at a constant hydraulic retention time 20 h and organic loading rate 0.50-2.16 g chemical oxygen demand (COD)/Ld. They reported that thermal pretreatment of sludge slightly improved H_2 production around 212 and 261 mL H_2/g for ABR1 and ABR2, respectively, and COD removal for both reactors. Furthermore, Maroušek et al. [30-33] have studied different methods to improve the biogas production from different residues such as pre-treatment methods (hot maceration, steam explosion, and pressure shockwaves) and found that the self-standing hot maceration did not significantly increase the methane yield, however, steam explosion has potential to significantly increase the methane yields, furthermore, the pressure shockwaves are capable of high methane yield.

From this background, the main objectives of this study are to investigate the effect of adding rice straw with different ratios to anaerobic digester of wastewater activated sludge (WWAS) on biogas production, analyze the chemical composition of the biogas production and study the effect of co-digestion on characteristics of digested material using batch and semi-continuous reactors.

Materials and Methods

Materials

Wastewater activated sludge samples were obtained from Altal-Alkabeer wastewater treatment plant, Ismailia Government, Egypt. Samples of WWAS were stored in cold (4°C) until their use. Rice straw samples were collected from El-Sharkia Government, Egypt with approximately length of 1.0 m. The rice straw was shredded to 2.0 mm and was stored at ambient temperature far from heat and moisture, for subsequent use.

Experimental methods

Batch experiment: The batch reactor consisted of a reactor connected to gas collector with P.V.C. tube and the gas collector was attached to an open jar by a tube with valve to measure the volume of water collected due to the pressure of the biogas production. The batch reactor was placed in a glass basin equipped with a heater and

thermostat to maintain a constant temperature (35°C) (Figure 2A). Five batch reactors were used in this experiment, each reactor contained 2.5 Kg WWAS and the first reactor was used as control contained sludge only. The other reactors contained WWAS mixed with rice straw at ratios 0.5, 1.0, 1.5, and 3.0% (straw to WWAS based on weight). The batch reactors were static except for daily mixing by hand after gas measurements.

Semi-continuous experiment: Figure 2B showed a schematic diagram for a semi-continuous reactor, it was similar to the batch reactor, in addition to influent and effluent tubes, moreover, shaking water bath was used to keep the reactor in continuous mixing and fixed temperature (35°C) during the reaction time. Two reactors were used in this experiment, the first reactor contained 2.5 Kg WWAS solo and operated as a control, meanwhile, the second one contained the same amount of WWAS mixed with rice straw (mixing ratio 1.5% based on weight) and retention time was 25 days for both reactors. Samples were withdrawn from effluent of both reactors around 100 mL and were replaced with the same volume of substrate prepared daily for each reactor.

Analytical methods: TS, total volatile solids (TVS), and COD samples were measured according to the procedure mentioned in the standard methods. Phosphorous was measured by the method described by Murphy and Riley [34] using spectrophotometer model UV-160A-SHIMADZU. Potassium was determined spectrophotometrically in the acid digested samples via Atomic Absorption Spectrophotometer model UNICAM 969a by the method of Nation and Robinson [35]. Finally, the organic carbon content was determined by the method of Walkley and Black [36]. The pH was measured using pH meter (pHep, HI 98107 pocket-sized pH Meter).

The biogas production was calculated by measuring the volume of the displaced water due to the pressure of biogas. Samples of the biogas produced were analyzed using gas chromatograph, LNG analyzer Varian 3800 cp (Liquefied natural gas analyzer) to determine its composition.

Results and Discussion

Characteristics of raw material

The physical and chemical characteristics of WWAS and rice straw are summarized in Table 1. The WWAS has a high COD ranged between 17.0 to 20.5g/L and a few-solid substrate, with a total solid content ranged from 1.29 to 1.39%, while the rice straw contains 93.63%. The majority of TS present were TVS, 64.74 to 69.09 and 74.10 for WAS and rice straw, respectively. The C/N ratio of WWAS was 6.59-9.40, however, it was 72.90 for rice straw. These values indicated the importance of rice straw in anaerobic co-digester due to its high carbon content that may adjust the optimum digestion C/N ratio to improve the biogas production. Furthermore, P and K were observed in both raw materials constituent and they are considered as macronutrients and required for functioning of many microorganisms in biological processes.

Batch experiment

Figure 3 showed the cumulative biogas produced from digestion of WWAS and mixture of WWAS with rice straw at ratios (0.5, 1.0, 1.5, and 3.0%), finding that the rate of biogas production was elevated during the first 20 days of reaction time, especially in the presence of rice straw with different mixing ratios. Then, the rate decreased with higher reaction time up to 60 days, and then it was negligible at higher reaction time.

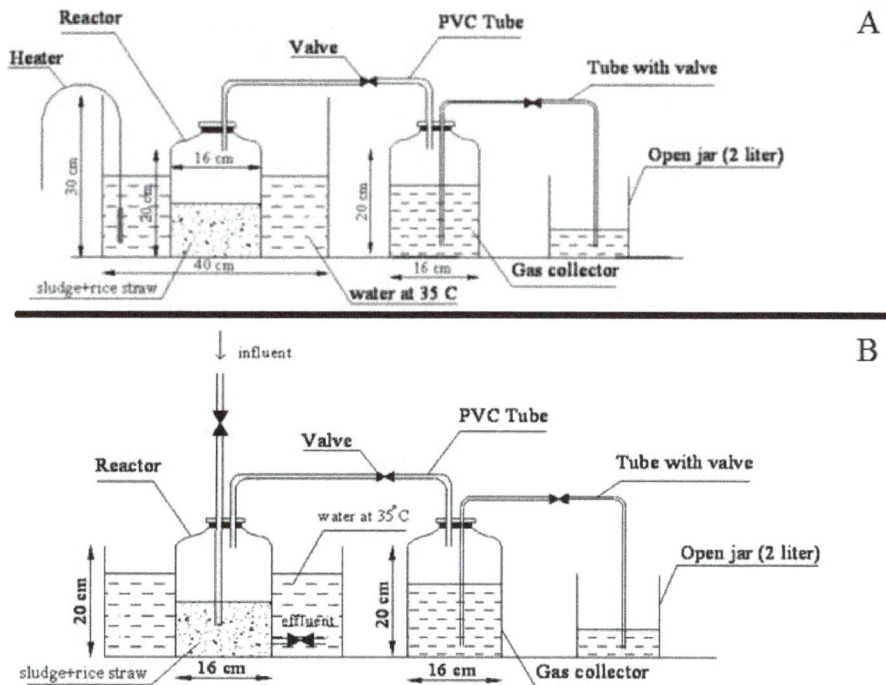

Figure 2: A schematic diagram of anaerobic co-digester reactors. A. The batch reactor and B. The semi-continuous reactor.

Parameter	Sewage sludge		Rice straw
	Batch test	Semi-continuous test	
pH	7.2	6.9	N.D.
COD (g/L)	17	20.5	N.D.
TS (%)	1.29	1.39	93.63
TVS (%)*	64.74	69.09	74.1
C/N	6.59	9.4	72.9
K (%)*	0.3	0.38	1.82
P (%)*	1.4	0.69	0.05
N.D. Not determined. *Percentages from TS.			

Table 1: Characterization of raw materials used in the experiments.

In general, co-digestion of WWAS with rice straw increased the amount of biogas produced which can be explained as the addition of rice straw improved C/N ratio and this ratio increased with increasing of straw content, (Figure 4), consequently enhanced biogas production by 150, 200, 250, 400% at mixing ratios 0.5, 1.0, 1.5, and 3.0%, respectively, in comparison to WWAS only. These results are similar to pervious works [5,9,10].

The relation between mixing ratio and biogas production has been proved by the following linear equation [11]:

$$Y = A \times X + B \tag{1}$$

Where, Y is the total amount of biogas predicted from anaerobic co-digester at equilibrium within boundary conditions (0.5 to 3.0% and 80 days), X is the mixing ratio, B is the total amount of biogas predicted from anaerobic digester at equilibrium, and A is the constant represented the relation between the mixing ratio and gas production. Equation coefficients have been found 8.10 and 7.51 for A and B, respectively, and the deviation percentage was calculated (1.64%) with

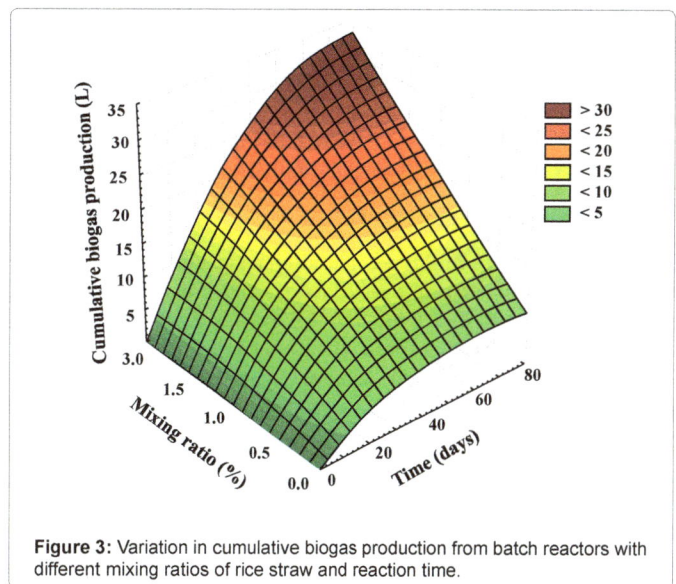

Figure 3: Variation in cumulative biogas production from batch reactors with different mixing ratios of rice straw and reaction time.

the following equation:

$$\%D = \frac{1}{N} \sum_{i=1}^{N} \left| \frac{V_{exp} - V_{pred}}{V_{exp}} \right| \times 100\% \tag{2}$$

where, V_{exp} is the value of experiment data and V_{pred} is the value of predicted data, Furthermore, C/N showed similar linear equation in relation to mixing ratio and the equation coefficients were 4.91 and 7.13 for A and B, respectively, and % D was 3.36 %, as shown in Figure 4.

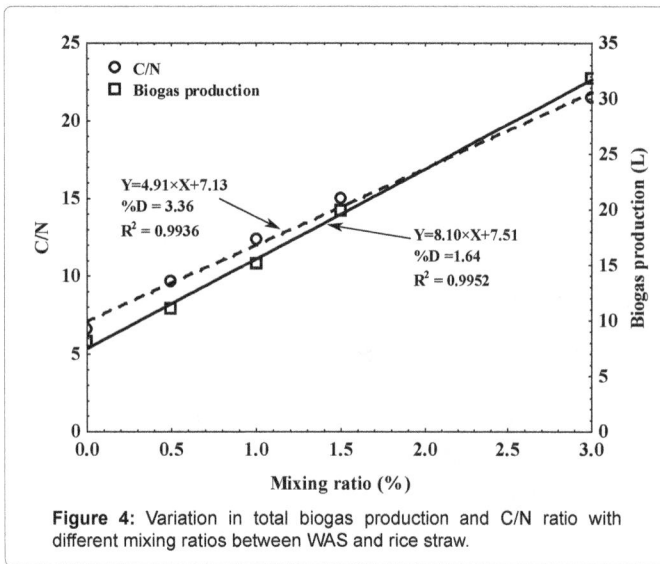

Figure 4: Variation in total biogas production and C/N ratio with different mixing ratios between WAS and rice straw.

Figure 5 presented the variation of pH, COD, TS, and TVS that occurred during the reaction time. The results showed that pH for all cases before and after digestion varied between 7 and 7.5 (Figure 5A), which nearly lied in the optimum pH range (6.5-7.5) for anaerobic digestion [37-39]. COD values decreased for all cases with about 50%, meanwhile, the reactor contained WWAS only recorded around 82 % reduction percent, which can be explained that WWAS reactor contained lower organic matter than co-digester reactors, due to the absence of rice straw (Figure 5B), that was completely degraded before the others co-digester reactors and it was confirmed by the produced biogas that almost stopped after 50 days of reaction time (Figure 3). The reduction in TS and TVS due to digestion and co-digestion varied in their percentages, however, the reduction percentages reached to about 40% and 53%, respectively (Figures 5C and 5D).

Semi-continuous experiment

Biogas production: From application stand point of view, semi-continues reactor was investigated. As a result obtained from batch reactor, co-digestion of 3.0% mixture of WWAS and rice straw recorded the highest amount of biogas production among the studied ratios; semi-continuous experiment for the same ratio was investigated, however, at this ratio, the reactor was stopped many times because clogging of tubes due to the high percent of rice straw, thus the ratio 1.5% was investigated and showed a flexibility in movement. Figure 6 showed the cumulative biogas production that was obtained from digestion co-digestion with 1.5% of rice straw. The amount of biogas released from co-digestion was 2.5-folded higher than that from digestion. This may due to higher C/N ratio in co-digester (16.0) than that in digester (9.4). Furthermore, the continuous increase in the total amount of biogas is attributed to the daily substrate feeding, consequently, continuous decomposition of organic compounds into biogas by anaerobic bacteria activation [40,41].

Biogas analysis: Samples of biogas produced from both reactors were collected and analyzed to determine its composition. Table 2 shows the composition of the biogas production from anaerobic digester and co-digester. It has been found that the main component in produced biogas was methane and followed by CO_2, then N_2. It was observed that there was no significant effect on methane content in produced biogas via co-digester. However, the released volume of methane based on the produced from co-digester was around 2.5-fold higher than that

anaerobic digester, similar results were found by Komatsu et al. [5].

Digested samples analysis: At constant period time (every 10 days), digested samples were taken from both reactors to measure TS, TVS, COD, TN, and pH during the digestion time. Figures 7A and 7B showed the variation of TS and TVS contents, they showed similar behavior for both reactors and were higher (2-fold) for co-digester compared to digester due to the presence of rice straw which has high contents of TS and TVS. It was observed that the decrease in their contents was higher in case of co-digester compared to digester only up to 40 days, may be due to the higher organic content in the co-digester than the digester which converted to biogas by anaerobic bacteria activation [40,41], and then remained virtually constant at a higher digestion time for both reactors. Figure 7C showed variations in COD during the reaction time for both reactors. The presence of rice straw in co-digester increased COD value by about 49% of that at digester due to the presence of rice straw, moreover, it was observed a similar trend for both reactors during digestion time and there was no difference between the both reactors for the removing of COD (around 12 g/L) during the reaction time.

Concentration of TN was found higher in the co-digester than that in digester, due to the presence of rice straw in co-digester (Figure 7D). However, a marked decrease in TN was detected for co-digester and digester up to 30 days of reaction time, and this is because the nutrients as nitrogen are conserved and mineralized to more soluble and biologically available forms [42] and can also be explained due to the conversion of some nitrogen into nitrogen gas during anaerobic decomposition, meanwhile, a slight increase was observed in co-digester at 40 days that may be related to the formation of ammonium ion (NH_4^+) during the biological processes [41,43].

Figure 7E showed that the pH was almost constant for digester during the digestion time, however, a marked decrease was observed for co-digester after ten days of digestion time. This may be attributed to acidogenic bacteria that produce organic acid which tend to lower the pH under normal conditions. The pH value increased at 40 days, reaching to the initial value. This increase may be explained due to the bicarbonate produced by methanogens [37] and the formation of ammonia, which usually formed in anaerobic processes as a result of mineralization of organic nitrogen [41], moreover, prolific methanogenesis may result in a higher concentration of ammonia [44]. Then pH values remained constant at higher digestion time. These results were similar to results obtained by Lei et al. [25].

However, the pH values through digestion process for both reactors were relatively in optimum range (6.5-7.5) which is favorable for the performance, stability, and to obtain maximal biogas yield [37-39].

Conclusions

Wastewater activated sludge and rice straw are produced in large quantities with serious effects on the environment. In this study, co-digestion of WWAS with grinded rice straw at different ratios (0.5-3.0%) was performed using batch model to investigate the effect of co-digestion of on biogas production. The results showed that the co-digestion of WWAS with rice straw improved biogas production compared to sludge digestion. Moreover, the biogas yield increased with the increasing of the rice straw content, it reached at 3.0% mixing ratio four times the amount released from the digested sludge only. The obtained results from semi-continuous reactor showed that the co-digestion increased total biogas amount continuously and no significant effect on the digested material characteristics which was

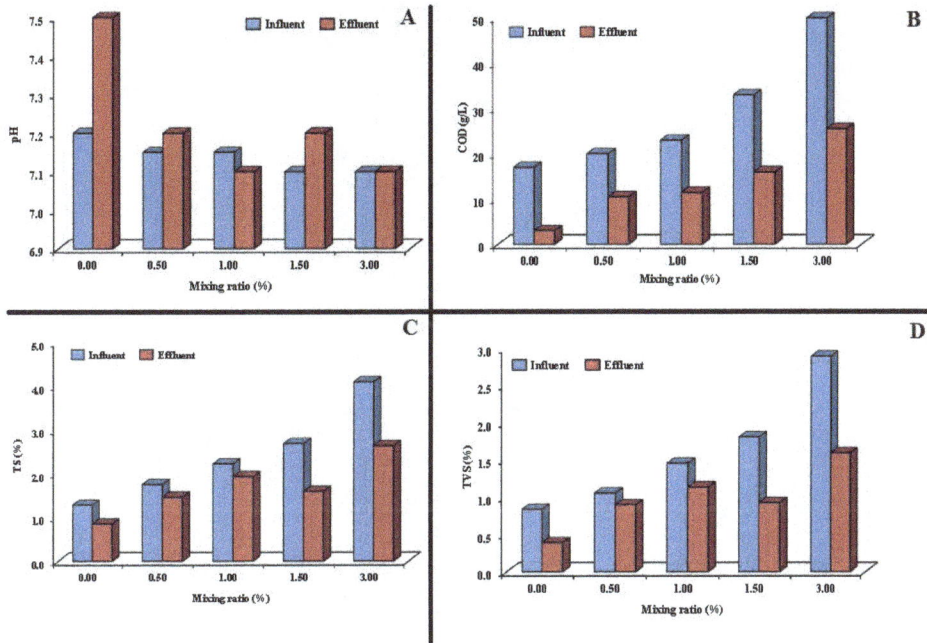

Figure 5: Characteristics of influent and effluent for the batch model with digestion time (80 days). A. pH. B. COD. C. TS. D. TVS.

Components	Composition %	
	Digester	Co-digester
Nitrogen	2.21	5.6
Methane	62.11	60.85
Carbondioxide	35.68	33.55

Table 2: The composition of the released biogas from digester and co-digester.

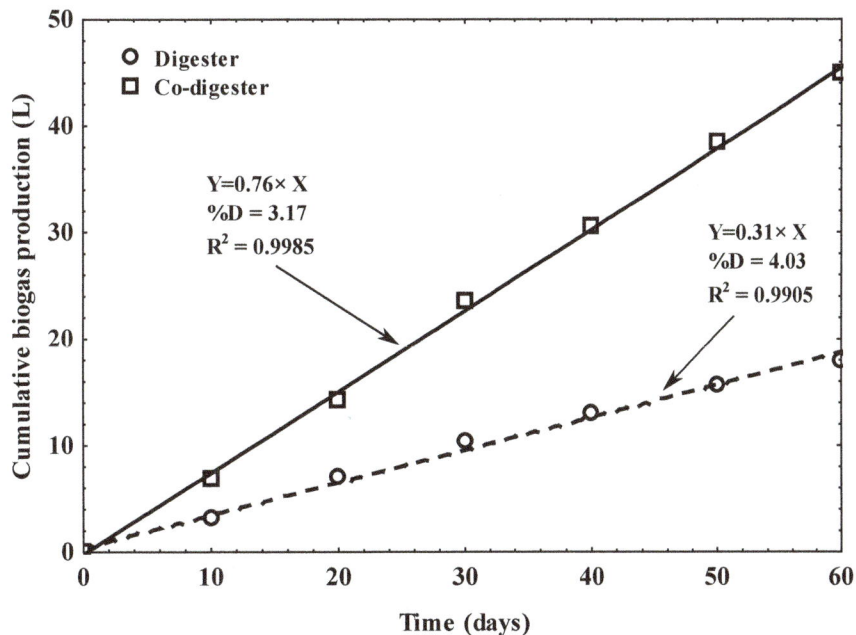

$Y=0.76 \times X$
$\%D = 3.17$
$R^2 = 0.9985$

$Y=0.31 \times X$
$\%D = 4.03$
$R^2 = 0.9905$

Figure 6: Cumulative biogas production via digestion and co-digestion for semi-continuous model.

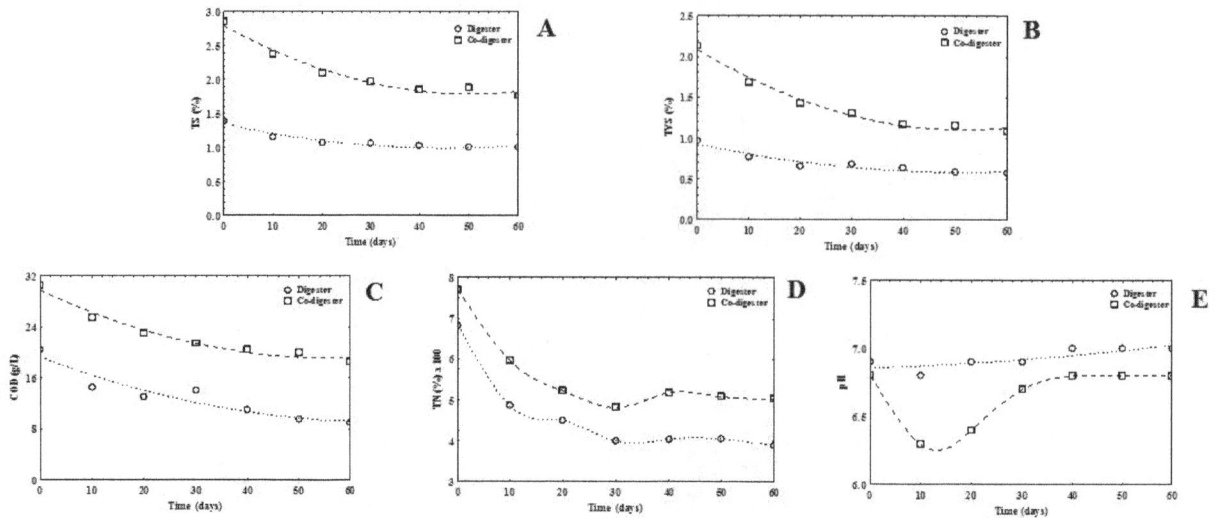

Figure 7: Characteristics variation of digested samples from digestion and co-digestion reactors during the digestion time in semi-continuous model. A. TS. B. TVS. C. COD. D. TN. E. pH.

noticed with the increase of the digestion time, although, no significant effect on methane percentage in the produced biogas was observed but its volume was increased due to the increase in the volume of produced biogas. Thus, the co-digestion of WWAS with rice straw using the existing digester is attractive option and could be beneficial to increase biogas production.

Acknowledgments

The authors are grateful to the team work at Environmental Engineering Department, Zagazig University for their valuable suggestions and encouragement through this study.

References

1. Luostarinen S, Luste S, Sillanpää M (2009) Increased biogas production at wastewater treatment plants through co-digestion of sewage sludge with grease trap sludge from a meat processing plant. Bioresour Technol 100: 79-85.

2. Kim M, Yang Y, Morikawa-Sakura MS, Wang Q, Lee MV, et al. (2012) Hydrogen prod uction by anaerobic co-digestion of rice straw and sewage sludge. Int J Hydrogen Energy 37: 3142-3149.

3. Ghazy M, Dockhorn T, Dichtl N (2009) Sewage Sludge Management in Egypt: Current Status and Perspectives towards a Sustainable Agricultural Use. World Acad Sci Eng Technol 57: 299-307.

4. El-Awady MH, Ali SA (2012) Nonconventional treatment of sewage sludge using cement kiln dust for reuse and catalytic conversion of hydrocarbons. Environmentalist 32: 464-475.

5. Komatsu T, Kudo K, Inoue Y, Himeno S (2007) Anaerobic co-digestion of sewage sludge and rice straw. pp: 495-501.

6. Tchobanoglous G, Theisen H, Vigil S (1993) Integrated solid waste management. New York: McGraw-Hill Inc 978.

7. Parkin G, Owen W (1986) Fundamentals of anaerobic digestion of waste water sludges. J Environ Eng 112: 867-920.

8. Hills DJ, Roberts DW (1981) Anaerobic digestion of dairy manure and field crop residues. Agr Wastes 3: 179-189.

9. Stroot PG, McMahon KD, Mackie RI, Raskin L (2001) Anaerobic co-digestion of municipal solid waste and biosolids under various mixing conditions-digester performance. Water Res 35: 1804-1816.

10. Yen HW, Brune DE (2007) Anaerobic co-digestion of algal sludge and waste paper to produce methane. Bioresour Technol 98: 130-134.

11. Serrano A, Siles JA, Chica AF, Martin MA (2014) Improvement of mesophilic anaerobic co-digestion of agri-food waste by addition of glycerol. J Environ Manage 140: 76-82.

12. Sosnowski P, Wieczorek A, Ledakowicz S (2003) Anaerobic digestion of sewage sludge and organic fraction of municipal solid wastes. Adv Environ Res 7: 609-616.

13. Edelmann W, Engeli H, Gradenecker M (2000) Co-digestion of organic solid waste and sludge from sewage treatment. Water Sci Technol 41: 213-221.

14. Lafitte-Trouqué S, Forster CF (2000) Dual anaerobic co-digestion of sewage sludge and confectionery waste. Bioresour Technol 71: 77-82.

15. Sekoai PT, Gueguim Kana EB (2013) A two-stage modelling and optimization of biohydrogen production from a mixture of agro-municipal waste. Int J Hydrogen Energy 38: 8657-8663.

16. Kim HW, Han SK, Shin HS (2003) The optimisation of food waste addition as a co-substrate in anaerobic digestion of sewage sludge. Waste Manag Res 21: 515-526.

17. Kim SH, Han SK, Shin HS (2004) Feasibility of biohydrogen production by anaerobic co-digestion of food waste and sewage sludge. Int J Hydrogen Energy 29: 1607-1616.

18. Nathao C, Sirisukpoka U, Pisutpaisal N (2013) Production of hydrogen and methane by one and two stage fermentation of food waste. Int J Hydrogen Energy 38: 15764-15769.

19. Einola JKM, Luostarinen SA, Salminen EA, Rintala JA (2001) Screening for an optimal combination of municipal and industrial wastes and sludges for anaerobic co-digestion. Proceedings of 9th World Congress on Anaerobic Digestion, Antwerpen, Belgium 1: 357-362.

20. Montañés R, Pérez M, Solera R (2013) Mesophilic anaerobic co-digestion of sewage sludge and a lixiviation of sugar beet pulp: optimisation of the semi-continuous process. Bioresour Technol 142: 655-662.

21. Cecchi F, Pava P, Mata-Alvrez J (1996) Anaerobic co-digestion of sewage sludge: application to the macroalgae from the Venice lagoon. Resour Conserv Recy 17: 57-66.

22. Pitk P, Kaparaju P, Palatsi J, Affes R, Vilu R (2013) Co-digestion of sewage sludge and sterilized solid slaughterhouse waste: Methane production efficiency and process limitations. Bioresour Technol 134: 227-232.

23. Zhou A, Guo Z, Yang C, Kong F, Liu W, et al. (2013) Volatile fatty acids productivity by anaerobic co-digesting waste activated sludge and corn straw:

effect of feedstock proportion. J Biotechnol 168: 234-239.

24. El-Bery H, Tawfik A, Kumari S, Bux F (2013) Effect of thermal pre-treatment on inoculum sludge to enhance bio-hydrogen production from alkali hydrolysed rice straw in a mesophilic anaerobic baffled reactor. Environ Technol 34: 1965-1972.

25. Lei Z, Chen J, Zhang Z, Sugiura N (2010) Methane production from rice straw with acclimated anaerobic sludge: Effect of phosphate supplementation. Bioresour Technol 101: 4343-4348.

26. Abou-Sekkina MM, Issa RM, Bastawisy AEM, El-Helece WA (2010) Characterization and Evaluation of Thermodynamic Parameters for Egyptian Heap Fired Rice Straw Ash (RSA). Int J Chem 2: 81-88.

27. Tewfik SR, Sorour MH, Abulnour AMG, Talaat HA, El Defrawy NM, et al. (2011) Bio-Oil from Rice Straw by Pyrolysis: Experimental and Techno-Economic Investigations. J Am Sci 7: 59-67.

28. Said N, El-Shatoury SA, Díaz LF, Zamorano M (2013) Quantitative appraisal of biomass resources and their energy potential in Egypt. Renew Sust Energy Rev 24: 84-91.

29. Mussatto SI, Roberto IC (2004) Optimal experimental condition for hemicellulosic hydrolyzate treatment with activated charcoal for xylitol production. Biotechnol Prog 20: 134-139.

30. Maroušek J, Kawamitsu Y, Ueno M, Kondo Y, Kolár L (2012) Methods for improving methane yield from rye straw. Applied Eng Agri 28: 747-755.

31. Maroušek J (2013) Prospects in straw disintegration for biogas production. Environ Sci Pollut Res Int 20: 7268-7274.

32. Maroušek J, Hašková S, Zaman R, Váchal J, Vanicková R (2015) Processing of residues from biobas plants for energy purposes. Clean Tech Environ Policy 17: 797-801.

33. Maroušek J, Hašková S, Zaman R, Váchal J, Vanicková R (2014) Nutrient management in processing of steam-exploded lignocellulose phytomass. Chem Eng Technol 37: 1945-1948.

34. Murphy J, Riley JH (1962) A modified single solution methods for the determination of phosphate in natural waters. Analytica Chimica Acta 27: 31-36.

35. Nation JL, Robinson FA (1971) Concentration of some major and trace elements in honeybee, royal jelly and pollen determined by atomic absorption spectrophotometer. J Apic Res 10: 35-43.

36. Walkley A, Black IA (1934) An examination of the Degt jareff method for determining soil organic matter and a proposed modification of the chromic acid titration method. Soil Sci 37: 29-38.

37. Anderson GK, Yang G (1992) pH control in anaerobic treatment of industrial wastewater. J Environ Eng 118: 551-567.

38. Yadvika, Santosh, Sreekrishnan TR, Kohli S, Rana V (2004) Enhancement of biogas production from solid substrates using different techniques: a review. Bioresour Technol 95: 1-10.

39. Liu CF, Yuan XZ, Zeng GM, Li WW, Li J (2008) Prediction of methane yield at optimum pH for anaerobic digestion of organic fraction of municipal solid waste. Bioresour Technol 99: 882-888.

40. Keshtkar A, Ghaforian H, Abolhamd G, Meyssami B (2001) Dynamic simulation of cyclic batch anaerobic digestion of cattle manure. Bioresour Technol 80: 9-17.

41. Zaher U, Cheong D, Wu B, Chen S (2007) Producing Energy and Fertilizer from Organic Municipal Solid Waste. Ecology Publication.

42. Wilkie AC (2005) Anaerobic digestion: Biology and Benefits. Dairy Manure Mangement Conference 176: 63-72.

43. Process Design Manual Sludge Treatment and Disposal.

44. Lusk P (1999) Latest progress in anaerobic digestion. Biocycle 40: 52-54.

3

Water Splitting Test Cell for Renewable Energy Storage as Hydrogen Gas

George Passas and Charles W Dunnill*

Energy Safety Research Institute (ESRI), Swansea University, College of Engineering, Bay Campus, United Kingdom

Abstract

The simple water splitting electrolysis cell has been shown that can easily be used to assess iteratively changed aspects of design and operation for the water splitting process and the design concepts for water splitting devices. The design characteristics and materials have been discussed such that a cheap and easy starting point for the assessment of design and process modification can be fully assessed. Concentration of electrolyte, and distance between electrodes have been shown to be key to the resistance of the cell and therefore to the efficiency of the process. This test cell will form the basis for comparison for future research regarding a number of aspects of potential improvements to the water splitting process.

Keywords: Renewable; Energy; Electrical energy; Hydrogen energy

Introduction

As the world turns more towards renewable for its energy supply, energy storage and transport become a more prevalent consideration. The lack of correlation between renewable sources of energy and energy demand create many challenges for the buffering of supply and demand. Simple water splitting devices can be used to convert spare (renewable) electrical energy into "Green Hydrogen" gas which can be transported, stored and used "On demand" for domestic, commercial and transport applications [1-3]. The electrolysis of water has been known for centuries [2,4] but only accounts for a very small proportion of worldwide hydrogen production [5] as the "Value" of hydrogen increases this is set to change, with large scale electrolysis becoming more prevalent given the added value of the "Green" agenda.

Alkaline water electrolysis could provide the key to low cost, sustainable and environmentally friendly hydrogen production worldwide [4,6]. This paper shows a simple and innovative water splitting cell that can be used to store renewable energy in the form of hydrogen gas. The paper goes on to discuss a number of design factors in the water splitting device and assess various improvements in design for their functionality, and the stacking of the cells into a useable devices.

A water splitting device operating through electrolysis consists of an anode, a cathode and an electrolyte solution. In an alkaline system, a DC charge is applied across the anode and the cathode and gas yielding reactions occur on both. Electrons on the cathode are consumed by the reduction of water into hydrogen with the resultant hydroxide ions passing through the electrolyte to the anode. At the anode the hydroxide ions deposit their electrons and are oxidized to form oxygen and water [4,7].

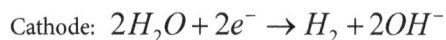

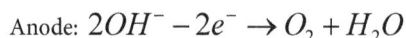

Cathode: $2H_2O + 2e^- \rightarrow H_2 + 2OH^-$

Anode: $2OH^- - 2e^- \rightarrow O_2 + H_2O$

A single cell water splitting device was prepared from acrylic and silicone using stainless steel electrodes as shown in Figure 1. This water splitting device was used as the starting point in the design improvements and a reference point for all of the characterizations, comparisons and discussions. In all cases the supply of electrolyte to the cell occurs from a large reservoir to ensure that the electrolytic concentration does not change over time between experiments.

The energy efficiency of water splitting devices is keys to the production of green hydrogen from renewable energy alongside the flexibility of such devices to absorb and handle spikes in political output. It is understood that a conventional industrial electrolyze requires between 4.5 and 5 kWh of energy in order to produce a single meter cubed of hydrogen [8]. Different forms of renewable energy have different constraints. Wind energy for instance suffers greatly from the creation of electrical spikes that correspond to spikes in the speed and wind, or gusts. The ability to smooth the spikes or indeed "Top slice" makes the storage of renewable energy in the form of hydrogen gas very versatile and potentially beneficial.

Experimental Methods

The simple water splitting cell set-up is shown and modifications and assessment procedures discussed. In this multi component water splitting device (Figure 1) the central chambers can be filled with electrolyte solution from the reservoir and a current passed between the electrodes. Hydrogen forms on the cathode, while oxygen forms on the anode. Both gasses rise through the electrolyte and pool at the top where they are separated by the gas trap before leaving the chamber due to buoyancy. The multi component water splitting device is held together using nylon threaded rod and nylon bolts so that there is no electrical short-circuit between the plates and the electrolyte supplied and balanced via an external reservoir. This represents a basic point of origin for a water splitting concept from which all future adaptations can be measured and assessed.

Water splitting device parameters

Dimensions: The test cells are 100 mm × 100 mm with varying distance between the electrodes. The spacer wall thickness was 20 mm giving an initial surface area for the accessible electrodes as 60 mm × 60 mm or 3600 mm². In addition there was the presence of the gas trap which reduced the face to face surface area by 300 mm² giving a total active surface area of 3300 mm².

***Corresponding author:** Dunnill CW, Energy Safety Research Institute (ESRI), Swansea University, College of Engineering, Bay Campus, United Kingdom, E-mail: C.Dunnill@Swansea.ac.uk

Electrodes: Much research has been carried out on the development of electrode materials by groups all over the world as seen in the review by Park et al. [9]. Many electrode materials have been tested by us, with medical grade stainless steel (316) performing well and with maximum cost to efficiency ratio. This material has been chosen as it is easily accessible and allows a reasonable starting point for the design and improvement of the electrolysis cell. 0.9 mm thick plates were used throughout these experiments and were cut to the same dimensions and design as the spacers with the addition of a 20 mm x 20 mm tab for connecting the electrodes on the end plates.

Spacer materials: Acrylic sheet is easy to procure, chemical resistant and easy to work with and was used as the spacer material. The cells were cut from single sheets using a laser cutter so each section was a unique piece rather than glued square rod. The spacers were cut from 20 and 12 mm thick sheet.

Gas trap: 1 mm thick acrylic sheet was laser curt to the same design as the spacers however with the upper internal edge lowered by 5 mm, so as to trap and separate the gas as it rises on the side electrodes. The gas trap allows the collection of gas at the top of the device and the channeling of the gas through the gas output.

Seals: 0.75 mm thick silicone sheeting was used with the same design as the spacers cut out using the laser cutter. Under compression the thickness was assumed to be 0.5 mm.

Electrolyte input: The electrolyte balance was made through 8 mm acrylic tubes affixed into the base of each of the oxygen side spacers and connected to the base of the reservoir. During electrolysis twice as much gas formation occurs on the hydrogen side as on the oxygen side so the electrolyte input is performed on the least busy side.

Gas output: Acrylic tube of 8, 10 or 12 mm can be attached to standard pneumatic push-fit tubing for easy sealing and hydrogen processing. Push-fit tubing allows for the easy connection of both wet and dry gas processing. The gas was passed into an expansion tank (Figure 2) prior to release to the atmosphere such that the electrolyte level was fixed at roughly the centre of the expansion tank.

End plates: In addition to the materials and pieces outlined in Figure 1, end plates consisting of 8 mm thick green acrylic, placed over the end electrodes. These plates had an additional 20 mm of height such that the cell was positioned 20 mm above the ground, preventing short circuiting of the electrode plates and spreading the compression

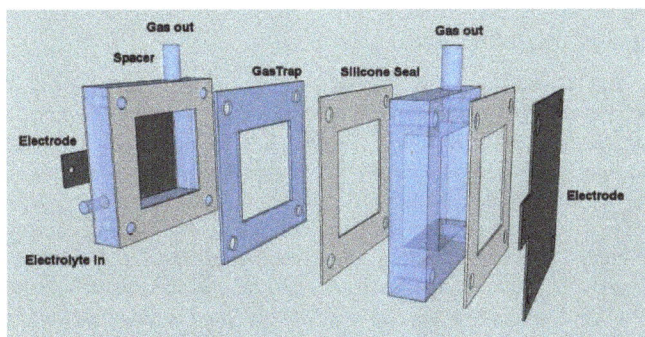

Figure 1: Schematic showing the makeup of a simple water splitting device for the storage of renewable energy. The multi component water splitting device consists of two electrodes, two spacer units and a central trap all held together with threaded nylon Rod and silicon seals to limit the leaks.

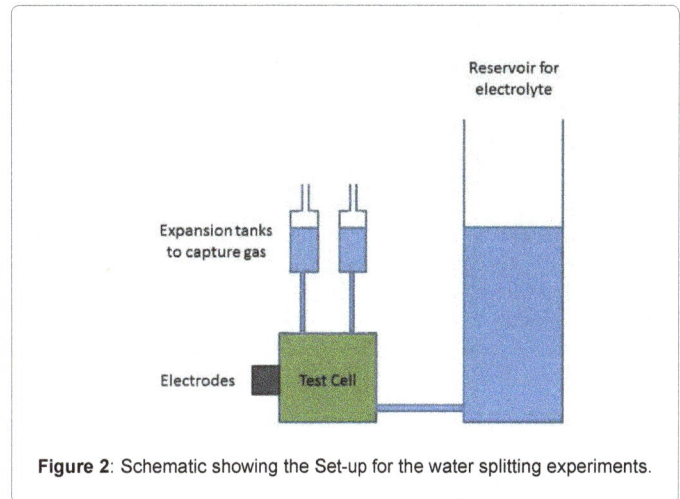

Figure 2: Schematic showing the Set-up for the water splitting experiments.

evenly across the spacer walls.

Fixing Screws: M8 size Nylon 66 threaded bar was used to apply a compressive force to hold the water splitting device together while maintaining electrical integrity between the electrodes.

Assessment process

The simple water splitting device was quantitatively assessed for a number of functional properties and assessment criteria, before modifications were performed in order to fully understand the impact of each modification as they occur. All measurements were taken using a large volume of electrolyte in the reservoir to maintain steady state concentration of the electrolyte throughout.

IV Curve: IV curves were performed using an Invium potentiometer at 0.5 volt increments from 0-10 V unless 2.5 A was reached. At each step the water splitting device was given 30 seconds to fully equilibrate and reach a static current value. In all of the analysis of the IV curve the raw data was processed through MatLab for interpretation. At each voltage setting the 60 measurements were collated, the first 10 readings, 5 seconds, were discarded and the final 25 seconds averaged (mean) to yield an operating current for that voltage setting. This was performed as there was a noticeable capacitance effect on the water splitting device as it charged and maintained steady state current, especially in the pre Ohmic region of the plot. This process invariably took less than 3 seconds to occur.

R_{opp}: The operating resistance of the cell (R_{opp}) is a direct measure of the efficiency of the cell and calculated from the IV curve $R_{opp} = 1/$ gradient of the IV line once it has stabilized. Matlab was programmed to calculate the Gradient based on the straight line between 3V and the maximum voltage reading in the calculation of the resistance.

V_{min}: The V_{min} value of the IV curve also yields information as to the efficiency of the cell. A low V_{min} implies a higher efficiency process. The lowest V_{min} possible theoretical value should be 1.23 V, corresponding to the potential for water splitting [7,8] in reality no gas evolution is observed below 1.65 to 1.7 V, [8] hence there is a need for the over potential. Industrial cells are often operated at between 1.8 and 2.6 V[8]. The V_{min} was assessed by extrapolation of the straight line part of the graph to I=0 value and is effectively the voltage at which the cell begins to operate as a water spitting device rather than a capacitor, when the minimum over potential has been achieved.

V_{min} =1.23 + Overpotential

Cell efficiency- thermal: The thermodynamic cell efficiency, $Cell_{Eff}$ is defined as:

$$Cell_{Eff} = \frac{HydrogenEnergy_{out}(KJ)}{ElectricalEnergy_{in}(KJ)} \times 100$$

Hydrogen gas has a calorific content of 286 KJ/Mol. This is regarded as the amount of energy that can be recovered from 1 mole of hydrogen by reaction with oxygen yielding only water as a by-product. Energy input is calculated by the multiplication of the voltage, current and time plots from the potentiostat. Measurements were performed at constant voltage for 5 minutes periods with a gas syringe measuring the volume of hydrogen produced. To lower errors the gas measurements were performed in triplicate and the cell efficiency values averaged.

Result and Discussion

The water splitting set-up is operated as a test rig, Figure 2. The cell was connected to a reservoir of electrolyte of greater volume than the cell cavity so that changes in volumes and concentration due to the electrolysis process are minimised. The gas output connections are connected to expansion tanks so that they don't overflow when gas bubbles are produced and the electrodes are connected to a potentiostat. Prior to each experiment the working electrolyte is returned to the ballast so that continued use does not lead to significant increasing concentration. The temperature was maintained at room temperature so as to avoid significant changes in electrolytic conduction [10] and buffered using the large ballast tank.

Baseline measurement

The setup as described in Figure 1 and Figure 2 was used to display the initial findings for the experiment sets. Two 20 ×20 square spacers were joined together with a gas trap and the IV line calculated. The raw data Figure 3a was then interpreted as set out in the experimental section to give the IV curve shown in Figure 3b. The values for the operational resistance and the minimum voltage required to drive the cell were calculated and are 6.8 Ohms and 2.03 Volts respectively with a concentration of 0.1 M NaOH.

Investigating the effect of distance between the electrode plates

It is understood that the distance between the plates has a significant effect on the overall efficiency of the process. There is a trade-off between gas separation and cell efficiency as higher distances giving better gas separation but lower efficiency.

To investigate the relationship between the spacer lengths and the resistance of the cells a 20 mm spacer with an electrolyte inlet was used as the cell, and sequentially increased in size by a 12 mm spacer. Measurements were taken accordingly.

There is a clear relationship between the distance between the electrodes and the resistance of the water splitting cell. The results show a linear response between spacer length and the resistance. The gradient of this line, inset in Figure 4 is 0.1515 ohms/mm and is resultant from the surface area of the electrodes, the concentration of the electrolyte and the temperature of the electrolyte. The closer the electrodes are together, the lower the resistance and therefore the lower the resistance and the higher the efficiency. However, as the water splitting cell has no diaphragm to separate the gasses across the face of the electrodes the closer the plates are to each other the less gas separation occurs leading

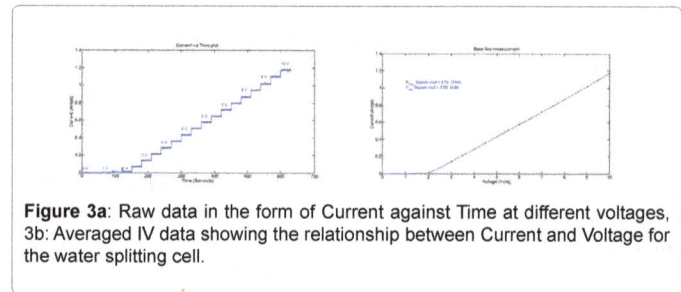

Figure 3a: Raw data in the form of Current against Time at different voltages, 3b: Averaged IV data showing the relationship between Current and Voltage for the water splitting cell.

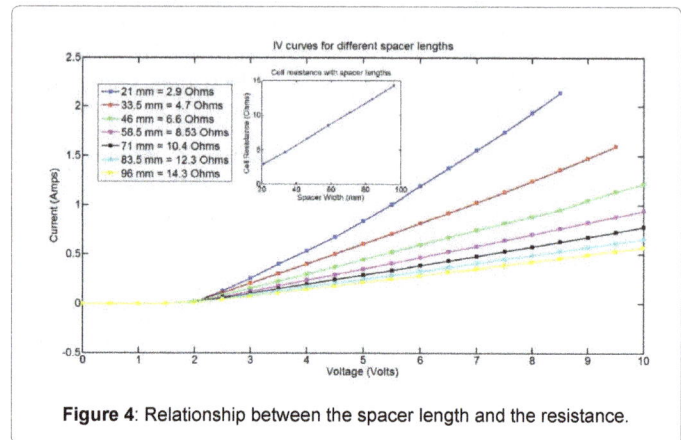

Figure 4: Relationship between the spacer length and the resistance.

to inefficiencies and safety concerns. At 96 mm there is no hydrogen / oxygen mixing whilst at 21 mm there would be approximately 50% loss for the hydrogen stream if the gas separator were not in place. As the hydrogen evolution is twice the volume of the oxygen evolution there is no need for the spacers to be of the same diameter. A 20 mm hydrogen spacer can easily be paired with a 12 mm oxygen spacer to give a cell, (20 × 12) of lower resistance than the 20 × 20, yet comparable gas separation capabilities.

Interior roof design of the spacer

A build-up of gas during the electrolysis process in the top of the cell is observed to purge over once a critical volume is reached. This had a detrimental effect on the cell operation as it sucks electrolyte solution though the entire system and results in the mixing of gas bubbles. Hydrogen contamination in the oxygen stream results in wasted efficiency whilst oxygen contamination in the hydrogen stream has safety implications for the long term storage. A redesign of the shape of the cell cavity roof, Figure 5, to channel the gas out of the cell, improved the gas separation characteristics.

The sloped inside roof appears to have had little effect on the resistance of the cell, Figure 6. The marginal improvement in performance, (lower R value) is likely due to the high effective surface area between the plates. There is no change in the face to face surface area of the electrodes, as the gas trap is still in place, however the sloped roof will allow for additional linear contact between the plates at non-90° angles. At a concentration of ~0.1 M the resistance of the cell will be very sensitive to changes in the effective face to face area. The change in the roof design will also allow for better gas handling, getting the bubbles of hydrogen and oxygen out of the system such that they do not interfere with the electrical processes occurring on the electrodes. The gas handling of the cell was considerably improved with gas passage from the cell into the expansion tank facilitated.

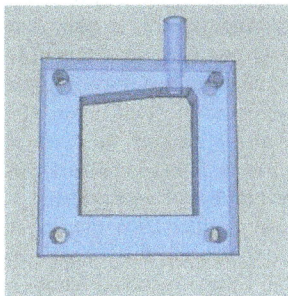

Figure 5: Diagram showing the design of the cell cavity roof- leading to improvements in gas flow dynamics.

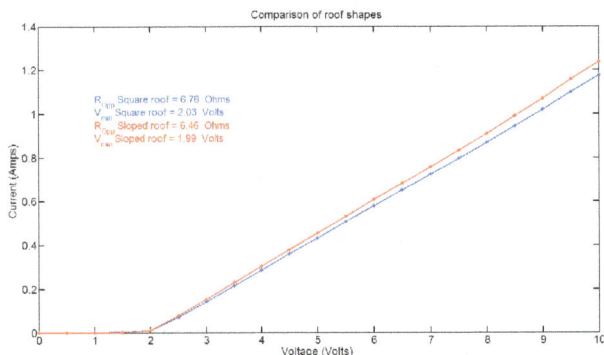

Figure 6: IV curves showing the effect of the different shape of the spacer roof.

Investigation of the concentration of the electrolyte solution

In all cases the 20 × 20, sloped roof setup was used to assess the different concentrations. Electrolyte mixtures from 0.1 M to 1 M, incrementing in 0.1 M intervals were prepared in a volume of 4 Litres. The same protocol as above was run with the samples denoted by their concentration of NaOH.

The concentration of electrolyte has a profound effect the resistance of the cell and therefore the efficiency of the overall water splitting process. Again a trade-off exists between the concentration of the electrolyte, the efficiency of the cell and the safety of the operators. A highly concentrated electrolyte whilst more efficient has more significant safety implications, which depending on the application of the water splitting cell poses a more or less significant threat, both to the operator and the environment. Alkali electrolysis is therefore limited in practice by the concentration of the electrolyte [11]. A 25-30% KOH solution has been widely reported [6,11]. One would expect that doubling the concentration of the electrolyte results in halving the resistance as a result of their being twice as many charge carriers available for conduction. This is evidenced by the numbers for the resistance measurements in-set, Figure 7. At 0.1 M the resistance was 7.3 Ω and reduced to 3.75 Ω at 0.2 M, approximately halving in resistance. Doubling the concentration from 0.2 M to 0.4 M, the resistance drops from 3.75 Ω down to 2.01 Ω again approximately halving, etc. In reality the efficiency gain in doubling the concentration does not quite halve the resistance with a lesser effect at the higher concentrations with the exponential decay dropping off slightly. Hence there is not

a 10 fold decrease in resistance associated with the 10 fold increase in concentration. Commercial alkaline water splitters running at a very high concentration of KOH have a very significant risk to operators and the environment in the event of a leak [4].

Figure 8 shows the straight lines formed in the ohmic region of the IV curves, extrapolated backwards to the 0 amps position. The inset shows the point on the X axis where the lines pass. The inset diagram shows that the position for the cell under all conditions is around 2V however there is a general trend that the higher the concentration the lower the Vmin number. Low V_{min} values would lead to the earlier onset of hydrogen gas and thus a higher efficiency for the water splitting process. The position of the Vmin relative to 1.23 V is down to be overpotential of the cell. This can be changed or tuned by a number of factors not just the concentration of the electrolyte. Primarily it is important to note that the stainless steel electrodes are considered to have middle over potential [8] compared to metals such as lead zinc and tin with a high over potential or indeed platinum and palladium with a very low over potential [8]. The lower the overpotential, the better the material is for electrolysis and the higher the cell efficiency (Figure 9).

The thermal cell efficiency was calculated at 2.5, 5 and 7 V as described in the experimental section, for a number of different cell configurations. In each case the cell was filled with 0.1M NaOH and the current measured at fixed voltage for 5 minutes. The hydrogen produced was passed into a gas syringe and the total volume of hydrogen recorded. Measurements were taken in triplicate to lessen the effect of inaccuracies in hydrogen volume measurement using a gas syringe.

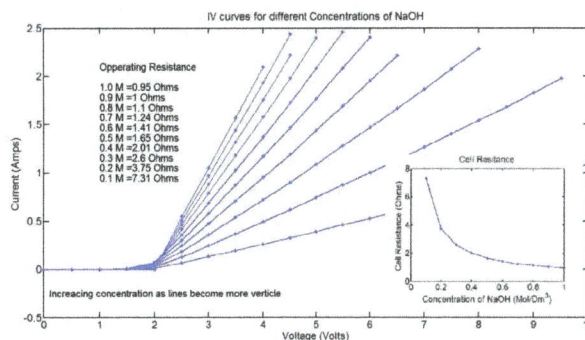

Figure 7: Graphical plot showing the IV curves for the varying concentrations, and the effect on the effective resistance of the cell.

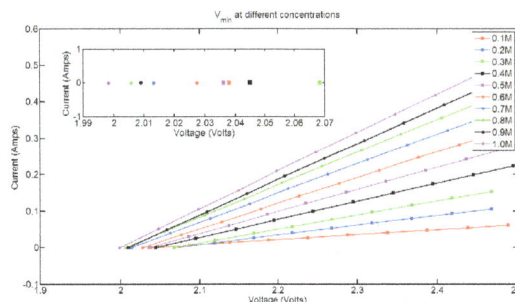

Figure 8: Graph showing the relationship between concentration and the V_{min} values.

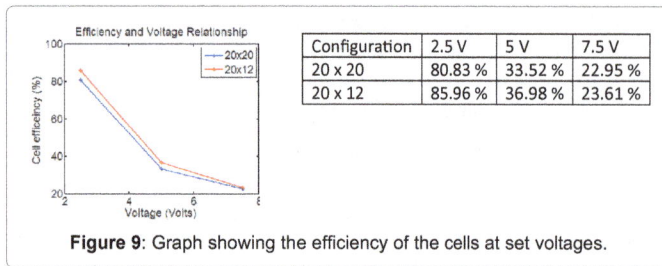

Figure 9: Graph showing the efficiency of the cells at set voltages.

Configuration	2.5 V	5 V	7.5 V
20 x 20	80.83 %	33.52 %	22.95 %
20 x 12	85.96 %	36.98 %	23.61 %

There is a clear relationship between the voltage of the cell and the thermal efficiency when calculated in this way. It is clear that the effect of resistance in the cell leads to a significantly more inefficient process and again leads to a trade-off. The most efficient process would occur at 2 V and produce almost no hydrogen however using almost no current. Whilst technically efficient this process is not useful for the storage of renewable energy.

As the measurement of hydrogen is directly related to the charge which is given by the integration of the current time graph it would seem obvious that any arrangement that raises the current whilst lowering the voltage would be beneficial to the efficiency. An increase in concentration of electrolyte would have this effect, as would increase the surface area of the electrodes. At optimum efficiency these cells would be operating at close to 2 V. The electrolyser working in this study is operating at low voltage and achieving approximately 81-86% thermal efficiency. This can be compared to commercial electrolysers having a nominal efficiency of 70% [11] or to standard industrial operating parameters whereby 4.5 to 5 kwh/m³ of hydrogen [4,8]. This equates to between 70 and 78% thermal efficiency. Improvements in design and the use of higher concentration of electrolyte will no doubt yield higher the efficiencies.

There are many ways of calculating the efficiency of an electrolysis cell [7]. In many ways the efficiency of the cell is also considered to be a moot point given that in real terms the efficiency of the cell should be a value-based metric rather than energy-based metric. In this regard the percentage efficiency should be calculated as follows:

$$Efficiancy_{\varepsilon\varepsilon\varepsilon} = \frac{Value(\varepsilon)\,output}{Value(\varepsilon)\,output} \times 100$$

It is obvious from the equation above that the value efficiency metric is thus largely skewed by the potential value of energy input, as well as the value of the energy output. Energy that would otherwise be wasted from a power station, wind farm or solar plant would therefore have a very low value energy input and therefore a large Efficiency$_{\varepsilon\varepsilon\varepsilon}$ value irrespective of the use for the energy output. Likewise there is a dilemma regarding the value of the output. Hydrogen gas could have a number of different uses and therefore a number of different values per unit volume. Hydrogen used to replace electricity would be valued against the price of electricity whilst hydrogen used to replace petrol in vehicles would have to be valued against the price of petrol. The additional cost of processing should also be taken into account as hydrogen direct from the electrolysis process would be suitable for some applications however unsuitable for others.

Stack configuration

The water splitting cells, once optimised can be stacked together such that the anode of one cell is the cathode of the next. In this way a 2V cell can be designed to fit any reasonable voltage input. Small scale renewable energy inputs, such as photovoltaic sheets and wind generators are mostly found to give an output of 12 or 24 V. A set of six cells can be stacked together in "series" configurations to provide water splitting device that operates most efficiently at 12 V. As seen before there is a direct relationship between the current passing through the cell and the hydrogen production, so in a stack configuration this needs to be multiplied by the number of cells available. 1 A of current passing through a single cell will produce 1/6 of the amount of hydrogen as 1 A of current passing through 6 cells, the voltage required to drive the six cells will of course be six times the voltage required to drive a single cell.

Conclusion

The simple water splitting electrolysis cell has been shown that can easily be used to assess iteratively changed aspects of design and operation for the water splitting process and the design concepts for water splitting devices. The design characteristics and materials have been discussed such that a cheap and easy starting point for the assessment of design and process modification can be fully assessed. Concentration of electrolyte, and distance between electrodes have been shown to be key to the resistance of the cell and therefore to the efficiency of the process. This test cell will form the basis for comparison for future research regarding a number of aspects of potential improvements to the water splitting process. These water splitting cells have the potential to be used as methods of storage of renewable energy.

References

1. Kreuter W, Hofmann H (1998) Electrolysis: The important energy transformer in a world of sustainable energy. Int J Hydrogen Energy 23: 661-666.

2. Trasatti S (1999) Water electrolysis: who first?. J Electroanalytical Chem 476: 90-91.

3. Mahrous AFM, Sakr LM, Balabel A, Ibrahim K (2011) Experimental Investigation of the Operating Parameters Affecting Hydrogen Production Process through Alkaline Water Electrolysis. Int J of Thermal Environ Engg 2: 113-116.

4. Rashid M, Mesfir MKA, Nasim H, Danish M (2015) Hydrogen Production by Water Electrolysis: A Review of Alkaline Water Electrolysis, PEM Water Electrolysis and High Temperature Water Electrolysis. Int J Engg Adv Technol 4: 2249-8958.

5. Dunn S (2002) Hydrogen futures: toward a sustainable energy system. Int J Hydrogen Energy 27: 235-264.

6. Williams JH, DeBenedictis A, Ghanadan R, Mahone A, Moore J, et al. (2012) The Technology Path to Deep Greenhouse Gas Emissions Cuts by 2050: The Pivotal Role of Electricity. Science 335: 53-59.

7. Zeng K, Zhang D (2010) Recent progress in alkaline water electrolysis for hydrogen production and applications. Progress Energy Combust Sci 36: 307-326.

8. Wang M, Wang Z, Gong X, Guo Z (2014) The intensification technologies to water electrolysis for hydrogen production – A review. Renew and Sustain Energy Rev 29: 573-588.

9. Park S, Shao Y, Liu J, Wang Y (2012) Oxygen electrocatalysts for water electrolyzers and reversible fuel cells: status and perspective. Energy Environ Sci 5: 9331-9344.

10. Nikolic VM, Tasic GS, Maksic AD, Saponjic DP, Miulovic SM, et al. (2010) Raising efficiency of hydrogen generation from alkaline water electrolysis – Energy saving. Int J Hydrogen Energy 35: 12369-12373.

11. Mazloomi K, Sulaiman NB, Moayedi H (2012) Electrical Efficiency of Electrolytic Hydrogen Production. Int J Electrochem Sci 7: 3314-3326.

4

Analysis of the Distribution of Onshore Sedimentary Basins and Hydrocarbon Potential in China

Zhenglong Jiang[1]*, **Yajun Li[2]**, **Yunfei Zhang[1]** **and Kangning Xu[1]**

[1]*School of Marine Sciences, China University of Geosciences, Beijing 100083, China*
[2]*School of Energy Resources, China University of Geosciences, Beijing 100083, China*

Abstract

Based on 20 years of onshore exploration in China, we compiled onshore sedimentary basin maps and oil and gas horizontal distributions for the major basins in China and analyzed the characteristics of basin distribution and hydrocarbon potential, which can be summarized in the following four points. (1) More than 400 basins have developed in China's onshore region. The development of large-scale basins is based on paleo-platforms (plates) or the micro-plate geological background. (2) The Tianshan-Xingmeng orogen and basin cluster along the North China/eastern Yangtze Platform host the majority of China's onshore oil reserves (66.64%). (3) The three major craton basins, Tarim, North China and Yangtze, host 70.19% of the total onshore natural gas reserves. (4) China's onshore oil reserves are concentrated in layers originating from the Cenozoic and Mesozoic, and the natural gas reserves are primarily distributed in layers originating from the Cenozoic, Mesozoic, and Upper Paleozoic.

Keywords: Onshore China; Sedimentary basin; Resource extent; Oil and gas; Layer series

Introduction

There are a large number of basins in China's onshore region. The type, distribution, formation, and development of sedimentary basins have been well studied by numerous scholars since the twentieth century, and numerous oil and gas fields have been detected since 2000 [1,2]. Our understanding of the geological characteristics of these basins has also increased; however, the exploration of unconventional resources (oil shale, oil sand, coalbed gas, shale gas, etc.) requires increased study of sedimentary basins.

China has a large reserve of oil and gas, and because of the rapid development of oil and gas exploration over the past 30 years, China's hydrocarbon potential has been increasingly evaluated [1-4]. The Ministry of Land and Resources proposed a new nation-wide evaluation of the hydrocarbon reserve resources and conducted a series of studies on the geological conditions of oil and gas formation, enrichment regularity, and potential areas to be explored in the future [3]. In the past ten years, the Ministry of Land and Resources strengthened the exploration of unconventional resources. Based on previous results, this paper will systematically study the characteristics of China's onshore sedimentary basins and distribution of oil and gas. The exploration of different fossil energy also has an important impact on the application of related technology.

Methods and Data

Sedimentary basin data

The distribution of basins is based on the following: "Map of exploration results of oil and gas basins in China" [5] "Distribution of Sedimentary Basins and Oil and Gas in China" [6] "Map of Oil and Gas Basins and Fields in China " [7], and "China Atlas of Oil and Gas Basin" [8]. The data from certain basins (Qiangtang, Songpan-Abei, South Yellow Sea, etc.) were selected from results within the national oil and gas strategic area. There are 410 sedimentary basins included in the distribution, and the total area is $421.8 \times 10^4 \text{ m}^2$ (Figure 1).

Figure 1 the sedimentary basins and their hydrocarbon potential in the onshore region of China. The boundaries of the basins were digitized by the MapGis software and the selected basin boundaries were compiled together according to their coordinates except the duplicate. The details are as follows: (a) All the large basins, the middle and small basins of the West, the Northwest and the Northeast of China are selected from "Distribution of Sedimentary Basins and Oil and Gas in China" [6]. (b) The middle and small basins of the Yangtze-Platform and the Southwest of China are selected from "Map of exploration results of oil and gas basins in China" [5]. (c) The middle and small basins of the Southeast of China are selected from "Map of Oil and Gas Basins and Fields in China" [6]. (d)The boundaries of the basins (Qiangtang, Songpan-Abei, South Yellow Sea) are selected from the results of MLRSCOG [3]. (e)The oil and gas fields are selected from the "China Atlas of Oil and Gas Basin" [8] and the new oil and gas fields discovered by Petro China and SINOPEC in the last 10 years.

Oil and gas data

Table 1 presents data on the exploited, detected, and show basins that contain major oil and gas reserves within China's onshore region. The data were collected from evaluations of national oil and gas resources [3], petroleum geology annals of China, and exploration results from PetroChina, SINOPEC.

Oil and gas resource basic data

Table 2 shows the oil and gas reserves of seven major basins. The data are from evaluations of national oil and gas [9].

The methods to evaluate the onshore conventional oil and natural gas resources are selected mostly on the basis of the exploration degrees and types of the basins [3]. The data of Tables 3, 4 and 5 are integrated from the results of the national oil and gas resources evaluation [3].

*****Corresponding author:** Jiang Z, School of Marine Sciences, China University of Geosciences, Beijing 100083, China, E-mail: jiangzl@ cugb.edu.cn

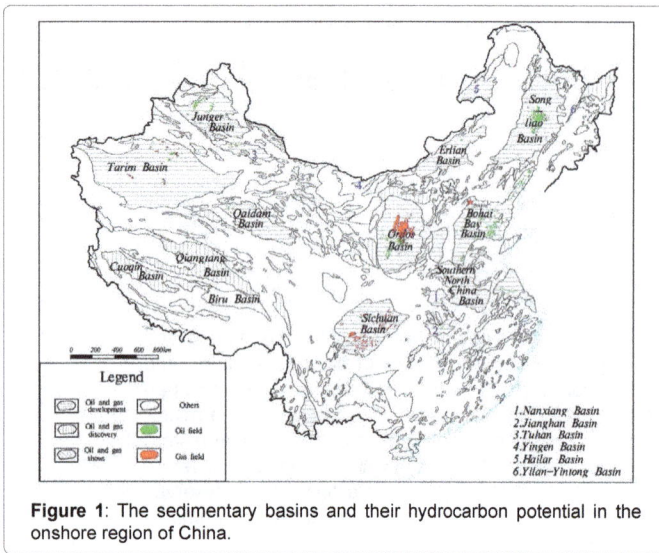

Figure 1: The sedimentary basins and their hydrocarbon potential in the onshore region of China.

Exploration Situation	Typical Basin
Exploitation	Songliao, Bohai Bay, Ordos, Junggar, Sichuan, Tarim, Qaidam, Turpan-Hami, Santanghu, Erlian, Hailar, Jiuxi, Jiudong, Jianghan
Detected	Yitong, Zhangwu, Fuxin, Huahai
Show	Dayangshu, Jiaolai, Wutang, Changdu, Cuoqin, Dongting, Hetao, Kumukuli, Liupanshan, Luoyang-Yichuan, Sanjiang

Table 1: Typical onshore petroleum basins in China.

Basin	Oil (10^8t)	Natural Gas (10^{12} m³)	Total (10^8t Oil Equivalent)
Bohai Bay	224.52	1.088	235.4
Songliao	113.07	1.403	127.1
Ordos	73.53	4.666	120.19
Junggar	53.19	6.514	118.33
Tarim	80.62	8.862	169.24
Qaidam	12.91	16	28.91
Sichuan	11.35	5.374	65.09
Total	569.19	29.507	864.26

Table 2: China's major onshore basin oil and gas resources.

Table 3 presents the extent of oil and gas resource in three petroliferous craton basins. The classification and statistical analysis are based on the petroleum geology conditions of the basins and the results of oil and gas exploration.

Table 4 presents the statistical results based on the distribution layers and areas of China's onshore petroleum resources. The layer series are Cenozoic, Mesozoic, Upper Paleozoic and Lower Paleozoic, and the areas are divided into East China, Central China, West China, South China and Qinghai-Tibet.

Table 5 presents the statistical results based on the distribution layers and areas of China's onshore natural gas resources. The layer series are Cenozoic, Mesozoic, Upper Paleozoic and Lower Paleozoic. Areas are divided as East China, Central China, West China, South China and Qinghai-Tibet.

Result and Discussion

Analysis of the sedimentary basin areas

The 410 basins (Figure 1) can be grouped into four categories according to area. There are 11 basins larger than 10×10^4 m²: Tarim, Songliao, Ordos,

Basin	Regional Tectonic Settings	Structural Feature	Source Rocks	Oil and Gas Formation	Resources Oil (10^8 t)	Resources Gas (10^{12} t m³)
Tarim	Paleo-plate since the Paleozoic, evolved into a craton basin in the Mesozoic	Active crust motion, strong vertical motion, developed faults	J-T, C-P,Є—O	K-E,T-J, C,Є-O	80.62	8.862
Ordos	Departed from the North China plate after Indo-China movement, evolved into a craton basin after Yanshan movement	Stable basin, several uplifts	J₁-T₃,P₁,C₃,Є-O	J₁,T₃, P₁,C₃,O₁	73.53	4.666
Sichuan	Craton basin formed from the Yangtze paleo-plate after Indo-China movement	Basin under the lateral extrusion after Mesozoic, formed a number of regional tectonic settings.	J₁,T₃, P,S,C	J₁,T₃, T₁₊₂,P,C,Z	11.35	5.374

Table 3: Characteristics of three petroliferous craton basins in China (reorganized according to the resource data of the database of MLRSCOG).

Layer Series	Resources (10^8 t)	East	Central	West	South	Qinghai-Tibet	Total
Cenozoic	Detected	159.60	2.24	28.86	0.74	5.40	196.84
Cenozoic	Available	40.83	0.61	8.56	0.17	0.81	50.98
Mesozoic	Detected	144.80	81.18	64.84	1.01	64.21	356.04
Mesozoic	Available	54.20	18.89	18.95	/	13.19	105.23
Upper Paleozoic	Detected	2.36	3.07	42.40	1.27	/	49.1
Upper Paleozoic	Available	0.58	0.73	11.61	0.23	/	13.15
Lower Paleozoic	Detected	17.64	/	39.03	/	/	56.67
Lower Paleozoic	Available	4.62	/	8.75	/	/	13.37
Total	Detected	324.41	86.48	175.13	2.02	69.61	657.65
Total	Available	100.24	20.23	47.87	0.40	14.00	182.74

Table 4: National Petroleum Resources Distribution (reorganized according to the resource data of the database of MLRSCOG).

Bohai Bay, Sichuan, Cuoqin, North Qiangtang, Junggar, Erlian, south of North China, and Qaidam. There are 51 basins with an area of approximately 10 to 1×10^4 m², including the following: southern Qiangtang, Gamba Tingri, southern Guizhou - Guangxi, BadanJilin, Nanpanjiang, Turpan-Hami, Hailar, Subei, and Biru. There are 66 basins with an area of approximately 1 to 0.5×10^4 m² and 46 with an area of approximately 0.5 to 0.2×10^4 m². The remaining 236 basins have an area smaller than 0.2×10^4 m².

The development of large-scale sedimentary basins in China's onshore region is based on the paleo-platform (plate) or micro-plate geological background [10]. Among the basins larger than 10×10^4 m², Tarim, Ordos, Bohai Bay, Sichuan and southern North China basins developed from the Paleozoic craton. Junggar and Qaidam basins had micro-plate background. Songliao and Erlian basins were the Mesozoic depression basins, and developed from the relative stable Late Paleozoic Jiangmeng

Layer Series	Resources (10^{12} m³)	East	Central	West	South	Qinghai-Tibet	Total
Cenozoic	Detected	0.62	/	4.77	0.01	/	5.4
	Available	0.36	/	2.92	0.01	/	3.29
Mesozoic	Detected	1.84	3.29	4.39	0.44	1.35	11.31
	Available	0.97	2.01	3.02	0.28	0.82	7.1
Upper Paleozoic	Detected	0.08	5.07	0.97	0.24	0.34	6.7
	Available	0.02	3.42	0.57	0.13	0.21	4.35
Lower Paleozoic	Detected	0.23	1.76	1.46	0.06	/	3.51
	Available	0.12	0.94	0.95	0.03	/	2.04
Total	Detected	2.77	10.11	11.6	0.76	1.69	26.93
	Available	1.47	6.37	7.46	0.44	1.03	16.77

Table 5: National Natural Gas Resources Distribution (reorganized according to the resource data of the database of MLRSCOG).

block.

Analysis of the basin hydrocarbon potential

Exploited oil and gas basins: There are 40 exploited oil and gas basins (Table 1 and Figure 1), and the Bohai Bay, Longliao, Ordos, Sichuan, Tarim, Junggar and Qaidam basins are the seven major production bases.

Detected oil and gas basins: The detected onshore oil and gas reserves are concentrated in Bohai Bay, Songliao, Ordoc, Sichuan, Tarim, Junggar and Qaidam basins (Table 1 and Figure 1).

Commercial hydrocarbon flows have been found in 45 basins in China. The oil reserves are over 10×10^8 t for Bohai Bay, Songliao, Ordos, Junggar and Tarim basins. Bohai Bay basin has the largest detected oil reserve at more than 1000×10^8 t. There are six basins (Ordos, Sichuan, Tarim, Songliao, Bohai Bay, and Qaidam) that have detected gas resources larger than 1000×10^8 m³. Among them, the Ordos and Sichuan basin contain more than 1×10^{12} m³ gas resources [10].

Oil and gas basins: There are 17 basins with detected oil and gas reserves that have not produced oil and gas. The Qiangtang and Cuoqin basins have large areas and a high density of resources, and each has an estimated reserve greater than 1 billion tons. The Changdu, Hetao, Sanjiang and Liupanshan basins each have an estimated reserve larger than 0.3 billion tons (Table 1 and Figure 1).

Other sedimentary basin: In addition to the previously mentions basins, 348 basins have no detectable oil and gas reserves (Figure 1).

Distribution of exploited oil and gas

China's petroleum basins can be characterized as oil enrichment rift basins and gas enrichment craton basins [11]. China's middle- and large-sized oil fields are mostly distributed in rift basins, whereas the middle- and large-sized gas fields are distributed in craton basins and foreland basins [1] (Figures 2 and 3).

Two oil enrichment zones: The oil reserve in the Tianshan-Xingmeng Mountains, North China and eastern Yangtze-Platform rift basin cluster is approximately 440×10^8 t, which is 66.64% of the total onshore oil reserve of China (Figure 2).

(1) The oil enrichment zone of Tianshan-Xingmeng orogeny: A 3000 km long orogenic belt located in a northern rift of the Zhongtian Mountains and Yanbula-Chifeng rift starting from Xingjiang and connecting Gansu, Ningxia, Inner Mongolia, and Heilongjiang. The Tianshan-Xingmen orogen includes the Junggar, Tuhan, Yingen, Erlian, Hailar and Songliao basins from west to east. The west central part contains the Jurassic basin in the northern rift of the Tian and Qilian Mountain. The basins are all located in the Junggar-Inner Mongolia-Songliao suture zone and along the ancient Asian Variscan fold belt. From west to east are pre-rift volcanic eruption events from the Carboniferous to Jurassic.

The rift basin structure is created by rifting and uplifting and mainly belongs to the Jurassic-Cretaceous period. The oil enrichment layers uplift from the Middle and Lower Jurassic to the Upper Cretaceous, and they form a major oil production belt from west to east. The total oil reserve is approximately 200×10^8 t, which represents 30.32% of the total onshore oil resources in China.

(2) The oil enrichment zone of eastern rift basin cluster in the North China and Yangtze-Platform: Belt located between the east fault of the Taihang Mountains and Tancheng-Lujiang fault and within most of the Bohai Bay basin. The northern part contains the Yilan-Yitong basin, and the southern part includes the southern North China, Nanxiang and Jianghan basins. It is composed of a tertiary oil enrichment rift basin belt [11]. The oil reserve in this area is approximately 240×10^8 t, which represents 36.32% of the total onshore reserve.

Three gas enrichment craton basins: The Tarim, North China and Yangtze paleo-plates were formed in the Early Paleozoic. They were under the Tethys Ocean, where they formed thick marine carbonate sedimentation. The plate tectonic process in the Mesozoic and Cenozoic created the Tarim, Ordos and Sichuan craton basins, and the hinterland preserved the entire Paleozoic strata. These gas enrichment craton basins contain 70.19% of the total onshore natural gas in China.

Distribution of oil and gas resources and layer series (or Era)

The long-term, multi-cycle and complex characteristics of China's tectonic evolution determined the sedimentation conditions of the multi-stage basins, multi-cycle sediments and multi-type and overlapping characteristics of origin basins that are controlled by different tectonic settings. Therefore, source rocks are widely distributed in China, and oil and gas are buried in layers from the Middle and Upper Proterozoic to the Quaternary (Figure 3).

Based on the results of recent evaluations of national resources, onshore conventional oil resources are estimated at approximately 657.65×10^8 t in China and conventional gas resources are estimated at approximately 26.93×10^{12} m³. The distribution layer series are shown

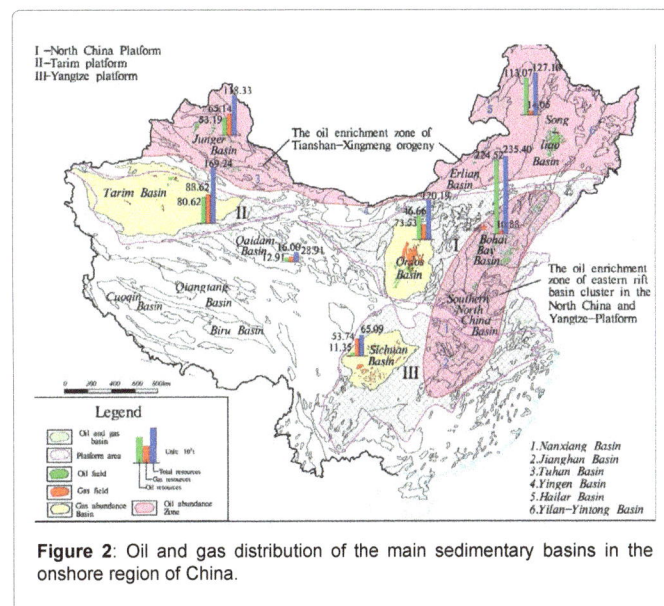

Figure 2: Oil and gas distribution of the main sedimentary basins in the onshore region of China.

below.

(1)   The oil resources in the Cenozoic layer is approximately 196.84×10^8 t (29.93% of the total reserve), whereas the Mesozoic layer contains 356.04×10^8 (54.14% of the total reserve) (Figure 4 and Table 4).

(2)   The natural gas resources are distributed in layers from the Cenozoic, Mesozoic and Upper Paleozoic, with reserves of [missing in origin draft] (20.05%), 5.4×10^{12} m^3 (41.99%) and 3.29×10^{12} m^3 (24.88%), respectively (Figure 5 and Table 5).

(3)   The distribution of the petroleum resources layer series is strongly uneven, and there are various oil and gas distribution features in different regions. In the eastern region, oil is primarily concentrated in Cenozoic and Mesozoic layers and gas is concentrated in Mesozoic layers. In the central region, oil occurs in Mesozoic layers, whereas gas occurs in Upper Paleozoic and Mesozoic layers. In the western region, oil is observed in all four layer series, with the most abundant in the Mesozoic, and gas occurs in Cenozoic and Mesozoic layers. In the southern region, rich oil deposits are found in the Upper Paleozoic and Cenozoic layers, whereas natural gas is mainly distributed in Mesozoic and Upper Paleozoic layers. The oil and gas in the Qinghai-Tibet region is mostly concentrated in the Mesozoic layers (Tables 4 and 5).

Conclusion

There are more than 400 onshore sedimentary basins in China, with 11 basins larger than 10×10^4 m^2 and 236 basins smaller than 0.2×10^4 m^2. The development of large-scale sedimentary basins is based on the paleo-platform (plate) or micro-plate geological background.

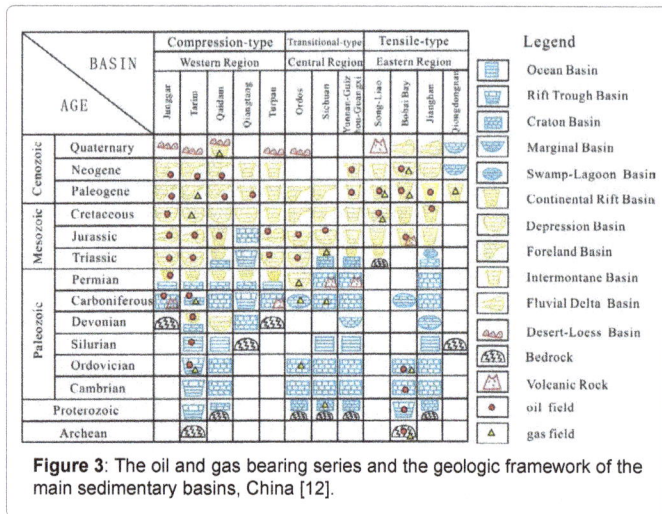

Figure 3: The oil and gas bearing series and the geologic framework of the main sedimentary basins, China [12].

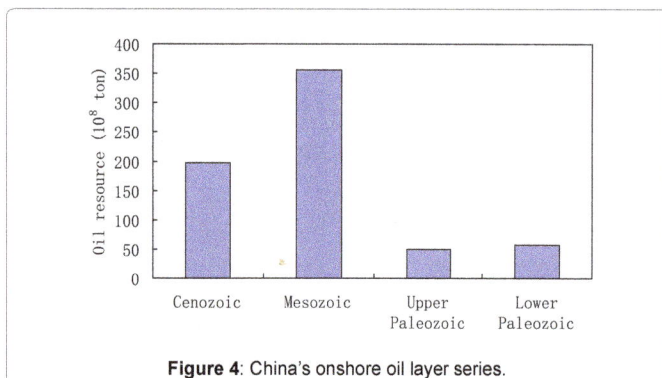

Figure 4: China's onshore oil layer series.

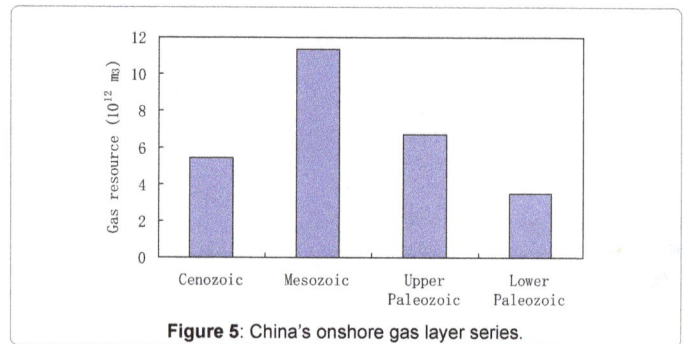

Figure 5: China's onshore gas layer series.

The Tianshan-Xingmeng orogen and basin cluster in the North China/eastern Yangtze Platform are the major distribution areas of China's onshore oil reserves (66.64%).

The three major craton basins, Tarim, North China and Yangtze, host 70.19% of the total onshore natural gas reserve in China.

The onshore oil resources in China are concentrate in Cenozoic and Mesozoic layers, and the natural gas reserves are mainly distributed in Cenozoic, Mesozoic, and Upper Paleozoic layers.

Acknowledgements

This work was supported by the project "Geological Map Compilation of Comprehensive National Petroleum" funded by the Oil & Gas Survey, the Geological Survey of China and the National Natural Science Foundation of China (Grant No. 41030853).

Software

The software used in this study is MpaGis Ver. 6.7. The base map references are the State Bureau of Surveying and Mapping's 1:4,000,000 base maps. The maps are displayed as Albers equal-area conic projections with a central longitude of 110° and standard latitudes of 25° and 47°.

References

1.   Jin ZJ (2008) Distribution and Structures of Large and Medium Oil-Gas Fields in China. Xinjiang Petroleum Geology 29: 385-388.

2.   Jin ZJ, Bai GP, Ali MG (2004) An introduction to petroleum and natural gas exploration and production research in China. J Petroleum Sci Engg 41: 1– 7.

3.   MLRSCOG Ministry of Land and Resources Strategic Research Center of Oil and Gas (2010) National Oil and Gas Resource Evaluation (Volume I, II and III). China Land Press, Beijing.

4.   Qu H, Zhao WZ, Hu SY (2006) Oil & Gas Resources Status and the Exploration Fields in China. China Petroleum Exploration 4: 1-4.

5.   SINOPEC Star Petroleum, Ltd. and Sinopec Exploration and Research Institute Jinzhou District (2004) Map of exploration results of oil and gas basins in China (1:4,000,000). Beijing Geological Publishing House, Beijing.

6.   PetroChina Company Limited and Institute of Petroleum Exploration and Development, Beijing (2001) Distribution of Sedimentary Basins and Oil and Gas in China (1:9,000,000) (unpublished).

7.   Hu JY (1995) Map of Oil and Gas Basins and Fields in China (1:4,000,000). Petroleum Industry Press, Beijing.

8.   Li GY, Lv MG, Zhao XZ (2002) China Atlas of Oil and Gas Basin. Petroleum Industry Press, Beijing.

9.   Zhou QF, Zhang L, Zhuang L (2009) Hydrocarbon Exploration, Production Status and Prospects in China's Major Petroliferous Basins. Sino-Global Energy 14: 41-48.

10. Zhao ZY (2003) The Geological Background and Tectonics of China's Sedimentary Basins. Marine Petrol Geol 8: 9-20.

11. Luo ZL (1998) Distribution and Outlook for Oil/Gas Exploration of Petroliferous Basin in China. Xinjiang Petroleum Geology 19: 441- 449.

12. Li DS (2012) Tectonics of polycyclic superimposed oil and gas basins in China. Science Press, Beijing.

5

Bio-Oil Production by Thermal Cracking in the Presence of Hydrogen

Renato Cataluña Veses[1]*, Zeban Shah[1], Pedro Motifumi Kuamoto[1], Elina B. Caramão[1], Maria Elisabete Machado[1] and Rosângela da Silva[2]

[1]*Instituto de Química, Federal University of Rio Grande do Sul, Brazil*
[2]*Pontifical Catholic University of Rio Grande do Sul, RS, Brazil*

Abstract

This paper describes the bio-oil production process of a mixture of agricultural wastes: discarded soybean frying oil, coffee and sawdust, by pyrolysis and thermal cracking in the presence of hydrogen. The fractions obtained in the pyrolysis and/or cracking processes were divided into a light fraction and a heavy one. All the fractions were analyzed by comprehensive two-dimensional gas chromatography with time-of-flight mass spectrometry detection (GC×GC/TOFMS). The characteristics of the fractions obtained in from the cracking process in the presence of hydrogen were similar to those of petroleum-based naphtha, while the fractions obtained by pyrolysis contained significant quantities of compounds such as furanmethanol, hexanol, and benzofuran, whose commercial value is high.

Keywords: Thermal cracking; Biomass pyrolysis; Hydrogen; Chromatography

Introduction

The recent environmental restrictions on the use of fossil fuels have intensified research into new alternative energy sources. Many alternative technologies to produce cleaner fuels have been developed, including the use of biomass, which offers a promising potential [1-4].

Biomass is a renewable source which has received attention due to various characteristics, particularly its low cost and wide availability. Biomass can be converted into bio-fuel by means of different processes, e.g., reductive combustion, liquefaction, pyrolysis and gasification [5]. The use of biomass is particularly interesting when it involves waste products such as waste vegetable oil, fruit seeds, sugarcane bagasse, sugarcane straw, rice husks, coconut fibers, and coffee grounds, which are also potential sources of energy [6-8].

Bio-oil from biomass pyrolysis, also known as pyrolysis oil, is a dark brown almost black liquid with a characteristic smoky odor, whose elemental composition is analogous to that of the biomass from which it derives. It is a complex mixture of oxygenated compounds with a significant amount of water originating from the moisture of the biomass and from cracking reactions. Bio-oil may also contain small coal particles and dissolved alkali metals coming from the ash. Its composition depends on the raw material and on the operating conditions used in its production. Pyrolysis oil is an aqueous microemulsion resulting from the products of fragmentation of cellulose, hemicellulose and lignin [9,10].

Much attention has focused on pyrolysis, a biomass thermal decomposition process, for which the literature describes numerous different reactors and conditions [11-13]. The presence of oxygen exerts a highly negative impact on the potential uses for bio-oil. For example, oxygen it lowers the heating value, gives rise to immiscibility with petroleum fuels, and leads to corrosiveness and instability during long-term storage and transportation [14]. The biomass pyrolysis process is an economically feasible option for producing chemicals and/or fuels [15,16]. The bio- oil resulting from the pyrolysis process consists of a mixture of more than 300 organic compounds [17], but its processing, separation and characterization pose technological challenges. In the thermal cracking process, the volatile compounds generated during pyrolysis also present a promising potential for energy generation [18]. Moreover, the upgrading process, which involves the reduction of oxygenates and is necessary to improve the quality of bio-oil, normally requires processes such as catalytic cracking, hydrogenation and steam reforming [19-22].

Hydropyrolysis is an important technique for improving the quality of bio-oil produced from biomass pyrolysis. Hydrogen is a reducing gas and cracking biomass in the presence of hydrogen can reduce the oxygen content in bio-oil [23]. This paper discusses the characterization of bio-oil generated from the pyrolysis of a mixture of wastes: discarded soybean frying oil, coffee grounds and sawdust. The thermal cracking process, which was performed in the presence of hydrogen in order to upgrade the bio-oil, resulted in lower molecular weight fractions and substantially reduced the content of oxygenated and nitrogenated species.

Experimental Materials

The bio-oil was obtained by pyrolysis of a mixture (1: 1: 1, in mass) of wastes: discarded soybean frying oil, coffee grounds and eucalyptus sawdust. The frying oil was mixed with the solids after their particle size was reduced to 0.21 mm. To this mixture were added calcium oxide (20 mass %) and sufficient water to produce a malleable mass that could be moulded into cylindrical samples (50 mm × 180 mm). The samples were allowed to dry at room temperature for a week.

Biomass pyrolysis and thermal cracking of the bio-oil

The bio-oil was produced by conventional pyrolysis of the cylindrical samples in an electrically heated stainless steel reactor. Before beginning the pyrolysis, the system was purged for 20 minutes with Argon containing 5% of hydrogen (100 mL min^{-1}). After purging, the pyrolysis started and the system was heated to 850°C at a heating rate of 15°C min-1. The volatiles produced during the process were

***Corresponding author:** Renato Cataluña Veses, Instituto de Química, Universidade Federal do Rio Grande do Sul, Porto Alegre, RS, Av. Bento Gonçalves, 9500, Brazil, E-mail: rcv@ufrgs.br*

treated by isothermal hydrocracking in another reactor (stainless steel, 20 mm in diameter and 600 mm in length) at 850°C. The final effluent was cooled to 100°C and the water phase was separated by decantation.

After phase separation, the effluent was condensed at 5°C and the aqueous phase separation process was repeated, while the gaseous phase was discarded.

For purposes of comparison, the pyrolysis was repeated without thermal cracking. Four samples were thus produced: bio-oil obtained at 100°C, and bio-oil obtained at 5°C, both with and without thermal cracking. These samples are hereinafter referred to as: OPH (Oil from Pyrolysis obtained at 100°C - High temperature) and OPL (Oil from Pyrolysis obtained at 5°C- Low temperature) for samples obtained only by pyrolysis, and OCH (Oil after Cracking obtained at 100°C-High temperature) and OCL (Oil after Cracking obtained at 5°C-Low temperature) for those obtained after thermal cracking. Figure 1 illustrates the production scheme of the four fractions produced, i.e., OPH and OPL by the pyrolysis process and OCH and OCL by the pyrolysis process followed by thermal cracking.2.2. Characterization of the products

The four fractions were analyzed by GC×GC/TOFMS using a LECO Pegasus IV (LECO, St Joseph, MI, USA) system. Experiments were performed in a conventional split/splitless injector (Agilent Technologies) at 320°C (1μL) with a split ratio equal to 1: 30. Helium

(99.999%, Linde Gases, Porto Alegre, RS, Brazil) was used as carrier gas, at 1 mL min-1. The oven temperature was programmed from 40°C to 300°C at 3°C min-1. The difference between ovens 1D and 2D was 15°C and the modulation period was 8 s (cryogenic quadjet modulator, cooled with liquid nitrogen). The transfer line and electron impact ionization source operated at 300°C and 250°C. The acquisition frequency of the detector was 100 Hz, using a mass range of 45 to 400 Daltons. Electron ionization was carried out at 70 eV. The data were processed on the Pegasus 4D platform of the ChromaTOF software. A DB-5 column was used as first dimension, and a DB-17 as second dimension column, using a cryogenic modulator.

The compounds were identified based on the following parameters: retention times, regions of spatial structuration, mass spectral match factor (NIST library), and spectral deconvolution. Given the spatial structure provided by GC×GC, some compounds with similarity below 700 were considered to be identified, since the elution region in the two-dimensional (2D) space, as well as other parameters, provide a higher degree of reliability in the identification of analytes. The data generated in the peak table were transferred to the Microsoft Excel™ program in order to build dispersion graphics to better visualize the distribution of compounds in 2D space.

Results and Discussion

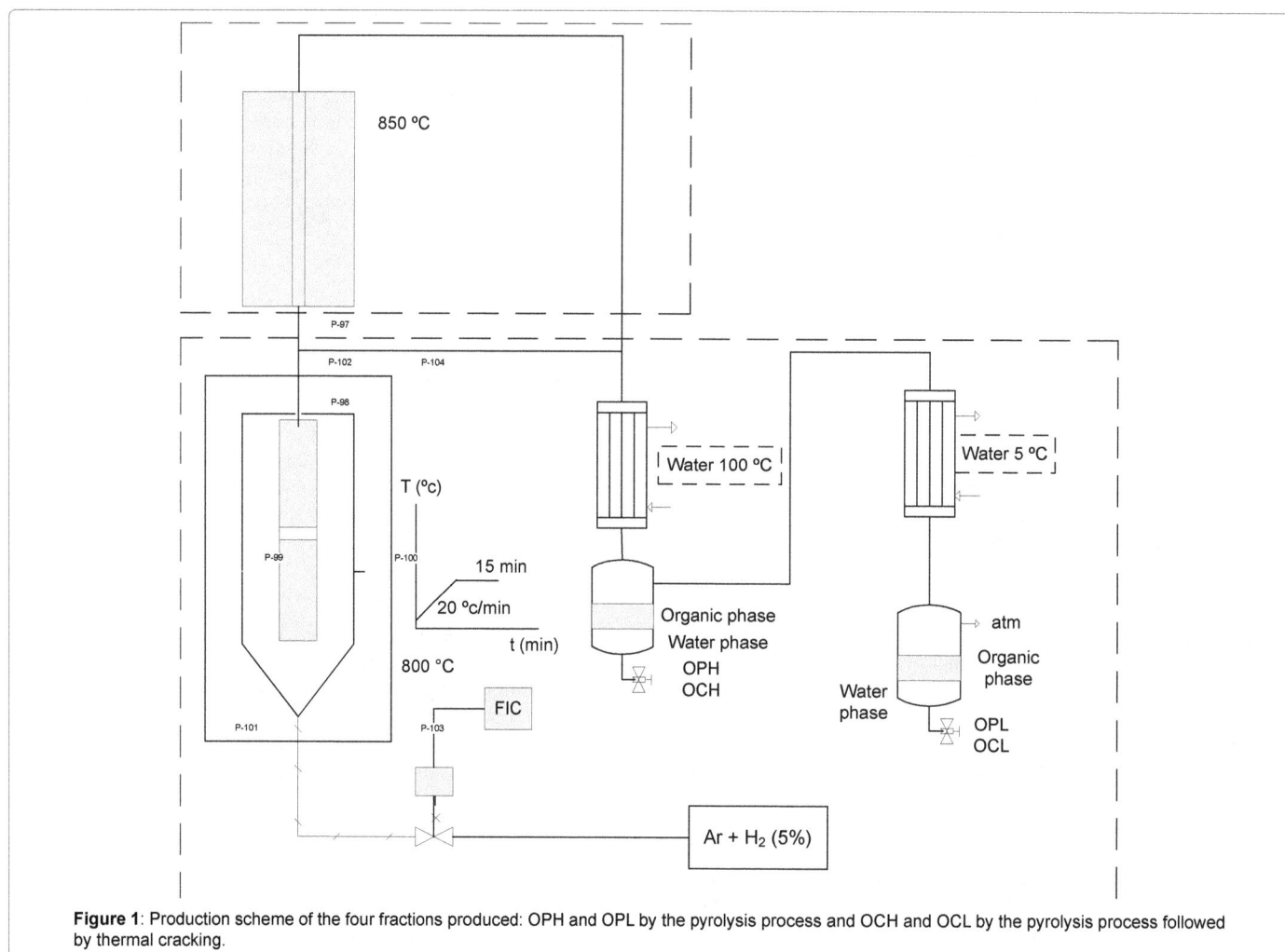

Figure 1: Production scheme of the four fractions produced: OPH and OPL by the pyrolysis process and OCH and OCL by the pyrolysis process followed by thermal cracking.

Product yields from pyrolysis and thermal cracking

Pyrolysis product distribution depends on reaction parameters such as temperature, heating rate and reactant particle size, as well as on the starting biomass.

The oil fractions obtained in this work came from the same raw materials and the same operational conditions, but from different production processes. The OCH sample was obtained by pyrolysis followed by thermal cracking, while the OPH sample was obtained solely by pyrolysis. The application of thermal cracking after pyrolysis led to a significant increase in the condensed fraction at the temperature of 5°C. The average yield of the pyrolysis process is approximately 30% oil fractions, 50% aqueous fractions, and 20% gas phase (uncondensed, obtained by difference). In the pyrolysis process, the oil fraction condensed at a temperature of 100°C corresponds to approximately 90% of the oil fraction. Pyrolysis process followed by thermal cracking results in a distribution of approximately 40% of the fraction condensed at a temperature of 5°C (OCL) and 60% of the oil fraction condensed at a temperature of 100°C (OCH).

Composition of bio-oil fractions

Given that the four bio-oil fractions are very complex mixtures of different chemical species derived from depolymerization and fragmentation of the main components of the biomass, which comprise a wide range of molecular weights, a GC×GC/TOFMS was used for their identification.

The compositions of the four bio-oil fractions shown in Figure 2 are grouped according to types of chemical compounds: acids, aldehydes, ketones, alcohols, phenols, aromatics, cyclic and aliphatic hydrocarbons, ethers and nitrogen compounds. The compounds were tentatively identified when the similarity between a sample's spectrum and that of the library was greater than 750. In total, 214 compounds in OCH, 324 in OCL, 84 in OPH and 312 in OPL were tentatively identified.

Some observations apply both to the bio-oils obtained from thermal degradation and to the light fraction of pyrolysis (OPL-Figure 2). For example, note that there is a high proportion of hydrocarbon compounds, the most important ones being aromatics and aliphatics, representing between 57 and 79 wt % of the products. On the other hand, the OPH sample obtained by pyrolysis and condensed at 100°C (fraction containing heavy compounds) is composed mainly of ketones and nitrogens, and smaller amounts of alcohols, ethers and phenols. This fraction does not contain hydrocarbons. Nitrogenous compounds in bio-oil originate from the thermal degradation of caffeine derivatives contained in coffee grounds.

As can be seen in Figure 2, the fractions obtained by pyrolysis and thermal cracking (OCH and OCL) consist mostly of aliphatic, aromatic and cyclic hydrocarbons. The OPH fraction is composed mainly of hydrocarbons with nitrogen (46% in area) and oxygen (47% in area) compounds. The oxygen content in pyrolysis bio-oils usually varies from 45 to 50 w/w%, and oxygen is present in most of the more than 300 compounds [10,24,25]. The distribution of these compounds depends mainly on the type of biomass and the process conditions. The presence of oxygenated compounds in bio-oil reduces its calorific value and renders it chemically unstable [9], limiting its use as fuel or in formulations for direct use in diesel cycle engines [9,26,27]. However, when separated, they present high added commercial value [28].

Table 1 lists the main identified compounds and their corresponding percentage area in the bio-oil and fractions (OPH, OPL, OCH and

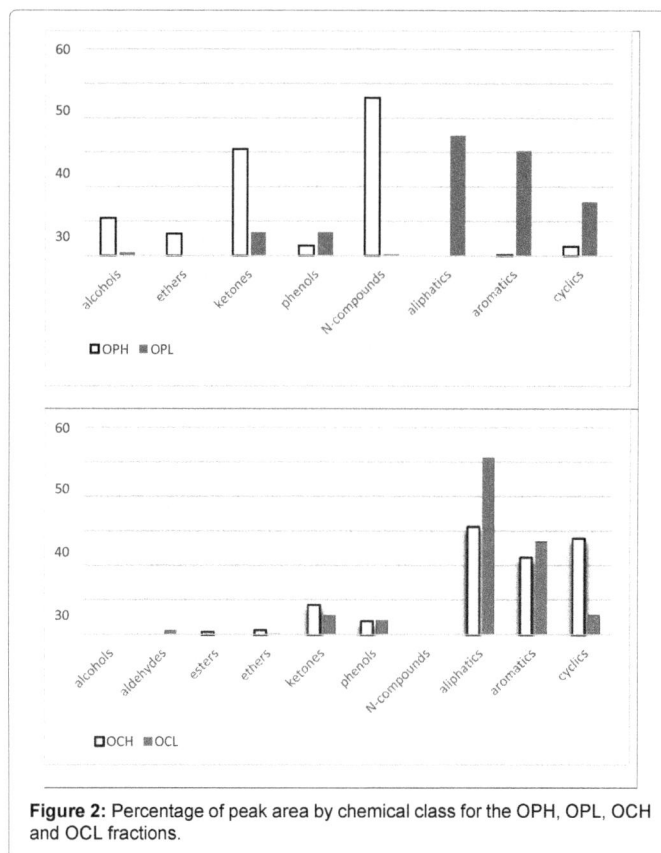

Figure 2: Percentage of peak area by chemical class for the OPH, OPL, OCH and OCL fractions.

OCL). The other compounds contained in the fractions of this study are listed in the Appendix.

As can be seen in Table 1, the oxygenated compounds in the OPH fraction alcohols include furanmethanol (8% in area) and hexanol (2% in area). These two alcohols are important raw materials for the preparation of a wide range of drugs and industrial products of high commercial and industrial value [29]. Benzofuran and dioxyethane ethers are also present in the OPH fraction in percentage areas of 4.0 and 2.0, respectively. Benzofuran is considered an important class of heterocyclic compounds which is present in numerous bioactive natural products and in pharmaceuticals and polymers. Benzofuran is one of the most important heterocyclic rings due to its broad microbiological range. Medicinal chemistry is widely involved in the synthesis of the benzofuran ring owing to its clinical importance. Benzofuran can be used as an enzyme activator and inhibitor, as an antimicrobial, anti-inflammatory, anti-cancer, antiviral, anti-tuberculosis, antioxidant agent, etc. [30].

The four fractions of this study contained phenolic compounds, namely, around 7% OPL, 4% OCH and OCL, and 3% OPH. These compounds are widely employed in the production of phenolic resins [31]. They also have antioxidant and antimicrobial properties that inhibit the proliferation of microorganisms, corrosion and deposits when added to diesel fuel formulations and/or biodiesel for use in engines (use of biomass-derived compounds) [32,33].

Moreover, chemical products containing oxygen are produced mainly from fossil fuels, through the oxidation or hydration of olefins to introduce oxygen containing functional groups. Fortunately, these functional groups are already present in bio-oil. Therefore, obtaining

Major compounds		Area %			
		OCH	OPL	OCL	OPH
Alcohol	Furanmethanol	n.d.	n.d.	n.d.	8.4
	Hexanol	n.d.	0.2	n.d.	2.3
Aldehyde	Propenal, phenyl	n.d.	0.1	1.1	n.d.
Ether	Benzofuran	0.5	n.d.	0.1	4.3
	Ethane, diethoxy-	n.d.	n.d.	n.d.	2.0
Ketone	Hexadecanone	1.5	n.d.	n.d.	n.d.
	Nonanone	n.d.	n.d.	1.4	n.d.
	Cyclopentenone, methyl-	n.d.	n.d.	n.d.	6.6
	Cyclopentenone, C3	0.2	0.5	0.2	6.2
	Cyclopentenone, C2	0.1	0.4	n.d.	5.6
	Cyclohexenone, methyl-	n.d.	0.2	n.d.	1.7
	Cyclopentanone	n.d.	0.1	0.1	1.2
	Cyclopentanone, methyl	n.d.	0.3	0.1	1.1
N-Compound	Pyrrole, methyl-	n.d.	n.d.	n.d.	9.3
	Pyrazine, C5	n.d.	n.d.	n.d.	8.1
	Pyrazine, C3	n.d.	n.d.	n.d.	2.8
	Pyrazole, C4	n.d.	n.d.	n.d.	1.7
	Imidazole, C5	n.d.	n.d.	n.d.	1.6
	Pyrazine, C4	n.d.	n.d.	n.d.	1.5
	Pyridine, methyl-	n.d.	n.d.	n.d.	1.4
	Pentanamide, methyl-	n.d.	n.d.	n.d.	1.0
Phenol	Phenol, ethyl	1.7	1.4	1.1	n.d.
	Phenol, methyl-	n.d.	1.9	n.d.	0.1
	Phenol	0.3	0.6	1.0	2.0

Class	Identification	Area %			
		OCH	OPL	OCL	OPH
HC Cíclicos-C5	Cyclopentadiene, C2	2.6	n.d.	n.d.	n.d.
	Cyclopentadiene, methyl	2.1	n.d.	n.d.	n.d.
	Cyclopentadiene, C3	1.2	n.d.	n.d.	n.d.
	Cyclopentene, C6	n.d.	1.2	n.d.	n.d.
	Cyclopentane, C8	n.d.	n.d.	1.1	n.d.
	Cyclopentadiene, C5	n.d.	n.d.	n.d.	2.2
HC Cíclicos-C6	Cyclohexene, C2	4.2	0.1	n.d.	n.d.
	Cyclohexene, methyl	5.1	0.2	0.5	n.d.
	Cyclohexadiene, C2	4.7	0.4	0.4	n.d.
	Cyclohexadiene, C4	n.d.	1.1	n.d.	n.d.

	Indene, methyl	1.3	1.2	1.3	n.d.
	Indene, C2	1.1	n.d.	n.d.	n.d.
	indene	n.d.	2.4	1.4	n.d.
	Naphthalene, dihydro-	n.d.	1.8	n.d.	n.d.
	Indane	n.d.	1.0	0.5	n.d.
	Naphthalene, methyl	n.d.	0.6	1.1	n.d.
	Heptadiene	2.4	n.d.	0.6	n.d.
	Dodecadiene	n.d.	1.3	n.d.	n.d.
	Toluene	6.2	1.0	0.1	0.2
	Benzene	2.1	n.d.	2.1	n.d.
	Benzene, C3	3.9	2.0	3.4	n.d.
HC diaromatico	Benzene, C4	1.0	3.4	1.9	n.d.
	Benzene, C5	0.1	1.3	0.3	n.d.
	Benzene, C2	n.d.	1.1	6.3	n.d.
	Octene	2.9	2.7	1.3	n.d.
	Decene	1.3	n.d.	1.0	n.d.
	Dodecene	n.d.	1.7	1.0	n.d.
	Undecene	0.6	1.2	0.6	n.d.
	Octane	6.9	1.7	0.4	n.d.
	Nonane	1.0	n.d.	1.5	n.d.
	Tridecane	0.4.	4.4	1.1	n.d.
	Pentadecane	0.6	1.2	0.8	n.d.
	Docecane	03	1.0	0.5	n.d.

Table 1: Main identified compounds and their corresponding percentage area in the bio-oil and fractions (OPH, OPL, OCH and OCL).

value-added chemicals from bio-oil is a potential approach for the efficient use of biomass energy.

With respect to the N-compounds present only in the OPH fraction, pyrazines corresponded to 25% in area. Pyrazine is an important product that participates together with benzene in the synthesis of quinoxaline, also known as benzopyrazine, which is rare in its natural state, but is easy to synthesize. Quinoxaline and its derivatives are very industrially important because of their ability to inhibit metal corrosion [34] in the preparation of porphyrins [35]. The pharmaceutical industry has a potential interest in them because of their broad spectrum of biological properties [36-38].

The composition of the OPH fraction contained practically no aromatic hydrocarbons. On the other hand, the OPL, OCH and OCL fractions each presented approximately 16% in area of alkylbenzenes, which could be isolated and, together with the pyrazines of the OPH

fraction, serve as raw material for the synthesis of quinoxaline and its derivatives.

The OCL and OCH fractions obtained by thermal cracking in the presence of a mixture of argon and 5% hydrogen resulted in the elimination of oxygen (deoxygenation) with the formation of water [39], and a stronger breakdown of the heavier organic compounds into lighter organic compounds, as well as the elimination of nitrogen from the nitrogenated compounds. This is illustrated in Figure 2 and in the supplementary material. Because thermal cracking produces various fragments of C-C, they may undergo oligomerization to form olefins, which in turn may undergo aromatization followed by alkylation and isomerization, producing aromatics.

The OCH and OCL fractions presented percentage areas of 82 and 84, respectively, of these hydrocarbons. The OPL fraction obtained by pyrolysis showed a profile similar to that of the OCH and OCL fractions with respect to hydrocarbons, with 81% in area, but with 8.0% of oxygenated compounds.

Conclusions

In the fractions obtained by pyrolysis, 84 compounds were tentatively identified in the heavy fraction and 312 in the light fraction. The vapors were subjected to thermal cracking in the presence of 5% hydrogen as a way to upgrade the bio-oil, and 214 compounds were identified in the heavy fraction and 324 in the light fraction. The thermal cracking process produced mainly aliphatic, aromatic and cyclic hydrocarbons, yielding approximately 80% in weight of these compounds with characteristics similar to those of naphtha derived from the atmospheric distillation of petroleum with potential applications as fuels. The fractions obtained solely by pyrolysis consisted predominantly of hydrocarbons with nitrogen (46% in area) and oxygen (47% in area) compounds. The oxygenated compounds included furanmethanol and hexanol alcohols, and benzofuran and dioxyethane ethers. All the analyzed fractions contained phenolic compounds. When isolated, these compounds are an excellent potential source of raw material for the preparation of pharmaceutical and industrial products of high commercial and industrial value.

Acknowledgements

The authors want to tkank CNPq

References

1. Arbogast S, Bellman D, Paynter JD, Wykowski J (2012) Advanced bio-fuels from pyrolysis oil: The impact of economies of scale and use of existing logistic and processing capabilities. Fuel Processing Technol 104: 121-127.

2. Arbogast S, Bellman D, Paynter JD, Wykowski J (2013) Advanced biofuels from pyrolysis oil... Opportunities for cost reduction. Fuel Processing Technol 106: 518-525.

3. Khan AA, Jong W, Jansens PJ, Spliethoff H (2009) Biomass combustion in fluidized bed boilers: Potential problems and remedies. Fuel Processing Technology 90: 21-50.

4. Cataluña R, Kuamoto PM, Petzhold CL, Caramão EB, Machado ME, et al. (2013) Using Bio-oil Produced by Biomass Pyrolysis as Diesel Fuel. Energy Fuels 27: 6831-6838.

5. Laksmono N, Paraschiv M, Loubar K, Tazerout M (2013) Biodiesel production from biomass gasification tar via thermal/catalytic cracking. Fuel Processing Technology 106: 776-783.

6. Dukua MH, Gu S, Hagan EB (2011) A comprehensive review of biomass resources and biofuels potential in Ghana. Renewable and Sustainable Energy Reviews 15: 404-415.

7. http://www.unep.org/ietc/Portals/136/Publications/Waste%20Management/ WasteAgriculturalBiomassEST_Compendium.pdf

8. Virmond E, Rocha JD, Moreira RFPM, José HJ (2013) Valorization of agroindustrial solid residues and residues from biofuel production chains by thermochemical conversion: a review, citing Brazil as a case study. Braz J Chem Eng 30: 197-229.

9. Bridgwater AV (2004) Biomass fast pyrolysis. Thermal science 8: 21-49.

10. http://www.ieabioenergy.com/

11. Mohan D, Pittman Jr CU, Steele PH (2006) Pyrolysis of wood/biomass for bio-oil: a critical review. Energy Fuel 20: 848-889.

12. Li S, Xu S, Liu S, Yang C, Lu Q (2004) Fast pyrolysis of biomass in free-fall reactor for hydrogen-rich gas. Fuel Processing Technol 85: 1201-1211.

13. Özbay N, Uzun BB, Varol EA, Pütün AE (2006) Comparative analysis of pyrolysis oils and its subfractions under different atmospheric conditions. Fuel Processing Technol 87: 1013-1019.

14. Girisuta B, Kalogiannis KG , Dussan K, Leahy JJ, Hayes MHB, et al. (2012) An integrated process for the production of platform chemicals and diesel miscible fuels by acid-catalyzed hydrolysis and downstream upgrading of the acid hydrolysis residues with thermal and catalytic pyrolysis. Bioresource Technology 126: 92-100.

15. Vispute TP, Zhang H, Sanna A, Xiao R, Huber GW (2010) Renewable chemical commodity feedstocks from integrated catalytic processing of oils. Science 330: 1222-1227.

16. Wright MM, Daugaard DE, Satrio JA, Brown RC (2010) Techno-economic analysis of biomass fast pyrolysis to transportation fuels. Fuel 89: 52-510.

17. Mythili R, Venkatachalam P, Subramanain P, Uma D (2013) Characterization of bioresidues for bioil production through pyrolysis. Bioresour Technol 138: 71- 78.

18. Mortensen PM, Grunwaldt JD, Jensen PA, Knudsen KG, Jensen AD (2011) A review of catalytic upgrading of bio-oil to engine fuels. Applied Catalysis A: General 407: 1-19.

19. Zhu XF, Zheng JL, Guo QX, Zhu QS (2005) Upgrading and utilization of bio-oil from biomass. Engineering Science 7: 83-88.

20. Zhang Q, Chang QJ, Wang TJ, Xu Y (2006) Progress on research of properties and upgrading of bio-oil. Petrochemical Technology 35: 493-498.

21. Guo XY, Yan YJ, Li TC (2004) Influence of catalyst type and regeneration on upgrading of crude bio-oil through catalytical thermal cracking. The Chinese Journal of Process Engineering 4: 53-58.

22. Kanaujia PK, Sharma YK, Garg MO, Tripathi D, Singh R (2014) Review of analytical strategies in the production and upgrading of bio-oils derived from lignocellulosic biomass. J Analyt Appl Pyrolys 105: 55-74.

23. Thangalazhy-Gopakumar S, Adhikari S, Gupta RB, Tu M, Taylor S (2011) Production of hydrocarbon fuels from biomass using catalytic pyrolysis under helium and hydrogen environments. Bioresour Technol 102: 6742-6749.

24. Meier D, Oasmaa A, Peacocke GVC (1997) Proper ties of Fast Pyrolysis Liquids: Status of Test Methods. Characterization of Fast Pyrolysis Liquids, in: Develop ments in Thermochemical Bio mass Conversion. Blackie Academic & Professional (Eds.), London.

25. Huber GW, Corma A (2007) Synergies between bio- and oil refineries for the production of fuels from biomass. Angew Chem Int Ed Engl 46: 7184-201.

26. Wang SR, Wang YR, Cai QJ, Wang XY, Jin H, et al. (2014) Multi-step separation of monophenols and pyrolytic lignins from the water-insoluble phase of bio-oil. Sep Purif Technol 122: 248-255.

27. Zhang Q, Chang J, Wang TJ, Xu Y (2007) Review of biomass pyrolysis oil properties and upgrading research. Energy Conversion and Management 48: 87- 92.

28. Bridgwater VA, Meier D, Radlein D (1999) An overview of past pyrolysis of biomass. Org Geochem 30: 1479-1493.

29. Liu W, Wang X, Hu C, Tong D, Zhu L, et al. (2014) Catalytic pyrolysis of distillers dried grain white solubles: An attempt towards obtaining value-added products. International Journal of Hydrogen Energy 39: 6371-6383.

30. Nevagi RJ, Dighe SN, Dighe SN (2015) Biological and medicinal significance of benzofuran. Eur J Med Chem 97: 561-581.

31. Migliorini MV, Moraes MSA, Machado ME, Caramão EB (2013) Caracterização de fenóis no bio-óleo da pirólise de caroço de pêssego por GC/MS e GCxGC/ TOFMS. Scientia Chromatographica 5: 47-65.

32. Almeida LR, Hidalgo AA, Veja ML, Rios MAS (2011) Utilização de compostos derivados da biomassa para solução problemas industriais do setor de biocombustíveis. Estudos Tecnológicos 7: 163-176.

33. Yang H, Zhao W, Norinaga K, Fang J, Wang Y, et al. (2015) Separation of phenols and Ketones from bio-oil produced from ethanolysis of wheat stalk. Sep Purif Technol 152: 238-245.

34. Zarrouk A, Zarrok H, Salghi R, Hammouti B, Al-Deyab SS, et al. (2012) A Theoretical Investigation on the Corrosion Inhibition of Copper by Quinoxaline Derivatives in Nitric Acid Solution. Int J Electrochem Sci 7: 6353-6364.

35. Kim J, Jaung JY (2008) The synthesis and optical properties of meso- substituted porphyrins bearing quinoxaline derivatives. Dyes and Pigments 77: 474-477.

36. Pereira JA, Pessoa AM, Cordeiro MN, Fernandes R, Prudêncio C, et al. (2015) Quinoxaline, its derivates and applications: A State of the Art rewiew. Eur J Med Chem 97: 664-672.

37. Vieira M, Pinheiro C, Fernandes R, Noronha JP, Prudencio C (2014) Antimicrobial activity of quinoxaline 1,4-dioxide with 2- and 3-substituted derivatives. Microbiol Res 169: 287-293.

38. http://onlinelibrary.wiley.com/book/10.1002/9780470187333

39. Adjaye JD, Bakhshi NN (1995) Production of hydrocarbons by catalytic upgrading of a fast pyrolysis bio-oil. Part II: comparative catalyst performance and reaction pathways. Fuel Processing Technol 45: 185-202.

6

Microalgae Harvesting Methods for Industrial Production of Biodiesel: Critical Review and Comparative Analysis

Mariam Al hattab[1], Abdel Ghaly[1]* and Amal Hammouda[1,2]

[1]Department of Process Engineering and Applied Sciences, Faculty of Engineering, Dalhousie University, Halifax, Nova Scotia, Canada
[2]Food Technology Research Institute, Agricultural Research Center, Giza, Egypt

Abstract

Microalgae biomass can be used to produce numerous value added products such as biodiesel, bioethanol, biogas and bio hydrogen, fish feed, animal feed, human food supplements and skin care products. Production of value added products from microalgae biomass requires growing and recovery of the algae biomass and extraction and downstream processing of the desired product. However, the major obstacle for using microalgae biomass on an industrial-scale for the production of biodiesel and other value added products is the dewatering step which accounts for 20-30% of the total costs associated with microalgae production and processing. The aim of this study was to review the current methods used for harvesting and concentrating microalgae and to perform a comparative analysis in order to determine the most efficient and economically viable dewatering methods for large scale processing of microalgae biomass. The harvesting techniques investigated included sedimentation, vacuum filtration, pressure filtration, cross flow filtration, disc stack centrifugation, decanter centrifuge, dispersed air floatation, dissolved air flotation, fluidic oscillation, inorganic flocculation, organic flocculation, auto-flocculation, bio-flocculation electrolytic coagulation, electrolytic flocculation and electrolytic floatation. Eight criteria were used for evaluation of these microalgae harvesting techniques: (a) dewatering efficiency (b) cost (c) toxicity (d) suitability for industrial scale (e) time (f) species specificity (g) reusability of media and (h) maintenance. Each criterion was assigned a score between 7 and 15 based on its degree of importance. Higher values were given to the criteria that were deemed most important for development of an efficient and economic large scale dewatering method for microalgae whereas lower values were given to criteria that were deemed necessary for determining a suitable method but were considered less important. The results indicated that of the 16 methods evaluated, 4 scored values of 80/100 and above and were deemed suitable for harvesting microalgae on an industrial scale. Three were physical techniques (disc stack centrifuge (87/100), cross flow filtration (84/100), decanter centrifugation (82/100)) and the forth was the organic flocculation (80) method. These techniques were deemed suitable for large scale use because of their effectiveness, low operational costs, suitability for numerous species, rapidness, minimal maintenance requirement and being environmentally friendly. The other methods were deemed unsuitable because they are not effective in dewatering a wide array of microalgae species, not suited for large volumes, costly and require high maintenance. Although each of the optimum techniques was deemed suitable for harvesting of microalgae on its merit, a combination of methods can also be used to enhance the recovery efficiency and improve the economics. The use of organic flocculation as an initial harvesting step to concentrate the algae suspension and the centrifugation (or filtration) as a secondary dewatering step will reduce the time and costs associated with dewatering. Flocculation allows for effective removal of algae from large amounts of liquid media and as such the costs associated with energy intensive centrifugation and filtration techniques (used individually) can be reduced by using them as secondary techniques since less volumes of microalgae suspension will undergo the secondary treatment.

Keywords: Microalgae; Harvesting; Dewatering; Physical treatment; Chemical treatment; Electrophoresis processes

Introduction

Microalgae are photosynthetic microorganisms that are abundant in nature and capable of growing in various environments [1]. Microalgae biomass can be used to produce numerous value added products such as biofuels (biodiesel, bioethanol, biogas and biohydrogen) [2], fish feed [3], animal feed [4], human food supplements such as vitamins A, B1, B2, B12, C, E, nicotinate, biotin, folic acid and pantothenic acid), Omega 3 fatty acids (Eicosapentaenoic acid (EPA), Docosahexaenoic acid (DHA)) and chlorophyll [5,6] and skin care products such as anti-aging creams, anti-irritant creams and skin regenerate creams [3,7,8]. Various microalgae strains contain high amounts of proteins (43-71% of dry matter) compared to meat (43%), soybeans (37%), milk (26%) and rice (8%). They synthesize a wide range of amino acids essential for humans and animals which make them great for use in food supplements [9,10]. Microalgae carbohydrates (10-30% of dry matter) are synthesized in the forms of sugars, starch and polysaccharides which are easy to digest [9]. The oil content in the cells can make up 25-77% of the dried biomass weight [11]. Microalgae are regarded as the best candidate for the production of biodiesel as they do not compete with edible crops [1,12] and can produce between 20,000 to 80,000 L of oil per acre per year which is 7-31 times greater than that produced by the best terrestrial crop (palm tree) [13]. Application of biorefinery concept to produce biodiesel and other value added products will enhance the economics of biodiesel production.

However, processing microalgae into biodiesel and other value added products requires culturing of the microalgae, recovery of the microalgae biomass and the extraction and downstream processing of the oil and other value added products [14]. However, the major

---

*Corresponding author: Ghaly A, Professor, Department of Process Engineering and Applied Sciences, Dalhousie University, Halifax, Nova Scotia, Canada
E-mail: abdel.ghaly@dal.ca

obstacle for using microalgae biomass on an industrial-scale for the production of value added products is the dewatering step [15,16]. Microalgae cultures need to be concentrated because they exist as a dilute suspension containing 0.1-2.0 g of dried biomass per litre [15,17]. Dewatering microalgae accounts for 20-30% of the total costs associated with microalgae production and processing [17,18]. The cost of the extraction, purification and extraction processes decrease with increased biomass concentration [15-17].

Therefore, in order to achieve economically viable biodiesel production, microalgae recovery needs to be made less costly. Different methods for solid-liquid separation can be employed to dewater/concentrate the microalgae culture to 10-450 g/L. Such methods include sedimentation, vacuum filtration, cross flow filtration, pressure filtration, decanter centrifugation, disc stack centrifugation, dissolved air flotation, dispersed air flotation, micro bubble generation organic flocculation, inorganic flocculation, bio-flocculation, auto-flocculation) and electrolytic coagulation, electrolytic flocculation and electrolytic flotation.

The aim of this study was to review the current methods used for harvesting and concentrating microalgae and perform a comparative analysis in order to determine the most efficient economically dewatering methods for large scale processing of microalgae biomass for production of biodiesel and value added products.

Physical Harvesting Methods

Numerous physical methods for microalgae dewatering processes have been used to retrieve the microalgae cells from their liquid suspension. These can be divided into four categories: sedimentation, filtration, centrifugation and flotation.

Sedimentation

In this technique, the solids and liquids are separated from one another by gravitational forces as shown in Figure 1 [19]. Different materials are separated from one another based on the density of the material and/or particle size. A larger difference in density would result in faster sedimentation rates while a smaller difference in densities and/or smaller particle size would require longer time to settle out by gravitational forces [20].

Type of Settling Tanks

Lamella separator (Figure 2a) and sedimentation tanks (Figure 2b) are used to separate solids from liquid [21,22]. Lamella separators offer a greater settling area than conventional thickeners as a result of plate orientation [23]. Lamella tanks work by inserting the microalgae biomass through the inlet. The liquid floats to the surface (effluent) and the biomass is caught onto the slanted plates. With time, the biomass settles down to the bottom of the tank and can be collected through

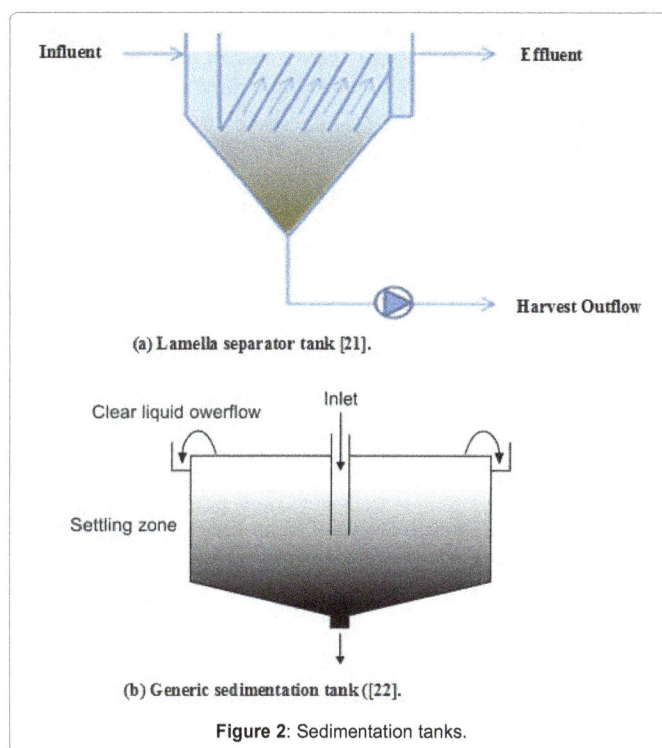

(a) Lamella separator tank [21].

(b) Generic sedimentation tank ([22].

Figure 2: Sedimentation tanks.

the harvest outflow. Sedimentation tanks are cylindrical with a funnel shaped bottom so that the settled microalgae are concentrated near the outlet. The outlet is placed at the bottom of the tank so that the collection of the settled microalgae can easily be recovered. The tank is equipped with a pump that carries the microalgae biomass from the cultivation tank into the sedimentation tank through the inlet. These tanks work by allowing the denser solids to settle to the bottom of the tank, leaving the clear water at the surface. Once the settling process is complete, the algae can be retrieved from the tank through the outlet.

Factors Affecting Sedimentation

The factors influencing the settlement rates of microalgae include: density and particle size, temperature, aging of the cells, light intensity and time [24-26].

Density and particle size: The density of marine microalgae varies from 1030 to 1100 kg/m^3 and the density of freshwater microalgae varies from 1040 to 1140 kg/m^3 [27-29]. Granados et al. [30] reported that the densities of fresh water (1000 kg/m^3) and salt water (1025 kg/m^3) are similar to that of microalgae and as a result the rate of settlement of algae is low. Murphy and Allen [31] stated that it is a challenge to remove microalgae biomass from the liquids because of the identical densities of the cells and media.

Cole and Buchak [32] indicated that the rate of settlement is dependent on the type of microalgae present and found the green microalgae to have an average settling rate of 0.1 m/d. Peperzak et al. [33] noted that the sedimentation rate of 24 different microalgae species (ranging in size from 10-1000 μm) varied from 0.4 to 2.2 m/d and there was no correlation between the size of the cells and the sinking rates. Milledge and Heaven [20] reported a settlement rate of 0.1 m/d for *Chlorella* species in freshwater. Yang et al. [34] reported an algae settling rate in the range of 0.1-0.3 m/d. Choi et al. [35] noted that the sedimentation rate of large and small sized algae were 2.6 cm/h and less than 1.0 cm/h, respectively.

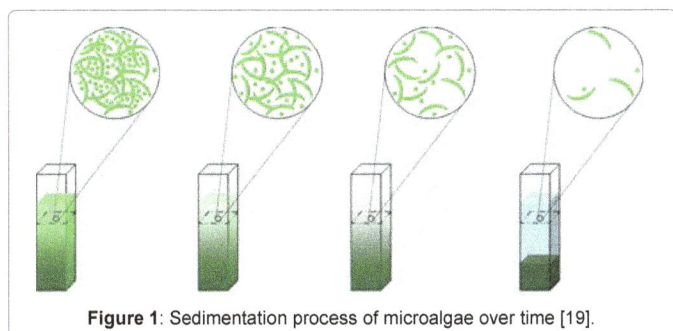

Figure 1: Sedimentation process of microalgae over time [19].

Temperature: Knuckey et al. [25] noted that the temperature of 4°C settled a wide array of microalgae species after 24 h when the pH was adjusted in the range of 8-8.5. Davis et al. [36] noted slower settling rates of microalgae in colder waters as a result of increased viscosity. Harith et al. [24] tested the effect of varying temperature (4-27°C) and the presence and absence of light on sedimentation rates of *Chaetoceros calcitrans* at a pH of 8. The highest efficiency (9%) was obtained at day 8 at a temperature of 27°C in the dark. Greenwell et al. [37] noted that harvesting microalgae by sedimentation in high temperature areas will deteriorate the cells.

Cell age: Danquah et al. [15] noted that the settling rate for microalgae harvested during the high growth phase (4-10 days) was lower than that harvested during low growth phase (10-12 days). Choi et al. [35] reported that the settling rate of algae significantly increased in the stationary growth phase of microalgae. Manheim and Nelson [38] noted that in the exponential growth phase (day 6) there was little to no settling in *Scenedesmus sp.* observed over 2 h period, but the greatest removal efficiency was noted in the stationary phase (day 15). They also noted that the settling rate for C. vulgaris species in the exponential phase was 6 times greater than the late stationary phase. Peperzak et al. [33] reported that the settling rate at 15 and 20 weeks for *Phaeocystis globosa* and *Eucampia zodiacus* were 0.5 and 1.0 m/day and 0.7 and 1.0 m/day, respectively.

Light: Danquah et al. [15] noted that the absence of light increased the settling rate during high growth and low growth phases. The supernatant obtained during the high growth phase contained 0.57 g/L of biomass in the presence of light and 0.39 g/L in the absence of light, while the supernatant obtained during the low growth phase contained 0.28 g/L in the presence of light and 0.17 g/L in the absence of light. Schlenk et al. [39] noted that the concentration of microalgae in the light and dark conditions were 1075 cells/mL and 775 cells/mL, respectively. On the other hand, Harith et al. [24] reported that the settling rates observed in the presence and absence of light in *Chaetoceros calcitrans* were similar.

Time: The concentration of microalgae by sedimentation requires long settling times that are greater than 24 h. Park et al. [40] noted long retention times of 1-2 days for algae recovery in large-scale settling tanks. Harith et al. [24] reported that increasing the settling time to 15 days increased the settling efficacy to 94%. Griffiths et al. [41] noted that the percentage of biomass recovery after 24 h of settling for *S. platensis, C. fusiformis, T. suecica, Nannochloropsis* and *Scenedesmus* were 95, 96, 80, 59 and 86%, respectively. Wang et al. [42] noted that the biomass recovery for the species *S. dimorphus* and *C. vulgaris* after 2 h of gravitational settling was 80 and 55%, respectively.

Advantages and Disadvantages

Although sedimentation tanks are effective in concentrating microalgae suspensions to 1.5% total suspended solids (TSS), they are not widely use in the industry. The costs associated with gravitational settling are low, but the reliability of this method without the use of flocculating agents is also low [16,43]. The settling time required is much longer than other processes [44] and energy is required for pumping the slurry [16]. Gonzalez-Fernandez and Ballesteros [45] stated that this method is time consuming and the composition of the cells can change. Mata et al. [46] stated that the cell concentrations obtained by sedimentation are low. Ras et al. [47] indicated that harvesting microalgae biomass by sedimentation alone is not the most efficient method since the cell recovery rates of 60-65% are low.

Filtration

This type of algae harvesting method uses a medium that is permeable so that it can retain the algae biomass while allowing the liquid to pass through. This technique requires a pressure difference across the filter which can be driven by vacuum, pressure or gravity. The membrane filters can be classified based on the size of the pores into macro filtration (greater than 10 μm), micro-filtration (0.1-10 μm), ultrafiltration (0.02-0.20 μm) and reverse osmosis (less than 0.001 μm) [48]. The pressure required to force the fluid across the membrane decreases as the pore size of the membrane is increased. Filtration

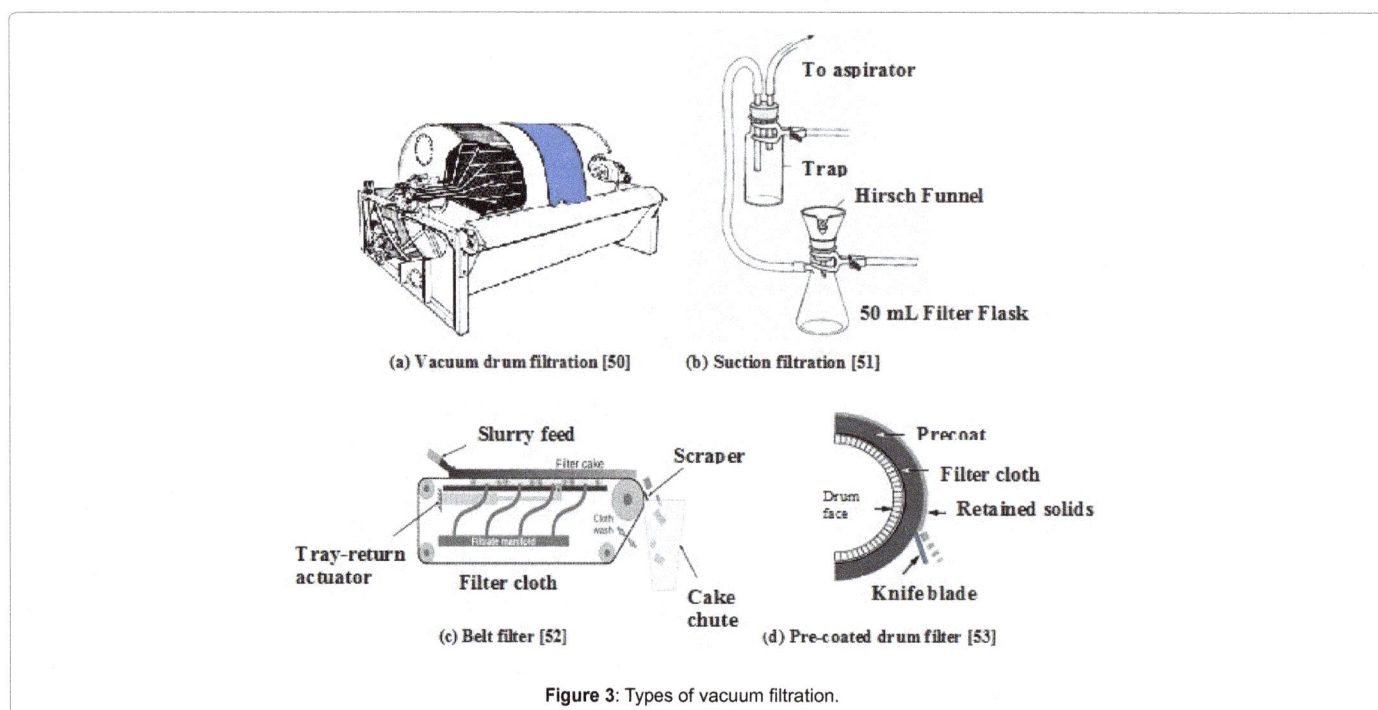

(a) Vacuum drum filtration [50]

(b) Suction filtration [51]

To aspirator

Trap

Hirsch Funnel

50 mL Filter Flask

Slurry feed

Scraper

Tray-return actuator

Filter cloth

Cake chute

(c) Belt filter [52]

Precoat

Filter cloth

Drum face

Retained solids

Knife blade

(d) Pre-coated drum filter [53]

Figure 3: Types of vacuum filtration.

techniques can concentrate microalgae cells in the suspension upto 5-18% and the operating costs vary from $10 to $20/gal. The harvesting efficiency using filtration methods ranges from 20% to 90% [49].

Vacuum Filtration

Vacuum filtration separates solids from liquid media by capturing the solid particles onto a filter while pulling the liquid through by suction from the filter Figure 3 [50-53]. Microalgae range in size from 2 to 30 μm indicating that a micro-filtration membrane is suitable for vacuum filtration [17,48]. Milledge and Heaven [20] stated that the macro-filtration membranes can be used for large microalgae cells or if the algae cells are flocculated together. Uduman et al. [16] reported that the vacuum filtration harvesting technique is most suited for large microalgae cells (greater than 10 μm). Stucki et al. [54] separated *Spirulina platensis* species using vacuum filtration equipped with regenerated cellulose membrane with a pore size of 0.45μm.

Type of vacuum filters: There are five different filter membranes that can be used in vacuum filtration. They are vacuum drum filter, suction filter, filter thickener, belt filter and starch precoated drum filter. Mohn [23] noted that the suction filter, starch precoated drum filter and belt filter were suitable for concentrating the *Coelastrum* microalgae species to a range of 5-37%. The author found the drum filters were not effective harvesting techniques as a result of clogging. Filter thickeners were not recommended as a result of low solid contents (3-7%) and

high energy requirements. Ferrentino et al. [55] noted that vacuum filtration equipped with a Buchner funnel and cellulose fiber filters were effective in the recovery of microalgae from solution.

Successful recovery of microalgae cells has been noted using filtration equipped with diatomaceous earth as a filter aid to avoid the clogging of the filter [17]. Gudin and Chaumont [56] reported that precoated drum filter with filter aid (diatomaceous earth) is effective in harvesting the microalgae *Chlamydomonas reinhardtii*. Molina Grima et al. [17] evaluated filters made of cellulose fibers and sand filters and obtained unsatisfactory results but found that diatomaceous earth filter effectively recovered the micro sized *Dunaliella* species. Uduman et al. [16] noted an exceptional recovery of *Dunaliella* species using diatomaceous earth aided filter. Brennan and Owende [48] stated that the use of diatomaceous earth as a filter aid can effectively remove microalgae cells from medium.

Energy consumption: Shelef et al. [43] reported energy consumption in the range of 0.1-5.9 kWh/m³ depending on the type of filter used. Mohn [23] noted that vacuum filtration consumed 5.9 kWh/m³ of energy in order to concentrate the suspended solids in solution to 18-27%. They also reported that the energy required to dewater the *C. proboscideum* using suction filter (8% SS), belt filter (9.5% SS) and filter thickener (5-7% SS) was 0.1, 0.45 and 1.6 kWh/m³, respectively. Umesh [57] noted that harvesting the microalgae strain *Spirulina fusiformis* under vacuum filtration with a coarse pores medium was low in cost

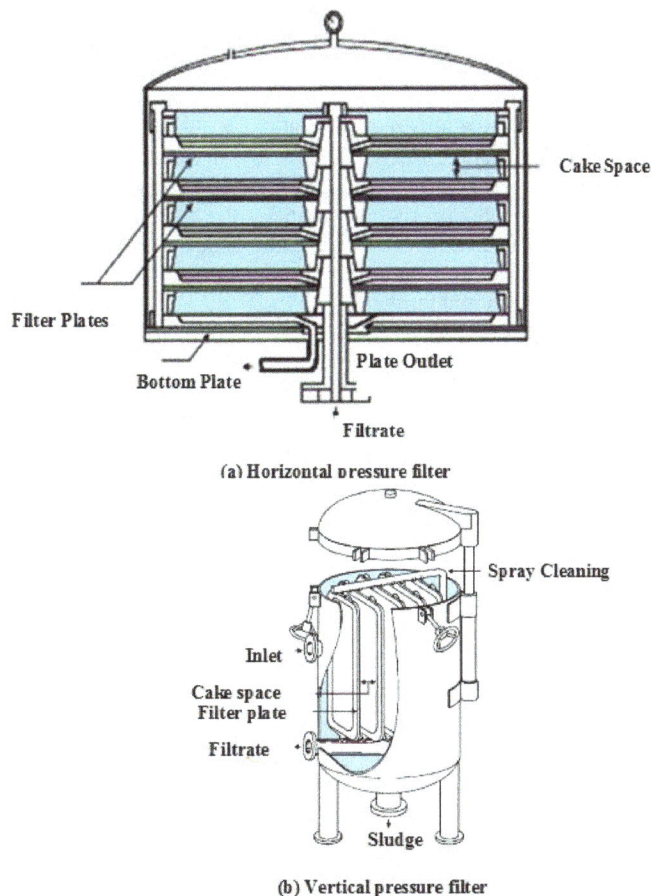

(a) Horizontal pressure filter

(b) Vertical pressure filter

Figure 4: Pressure filter [50].

($83.3/ m²) and capable of harvesting 23 kg/m² kwh. Shelef et al. [43] reported that the pressure drop required for vacuum operations is in the range of 70-80 kPa. Milledge and Heaven [20] reported a power consumption of 0.25 kWh/m³ for microalgae harvest using vacuum belt filter. Mohn [58] reported an energy consumption value of 3 kWh/m³ for microalgae harvesting, using vacuum drum filtration.

Advantages and disadvantages: The advantages of using vacuum filtration technique for harvesting microalgae are the preservation of the cells after the recovery process [59]. The effectiveness of the filtration process is dependent on the membrane size and the microalgae cell size. Harvesting microalgae by filtration is more efficient than the sedimentation technique, but drawbacks of this method include membrane replacement and/or periodical washing of the membrane to avoid clogging the membrane pores [45]. However, drawbacks are associated with large energy requirements and costs associated with periodic replacement of membrane as a result of clogging [17,46,60-62]. Arar and Collins [63] recovered microalgae for chlorophyll extraction using vacuum filtration at 6 in. Hg (20 kPa) and noted that higher pressures and prolonged filtration (beyond 10 min) may damage the cells. Uduman et al. [16] noted that filtration technique is suitable for larger microalgae cells, but inadequate for recovery of microalgal species. Rossi et al. [64] noted that rapid clogging of the membrane resulted using the ultrafiltration membrane technique. Flocculation assisted filtration processes would lower the energy requirements, but additional costs for the flocculent would be encountered [20]. Molina Grima et al. [17] recovered microalgae biomass using filtration method and concluded that this harvesting method is not economically viable for large scale production.

Pressure Filtration

Pressure filtration is a technique used for separating particles form a liquid suspension into a compacted form. It works by separating the liquid from the particles (that are collected onto the filter) by means of pressure [65]. The flow of fluids through the filter is created by raising the pressure above the atmospheric pressure to create a pressure differential across the filter [66]. This process is operated in batches that are most often fed from and discharged to a continuous process. A surge tank is required upstream to the filter and one is required for the collection of the filtrate [67].

Type of pressure filtration: Pressure filtration harvesting can be achieved by plate-and-frame filter presses or by using a pressure vessel that is equipped with filters as shown in Figure 4 [50]. The plate-and-frame filter presses works by forcing the liquid in the microalgae suspension through the filter using high pressure. A series of rectangular plates that are mounted in a vertical position, face to face, make up the

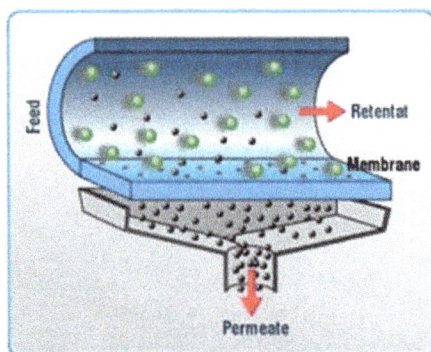

press system. A fitted filter cloth is applied to each of the plates and they are held together with one another by force under pressure. The fluid that contains the algae is pumped into the gaps between the plates and the pressure is applied in order to force the liquid through the plate outlets and filter cloths. After separation, the dewatered microalgae cake is recovered [68].

Energy consumption: This method can be considered energy efficient since a minimal amount of energy is required upon assessment of the output product and the amount of initial feedstock added [15]. However, the effectiveness of the method is dependent on the type of algae species.

Molina Grima et al. [17] noted that the amount of energy required to harvest 22-27% (w/v) of C. *paroboscideum* species using pressure filters is 0.88 kWh/m³. Nagle and Lemke [69] noted an 8% concentration of microalgae (up to 0.5%) using a filter press that has 20 plates and frame (30 cm in diameter), equipped with filter paper that has a pore size of 5 μm. Harun et al. [8] noted that the microalgae *Dunaliella* and *Chlorella* species were too small to be recovered by pressure filtration. Mohn [23] reported that the energy consumed for harvesting C. *proboscideum* using cylindrical sieve (7.5% suspended solid concentration) and filter basket (5% suspended solid concentration) was 0.3 kWh/m³ and 0.2 kWh/m³, respectively. He also found that pressure filtration was not suitable for the species *Scenedesmus, Dunaliella* and *Chlorella*, but was satisfactory for other larger microalgae species such as *Coelastrum proboscideum* and *Spirulina platensis*.

Advantages and disadvantages: Some of the advantages of using pressure filtration are: the cakes collected (composed of the particles in the liquid suspension) have low moisture content, the soluble recovery from the cake is high, re-circulating the filtrate for 1-2 min will clean the filter, high degree of clarity in solutions can be achieved and alloy and synthetic materials can be used to construct the filters and the internal parts [66,70]. The disadvantages of using this technique include: the difficulty in washing the filter medium which increases when the solid is sticky, the internals are difficult to clean in food-grade applications and the difficulty in viewing the condition of the filter due to vessel encapsulation [70].

Cross Flow Filtration

Harvesting microalgae cells in large volumes can be effectively done using cross flow filtration a shown in Figure 5 [71]. In this technique, the sample flows tangentially across a membrane. The particles larger in size than the membrane pores are retained and referred to as the retentate. Smaller particles pass through the membrane with the liquid solution and are referred to as the permeate.

Membrane type: Ultrafiltration or microporous membranes are the type of filter membranes used in this technique. These membranes are available with a wide range of pore sizes and molecular weight retentions. Polymer membranes have a long operating life when used at suitable cross flow velocity conditions and low transmembrane pressures. Petrusevski et al. [59] used a cross filtration with a membrane pore size of 0.45 μm and achieved a biomass recovery efficiency of 70-89%. Rossignol et al. [72] found that polymer membranes were effective in recovering the marine microalgae species *Haslea ostraria* and *Skeletonema costatum*, but the performance depended on the hydrodynamic conditions,properties of the microalgae and the concentration of the microalgae cells.

Uduman et al. [16] reported that the initial flux for microfiltration membranes were much higher than those of ultrafiltration, but they

Figure 5: Cross flow filtration diagram [71].

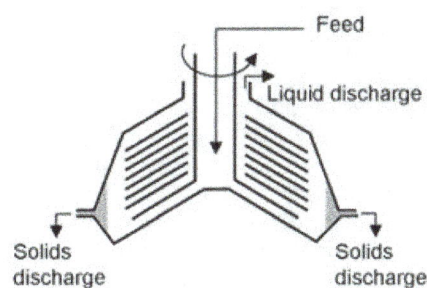

(a) Borowitzka and Moheimani [79]

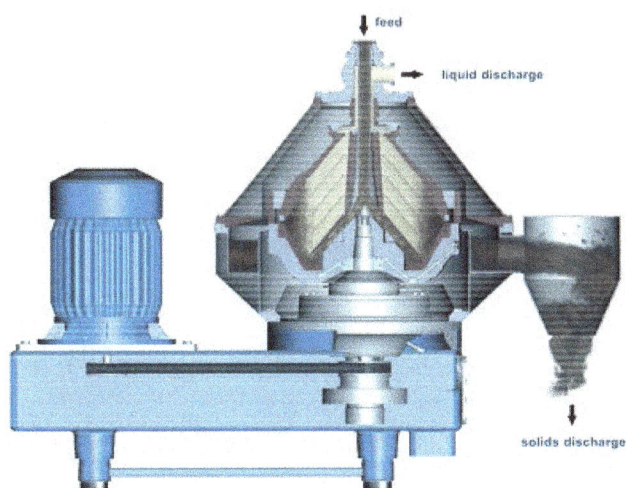

(b) Enviropro [80]

Figure 6: Disc stack centrifuge.

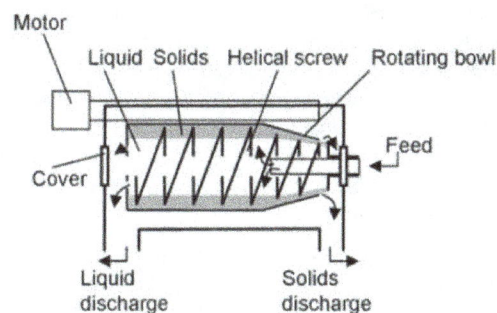

Figure 7: Decanter centrifuge [79].

clogged more easily. Zhang et al. [73] used a cross-flow ultrafiltration membrane with a cross-flow velocity of 0.17 m/s and noted an increase in algae concentration from 0.104% to 92.5% at the membrane surface with a harvesting efficiency value of 46.01 g/m²/h. Rossi et al. [74] tested 14 various inorganic membranes and noted that the ultrafiltration membrane ATZ-50 kDa illustrated the best performance and concentrated the *Arthrospira platens* species by a factor of 20. Rossi et al. [64] used a cross-flow filtration technique equipped with an organic ultrafiltration membrane (polyacrylonitrile, 40kDa) and concentrated *Arthrospira platens* by a factor of 10. Rossignol et al. [72] concentrated the species *Skeletonema costatum* using cross-flow ultrafiltration with a flux of 30 l/h for 12 h. Rose et al. [75] effectively concentrated the species *Dunaliella salina* by cross-flow ultrafiltration with flux rates of 30-40 l/h. Walsh et al. [76] concentrated the species *Thalassiosira pseudonana* to 2.3 L from 2840 L which was composed of 2.33×10^{12} cell/L using microfiltration membrane system. Ahmed et al. [77] noted that the resistance of the cross-flow microfiltration decreased as the cross-flow velocity increased from 0.13 to 4 m/s while harvesting *Chlorella* sp. species.

Energy consumption: Rossignol et al. [72] reported that the energy consumption using cross-flow filtration techniques can range from 3kWh/m³ to 10kWh/m³ depending on the feed characteristics, the system design and the pressure used. Danquah et al. [15] noted that cross-flow filtration can also be used in sensitive suspensions and is a

cheap method for concentrating suspended solids in the range of 2.5-8.9 % with an energy consumption in the range of 0.38-2.06 kWh/m³. Crittenden et al. [78] reported that the energy consumption for a cross-flow filtration technique was 5kWh/m³. Danquah et al. [15] reported that the amount of energy required for dewatering microalgae to a concentration of 8.88% (w/v) was 2.06kWh/m³.

Advantages and disadvantages: Cross flow microalgae filtration is advantageous over other conventional harvesting methods such as sedimentation, flocculation and centrifugation because it results a in complete removal of debris and microalgae cells [16]. The equipment are considered to be cheap because the costs are only associated with pumping and replacement of membranes [72]. The structure and properties of the recovered microalgae are preserved using this filtration technique [59]. However, large scale recovery of algae cells using this method can be limited due to fouling and frequent replacement of the membrane [16].

Centrifugation

This type of removal mechanism is widely used in beverage, food and pharmaceutical industries. Centrifugation is a process in which a centrifugal force is used to enhance the separation of solids. Spinning the suspension creates the pressure differential necessary for particle separation from the liquid suspension. Thus, the efficiency of the recovery process is dependent on the centrifugal force [17].

Types of Centrifuges

The two types of centrifugation used for harvesting microalgae are: disc stack and decanter centrifuges.

Disc stack centrifuge: The most common industrial centrifuge used today in commercial plants producing high value products and algal biofuel is the disc stack type centrifuge. It consists of a shallow cylindrical bowl that has numerous stacks of metal cones (discs) which are closely spaced together as shown in Figures 6 [79,80]. Separation of the materials is based on densities. The mixture is placed on the centre of a stack of discs and the lighter phase of the mixture remains on the inside towards the centre while the denser phase is displaced outwards to the underside of the discs. This technique separates materials of different densities by layering them [81]. It is most suited for separating materials with particle sizes in the range of 3-30 μm and for concentrations of suspensions that has solid content ranging from 2 to 25% [20].

Heasman et al. [82] evaluated the cell recovery efficiency of nine different microalgae species using a disk stack centrifuge and noted a recovery efficiency greater than 95% at a force of 13,000g. They also noted that the recovery efficiency declined with a decrease in the gravitational force to 60% and 40% at gravitational forces of 6,000g and 1,300g, respectively. Sim et al. [83] noted a 90 % microalgae removal efficiency using a disc stake type of centrifuge. Vasudevan et al. [84] achieved an 18% microalgae concentration using a disc centrifuge. Mackay [85] used a disc centrifuge operating at a force of 4,000-15,000g for a biomass suspension containing 0.2-20% v/v algae cells. Chojnacka et al. [86] harvested the *Spirulina* sp. using a disc type centrifuge operating at 6,000 rpm for 5 min.

Decanter centrifuge: Decanting centrifugation is based on the concept of using a special settling tank in which the solids in suspension are forced to fall down due to the gravitational forces [87]. The decanter centrifuge (Figure 7) operates continuously by pumping the cultivated microalgae biomass into the centrifuge bowl whereby the suspended particles in solution are forced to the bottom of the bowl. The liquid left after the particles have been extracted is passed through the overflow pipe [88].

Molina-Grima et al. [17] noted that concentration of microalgae biomass using a decanter centrifuge is better than other harvesting methods. Dassey and Theegala [89] achieved a harvesting efficiency of 28.5% for microalgae at a flow rate of 18 L/min using continuous flow decanter centrifuge. Smith and Charter [87] reported that the clarity of the liquid produced after separation was not as great as that achieved using disc-stack centrifugation. Vasudevan et al. [84] reported that a 12% microalgae concentration was achieved using a decanter centrifuge. Mackay [85] reported that the decanter centrifuge operates using a force of 4000-10000 g and is effective for slurries with a biomass content of 5-80%. Vasudevan et al. ([84] stated that microalgae biomass needs to undergo an initial thickening step such as dissolved air flotation in order to concentrate microalgae suspensions (0.02-0.05 weight %) to 2-3% before using decanter centrifugation.

Energy Consumption

Disc-stack centrifuge: The energy consumption reported in the literature for the disc stack centrifuge varied from 0.53 kWh/m³ to 5.5 kWh/m³. Alfa Laval [90] used a disc type centrifuge for dewatering microalgae and achieved a 16% with a power consumption of 0.53kWh/m³. Mohn [23] noted a 12% suspended solids concentrate of the microalgae species *Scendesmus* using a disc-stack centrifuge with an energy consumption of 1 kWh/m³. Goh [91] harvested microalgae grown in pig waste using disc centrifugation with an energy

consumption of 1.4kWh/m³. Sharma et al. [92] noted that the disc stack centrifuge consumed 5.5kWh/m³ for *Chlorella sp*. harvesting.

Decanter centrifuge: The energy consumption reported in the literature for decanter centrifuge varied from 1.3kWh/m³ to 8kWh/m³. Sim et al. [83] noted that an energy consumption of 1.3kWh/m³ was required for concentrating microalgae biomass from 0.04% to 4.00% using a decanter type centrifuge. Molina Grima et al. [17] achieved a microalgae concentration of 22% (w/v) using decanter centrifuge with an energy consumption of 8kWh/m³. Mohn [58] reported that the energy consumption for harvesting microalgae (20% DS) using decanter centrifugation was 4 kWh/m³.

Advantages and Disadvantages

Disc-stack centrifuge: The advantages of using disc-stack centrifuge for harvesting microalgae is their high removal efficiency compared to other industrial centrifuges. The concentration of the feed for these units is typically in the range of 0.5-10% w/w. This type centrifuge handles high flow rates and is capable of separating fine (0.1-100 µm) particles [93]. This device can be used to separate solid from liquid in continuous, semi-continuous and batch operation. Some of the disadvantages of this type of centrifuge include: low dry substance content in the discharge system, mechanically complex, costly and the small space between the closely stacked discs makes it harder to clean and may require chemicals for cleaning [92,93].

Decanter centrifuge: The dewatered biomass using the decanter centrifuge is much more concentrated than that achieved using the disc centrifuge. However, the decanter centrifuge is more suited for suspensions with higher solid particles and is unsuitable for microalgae suspensions [58]. This type of centrifuge is most suited for separating materials with particle sizes greater than 15 µm and solid suspensions containing higher than 15% [20]. It operates at inertial forces that are less than 6000 g. The disadvantages of using this method for microalgae harvesting are: highly concentrated feeds (typically in the range of 4-40% w/w) is required, the liquid leaving the system may not be clear due to the presence of fines, processing finer particles may result in poor flow properties of the thickened solids and cause mechanical difficulties, much more energy intensive than disc centrifuges and the high costs associated with the equipment required for processing large volume [17,20,93-95]. This type of centrifuge has been estimated to consume 3000 kWh/ton of dry alga biomass.

Flotation

The flotation technique for microalgae harvesting takes advantage of the low density of microalgae [28]. This technique is classified as a physiochemical gravity separation process in which gas bubbles pass

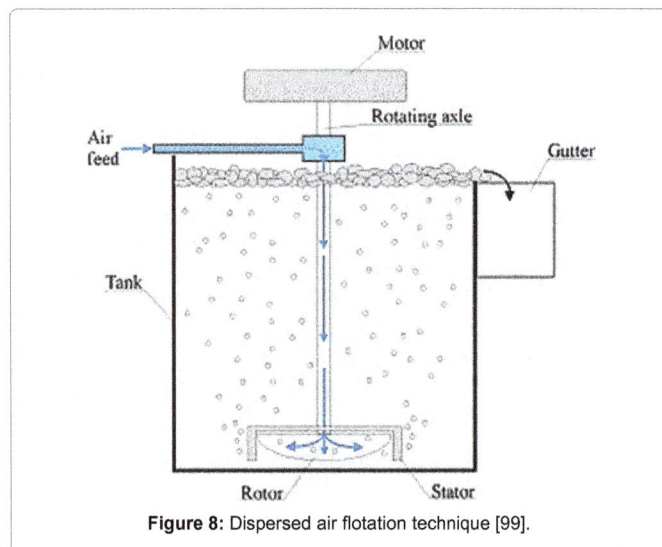

Figure 8: Dispersed air flotation technique [99].

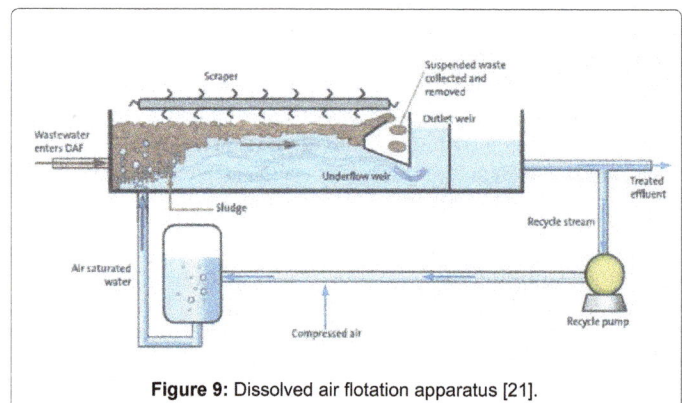

Figure 9: Dissolved air flotation apparatus [21].

through a liquid-solid suspension causing the microalgae to float to the surface by adhering to the gaseous bubbles [43,96]. The aeration also assists in removing the volatile organic compounds that are in the solution which provide cleaner residual water [43]. The efficiency of this method depends on the suspended particles instability, higher air-particle contact corresponds to a lower instability [97]. In flotation technique, the size of the particle is of importance, the smaller the particle size the more likely it will be lifted to the top of the medium by the bubbles. Solutions with particle sizes of less than 500 μm can be used in flotation [98].

Types of Air Flotation

The flotation processes are grouped by the method that is used for the bubble formation into: dispersed air flotation, dissolved air flotation, microbubble generation and electrolytic flotation.

Dispersed air flotation: This technique requires the use of a high speed mechanical agitator for bubble formation and an air injection system as shown in Figure 8 [99]. The gas mixes with the liquid as it is introduced at the top of the vessel and is allowed to pass a disperser that creates bubbles ranging in diameter from 700 to 1500 μm [100]. These bubbles are a magnitude larger (1000 μm in diameter) than those produced using dissolved air flotation technique.

Chen et al. [101] used dispersed air flotation for dewatering of microalgae *Scenedesmus quadricauda*. They used three varying surfactants in order to remove the microalgal cell: non-ionic X-100, cationic *N-Cetyl-N-N-N*-trimethulammonium bromide and anionic sodium dodecylsulfate. They also found that surfactants played a role in increasing the integrity of the bubble avoiding rupturing, and noted that this removal method was most successful with the use of cationic *N-Cetyl-N-N-N*-trimethulammonium bromide surfactant.

Yan and Jameson [102] noted that the dispersed air flotation technology resulted in 98% microalgae removal efficiency. Xu et al. [103] reported a 93.6% recovery efficiency of *B. braunii* using dispersed air flotation technique. Kurniawati et al. [104] harvested *Chlorella vulgaris* and *Scenedesmus obliquus* using dispersed air flotation assisted with chitosan and achieved a recovery efficiency greater than 93%.

Dissolved air flotation: The dissolved air flotation technique is the most widely used flotation technique in the treatment of industrial effluent [98,100]. This method requires a reduction in water pressure that is presaturated with air. The liquid is then injected into the flotation tank at atmospheric pressure. Bubbles are generated from the diffuser nozzles and rise through the liquid carrying microalgae cells in the suspended media to the surface of the tanks as shown in Figure 9

[21]. The cumulated biomass at the surface can be skimmed off and collected. The clarified liquid portion is saturated with air and recycled back into the flotation tank [100]. The supply of air into the system can be controlled by changing the saturator pressure or by changing the ratio that is recycled back into the tank. The size of the bubbles formed can be controlled by the saturator, operated above atmospheric pressure and by the injection flow rate [16,105]. The flow rate must be great enough to prevent backflow, provide a pressure drop and allow for bubble growth on the pipes surface [16]. Small bubbles ranging in size from 10 to 100 μm (with a mean size of 40 μm) are desirable [105].

The dissolved air floatation microalgae separation is usually coupled with the use of a chemical flocculation process. Edzwald [28] investigated the use of dissolved air flotation for microalgae recovery and noted that this method required pretreatment by flocculation, but was more successful than the settling technique. Wiley et al. [106] used dissolved air flotation for microalgae harvest and noted a suspended solids concentrate of 5% with an energy consumption of 7.6 kWh/m³. Goh [91] noted that this method was effective in harvesting microalgae from pig slurry when coupled with the alum flocculent with a high dosage of 0.3 g/L.

Fluidic oscillation: The recovery of microalgae can also be achieved by micro-bubble generation through fluidic oscillation as shown in Figure 10 [107]. This method works by converting a continuous air supply into oscillatory flow with a regular frequency, generating bubbles that are the size of the exit pores [108]. The miniature bubbles are formed by fitting a diffuser to the bi-stable valve which ensures that the bubbles formed are approximately 10 times smaller than those originally dispersed in flotation methods [107]. Fine bubbles that are the size of the exit pores are generated by use of a fluidic oscillator [108]. The bubbles formed attach to the hydrophobic cells suspended in solution and carry them to the surface. At the surface the bubble raptures leaving the cells behind [109].

Hanotu et al. [108] reported a microalgae recovery efficiency of 99.2% using microbubble generation at a pH of 5 with the aid of ferric chloride coagulant (150 mg/L). Elder [110] noted a removal efficiency greater than 95% using micro-bubble flotation. Yap et al. [111] noted a removal efficiency greater than 95% for removal of *Microcystis* and filamentous *Cylindrospermospsis* using micro-bubble flotation. However, not much research has been performed using this technique for microalgae recovery to deem it economically suitable for large scale recovery of microalgae cells [20].

Factors Affecting Air Flotation

The factors affecting the efficiency of harvesting microalgae using the air flotation technique include: pH, air flow rate, alkalinity, recycle rate, hydraulic loading and time. The velocity of the skimmer determines the concentration of the slurry formed and the height above the surface of the water [16].

pH: One of the most important parameters that affects the flotation processes is the pH. Lin and Liu [112] stated that the pH affects the reaction routes and the interfacial properties. Chen et al. [101] noted that maintaining the pH in the range of 5-8 using anionic sodium dodecylsulfate had little to no effect on the removal efficiency (95%) of microalgae *Scendesmus quadricauda*. They attributed this to the electrostatic interactions between the positively charged microalgae surface and the surfactant that are facilitated by pH values less than 8. However, at pH values above 8 the interaction between the algae surface and surfactant was weak, which did not allow for efficient removal. Kwon et al. [113] reported that flocculent assisted air flotation

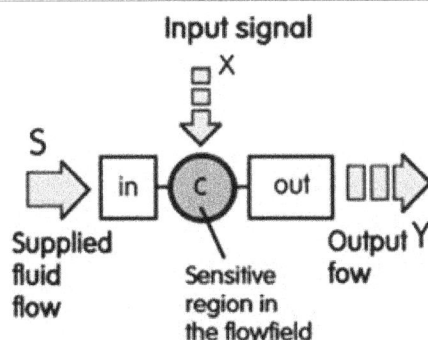

Figure 10: Principle of fluidic oscillation - The continuous flow of fluid (S) is integrated with a weak input signal (X) which causes the change in output flow (Y) [107].

technique required a pH of 7-8 for organic and 5-6 for inorganic flocculants in order to reach a removal efficiencies greater than 90%. Zhang et al. [114] reported a harvesting efficiency greater than 90% for *Chlorella zofingiensis* using an air flotation technique by adjusting the pH to 6.2. Hanotu et al. [108] reported that acidic conditions were optimal for effective removal of microalgae.

Alkalinity: Besson and Guiraud [115] noted that the microalgae flotation recovery efficiency was improved by using sodium hydroxide at a concentration of 0.0085 mol/L (over the tested range of 0.000 to 0.025 mol/L). Schlesinger et al. [116] noted that calcium hydroxide is most effective in flocculating microalgae that can be recovered using flotation techniques. Thangavel and Sridevi [117] reported that carbonate salts can be used for microalgae flocculation to enhance the recovery efficiency of the microalgae. Chen et al. [101] reported that the change in alkalinity from 0 to 50 mg/L using $NaHCO_3$ did not have any effect on the removal efficiency (95%) of *Scenedesmus quadricauda* species.

Air flow rate: Reports in the literature show that varying the air flow rate from 68 to 206 ml/min did not have any effect on the removal efficiency (92%) of the *Chlorella* sp. species (Liu et al. [118]. Coward et al. ([119] harvested *Chlorella* sp. using flotation technique at a flow rate of 100 L/h. Dassey and Theegala ([120] reported that an increase in the air flow rate beyond 160 ml/min did not significantly increase the microbubble production and as such no increase in harvesting efficiency was noted. Hanotu et al. [108] noted that the air flow rate of 85 L/min was effective in the separation of microalgae form the media.

Medium: Zhang et al. [121] found that the algal media influences the rate of harvesting efficiency. They noted that the algal cells that were nitrogen deprived had lower concentrations of surface functional groups as they went from the exponential to the stationary and declining growth phases. They noted that these functional groups are made up of proteins and polysaccharides that are important for stabilizing the surface charge which affects the adhesion onto the bubbles in dissolved air flotation technique. They also noted that more Alum coagulant was required to achieve higher harvesting efficiency (90%) in the exponential phase compared to the stationary and declining phases as more alum is required to destabilize the surface of the generated cells in the exponential phase. There was a linear correlation between the concentration of surface functional groups and alum dosage for a given harvesting efficiency. Coward et al. [119] stated th cells at the highest flotation harvesting efficiency was observed during high growth phase as a result of higher zeta potential. Schenk et al. [11] noted that nutrient limitation in the media can improve the harvesting efficiency of microalgae flotation methods.

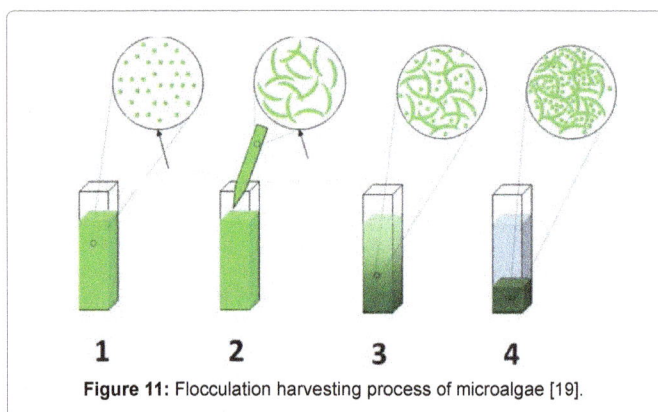

Figure 11: Flocculation harvesting process of microalgae [19].

Recycle rate: The recycle rate reported in the literature varied from 5 to 10%. Edzwald and Wingler [122] obtained a 97-99% removal efficiency of *Chlorella vulgaris* species with a recycle rate of 8%. Vlaski et al. [123] reported a removal efficiency of 94.5% for *Microcystics* with a recycle rate of 5-10%. Kempeneers et al. [124] achieved a removal efficiency of 80% for *Melosira cyclotella* with a recycle rate of 6%. Teixeira and Rosa [125] reported a 92-98% recovery efficiency of blue-green microalgae using a recycle rate of 8% and stated that the recycle system was vital for effective particle recovery but recycle rates past 8% illustrated little improvements. They also found that the addition of pressurized recycle system did not improve the recovery rate of the cells. This phenomena is attributed to the lack of particle destabilization since particle destabilization is vital to the effectiveness of dissolved air flotation as opposed to floc size. Gregory and Edwald [126] reported a recovery efficiency of 90-99% using dissolved air flotation with a recycle rate of 10%.

Hydraulic loading: The hydraulic loading rate for industrial air flotation applications ranges from 0.504 m/h to 40 m/h [127, 128]. Edzwald [129] reported that high rate dissolved air flotation techniques can be performed at hydraulic loadings of 20-40 m/h. Haarhoff and Rykaart [130] noted that increasing the hydraulic loading lowered bubbles formation. Dassey and Theegala [120] stated that increased hydraulic loadings decreased the time for air to dissolve which results in poor bubble productions.

Time: The time required for effective air floatation reported in the literature varied from 3 to 30 min. Edzwald and Wingler [122] noted that the time required to remove 96-99% of *Chlorella vulgaris* using air flotation was 5 min. Vlaski et al. [123] used an air flotation technique to concentrate the species *Microcystis aeruginosa* to 94.5% in 8 min. Kempeneers et al. [124] achieved an 80% removal efficiency of *Melosira cyclotella* using air flotation in 3 min. Xu et al. [103] noted a 93.6% recovery efficiency for *B. braunii* using air floatation in 14 min. Coward et al. [119] noted that the air flotation technique effectively harvested microalgae within 30 min.

Energy Consumption

The flotation technique is an energy intensive process as a result

Chemical Type	Flocculent	Reference
Inorganic	Ferric sulphate	Kown et al. [113]
	Ferric chloride	Papazi et al. [137]
	Aluminium chloride	Molina-Grima et al. [17]
	Aluminium sulfate	Oh et al. [136]
Organic	Magnafloc LT 25	Knuckey et al. [25]
	Zetag	t'Lam et al. [138]
	Praestol	Pushparaj et al. [139]
	Chitosan	Chen et al. [140]

Table 1: Chemical flocculent type and the effect on microalgae.

Flocculent Type	Effect on Microalgae	Reference
Inorganic	Toxic to the cells	Papazi et al. [137]
	Alter the color of the medium	Schenk et al. [11]
	Alter the pH of the media which may make it unsuitable for reuse	Molina Grima et al. [17]
Organic	High cell viability (>75%)	Harith et al. [24]
	No inhibitory effect the cells	Pushparaj et al. [139]
	Non-toxic	Vandamme et al. [141]

Table 2: Chemical flocculent type and the effect on microalgae.

of the high pressure required [108]. Algae harvest using dissolved air flotation is an efficient method but has high operational costs that are associated with the use of energy intensive compressor that function at pressures of 390 kPa [131-133]. However, the fluidic oscillator does not require much energy for operation and has been noted to consume 2-3 orders less energy than dissolved and dispersed air flotation methods [107,108].

Advantages and Disadvantages

Compared to sedimentation, flotation method is much faster and more effective for harvesting of microalgae [96]. Mohn [58] reported that dewatering microalgae using flotation methods (7% concentration) is much more rapid and efficient than the use of sedimentation (1.5% concentration). However, the flotation method has only been reported to be effective in microalgae harvesting on a bench scale and is not suitable for recovering microalgae cells on large scale [43]. In addition, this method is an energy intensive [16], has high operational costs and high energy is required for small bubble generation [58]. The cost of flotation has been reported to be as high as or even higher than the centrifugation method.

Chemical Harvesting Methods

On the basis of energy requirement, chemical flocculation as a dewatering method seems to be the most promising for large scale utilization [16,134, 135]. These flocculation methods work by concentrating the cells (coagulation) followed by settlement to the bottom of the cultivating apparatus due to the increased density of the concentrate as shown in Figure 11 [19].

Type of Chemical Flocculation

There are two types of chemicals (Table 1) that can induce flocculation: inorganic and organic polymers [136]. The effects of these polymers on microalgae are shown in Table 2 [11,17,24,137,139,141]. Typically, cationic, anionic and non-ionic polyelectrolytes are used to flocculate the microalgae cells [16].

Inorganic Compounds: Microalgae flocculation using inorganic compounds works by charge neutralization [142]. The flocculation process using these compounds works in low pH environments in order to form cationic hydrolysis products [143]. Under optimal pH, these flocculants form polyhydroxy complexes. A large number of chemicals (ferric sulphate, ferric chloride, aluminum chloride, aluminum sulfate) have been tested with the microalgae inorganic flocculation process, the most effective of which was aluminum sulfate [136]. In wastewater treatment, multi-valent metal salts such as ferric sulphate, ferric chloride and aluminium chloride have been used to remove algae [16].

Papazi et al. [137] achieved a harvesting efficacy for *Chlorella minutissma* species greater than 85% using ferric salts. Kown et al. [113] reported a flocculation efficiency of 85.6% for *Tetraselmis sp.* using ferric sulfate (a dose of 0.7 g/L) at a pH of 4-8. Wyatt et al. [144] reported a harvesting efficacy greater than 90% for the *Chlorella zofingiensis* species using a ferric chloride concentration of 40% (w/v) at a pH of 4.0. Xuan [145] achieved a 90% removal efficacy for *Nannochloropsis sp.* using ferric chloride administered at 0.18 mg/l. Sukenik et al [146] achieved a flocculation efficiency greater than 80% for marine microalgae using ferric chloride. Bintisaarani [147] found the ferric chloride flocculation to be the most effective for harvesting *Nannochloropsis* species and reported a removal efficiency of 89% using a ferric chloride concentration of 0.9 M at a pH of 7.5.

Aluminium (alum) salts have been noted to effectively flocculate the microalgae species *Chlorella* and *Scenedesmus* [17]. Papazi et al. [137] found aluminum salts to be more effective in flocculating *Chlorella* species than ferric salts. Shelef et al. [43] noted that alum was a superior flocculating agent compared to ferric sulfate in terms of pH, amount of flocculent and the quality of the final water slurry. Kown et al. [113] reported a flocculation efficiency of 92.6% for *Tetraselmis sp.* using aluminium sulfate dose of 1.2 g/L at a pH of 5-6. Millamena et al. [148] stated that alum was effective in flocculating *Chaetoceros calcitrans, Tetraselmis chui, Skeletonema costarum* and *Isochrysis goibana* species at a pH of 6.5. Aragon et al. [149] used aluminium sulfate to harvest a culture made up of *Scenedesmus acutus* (80%) and *Chlorella vulgaris* (20%) using a dosage of 30-50 mg/L at a pH of 6-6.5.

Organic Compounds

Organic polymers (chitosan) or polyelectrolyte (polyelectolyte polyamine) flocculants are known as polymeric flocculants (synthetic and natural) that consist of both ionic and non-ionic species. The use of organic compounds for flocculation works by combining both particle bridging and charge neutralization. The charge density and polymer chain length determines the extent to which each is used. The process begins by the attachment of the polymer onto the microalgal surface through chemical or electrostatic forces.

The polymer is able to attach to the surface of the cells through Coulombic (charge-charge), dipole-dipole, van der Waals or hydrogen interactions as shown in Figure 12 [150-153]. Coulmbic force attraction works by having unlike charges on the surface of the polymer and the microalgae attach to one another, following the notion that like charges repel one another and unlike charges attract one another. Dipole-dipole interactions occur when two polar molecules approach one another and the partially negative portion bonds to the partially positive one. Van der Waals forces are the attraction of intermolecular forces between molecules. Hydrogen bonding is a type of dipole-dipole attraction in

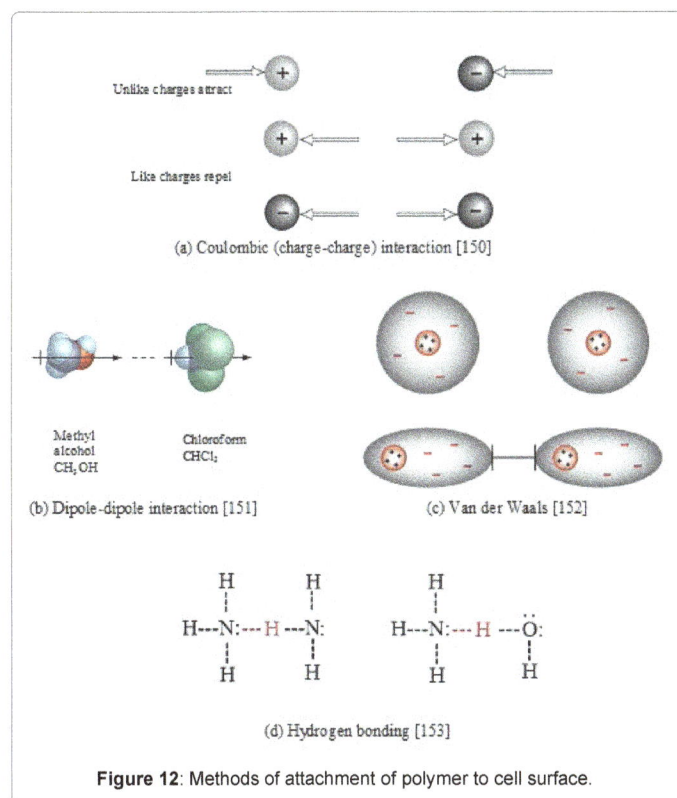

Figure 12: Methods of attachment of polymer to cell surface.

which a hydrogen atom is bonded to a highly electronegative atom nearby [154]. In this manner, the polymer attaches to the surface leaving its tail out into the solvent forming loops. The loops and tails of the polymer allow it to attach to other cells to from bridges between them [155].

The efficiency of this flocculation process depends on the degree in which the microalgae cells cover the polymer. If the attachment of the polymer to the cell surface is less than the optimum amount, then it may not be able to withstand shear forces as a result of agitation. On the other hand, excess coverage of the polymer onto the cell surface can cause static hindering of the bridging process [156].

Recent studies have revealed that cationic polyelectrolytes flocculent agents are the most effective for microalgae recovery [16,30]. Granadoes et al. [30] noted that inorganic flocculants were less efficient in the flocculation of *Muriellopsis sp.* species than organic agents. Knuckey et al. [25] reported that adding 0.5 mg/L of non-ionic polymer Magnafloc LT25 (anionic polyacrylamide from BASF chemical company) to a medium with a pH adjusted to 10-10.6 effectively concentrated and settled a wide range of microalgae species at rates that are 200-800 times higher than the control. Harith et al. [24] maintained a high microalgae cell viability (75%) for the *Chaetoceros* species using Magnafloc LT25 flocculent at a dosage of 1 mg/L. They stated that increasing the Magnafloc LT 25 and Magnafloc LT 27 dosage did not increase the flocculation efficiency but increased the settling rates.

t'Lam et al. [138] reported a 98% flocculation efficiency for the species *P. tricorntum* using Zetag flocculent at 10 ppm, but only achieved a 52% recovery for the *N. oleoabundans* species. Udom et al. [157] found that using Zetag at a dosage of 34 mg/l flocculated microalgae and resulted in a 98% recovery efficiency. Buelna et al. [158] noted a 95-100% removal efficiency for *Chlorella* culture using 5 mg/L Zetag 63 at a pH of 6-9.

Pushparaj et al. [139] flocculated *Teraselmis* and *Spirulina* with a 70% biomass recovery efficiency using praestol (a cationic organic polyacrylamide based flocculent) with no inhibitory effects on recycled and reused media. However, inhabitation of flocculation has been noted for organic cationic polymers in environments with salinity above 5 g/L [17,25]. Sukenik et al. [146] found that the amount of flocculent required to remove 90% of microalgae from liquid suspension increased linearly with increased salinity. Danquah et al. [15] noted that the amount of energy required to achieve 15% (w/v) microalgae concentration using polymer flocculation was 14.81 kWh/m^3.

Chen et al. [140] reported that the general dosage range of chitosan required to effectively flocculate microalgae species was 5-200 mg/L. Xu et al. [159] noted a 99% clarification efficiency for *Chlorella sorokiniana* using chitosan at an optimal dosage of 10 mg/g dried microalgae and pH values below 7. Ahmed et al. [160] achieved a 99% flocculation efficiency in 20 minutes for *Chlorella* sp. with a chitosan dosage of 10 ppm. Chang and Lee [161] reported a flocculation efficiency of 99% for *Chlorella vulgaris* using chitosan at a dosage of 200 mg/L and a pH of 8.7. Sirin et al. [162] reported a flocculation efficiency of 92% in 10 min for the *Phaeodactylum tricornutum* species using chitosan (20 mg/L). Morales et al. [163] noted a 100 % flocculation efficiency for *Chlorella* sp. using chitosan at a concentration of 40 mg/L. Beach et al. [164] compared the chitosan flocculation, centrifugation and filtration methods for microalgae harvesting and noted that chitosan flocculation was the least energy consuming method of the three.

Factors Affecting Chemical Flocculation

Inorganic Flocculation

The factors affecting inorganics flocculation include: concentration of the flocculent, pH and the surface charge of the flocculent.

Flocculent concentration: The concentration at which the flocculent is administered into the system has been noted to affect the efficiency of the microalgae recovery. Rakesh et al. [165] used aluminium sulphate concentrations ranging from 50 to 300 mg/L for the recovery of *Chlorella sp.*, *Chlorococcum sp.* and *Chlorella sorokiniana* and found 50 mg/L to be the most effective dose. Garzon-Sanabria et al. [166] evaluated the recovery efficiency of *N. oculata* using aluminum chloride at concentrations in the range of 50-100 mg/L and found 50 mg/L to be the most effective dose. Ferriols and Aguilar [167] reported on the use of calcium chloride and sodium hydroxide at concentrations of 100-200 mg/L for the recovery of *Tetraselmis terrahele* and achieved the highest recovery efficiency at 200 mg/L. Wyatt et al. [144] noted that in media with low algae concentrations, the concentration of ferric chloride required to flocculate *Chlorella zofingiensis* increases linearly with cell concentration.

pH: Varying the pH of the medium using inorganic flocculants can promote cell aggregation. Knuckey et al. [25] noted effective flocculation (>80%) of *Chaetoceros calcitrans*, *Chlorella muelleri*, *Thalassiosira pseudonana*, *Attheya septentrionalis*, *Nitzschia closterium*, *Skeletonema* sp., *Tetraselmis suecica* and *Rhodomonas salina* by altering the pH with the addition of sodium hydroxide. Garzon-Sanabria et al. [166] used aluminum chloride (50-100 mg/L) to modify the pH (4-7) and achieved the highest recovery efficiency of *N. oculata* using a dosage of 50 mg/L (pH =5.3). Lee et al. [168] noted that changing sodium hydroxide concentration affected the flocculation efficiency of *Botryococcus braunii* as a result of pH change in the medium. Sanyano et al. [169] successfully flocculated *Chlorella* sp. using ferric chloride at a pH of 8.1.

Surface charge: Microalgae surface cells are negatively charged, indicating that a positively charged flocculent would be required to bond the cells to one another [144,169]. Algal coagulation is induced by the attraction of the positively charged flocculent onto the negatively charged cell surface and the attachment of another algal cell onto the positively charged flocculent [161]. The efficiency of the flocculation is depended on the amount of flocculent available to bridge the algae to one another [144,170]. Wyatt et al. [144] noted that the positive nature of ferric chloride induced microalgae flocculation with a recovery efficiencies of 90% at a pH above 4.1 and below 8. Knuckey et al. [25] flocculated microalgae with an efficiency of 80% using Fe^{3+} ions. Garzon-Sanabria et al. [166] recovered *Nannocloris oculata* with a 90% efficiency using aluminium chloride to counteract the surface charge of the microalgae cells at a pH of 5.3. Sanyano et al. [169] successfully flocculated *Chlorella* sp. using ferric chloride. Lee et al. [171] achieved 100% flocculation efficiency in *Chlorella* sp. using synthesized cationic aluminum and magnesium organoclays.

Organic Flocculation

The factors affecting organic flocculation include: pH, charge on polymer, dosage and salinity.

pH: Some microalgae species can flocculate together by adjusting the pH [17]. Uduman et al. [16] stated that the pH and the chemical composition of the microalgal medium impact the amount of flocculent required. They noticed less electrostatic repulsion between colloids at low pH levels resulting in increased amounts of bridging since the polymer chains are longer. They also found the dose of polymeric flocculent required to vary with microalgae concentration, because of the charge in surface area of algae. Knuckey et al. [25] reported that the non-ionic polymer Magnafloc LT25 settled a wide range of microalgae species effectively using a rate of 0.5 mg/L in a pH adjusted media to 10-10.6. Tenney et al. [172] noted the most effective flocculation resulted when using cationic polyelectrolytes at low pH levels. Ras et al. [47] noted that the *Chlorella* species flocculated when the pH was increased to 11-12. Lee et al. [173] stated that extreme pH levels can result in cell death or impairment.

Charge on polymer: The polyelectrolyte charge plays an important role in the flocculation process of microalgae. Anionic polyelectrolytes are not effective flocculent agents on their own due to the negatively charged microalgae cell surface because like charges repel one another and/or the length of the polymer is not sufficient enough to bridge the particles together [142,172]. It is for this reason that cationic polyelectrolytes are found to be much more effective in the flocculation of microalgae. Morrissey et al. [174] noted that the N,N-dimethylaminopropyl acrylamid polymer (positive character) resulted in recovery efficiencies of microalgae greater than 99% at a pH of 7 and increasing the pH to 13 (activating negative functionality) resulted in flocculation efficiencies of less than 12%. Chang et al. [161] noted that the positively charged surface of chitosan resulted in a 99% removal efficiency for *Chlorella vulgaris*.

Flocculent concentration: The amount of cationic flocculent required for effective bridging between the cells depend on the amount of negative charges present in the system, the surface charge density, the cell counts per volume, the total cell surface area and the charge density of the positively charged polyelectrolyte. The negative charge on the surface is induced by the functional groups (carboxyl groups) present on the microalga cells which have been noted to affect the isoelectric point of the cells [142]. Uduman et al. [16] reported that the growth phase and the metabolic conditions of the microalgal cells dictate the concentration and the reactivity of these functional groups. Granados et al. [30] showed that the species *Chlorella* and *Scenedesmus* were effectively flocculated to a concentration of 2 g/L after 15 min using polyelectrolyte dosages of 2-25 mg/g. Tenney et al. [172] stated that the cationic polyelectrolyte polyamine was effective in flocculating the microalgae cells at a dosage of 2.5 mg/l. Sukenik et al. [146] reported that marine microalgae *Isochrysis galbana* and *Chlorella stigmatophora* require 5-10 times more flocculent dosages than those required by freshwater microalgae.

Salinity: The salinity level affects the organic flocculation of microalgae. Bilanovic and Shelef [175] noted that the polyelectrolyte flocculent was inhibited in the marine medium due to its high salinity and observed effective flocculation at salinity levels below 5 g/l. This was attributed to the fact that high ionic strength causes the polymer to shrink in dimension, thus failing to form a bridge to link the microalgal cells. Schlesinger et al. [116] reported that the addition of alkali to *Chlamydomonas* did not result in rapid flocculation in the saline medium.

Advantages and disadvantages

In comparison with other methods, chemical flocculation is considered to be one of the best methods for cell harvesting because it can handle large amounts of microalgae, it can be used with a wide range of species, it is reliable and it is cost-effective [16,176].

The costs of inorganic flocculants are much less than those of organic ones [138,177,178]. However, the higher amounts required using inorganic flocculants can result in higher costs per unit of microalgae than the more expensive organic flocculants [58]. Sukenik et al. [146] reported that the optimal dosage of inorganic flocculent required to flocculate marine microalgae was 5-10 times higher than that required to flocculate freshwater microalgae. Shammas [177] also noted that the higher cost of organic coagulants can be offset by the low dosages required compared to those of inorganic flocculants.

Microalgae harvesting techniques using chemical flocculation are not environmentally friendly because they introduce chemicals into the system which increase the dissolved solids and change color [179]. Inorganic flocculants can be toxic and can also have negative effects on microalgae by modifying their growth media and changing their color which prevents the reuse and recycling of water [11,17,137]. Hee-Mock et al. [180]) and Vandamme et al. [181] stated that chemical flocculants that are toxic and carcinogenic and are not suitable for harvesting microalgae biomass that is being processed for food supplements and food additives. Therefore, selection of flocculent should be based on cost, toxicity and reusability of the media [17].

Auto-flocculation Harvesting Methods

Some microalgae species can flocculate spontaneously, a process known, in a response to certain environmental stresses. This phenomenon is known as autoflocculation.

Types of Environmental Stress

There are several factors that affect the efficiency of autoflocculation, which include: pH, dissolved oxygen content, nitrogen concentration and the amount of calcium and magnesium ions in solution.

pH

When the pH of the medium is increased the cells come together and settle by gravitational force. The addition of more bases into the medium increased the formation of dense flocs which result in less settling times. However, it is important to note that not all species flocculate with increased pH levels [17,182].

Harith et al. [24] noted that at pH values less than 10 only slight separations between the microalgae and liquid media occurred after 4 h and increasing the pH from 8 to 10 using NaOH and KOH increased the flocculation efficiency from 13 to 82% and from 35 to 78% in 4 h, respectively. Wu et al. [182] noted that a pH of 10.5 resulted in 90% flocculation efficiency for the freshwater species *Chlorella vulgaris*, *Scenedesmus* sp. and *Chlorococcum* sp. and a pH of 9.0-9.3 resulted in 90% flocculation efficiency for the marine species *Nannochloropsis* sp. and *Phaeodactylum tricornutum*. Horiuchi et al. [183] noted a 96% flocculation efficiency in the marine species *Dunaliella tertiolecta* when the pH was adjusted to 8.6. Millamena et al. [148] also noted effective flocculation of microalgae when the pH was maintained above 10 in salt water.

Dissolved Oxygen

Uduman et al. [16] noted that flocculation in some microalgae species can occur naturally with changes in dissolved oxygen concentration. Schenk et al. [11] reported that dissolved oxygen stress can result in microalgae flocculation. Liao et al. [184] reported that increased dissolved oxygen in solution triggers autoflocculation by increasing the binding sites available on the cell surface. Greater binding sites result in bulk formation of the cells which increases the weight of

Microorganism	Bio-Flocculated Microalgae	Reference
Bacteria		
Bacillus licheniformis	*Desmodesmus sp.*	Ndikubwimana et al. [197]
P. stutzeri & B.cereus	*Pleurochrysis carterae*	Lee et al. [173]
Paenibacillus sp.	*Chlorella vulgaris*	Oh et al. [136]
Paenibacillus polymixa	*Scenedesmus sp.*	Kim et al. [198]
Bacillus subtilis	*Chlorella vulgaris*	Zheng et al. [114]
Fungi		
Ankistrodesmus falcatus	*Chlorella vulgaris*	Salim et al. [19]
Scenedesmus obliquus	*Chlorella vulgaris*	Salim et al. [19]
Tetraselmis suecica	*Nannochloropsis oleabundans*	Salim et al. [19]
Skeletonema	*Nannochloropsis*	Schenk et al. [11]

Table 3: Bio-flocculation of microalgae by use of fungi and bacteria microorganisms.

the flocs and increases the settling rate). They also noted that increased photosynthetic activity by the microorganisms increases the dissolved oxygen content and the formation of dense flocs. Wilen and Balmer [185] noted that large flocs can be generated when the dissolved oxygen concentration is high in the media. Koopman et al. [186] noted that dissolved oxygen concentrations of 14-16 mg/l promoted flocculation in the system.

Nitrogen

Auto-flocculation in microalgae cells may be triggered naturally as a result of environmental stress caused by nitrogen concentration [11.16]. Sukenik and Shelef [187] noted that certain species of microalgae flocculate with one another as a result of nitrogen stress in the media. Becker [188] noted that microalgae cells can aggregate with one another as a result of nitrate assimilation. Assimilation of nitrate nitrogen increases the pH of the medium and promotes cell flocculation [182,189]. Nurdogen and Oswald [190] also noted that nitrate assimilation resulted in auto-flocculation in microalgae species.

Nguyen et al. [191] noted a nitrate concentration of 840.4 mg/L was sufficient in flocculating *Chlorella vulgaris* in mBB medium.

Addition of Ca^{2+} and Mg^{2+}

Autoflocculation occurs in the culture media spontaneously as a result of coprecipitation of calcium and magnesium salts present in the media which results in change of the pH of the medium [43]. Smith and Davis [192] evaluated Mg^{2+}, Ca^{2+} and CO_3^{2-} ions for their effectiveness in flocculating and settling of microalgae cells and found that Mg^{2+} ion with high pH levels resulted in effective flocculation and rapid sedimentation. They achieved settling rates that were 100-fold higher than those achieved with sedimentation. The reason for this phenomenon is that magnesium hydroxide flocs are positively charged while calcium carbonate flocs are negatively charged [193]. Thus, destabilization of the negatively charged microalgae cells is greater using magnesium as opposed to calcium. The optimal pH for autoflocculation is strain specific [187]. Nguyen et al. [191] reported that the species *Chlorella vulgaris* autoflocculated with and efficiency of 90% by addition of Ca^{2+} and Mg^{2+} at concentrations of 120 mg/l and 1000 mg/l, respectively. Vandamme et al. [194] noted that addition of Mg^{2+} in *Chlorella vulgaris* culture induced autoflocculation.

Advantages and Disadvantages

The advantages of this harvesting technique are the simplicity and low costs. The process can be reverted by pH adjustment using HCl to decrease the pH back to 7.5-8 [25]. However, autoflocculation does not occur in all species making it an unreliable process [39].

Using pH flocculation is beneficial since pH induced flocculated cells were identical to non-flocculated microalgae cells. This means that this autoflocculation technique has low-shear force on the cells when compared to centrifugation [195]. Knuckey et al. [25] found that the chlorophyll a from *T. pseudonana* cultures were intact using pH-induced methods, but centrifuged microalgae cells only has slight chlorophyll a peaks. This is necessary when the microalgae biomass is required for

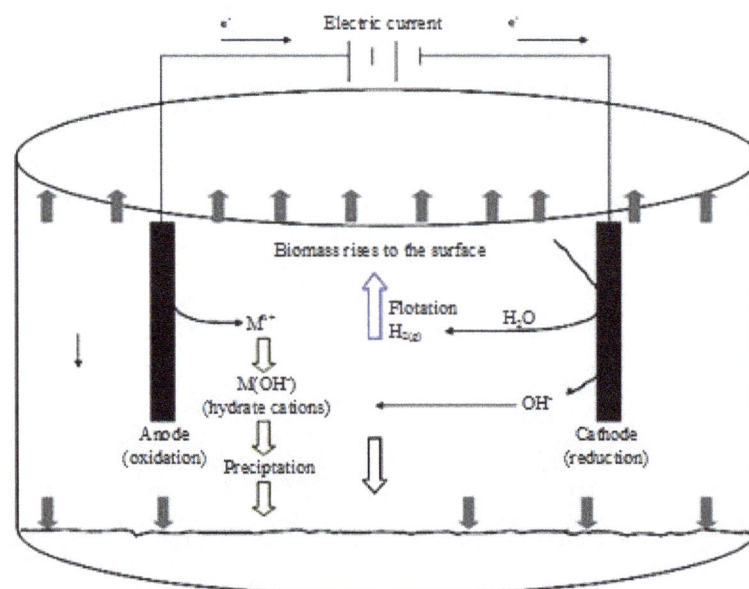

Figure 13: Dewatering microalgae through electrolytic coagulation [205].

use as feed diets or for use of extraction of certain compounds (such as chlorophyll a) from the cell. They also noted that the harvested microalgae species (used for oyster feed) using pH induced flocculation were a better diet choice compared to those harvested by centrifugation and much better than those harvested using ferric chloride flocculation processes. D'Souza et al. [196] also reported that pH induced harvesting of *C. muelleri* for feed to tiger prawn *P. mondon* were only slightly slower in developmental rate compared to those using fresh *C. muelleri*.

Bio-Flocculation Harvesting Methods

The use of microorganisms for the recovery of microalgae biomass has been investigated (Table 3) [11,19,114,136,173,197,198]. This method works by the addition of microorganisms to the culture which adhere to the microalgae cells causing the weight to increase and resulting in settlement of the cells to the bottom of the vessel. The supernatant containing the culture medium is decanted and washed with water in order to reduce the salinity [145,199].

Molina Grima et al. [17] noted the effective flocculation of *Chlorella* using bio-flocculent from bacteria species. Oh et al. [136] successfully harvested *Chlorella vulgaris* using the bio- flocculent bacterium *Paenibacillus* sp. Kim et al. [200] noted effective flocculation of the species *Scenedesmus* sp. using the bio-flocculent *Paenibacillus polymixa*. Ndikubwimana et al. [197] harvested the microalgae species *Desmodesmus sp.* using the bacterium *Bacillus licheniformis* with a 98% removal efficiency. Zhang and Hu [201] co-cultured the species *Chlorella vulgaris* with different filamentous fungi and extracted the microbial oil for transesterification into biodiesel.

Factors Affecting Bio-flocculation

The factors affecting bio-flocculation include: concentration of the bio-flocculent, pH and the selectivity of the microorganism.

Bio-flocculent Concentration

The rate at which bio-flocculation is achieved depends on the ratio of the bio-flocculent to the non-flocculating microalgae species. Bio-flocculent concentrations that are greater than the concentration of non-flocculating microalgae increase the rate of sedimentation [19]. Lee et al. [173] found that the addition of bacteria to the non-flocculating microalgae culture increased the rate of sedimentation. Salim et al. [19] successfully harvested non flocculating microalgae by addition of

bioflocculating species and noted that the addition of bioflocculating microalgae induced the microalgae sedimentation and increased the efficiency of harvesting. Oh et al. [137] reported that the flocculation efficiency of *C. vulgaris* using the bacterium *Paenibasillus* sp. decreased with increasing dilutions of the bacterium. Zheng et al. [114] reported that the flocculation efficiency of *C. vulgaris* using the bio-flocculent *B. subtilis* increased with increasing concentrations of *C. vulgaris* biomass. Lee et al. [168] noted that *C. vulgaris* flocculation efficiency increased with increasing bacteria (*Flavobacterium, Terrimonas* and *Sphingobacterium*) concentration in the culture.

pH

The efficiency of bio-flocculation was noted to be affected by the pH of the medium. The pH alters the surface charge of the molecules in the medium which dictate the degree of attraction/repulsion. Oh et al. [136] reported that the flocculation efficiency of *C. vulgaris* using the bacterium *Paenibasillus* sp. increased with increasing pH from 5 to 11. Ndikubwimana et al. [197] noted that the flocculation efficiency of *Desmodesmus* sp. increased from 43 to 98% using the bacterium *Bacillus licheniformis* as the pH decreased from 7.2 to 3. Lee et al. [168] noted that pure *C. vulgaris* cultures showed no flocculation as the pH was increased from 3-11, but cultures with bacteria demonstrated increased flocculation efficiencies (from 43 to 94%) with increases in the pH over the range of 3-11. However, Zheng et al. [114] noted that the bio-flocculation efficiency using *B. subtilis* with microalgae species *Chlorella vulgaris* and *Chlorella protothecoides* were noted effected by pH.

Species Selectivity

Bio-flocculants are species specific which indicates that not all bio-flocculants will flocculate varying types of microalgae species. Oh et al. [136] reported that the bio-flocculent bacterium *Paenibasillus* sp. resulted in flocculation efficiencies in the range of 91-95% for *Botryococcus braunii, Scenedesmus quadricauda, C. vulgaris* and *Selenastrum capricornutum*, but efficiency in the range of 38 to 49% was noted for *Anabaena flos-aquae* and *Microcystis aeruginosa*. Grossart et al. [202] reported that bacteria were successful in aggregate formation of *Thalassiosira weissflogii* but has little effect on flocculation of *Navicula* sp. Oh et al. [136] and Kim et al. [198] noted that the flocculating with

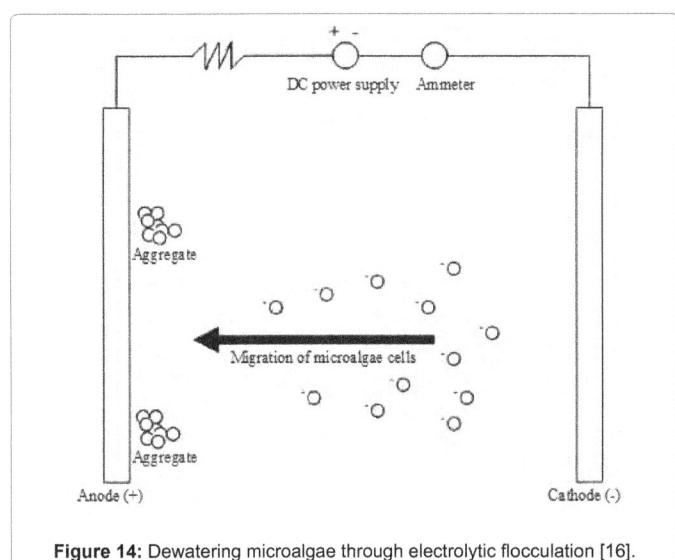

Figure 14: Dewatering microalgae through electrolytic flocculation [16].

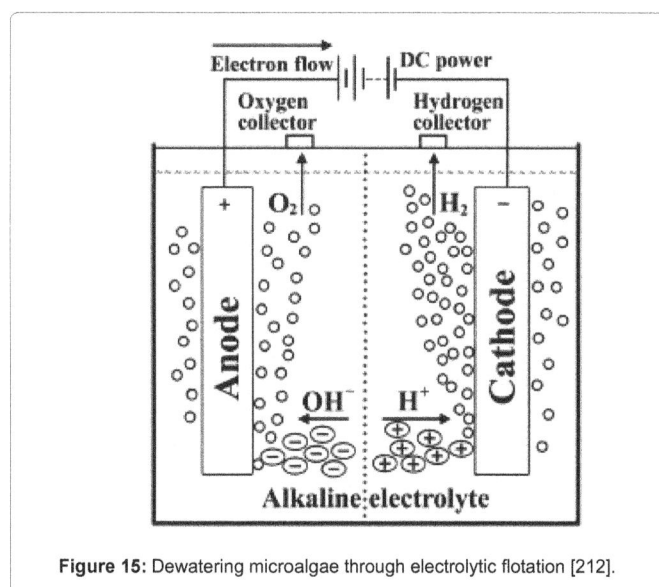

Figure 15: Dewatering microalgae through electrolytic flotation [212].

the bacterium *Paenibasillus* resulted in a flocculating efficiency of 83% and 95% with *Chlorella vulgaris* and *Scenedesmus* sp., respectively.

Advantages and Disadvantages

The advantages of using bio-flocculants include their biodegradability, non-toxic nature and the intermediates formed during degradation are not secondary pollutants [203]. Salim et al. [19] noted that a two step harvesting process using naturally flocculating microalgae to induce non-flocculating microalgae followed by centrifugation reduced the energy use from 13.8 MJ/kg (dry weight) to less than 2 MJ/kg (dry weight).

The disadvantages of this technique are that the microorganisms used to flocculate the algae are species-specific, and the recycling and recovery of these organisms can be difficult [17,204]. Oh et al. [136] stated that the bio-flocculants used to dewater the microalgae cells should be tested for acute oral toxicity in order for the retrieved biomass to be used in food additives and feed supplement. Vandamme et al. [181] indicated that the use of fungi or bacteria as flocculating agents results in microbiological contamination of the microalgae biomass, which needs to be assessed before use in feed or food products.

Electrophoresis Harvesting Methods

The electrolytic methods is used to eliminate the use of costly and toxic chemicals since microalgae behave much the same as colloid particles which allows for their separation from a water based medium by electric flied [16].

Types of Electrophoresis

The main electrophoresis methods that can be used for harvesting microalgae are: electrolytic coagulation, electrolytic flocculation and electrolytic flotation.

Electrolytic Coagulation

This type of electrophoresis method requires the use of both physical and chemical stimuli for the effective separation of microalgae biomass. The coagulation process is induced by the generation of current from an iron or aluminum electrode as shown in Figure 13 [205]. The amount of electrical current passing through the water medium dictates the amount of metal ions dissolved into the liquid suspension [206]. The metal ions released into the solution are metal hydroxides that contribute to the destabilization of colloid suspension and coagulate the biomass. Flocculation is achieved by the linking of the positively charged metal to the negatively charged microalgae cell and the movement toward the anode as a result of electrophoretic motion [179,206]. Rapid coagulation results from high current densities but the cost associated with the method is high.

Uduman et al. [207] used an aluminium electrode set at 5 V and achieved electrocoagulation efficiencies of 93.3% and 87.3% in 600 s for the species *Tetraselmis* sp. and *Chloroccum* sp., respectively. Azarian et al. [208] recovered 99.5% of total suspended solids in 15 minutes by electrocoagulation using a power source of 550 W with aluminium anode. They also noted that a power supply of 100 W required 30 minutes to achieve similar results. Ghernaout et al. [209] achieved a 99% removal of *Escherichia coli* in 20 min by electrocoagulation using aluminium electrode with a power supply of 12 W.

Electrolytic Flocculation

Electrolytic flocculation works by movement of negatively charged cells toward the anode as shown in Figure 14 [16]. At the anode, the cells lose their charge forming flocs that can be lifted to the surface by adhering to the bubbles formed by the electrolysis of the water [210].

Poelman et al. [210] tested the effectiveness of electrolytic flocculation in a 100 L vessel equipped with vertical electrodes and noted a removal efficiency of 80-95% in 35 min. They also noted that the rate of microalgae removal decreased with decreasing voltages and less energy was consumed when the total surface area of the electrodes was decreased and/or the distance between the electrodes was decreased.

Xu et al. [103] achieved a 93.6% recovery efficacy of *Botryococcus branii* after 30 minutes using electrolytic flocculation technique with a power supply of 6 W. Lee et al [168] noted a marine microalgae recoveries of 85% and 95% after 60 minutes using electrolytic flocculation with a power supply of 5 V and 5.2 V, respectively. Zenouzi et al. [211] reported a 97.4% removal efficiency of *Dunaliella salina* after 3 min using electrolytic flocculation.

Electrolytic Flotation

This technique is similar to electrolytic coagulation in that active metal anodes are used to flocculate the microalgae cells. The difference between the two techniques is that the cathode in electrolytic flotation is made from an inactive metal (steel) that is electrochemically nondepositing as shown in Figure 15 [212]. The inactive metal forms hydrogen bubbles from the electrolysis of the water. The particulates in the suspension attach to the gaseous bubbles and are lifted to the surface of the vessel [208,213].

Alfafara et al. [213] investigated the use of electrolytic floatation of microalgae cells using a polyvalent aluminium anode and titanium alloy cathode and found that increasing the electrical power decreased the electrolysis time and increased the rate of chlorophyll a removal. They also noted that the amount of chlorophyll measured was related to the concentration of microalgae removed by electrolysis. The usage of

Criteria	Importance	Description
Dewatering Efficiency	15	The system should be able to effectively concentrate and remove high percentage of the cells from their surrounding liquid media
Cost	15	The operational costs of the process should be low in order to reduce the total processing costs associated with microalgae recovery
Toxicity and health and environmental impact	15	The method should be non-toxic so that the retrieved algae biomass maybe processed for a number of value added products including ones for human consumption It should also be environmentally friendly in order to reduce the amount of toxic wastes produced
Suitability for Large Scale Use	15	The method should effective in handling large volumes for industrial production
Time	15	The rate of harvest should be quick to ensure the sustainability purposes
Species Specificity	10	The method should not be species or strain specific
Reusability of Media	8	The media should be recycled for reuse in order to minimize costs
Maintenance	7	Costs for maintaining the method should be low

Table 4: Criteria used for the comparative analysis of different harvesting techniques.

high electrical power is limited by the increase in heat and the increase in pH.

De Carvalho Neto et al. [214] reported a chlorophyll a removal efficiency of 99% using an electroflotation method running for 140 min with a power supply of 60 W. Ghernaout et al. [215] reported a microalgae removal efficiency of approximately 100% after 140 minutes in Ghrib Dam water using electroflotation method. Shelef et al. [43] noted that electroflotation technique resulted in total suspended solids in the range of 3-5%. Brennan and Owende [48] noted that there is little proof of the economic feasibility of this recovery method.

Energy Consumption

Dewatering microalgae by electrophoresis techniques requires less energy (0.2-2.1 Wh/g) compared to other harvesting methods (0.18-35.62 Wh/g). Vandamme et al. [141] reported that the amount of energy required to flocculate the freshwater microalgae *C. vulgaris* and the marine *P. tricornutum* species was 2.1 Wh/g and 0.2 Wh/g (dry weight), respectively. Gonzalez-Fernandez and Ballesteros [45]

Criteria	Description	Score
Dewatering Efficiency (15)	Settlement is based on density and since microalgae density is similar to that of water media, efficiency is low without the use of flocculants	5
Cost (15)	Minimum energy costs are required for this technique as gravitational forces are cost free	15
Toxicity and health and environmental impact (15)	This method is nontoxic to the cells, since it works by gravitational forces	15
Suitability for Large Scale (15)	Unsuitable for large scale use because of the long periods required for the process and it only works for microalgae cells with higher densities	5
Time (15)	Long periods of time are required to achieve settlement of microalgae cells through gravitational forces	2
Species Specificity (10)	Highly dependent on the type of species used. Species should have a higher density than that of water	4
Reusability of Media (8)	Method does not introduce any chemicals or alter the composition of the species/media	8
Maintenance (7)	No maintenance costs are required	7
Total (100)		61

Table 5: Evaluation of sedimentation.

Criteria	Description	Score
Dewatering Efficiency (15)	Effective recovery of microalgae cells Depends on filter size and the size of microalgae cells	13
Cost (15)	Costs associated with pump and replacement of filters	9
Toxicity and health and environmental impact (15)	Cell composition remains intact and toxic chemicals are not required	15
Suitability for Large Scale (15)	Large pump and large filters are required for large scale	10
Time (15)	Rapid cell recovery	12
Species Specificity (10)	Dependent on the microalgae cell size	6
Reusability of Media (8)	Liquid media can be recycled	8
Maintenance (7)	Frequent filter replacement as a result of clogging	2
Total (100)		75

Table 6: Evaluation of vacuum filtration.

Criteria	Description	Score
Dewatering Efficiency (15)	Effective in dewatering the microalgae. Suspended solids in the filtrate are low	13
Cost (15)	Costs associated with pump to create pressure and with filter replacements	9
Toxicity and health and environmental impact (15)	This method is non-toxic and cell composition is not altered	15
Suitability for Large Scale (15)	Suitable for large volumes but requires large filters and large pump	10
Time (15)	Relatively rapid cell recovery	12
Species Specificity (10)	Dependent on the size of the species	5
Reusability of Media (8)	Filtrate can be recycled and reused again for microalgae growth	8
Maintenance (7)	Costs associated with filter replacement	2
Total (100)		74

Table 7: Evaluation of pressure filtration.

Criteria	Description	Score
Dewatering Efficiency (15)	Complete removal of microalgae cells from the media	15
Cost (15)	Costs associated with pump and membrane	12
Toxicity and health and environmental impact (15)	This method is non-toxic	15
Suitability for Large Scale (15)	Suitable for large volumes of microalgae Smaller microalgae result in membrane clogging	10
Time (15)	Rapid cell recovery	12
Species Specificity (10)	Wide range of cell sizes can be used	8
Reusability of Media (8)	Filtrate can be recycled and reused again for microalgae growth	8
Maintenance (7)	Costs associated with filter replacement	4
Total (100)		84

Table 8: Evaluation of cross flow filtration.

Criteria	Description	Score
Dewatering Efficiency (15)	Effective separation of solid particles from liquid suspensions	13
Cost (15)	Large amounts of energy are required for operation	8
Toxicity and health and environmental impact (15)	The use of toxic materials is not required Cell composition is not altered	15
Suitability for Large Scale (15)	Suitable for large volumes of microalgae	12
Time (15)	Rapid	15
Species Specificity (10)	No dependence on the type of species	10
Reusability of Media (8)	Supernatant can be easily recovered and recycled	8
Maintenance (7)	Not much maintenance is required	6
Total (100)		87

Table 9: Evaluation of disc stack centrifuge.

Criteria	Description	Score
Dewatering Efficiency (15)	Effective separation of solid particles from liquid suspensions Solid concentrates are much more dense than those recovered using disc type	15
Cost (15)	More energy is required for operation compared to disc type	6
Toxicity and health and environmental impact (15)	The use of toxic materials is not required Cell composition is not altered	15
Suitability for Large Scale (15)	Suitable for large volumes of microalgae	12
Time (15)	Rapid	15
Species Specificity (10)	Suitable for larger species only	5
Reusability of Media (8)	Supernatant can be easily recovered and recycled	8
Maintenance (7)	Not much maintenance is required	6
Total (100)		82

Table 10: Evaluation of decanter centrifuge.

Criteria	Description	Score
Dewatering Efficiency (15)	Effectiveness depends on the likelihood of the cells coming into contact with the air bubbles in order to float to the surface Cell may rupture	12
Cost (15)	High costs are required for high speed agitation in order to produce the bubbles. Additional surfactants increase the costs	10
Toxicity and health and environmental impact (15)	The use of toxic materials is not required	12
Suitability for Large Scale (15)	Large volumes of microalgae can be used	13
Time (15)	The time required is dependent on the rate of agitation	10
Species Specificity (10)	Species should have high tolerance to avoid rupturing	5
Reusability of Media (8)	Media may be recycled for further use	8
Maintenance (7)	Not much maintenance is required	7
Total (100)		77

Table 11: Evaluation of dispersed air flotation.

Criteria	Description	Score
Dewatering Efficiency (15)	Effective with the use of additional flocculants	10
Cost (15)	Operational costs are high, large amounts of energy would be required and the cost of flocculants for effective recovery is also high	9
Toxicity and health and environmental impact (15)	Inorganic flocculants are toxic	8
Suitability for Large Scale (15)	Large volumes of microalgae can be harvested	13
Time (15)	Dependent on the likelihood of the cells interacting with air bubble	10
Species Specificity (10)	Wide range of species can be used but species ability to adhere onto gas bubble is key	8
Reusability of Media (8)	Chemicals are not used and the medium can be recycled after it is saturated with air	5
Maintenance (7)	Not much maintenance is required	7
Total (100)		70

Table 12: Evaluation of dissolved air flotation.

noted that the marine microalgae required less energy for harvesting, because the marine medium allows for a higher conductivity that favors the electrocoagulation process. Kim et al. [200] reported that the electrical energy consumption of polarity exchange using two types of electrodes ranged from 1.19 to 1.23 Wh/g for 99% harvesting recovery of microalgae after 15 min. Bektas et al. [216] noted that dewatering of microalgae using electrocoagulation by 0.8-1.5 Wh/g of microalgae

culture whereas cross flow filtration, pressure filters, vacuum filters, centrifugation and flocculation using polymers have consume 3.47, 0.18, 1.19, 1.67 and 35.62 Wh/g, respectively [15,17].

Advantages and Disadvantages

The advantages of the electrophoresis harvesting technique include: versatility, energy efficiency, safety, selectivity, environmental

Criteria	Description	Score
Dewatering Efficiency (15)	Effective with the use of additional flocculants	9
Cost (15)	Large amounts of energy would be required, but operational costs are much lower than those of dispersed and dissolved air techniques	10
Toxicity and health and environmental impact (15)	Coagulants are required for improving recovery effectiveness	10
Suitability for Large Scale (15)	Large volumes of microalgae can be harvested,	13
Time (15)	Dependent on the likelihood of the cells interacting the with air bubble	10
Species Specificity (10)	Can be used on a wide range of species but species ability to adhere onto gas bubble is key	8
Reusability of Media (8)	Recycled after it is saturated with air	6
Maintenance (7)	Not much maintenance is required	7
Total (100)		73

Table 13: Evaluation of fluidic oscillation.

Criteria	Description	Score
Dewatering Efficiency (15)	Cell concentration in liquid is low and depends on the position of the flocculent on the cell	10
Cost (15)	Large amounts of flocculants are required Does not require high amounts of energy	11
Toxicity and health and environmental impact (15)	Flocculating agents are toxic and not suitable for food additive and pharmaceutical products	5
Suitability for Large Scale (15)	Large volumes of microalgae suspensions can be used	15
Time (15)	Relatively fast	10
Species Specificity (10)	Wide range of species can be used but the process is dependent on the type of species used and how well the flocculent attaches to the cells	5
Reusability of Media (8)	The pH of the media left after harvest of microalgae is low which is not suitable for some microalgae species	2
Maintenance (7)	No maintenance required	7
Total (100)		65

Table 14: Evaluation of inorganic flocculation.

Criteria	Description	Score
Dewatering Efficiency (15)	Effectiveness is dependent on the position of the flocculent on the cell and the cell surface charge	11
Cost (15)	Expenses are associated with cost of flocculent Less amounts are required when compared to inorganic agents	11
Toxicity and health and environmental impact (15)	Organic compounds are non-toxic and can be used in the formation of addible and cosmetic by-products	15
Suitability for Large Scale (15)	Large volumes of microalgae suspensions can be used	15
Time (15)	Relatively fast	10
Species Specificity (10)	Process is dependent on the type of species used and how well the flocculent agent attaches to the cells	6
Reusability of Media (8)	Organic agents are non-toxic and the media can be recycled Changes in pH of media occur with flocculent addition and can affect the microalgae species	5
Maintenance (7)	No maintenance required	7
Total (100)		80

Table 15: Evaluation of organic flocculation.

Criteria	Description	Score
Dewatering Efficiency (15)	Flocculation is induced by pH change and separation depends on the response of the species to the environment	11
Cost (15)	Cost of chemicals purchased is relatively reasonable	12
Toxicity and health and environmental impact (15)	Chemicals used to induce flocculation can be toxic	8
Suitability for Large Scale (15)	Large volumes of microalgae suspension can be used Prolonged periods of time for sufficient settling is not suitable for large scale use	8
Time (15)	Long periods are required for the settling of the cells	8
Species Specificity (10)	Dependent on the response of the cell to the pH altered environment Is not suitable for all species	4
Reusability of Media (8)	The pH of the media is altered to induce the flocculation making the media unsuitable for reuse	2
Maintenance (7)	No maintenance required	7
Total (100)		60

Table 16: Evaluation of auto-flocculation.

compatibility and cost effectiveness [206]. Minimum energy is consumed when using optimum potential difference (0.331 kWh/m³) by controlling the electrode surface area and distance between the electrodes. There are no added costs associated with flocculent products [16]. The costs associated with dewatering microalgae via electrolytic methods were significantly less than other harvesting methods such

Criteria	Description	Score
Dewatering Efficiency (15)	Effectiveness is dependent on the linkage of the bio-flocculent to the cells in order to increase their density	11
Cost (15)	Costs are associated with the purchasing the microorganisms and the maintenance of the culture	10
Toxicity and health and environmental impact (15)	Bio-flocculent species are non-toxic	15
Suitability for Large Scale (15)	Effective on large volumes of microalgae species	15
Time (15)	Based on the ability of the cells to link onto the microorganisms in order be more dense and improve settling time	10
Species Specificity (10)	Wide array of species can be used	8
Reusability of Media (8)	Microorganisms can be harvested and the medium can be recycled	3
Maintenance (7)	Maintenance of microorganism culture is required	4
Total (100)		76

Table 17: Evaluation of bio-flocculation.

Criteria	Description	Score
Dewatering Efficiency (15)	Effective concentration of cells in liquid suspension	13
Cost (15)	High electrical power is required	11
Toxicity and health and environmental impact (15)	The addition of toxic chemicals is not required Cell composition can be altered	12
Suitability for Large Scale (15)	Large volumes require large power inputs which deem this method unsuitable for large scale	7
Time (15)	Rapid	15
Species Specificity (10)	Dependent on the charge of the species Conductivity of the water (marine water requires less energy)	3
Reusability of Media(8)	Microalgae can be harvested and the media can be recycled	5
Maintenance (7)	Effectiveness of electrode is reduced with continued use Frequent replacement of electrode maybe necessary	3
Total (100)		69

Table 18: Evaluation of electrolytic coagulation.

Criteria	Description	Score
Dewatering Efficiency (15)	Effective in cumulating the cells together with one another	13
Cost (15)	Energy is required for rapid accumulation of the cells	11
Toxicity and health and environmental impact (15)	The addition of toxic chemicals is not required Cell composition can be altered	12
Suitability for Large Scale (15)	Unsuitable for large scale production because of energy requirement and the alteration of cell composition	8
Time (15)	Rapid	15
Species Specificity (10)	Dependent on the charge neutralization of the cell Conductivity of the water (marine water requires less energy)	5
Reusability of Media (8)	Media can be reused Cell composition can be altered	4
Maintenance (7)	Effectiveness of electrode is reduced with continued use Frequent replacement of electrode maybe necessary	3
Total (100)		71

Table 19: Evaluation of electrolytic flocculation.

Criteria	Description	Score
Dewatering Efficiency (15)	Effective in accumulating the cells together with one another Effectiveness is based on the likelihood that the cell comes in contact with the bubble in order to float to the surface	13
Cost (15)	High costs are associated with power supple	11
Toxicity and health and environmental impact (15)	The addition of toxic chemicals is not required Cell composition can be altered	12
Suitability for Large Scale (15)	Unsuitable for large scale production because of the energy required and the alteration of cell composition	8
Time (15)	Dependent on the likelihood that the cell adheres to a bubbles in order to float to the surface	10
Species Specificity (10)	Dependent on the charge neutralization of the cell Conductivity of the water (marine water requires less energy)	4
Reusability of Media (8)	Media can be reused	4
Maintenance (7)	Effectiveness of electrode is reduced with continued use Frequent replacement of electrode maybe necessary	3
Total (100)		65

Table 20: Evaluation of electrolytic flotation.

Criteria	Physical									Chemical		AF	BF	Electrophoresis		
	S	VF	PF	CFF	DSC	DC	DAF	DVF	FO	IF	OF			EC	EFC	EFT
Dewatering Efficiency (15)	5	13	13	15	13	15	12	10	9	10	11	11	11	13	13	13
Cost (15)	15	9	9	12	8	6	10	9	10	11	11	12	10	11	11	11
Toxicity and health and environmental impact (15)	15	15	15	15	15	15	12	8	10	5	15	8	15	12	12	12
Suitability for large scale (15)	5	10	10	10	12	12	13	13	13	15	15	8	15	7	8	8
Time (15)	2	12	12	12	15	15	10	10	10	10	10	8	10	15	15	10
Species specificity (10)	4	6	5	8	10	5	5	8	8	5	6	4	8	3	5	4
Reusability of Media (8)	8	8	8	8	8	8	8	5	6	2	5	2	3	5	4	4
Maintenance (7)	7	2	2	4	6	6	7	7	7	7	7	7	4	3	3	3
Total (100)	61	75	74	84	87	82	77	70	73	65	80	60	76	69	71	65

S: Sedimentation
VF: Vacuum filtration
PF: Pressure filtration
CFF: Cross flow filtration
DSC: Disc stack centrifugation
DC: Decanter centrifugation
DAF: Dispersed air flotation
DVF: Dissolved air flotation

FO: Fluidic oscillation
IF: Inorganic flocculation
OF: Organic flocculation
AF: Auto-flocculation
BF: Bio-flocculation
EC: Electrolytic coagulation
EFC: Electrolytic flocculation
EFT: Electrolytic flotation

Table 21: Comparative analysis.

as sedimentation with flocculants, centrifugation and flotation with flocculants [210]. This indicates that although higher electrical energy is consumed using electrolytic methods, the cost to harvest is much lower than other harvesting techniques.

Some of the drawbacks associated with this harvesting technique include: cathode fouling and change in cell composition [210]. The current intensity decreases by 5-10% upon reuse of the cathode due to internal resistance [141] and changes in cell composition can be induced using high current densities [45].

Combination of Methods

Cost effective methods for cell harvesting are vital for the economics of biodiesel production. Harvesting processes account for 20-30 % of the total biomass production costs [17]. Selection of harvesting technique is dependent on the size and density of the microalgae cells, conditions of the culture and concentration of biomass and target product value [13].

Schenk et al. [11] reviewed various harvesting methods and noted that the combination of flocculation with sedimentation or flotation with filtration or flotation with centrifugation to be the most economical alternatives. Brennan and Owende [48] and Uduman et al. [16] reported that energy can be conserved and cost can be reduced by harvesting the microalgae in a two-step process using two techniques. Initially, the microalgae are concentrated to 2-7% total suspended solids by the process of flocculation and the cells are further concentrated into a paste (suspended solid concentration of 15-25%) by a secondary harvesting step such as filtration or electrophoresis. Funk et al. [217] integrated dissolved air flotation with chemical flocculation (ferric sulfate) and noted increased recovery efficiency from 88% to 95% for *Chlorella vulgaris*.

Kim et al. [200] reported that a new and innovative technique for improving the economics of the electrophoresis processes is the combination of electrolytic coagulation and electrolytic flotation into continuous electrolytic microalgae. In this method, the current direction (polarity exchanges) is exchanged for the continuous harvest of microalgae and their cultivation. The current direction creates two phases by using a pair of electrodes. The first phase works to destabilize the negatively charged microalgae cells forming flocs. The formation of flocs is mediated by metal ions that are released from the electrode

dissolving in solution. In the second phase, the metal ion generation is halted and the bubbles formed from both electrodes lift the flocs to the top of the solution causing them to float.

Xu et al. [131] used electroflocculation integrated with dispersed air flotation and noted a harvesting efficiency of 98.9% in 14 min. They noted that the cell aggregate increased with the integrated system as opposed to those observed with electroflocculation. The use of dispersed air flotation increased the rate of aggregate formation. However, the stress from continued air supplementation into the system disturbed the up-floated flocs into algal aggregates. Thus disturbance was avoided by halting the supplementation of air into the system once the aggregate size reached its peak value.

Comparative Analysis

Selection of Criteria

Eight criteria (Table 4) were used for the evaluation of microalgae harvesting techniques: (a) dewatering efficiency, (b) cost, (c) toxicity (d) suitability for large scale use, (e) time, (f) species specificity, (g) reusability of media and (h) maintenance. These criteria were selected based on the information reported in the literature about these microalgae harvesting methods. The comparative analysis was performed using these criteria to determine the most efficient, cost effective and environmentally friendly dewatering technique for a wide array of microalgae species that is suitable for large scale application.

Assigning Score to Each Criterion

Each of the selected criteria was assigned a score from 7 to 15 which was determined by the degree of importance of the criterion (Table 4). Higher values were given to the criteria that were deemed most important for development of an efficient and economic large scale dewatering method for microalgae. Lower values were given to criteria that were deemed necessary for determining a suitable method but were considered less important. These values were then used to determine the effectiveness of each harvesting method as shown in Tables 5-20.

Analysis

The sum of the scores obtained for each method are presented in Table 21. The results indicated that of the 16 methods used for microalgae harvesting evaluated in this study, 4 had scores of 80/100 or above and are, therefore, deemed suitable for harvesting a wide array of microalgae species at the industrial scale. These methods are disc stack centrifuge (87/100), cross flow filtration (84/100), decanter centrifugation (82/100) and organic flocculation (80/100).

Three of the nine physical harvesting methods (disk stack centrifugation, cross flow filtration and decanter centrifugation) were considered suitable for large scale harvesting of microalgae because they are highly effective in removing microalgae biomass from the liquid medium, non-toxic and rapid and the medium can be reused. The organic flocculation is suitable for microalgae harvesting on a large scale because it can be used with a wide range of microalgae, the organic chemicals are not toxic and the medium can be recycled.

The other 12 harvesting methods were deemed unsuitable for harvesting microalgae at the industrial scale because they did not meet the evaluation criteria (suitability for dewatering a wide array of microalgae species, suitability for large volumes, low operation costs and low maintenance). The other 6 physical methods were not as effective in removing the algae biomass, required long time, were not suitable for large scale, required high maintenance and were not effective for a wide array of microalgae. Autoflocculation techniques are unsuitable for large scale use because the chemicals used for pH change are toxic, are species specific, require long time and recycling of the medium requires additional costly treatment. The bio-flocculation technique depends on the desirable end product, since the microorganisms used for flocculation are harvested with the cells, the flocculating cultures must be adjusted for viable growth and the process is costly. Electrophoresis methods were deemed unsuitable for large scale microalgae harvesting because of difficulty in scaling up and the disruption of the cells can affect the quality and yield of the desired end product.

Conclusions

The major obstacle for using microalgae biomass on an industrial-scale for production of value added products is the dewatering step which accounts for 20-30% of the total costs associated with the process. A comparative analyses of 16 harvesting techniques that included 9 physical methods (sedimentation, vacuum filtration, pressure filtration, cross flow filtration, disc stack centrifugation, dispersed air floatation, dissolved air flotation, fluidic oscillation), 2 chemical methods (inorganic flocculation, organic flocculation), auto-flocculation method, bio-flocculation method and 3 electrophoresis methods (electrolytic coagulation, electrolytic flocculation, electrolytic floatation) were undertaken. Selection of the most suitable harvesting methods was based on the effectiveness, cost, toxicity, processing time, species specificity, maintenance and suitability for operating on large scale. Each of the selected criteria was assigned a score ranging from 7 to 15 depending on the degree of importance of the criterion.

The results indicated that of the 16 methods evaluated, 4 scored values of 80/100 or above and were deemed suitable for harvesting microalgae on an industrial scale. Three of which were physical techniques (disc stack centrifuge (87/100), cross flow filtration (84/100), decanter centrifugation (82/100)) and the forth was the organic flocculation (80/100) method. These techniques were deemed suitable for large scale use because of their effective dewatering ability, low operational costs, suitability for numerous species, rapidness, require minimal maintenance and being environmentally friendly. The other methods were deemed unsuitable because they are not suitable for dewatering a wide array of microalgae species, not suited for large volumes, costly and require high maintenance.

Although any of these four techniques is deemed suitable for harvesting of microalgae and can be used alone, a combination of methods can also be used to further enhance the recovery efficiency and improve the economics. The use of flocculation as an initial harvesting step to concentrate the algae suspension allows for effective removal of algae biomass from large liquid media. The costs associated with energy intensive centrifugation or filtration techniques can be reduced by using these methods as secondary techniques since less volume of microalgae suspension will be required to undergo the secondary treatment. It is, therefore, recommended that the use of centrifugation or filtration microalgae harvesting techniques be coupled (as secondary techniques) with organic flocculation (as an initial dewatering step) in order to improve the economics of the overall process.

References

1. Wahlen BD, Willis RM, Seefeldt LC (2011) Biodiesel production by simultaneous extraction and conversion of total lipids from microalgae, cyanobacteria, and wild mixed-cultures. Bioresourc Technol 102: 2727-2730.

2. Slade R, Bauen A (2013) Microalgae cultivation for biofuels: Cost, energy balance, environmental impacts and future prospects. Biomass Bioenergy 53: 29-38.

3. Spolaore P, Joannis-Cassan C, Duran E, Isambert A (2006) Commercial applications of microalgae. J Biosci Bioeng 101: 87-96.

4. Bishop WM, Zubeck HM (2012) Evaluation of microalgae for use as nutraceuticals and nutritional supplements. J Nutrition Food Sci 2: 147-183.

5. Luiten EEM, Akkerman I, Koulman A, Kamermans P, Reith H, et al (2003) Realizing the promises of marine biotechnology. Biomol Engg 20: 429-39.

6. Soontornchaiboon W, Joo SS, Kim SM (2012) Anti-inflammatory effects of violaxanthin isolated from microalga Chlorella ellipsoidea in RAW 264.7 macrophages. Biological Pharmaceutical Bulletin 35: 1137-1144.

7. Chen P, Min M, Chen Y, Wang L, Li Y, et al (2009) Review of the biological and engineering aspects of algae to fuels approach. Int J Agri Biol Engg 2: 1-30.

8. Harun R, Singh M, Gareth MF, Michael KD (2010) Bioprocess engineering of microalgae to produce a variety of consumer products. Renew Sustain Energy Rev 14: 1037-1047.

9. Becker W (2004) Microalgae in human and animal nutrition in Handbook Microalgal Culture: Biotechnology and Applied Phycology. (edn A), Blackwell Publishing Ltd, Oxford, UK.

10. Guil-Guerrero JL, Navarro-Juárez R, López-Martínez JC, Campra-Madrid P, Rebolloso-Fuentes MM (2004) Functional properties of the biomass of three microalgal species. J Food Engg 65: 511-517.

11. Schenk P, Thomas-Hall S, Stephens E, Marx U, Mussgnug J, et al (2008) Second generation biofuels: high-efficiency microalgae for biodiesel production. Bioenerg Res 1: 20-43.

12. Demirbas MF, Balat M, Balat H (2011) Biowastes-to-biofuels. Energy Convers Manag 52: 1815-1828. ISSN: 0196-8904.

13. Demirbas A (2010) Use of algae as biofuel sources. Energy Convers Manag 51: 2738-2749.

14. Pulz O (2001) Photobioreactors: production systems for phototrophic microorganisms. Appl Microbiol Biotechnol 57: 287-293.

15. Danquah MK, Gladman B, Moheimani N, Forde GM (2009) Microalgal growth characteristics and subsequent influence on dewatering efficiency. Chem Engg J 51: 73-78.

16. Uduman N, Qi Y, Danquah MK, Forde GM, A. Hoadley A (2010) Dewatering of microalgal cultures: a major bottleneck to algae-based fuels. J Renew Sustain Energy 2: 012701–012715.

17. Molina Grima E, Belarbi EH, Acien-Fernandez FG, Robles-Medina A, Yusuf C (2003) Recovery of microalgal biomass and metabolites: process options and economics. Biotechnol Adv 20: 491-515.

18. Zitelli GC, Rodolfi L, Biondi N, Tredici MR (2006) Productivity and photosynthetic efficiency of outdoor cultures of Tetraselmis suecica in annular columns. Aquaculture 261: 932-943.

19. Salim S, Bosma R, Vermuë MH, Wijffels RH (2011) Harvesting of microalgae by bio-flocculation. J Appl Phycol 23: 849-855.

20. Milledge JJ, Heaven S (2013) A review of the harvesting of micro-algae for biofuel production. Rev Environ Sci Biotechnol 12: 165-178.

21. Trevi (2014) Dissolved air flotation (DAF). Trevi nv, 9050 Gentbrugge, Belgium. 2014.

22. Generalic E (2014) Sedimentation. Croatian-English Chemistry Dictionary & Glossary.

23. Mohn FH (1980) Experiences and strategies in the recovery of biomass from mass cultures of microalgae. In Algae Biomass. (G. Schelef and C. J. Soeder ed.), Elsevier, Amsterdam: 547-571.

24. Harith ZT, Yusoff FM, Mohammed MS, Shariff M, Din M, et al (2009) Effect of different flocculants on the flocculation performance of microalgae, Chaetoceros calcitrans, cells. African J Biotechnol 8: 5971-5978.

25. Knuckey RM, Brown MR, Robert R, Frampton DMF (2006) Production of microalgal concentrates by flocculation and their assessment as aquaculture feeds. Aquaculture Engg 35: 300-313.

26. Waite AM, Thompson PA, Harrison PJ (1992) Does energy control the sinking rates of marine diatoms. Limnol Oceanography 37: 468-477.

27. Van Lerland ET, Peperzak L (1984) Separation of marine seston and density determination of marine diatoms by density gradient centrifugation. J Plankton Res 6: 29-44.

28. Edzwald JK (1993) Algae, bubbles, coagulants, and dissolved air flotation. Water Sci Technol 27: 67-81.

29. Reynolds SC (1994) The ecology of freshwater phytoplankton. Cambridge University Press, Cambridge, United Kingdom.

30. Granados MR, Acién FG, Gómez C, Fernández-Sevilla JM, Grima EM (2012) Evaluation of flocculants for the recovery of freshwater microalgae. Bioresource Technol 118: 102-110.

31. Murphy CF, Allen DT (2011) Energy-water nexus for mass cultivation of algae. Environ Sci Technol 45: 5861-5868.

32. Cole TM, Buchak EM (1995) CE-QUAL-W2: A two dimensional, laterally averaged, hydrodynamic and water quality model, version 2.0: user manual. Instruction Report EL-95-1, US Army Engineer Waterways Experiment Station, Vicksburg, Mississippi.

33. Peperzak LF, Koeman CR, Gieskes WWC, Joordens JCA (2003) Phytoplankton sinking rates in the Rhine region of freshwater influence. J Plankton Res 25: 365-368.

34. Yang MD, Sykes RM, Merry CJ (2000) Estimation of algal biological parameters using water quality modeling and SPOT satellite data. Ecological Modelling, 125: 1-13.

35. Choi SK, Lee JY, Kwon DY, Cho KJ (2006) Settling characteristics of problem algae in the water treatment process. Water sci Technol: A j Int Association Water Res 53: 113-119.

36. Davis EM, Downs TD, Shi Y, Ajgaonkar AA (1995) Recycle as an alternative to algal TSS and BOD removal from an industrial waste stabilization pond system. Industrial Waste Conference 50: 65-73.

37. Greenwell HC, Laurens LML, Shields RJ, Lovitt RW, Flynn KJ (2010) Placing microalgae on the biofuels priority list: a review of the technological challenges. J Royal Society Interface 7: 703-726.

38. Manheim D, Nelson Y (2013) Settling and bioflocculation of two species of algae used in wastewater treatment and algae biomass production. Environ Prog Sustain Energy 32: 946-954.

39. Schlenk D, Batlet G, King C, Stauber J, Adams M, et al (2007) Effects of light on microalgae concentrations and selenium uptake in bivalves exposed to selenium-amended sediments. Arch Environ Contamination Toxicology 53: 365-370.

40. Park JBK, Craggs RJ, Shilton AN (2011) Recycling algae to improve species control and harvest efficiency from a high rate algal pond. Water Res 45: 6637-6649.

41. Griffiths MJ, Van Hille RP, Harrison STL (2012) Lipid productivity, settling potential and fatty acid profile of 11 microalgal species grown under nitrogen replete and limited conditions. J Appl Phycol 24: 989-1001.

42. Wang ZJ, Hou, Bowden D, Belovich JM (2014) Evaluation of an inclined gravity settler for microalgae harvesting. J Chem Technol Biotechnol 89: 714-120.

43. Shelef G, Sukenik A, Green M (1984) Microalgae harvesting and processing: a literature review. Solar Energy Res Inst Golden, Colorado.

44. Taher S, Al-Zuhair S, Al-Marzouqi, Haik Y, Farid MM (2011) A review of enzymatic transesterification of microalgal oil-based biodiesel using supercritical technology. Enzyme Research 468292: 25. doi: 10.4061/2011/468292

45. Gonzalez-Fernandez C, Ballesteros M (2013) Microalgae Autoflocculation: an alternative to high-energy consuming harvesting methods. J Applied Phycology, 25: 991-999.

46. Mata TM, Martins AA, Caetano NS (2010) Microalgae for biodiesel production and other applications: a review. Renew Sustain Energy Rev 14: 217-232.

47. Ras M, Lardon L, Bruno S, Bernet N, Steyer JP (2011) Experimental study on a coupled process of production and anaerobic digestion of Chlorella vulgaris. Bioresourc Technol 102: 200-206.

48. Brennan L, Owende P (2010) Biofuels from microalgae a review of technologies for production, processing, and extractions of biofuels and co-products. Renew Sustain Energy Rev 14: 557-577.

49. Green FB (2008) Harvesting microalgae: challenges and achievements. Microalgae Biomass Summit, Algal Biomass Organization, Seattle, Washington, USA.

50. Dicalite (2012) Filtration. Dicalite, Bala Cynwyd, Pennsylvania. Accessed on October 21, 2014 form http://www.dicalite-europe.com/filtrat.htm

51. Science UA (2014) Filtration. Deparment of Chemistry and Biochemistry, University of Arizona, Tucson, Arizona. Accessed on October 22, 2014 from http://www.chem.arizona.edu/

52. Sparks T (2012) Solid-liquid filtration: understanding filter presses and belt filters. Filtration + Separation, 49: 20-24.

53. Koline-Sanderson (2014) Rotary drum vacuum filters. Peapack, New Jersey. Accessed on October 23, 2014 from http://www.komline.com/docs/rotary_drum_vacuum_filter.html

54. Stucki S, Vogel F, Ludwig C, Haiduc AG, Brandenberger M (2009) Catalytic gasification of algae in supercritical water for biofuel production and carbon capture. Energy Environ Sci 2: 535-541.

55. Ferrentino JM, Farag IH, Jahnke LS (2006) Microalgae oil extraction and in-situ transesterification. AIChE Annual Meeting, Conference Proceedings, San Francisco, California, USA.

56. Gudin C, Chaumont D (1991) Cell fragility, the key problem of microalgae mass production in closed photobioreactors. Bioresourc Technol 38: 141-51.

57. Umesh BV (1984) Performance of two new filtration devices for harvesting of Spirulina alga. Biotechnol Letters 6: 309-312.

58. Mohn F (1988) Harvesting of micro-algal biomass. In: Micro-algal biotechnology (LJ Borowitzka and MA Borowitzka edn.), Cambridge University Press, Cambridge, UK. ISBN: 0521323495

59. Petrusevski B, van Breemen AN, Alaerts GJ, Bolier G (1995) Tangential flow filtration: a method to concentrate freshwater algae. Water Research 29: 1419-1424.

60. Amaro HM, Guedes AC, Malcata FX (2011) Advances and perspectives in using microalgae to produce biodiesel. Appl Energy 88: 3402-3410.

61. Pittman JK, Dean AP, Osundeko O (2011) The potential of sustainable algal biofuel production using wastewater resources. Bioresourc Technol 102: 17-25.

62. Wang B, Li Y, Wu N, Lan CQ (2008) CO_2 bio-mitigation using microalgae. Appl Microbiol Biotechnol 79: 707-718.

63. Arar EJ, Collins GB (1997) In-vitro determination of chlorophyll a and pheophytin a in marine and freshwater algae by fluorescence. National Exposure Research Laboratory, US Environmental Protection Agency, Cincinnati, Ohio .

64. Rossi N, Jaouen O, Legentilhomme P, Petit I (2004) Harvesting of cyanobacterium Arthrospira platensis using organic filtration membranes. Food and Bioproducts Processing 82: 244-250.

65. Velamakanni BV, Lange FF (1991) Effect of interparticle potentials and sedimentation on particle packing density of bimodal particle distributions during pressure filtration. J American Chem Society 74: 166-172.

66. Rushton A (1996) Solid liquid filtration and separation technology (Rushton A, ASWard, RG Holdrich edn.) Wiley-VCH, Verlag GmbH, Weinheim, Germany: 1-32.

67. Raghvan GSV, Sanga ECM (2003) Encyclopedia of Agricultural, Food, and Biological Engineering. (DR Heldman edn.) CRC Press, Boca Raton, Florida: 329

68. Show KY, Lee DJ, Chang GS (2013) Algal biomass dehydration. Bioresourc Technol 135: 720-729.

69. Nagle N, Lemke P (1990) Production of methyl ester fuel from microalgae. Appl Biochem Biotechnol 25: 355-361.

70. Spellman FR (2008) Handbook of water and wastewater treatment plant operations. (2nd edn.) CRC Press, Boca Raton, Florida.

71. Novasep (2014) Novasep technologies for bio-based chemical manufacturing. Novasep, Pompey, France. Accessed on October 21, 2014 form http://www.novasep.com/biomolecules/BioIndustries/Bioindustries-technologies.asp

72. Rossignol N, Vandanjon L, Jaouen P, Quemeneur F (1999) Membrane technology for the continuous separation microalgae/culture medium: compared performances of cross-flow microfiltration and ultrafiltration. Aquacultural Engineering 20: 191-208.

73. Zhang X, Hu Q, Sommerfeld M, Puruhito E, Chen Y (2010) Harvesting algal biomass for biofuels using ultrafiltration membranes. Bioresource Technol 101: 5297-5304.

74. Rossi N, Petit I, Jaouen P, Legentilhomme P, Derouiniot M (2005) Harvesting of cyanobacterium Arthrospira platensis using inorganic filtration membrane. Separation Sci Technol 40: 3033-3050.

75. Rose PD, Maart BA, Phillips TD, Tucker SL, Cowan AK, et al. (1992) Cross-flow ultrafiltration used in algal high rate oxidation pond treatment of saline organic effluents with the recovery of products of value. Water Science Technol 25: 319-327.

76. Walsh DT, Petrovits EJ, Kraus RA, Wuetrich LM, Withstandley CA, et al. (1986) An evaluation of tangential flow filtration for microalgal cell harvesting. National Science Foundation, Washington.

77. Ahmed AL, Yasin NHM, Derek CJC, Lim JK (2012) Cross flow microfiltration of microalgae biomass for biofuel production. Desalination 302: 65-70.

78. Crittenden JC, Trussell RR, Hand DW, Howe KJ (2012) MWH water treatment: principles and design. Hoboken, New Jersey, USA.

79. Borowitzka MA, Moheimani NR (2013) Algae for biofuels and energy. Springer Dordrecht Heidelberg, New York, London.

80. Enviropro (2014) Disc stack centrifuges with self-cleaning bowl. Enviropro. Stirling, Scotland.

81. Mannweiler K, Hoare M (1992) The scale-down of an industrial disk stack centrifuge. Bioprocess Engg 8: 19-25.

82. Heasman M, Diemar J, O'Connor W, Sushames T, Foulkes L, et al. (2000) Development of extended shelf-life microalgae concentrate diets harvested by centrifugation for bivalve molluscs—a summary. Aquaculture Res 31: 637-659.

83. Sim T, Goh A, Becker E (1988) Comparison of centrifugation, dissolved air flotation and drum filtration techniques for harvesting sewage-grown algae. Biomass 16: 51-62.

84. Vasudevan V, Stratton RW, Pearlson MN, Jersey GR, Beyene AG, et al. (2012) Environmental performance of algal biofuel technology options. Environ Sci Technol 46: 2451-2459.

85. Mackay D (1996) Downstream processing of natural products: A practical handbook (MS Verral edn.) John Wiley and Sons: 11-40.

86. Chojnacka K, Chojnacki A, Górecka H (2005) Biosorption of Cr³⁺, Cd²⁺ and Cu²⁺ ions by blue-green algae Spirulina sp.: kinetics, equilibrium and the mechanism of the process. Chemosphere 59: 75-84.

87. Smith J, Charter E (2009) Functional food product development. Wiley-Blackwell, Etobicoke, Ontario.

88. Rees F, Leenheer J, Ranville J (1991) Use of a single-bowl continuous-flow centrifuge for dewatering suspended sediments: effect on sediment physical and chemical characteristics. Hydrological Processes 5: 201-214.

89. Dassey AJ, Theegala CS (2013) Harvesting economics and strategies using centrifugation for cost effective separation of microalgae cells for biodiesel applications. Bioresource Technol 128: 241-245.

90. Alfa Laval (2010) High-capacity Disc Stack Centrifuge for Fats and Oil Refining. Accessed August 23, 2010 from http://www.alfalaval.com/solution-finder/products/px-series/pages/documentation.aspx

91. Goh A (1984) Production of microalgae using pig waste as a substrate. Algal Biomass Workshop, University of Colorado, Boulder, USA.

92. Sharma KK, Garg S, Li Y, Malekizadeh A, Schenk PM (2013) Critical analysis of current microalgae dewatering techniques. Biofuels, 4: 397-407.

93. Tarleton S, Wakeman R (2006) Solid/liquid separation: equipment selection and process design. Elsevier Science, Sydney, Australia.

94. Metcalf & Eddy (2003) Wastewater Engineering: Treatment and Reuse, (Tchobanoglous G, Burton FL, Stensel HD, edn.) McGraw-Hill Education, New York, USA.

95. Packer M (2009) Algal capture of carbon dioxide; biomass generation as a tool for greenhouse gas mitigation with reference to New Zealand energy strategy and policy. Energy Policy 37: 3428-3437.

96. Singh A, Nigam PS, Murphy JD (2011) Mechanism and challenges in commercialisation of algal biofuels. Bioresource Technol 102: 26-34.

97. Schofield T (2001) Dissolved air flotation in drinking water production. Water Sci Technol 43: 9-18.

98. Matis KA, Gallios GP, Kydros KA (1993) Separation of fines by flotation techniques. Separations Technol 3: 76-90.

99. Kaartinen J, Koivo H (2014) Machine vision based measurement and control of zinc flotation circuit. Helsinki University of Technology, Helsiniki, Finland.

100. Rubio J, Souza ML, Smith RW (2002) Overview of flotation as a wastewater treatment techniques. Minerals Engineering, 15: 139-155.

101. Chen YM, Liu JC, Ju YH (1998) Flotation removal of algae from water. Colloids and Surfaces, B: Biointerfaces 12: 49-55.

102. Yan YD, Jameson GJ (2004) Application of the Jameson Cell technology for algae and phosphorus removal from maturation ponds. Int J Mineral Process 73: 23-28.

103. Xu L, Wang F, Li HZ, Hu ZM, Guo C, et al. (2010) Development of an efficient electroflocculation technology integrated with dispersed-air flotation for harvesting microalgae. J Chem Technol Biotechnol 85: 1504-1507.

104. Kurniawati HA, Ismadji S, Liu JC (2014) Microalgae harvesting using natural saponin and chitosan. Bioresourc Technol 166: 429-434.

105. Edzwald JK (2010) Dissolved air flotation and me. Water Res 44: 2077-2106.

106. Wiley PE., Brenneman KJ, Jacobson AE (2009) Improved algal harvesting using suspended air flotation. Water Environ Res 81: 702-708.

107. Zimmerman WB, Hewakandamby BN, Tesar V, Bandulasena HCH, Omotowa OA (2009) On the design and simulation of an airlift loop bioreactor with microbubble generation by fluidic oscillation. Food Bioproducts Processing, 87: 215-227.

108. Hanotu J, Bandulasena HCH, Zimmerman WB (2012) Microflotation performance for algal separation. Biotechnol Bioengg 109: 1663-1673.

109. Dai Z, Fornasiero D, Ralston J (2000) Particle-bubble collision models--a review. Adv Colloid Interf Sci 85: 231-256.

110. Elder AR (2011) Optimization of dissolved air flotation for algal harvesting at the Logan, Utah Wastewater treatment plant. Utah State University, Logan, Utah. Accessed on October 16, 2014 from http://digitalcommons.usu.edu/cgi/viewcontent.cgi?article=2050&context=etd

111. Yap RKL, Whittaker M, Diao M, Stuetz RM, Jefferson B, et al. (2014) Hydrophobically-associating cationic polymers as micro-bubble surface modifiers in dissolved air flotation for cyanobacteria cell separation. Water Res 61: 253-262.

112. Lin MC, Liu JC (1996) Adsorbing colloid flotation of as (V)—feasibility of utilizing streaming current detector. Separation Sci Technol 31: 1329-1641.

113. Kown H, Lu M, Lee EY, Lee J (2014) Harvesting microalgae using flocculation combined with dissolved air flotation. Biotechnol Bioprocess Engg 19: 143-149.

114. Zheng H, Gao Z, Yin J, Tang X, Ji X, et al. (2012) Harvesting of microalgae

by flocculation with poly (gamma-glutamic acid). Bioresource Technol 112: 212-220.

115. Besson A, Guiraud P (2013) High-pH-induced flocculation-flotation of the hypersaline microalga *Dunaliella salina*. Bioresourc Technol 147: 464-470.

116. Schlesinger A, Eisenstadt D, Bar-Gil A, Carmely H, Einbinder S, et al. (2012) Inexpensive non-toxic flocculation of microalgae contradicts theories; overcoming a major hurdle to bulk algal production. Biotechnol Adv 30: 1023-1030.

117. Thangavel P, Sridevi G (2012) Environmental Sustainability. Springer, New Delhi, India.

118. Liu JC, Chen YM, Ju YH (1999) Separation of algal cells from water by column flotation. Separation Sci Technol 34: 2259-2272.

119. Coward T, Lee JGM, Caldwell GS (2014) Harvesting microalgae by CTAB-aided foam flotation increases lipid recovery and improves fatty acid methyl ester characteristics. Biomass Bioenergy 67: 354-362.

120. Dassey A Theegala C (2012) Optimizing the air dissolution parameters in an unpacked dissolved air flotation system. Water 4: 1-11.

121. Zhang S, Amendola P, Hewson JC, Sommerfeld M, Hu Q (2012) Influence of growth phase on harvesting of *Chlorella zofingiensis* by dissolved air flotation. Bioresourc Technol 116: 447-484.

122. Edzwald JK, Wingler BJ (1990) Chemical and physical aspects of dissolved air-flotation for the removal of algae. Aqua, 39: 24-35.

123. 123. Vlaski A, van Breemen AN, Alaerts GJ (1996) Optimisation of coagulation conditions for the removal of cyanobacteria by dissolved air flotation or sedimentation. Aqua, 45: 253-261.

124. Kempeneers S, Menxel FV, Gille L (2001) A decade of large scale experience in dissolved air flotation. Water Sci Technol 43: 27-34.

125. Teixeira MR, Rosa MJ (2006) Comparing dissolved air flotation and conventional sedimentation to remove cyanobacterial cells of Microcystis aeruginosa: Part I: The key operating conditions. Separation Purif Technol 52: 84-94.

126. Gregory R, Edzwald JK (2010) Water Quality and Treatment (6th edn.) McGraw Hill, New York.

127. Metcalf and Eddy (1991) Wastewater Engineering: Treatment, Disposal, and Reuse. (3rd edn.) Mcgraw-Hill College Division New York.

128. Corbitt RA (1999) Standard handbook of environmental engineering. (2nd edn.) McGraw-Hill, New York, New York.

129. Edzwald JK (2007) Development of high rate dissolved air flotation for drinking water treatment. J Water Supply 56: 399-409.

130. Haarhoff J, Rykaart E (1995) Rational design of packed saturators. Water Sci Technol 31: 179-190.

131. Féris LA, Rubio J (1999) Dissolved air flotation (DAF) performance at low saturation pressures. Filtra Separat 36: 61-65.

132. Haarhoff J, Steinbach S (1996) A model for the prediction of the air composition in pressure saturators. Water. Res 30: 3074-3082.

133. Green FB, Lundquist TJ, Oswald WJ (1995) Energetics of advanced integrated wastewater pond systems. Water Sci Technol 31: 9-20.

134. Araujo GS, Matos LJBL, Fernandes JO, Cartaxo SJM, Goncalves LRB, et al. (2013) Extraction of lipids from microalgae by ultrasound application: prospection of the optimal extraction method. Ultrasonics Sonochemistry 20: 95-98.

135. Wijffels RH, Barbosa MJ (2010) Microalgae for the production of bulk chemicals and biofuels. Biofuels Bioproducts Biorefining, 4: 287-295.

136. Oh HM, Lee SJ, Park MH, Kim HS, Kim HC, et al. (2001) Harvesting of *Chlorella vulgaris* using a bio-flucclant form *Paenibacillus* sp. AM49. Biotechnol Letters 23: 1229-1234.

137. Papazi A, Makridis P, Divanach P (2010) Harvesting *Chlorella minutissima* using cell coagulants. J Appl Phycol 22: 349-355.

138. t'Lam GP, Vermue MH, Olivieri G, van den Broek LAM, Barbosa MJ, et al. (2014) Cationic polymers for successful flocculation of marine microalgae. Bioresource Technol 169: 804-807.

139. Pushparaj B, Pelosi E, Torzillo G, Materassi R (1993) Microbial biomass recovery using a synthetic cationic polymer. Bioresource Technol 43: 59-62.

140. Chen G, Zhao L, Qi Y, Cui YL (2014) Chitosan and its derivatives applied in harvesting microalgae for biodiesel production: an outlook. J Nanomaterials 2014: 1-9.

141. Vandamme D, Vieira Pontes SC, Goiris K, Foubert I, Jozef L, et al (2011) Evaluation of electro-coagulation-flocculation for harvesting marine and freshwater microalgae. Biotechnol Bioengg, 108: 2320-2329.

142. Bernhardt H, Clasen J (1991) Flocculation of micro-organisms. J Water Supply: Res Technol-Aqua 40: 76-87.

143. Briley DS, Knappe DRU (2002) Optimizing ferric sulphate coagulation of algae with streaming current measurements. American Water Works Association 94: 80-90.

144. Wyatt NB, Gloe LM, Brady PV, Hewson JC, Grillet AM, et al. (2012) Critical conditions for ferric chloride-induced flocculation of freshwater algae. Biotechnol Bioengg 109: 493-501.

145. Xuan DTT (2009) Harvesting marine algae for biodiesel feedstock. Department of Environmental Engineering, National University of Singapore. Accessed on November 16, 2014 from www.nus.edu.sg/nurop/2009/FoE/U067436X.PDF

146. Sukenik A, Bilanovic D, Shelef G (1988) Flocculation of microalgae in brackish and sea waters. Biomass 15:187–199.

147. Bintisaarani NA (2012) Screening and optimising metal salt concentration for marine microalgae harvesting by flocculation. Unpublished degree thesis for bachelor degree, Universiti Malaysia Pahang, Pahang, Malaysia. Accessed on September 21, 2014 from

148. Millamena O, Aujero E, Borlongan I (1990) Techniques on algae harvesting and preservation for use in culture as larval food. Aquacultural Engg 9: 295-304.

149. Aragon AB, Prdilla RB, Ros de Ursinos JAF (1992) Experimental study of the recovery of algae cultured in effluents from the anaerobic biological treatment of urban wastewaters. Resourc Conserv Recycl 6: 293-302.

150. Wikipremed (2014) Electrostatic Force- Coulomb's Law. WikiPremed. Atlanta, Georgia. Accessed on October 15, 2014 from http://www.wikipremed.com/01physicscards.php?card=628

151. Eremin VV (2011) Van der Waals interaction. RUSANANO, Menlo Park, California. Accessed on October 19, 2014 from http://eng.thesaurus.rusnano.com/wiki/article619

152. Puiu T (2014) Direct measurement of van der Waals force made for the first time. ZME Science. Accessed on October 15, 2014 from http://www.zme-science.com/science/physics/direct-measurement-of-van-der-waals-force-made-for-the-first-time/

153. Pietri J (2014) Hydrogen Bonding. UC Davis ChemWiki, Sacramento, Calfornia. Accessed on October 19, 2014 from http://chemwiki.ucdavis.edu/Physical_Chemistry/Intermolecular_Forces/Hydrogen_Bonding

154. Bondi A (1964) Van der Waals volumes and radii. J Phys Chem 68: 441-451.

155. Hunter RJ (1993) Introduction to Modern Colloidal Science. Oxford University Press, New York.

156. Hogg R (1999) The role of polymer adsorption kinetics in flocculation. Colloids and Surfaces A: Physicochemical and Engineering Aspects 146: 253-263.

157. Udom I, Zaribaf BH, Halfhide T, Gillie B, Dalrymple O, et al. (2013) Harvesting microalgae grown on wastewater. Bioresourc Technol 139: 101-106.

158. Buelna G, Bhattarai KK, Delanoue J, Taiganides EP (1990) Evaluation of various flocculants for the recovery of algae biomass grown on pig-waste. Biological Wastes 31: 211-222.

159. Xu Y, Purton S, Baganz F (2013) Chitosan flocculation to aid the harvesting of the microalga *Chlorella sorokiniana*. Bioresourc Technol 129: 296-301.

160. Ahmad AL, Mat Yasin NH, Derek CJC, Lim JK (2011) Optimization of microalgae coagulation process using chitosan. Chem Engg J 173: 879-882.

161. Chang, YR, Lee DJ (2012) Coagulation-membrane filtration of *Chlorella vulgaris* at different growth phases. Drying Technol 30: 1317-1322.

162. Sirin S, Trobazo RC, Ibanez, Salvado J (2012) Harvesting the microalgae *Phaeodactylum tricornutum* with polyaluminum chloride, aluminium sulphate, chitosan and alkalinity-induced flocculation. J Appl Phycol 24: 1067-1080.

163. Morales J, de la Noue J, Picard G (1985) Harvesting marine microalgae species by chitosan flocculation. Aquacultural Engg 4: 257-270.

164. Beach ES, Eckelman MJ, Cui Z, Brentner L, Zimmerman JB (2012) Preferential technological and life cycle environmental performance of chitosan flocculation for harvesting of the green algae Neochloris oleoabundans. Bioresourc Technol 121: 445-449.

165. Rakesh S, Saxena S, Dhar DW, Prasanna R, Saxena AK (2014) Comparative evaluation of inorganix and organix amendments for their flocculation efficiency of selected microalgae. J Appl Phycol 26: 399-406.

166. Garzon-Sanabria AJ, Davis RT, Nikolov ZL (2012) Harvesting Nannochloris oculata by inorganic electrolyte flocculation: effect of initial cell density, ionic strength, coagulant dosage, and media pH. Bioresoure Technol 118: 418-424.

167. Ferriols VNE, Aguilar RO (2012) Efficiency of various flocculants in harvesting the green microalgae Tetraselmis tetrahele (Chlorodendrophyceae: Chlorodendraceae). Aquaculture, Aquarium, Conservation and Legislation, 5: 265-273.

168. Lee YC., Kim B, Farooq W, Chung J, Han JI, et al. (2013) Harvesting of oleaginous Chlorella sp by organoclays. Bioresourc Technol 132: 440-445.

169. Sanyano N, Chetpattananondh P, Chongkhong S (2013) Coagulation-flocculation of marine Chlorella sp for biodiesel production. Bioresourc Technol 147: 471-476.

170. Ching HW, Tanaka TS, Elimelech M (1994) Dynamics of coagulation of kaolin particles with ferric chloride. Water Res 28: 559-569.

171. Lee SJ, Kim SB, Kim JE, Kwon GS, Yoon BD, et al. (1998) Effects of harvesting method and growth stage on the flocculation of the green alga Botrycococus braunii. Letters in Appl Microbiol 27: 4-18.

172. Tenney MW, Echelberger WF, Schuessler RG, Pavoni JL (1969) Algae flocculation with synthetic organic polyelectrolytes. Appl Microbiol 18: 965-971.

173. Lee A, Lewis D, Ashman P (2009) Microbial flocculation, a potentially low-cost harvesting technique for marine microalgae for the production of biodiesel. J Appl Phycol 21: 559-567.

174. Morrissey KI, He C, Wong MH, Zhao X, Chapman RZ, et al. (2014) Charge-tunable polymer as reversible and recyclable flocculants for the dewatering of microalgae. Biotechnol Bioengineering 112: 74-83.

175. Bilanovic D, Shelef G (1988) Flocculation of microalgae with cationic polymers: effects of medium salinity. Biomass 17: 65-76.

176. Benemann J, Koopman B, Weissman J, Eisenberg D, Goebel R (1980) Development of microalgae harvesting and high-rate pond technologies in California, in Algae Biomass: production and use. Elsevier/North-Holland Biomedical Press, Amsterdam, Netherlands.

177. Shammas NK (2005) Coagulation and flocculation. In: Physicochemical treatment processes, Handbook of environmental engineering. The Humana Press, Totowa, New Jersey, 3: 103-139.

178. Zhou FS, Wang X, Zhou L, Liu Y (2014) Preparation and coagulation behavior of a novel multiple flocculent based on cationic polymer-hydroxy aluminium-clay minerals. Adv Materials Sci Engg, Article ID: 581051.

179. Bukhari AA (2008) Investigation of the electrocoagulation treatment process for the removal of total suspended solids and turbidity from municipal wastewater. Bioresourc Technol 99: 914-921.

180. Hee-Mock O, Lee SJ, Park MH, Kim HS, Kim HC, et al. (2001) Harvesting of Chlorella vulgaris using a bioflocculant from Paenibacillus sp. AM49. Biotechnol Letters 23: 1229-1234.

181. Vandamme D, Foubert I, Muylaert K (2013) Flocculation as a low-cost method for harvesting microalgae for bulk biomass production. Trades in Biotechnol 31: 233-239.

182. Wu Z, Zhu Y, Huang W, Zhang C, Li T, et al (2012) Evaluation of flocculation induced by pH increase for harvesting microalgae and reuse of flocculated medium. Bioresourc Technol 110: 496-502.

183. Horiuchi J, Ohba I, Tada K, Kobayashi M, Kanno T, et al.(2003) Effective cell harvesting of the halotolerant microalga Dunaliella tertiolecta with pH Control. J Biosci Bioengg, 95: 412-415.

184. BQ, Lin HJ, Langevin SP, Gao WJ, Leppard GG (2011) Effects of temperature and dissolved oxygen on sludge properties and their role in bioflocculation and settling. Water Res 45: 509-520.

185. Wilén BM, Balmér P (1999) The effect of dissolved oxygen concentration on the structure, size and size distribution of activated sludge flocs. Water Res 33: 391-400.

186. Koopman B, Lincoln EP, Nordstedt RA (1982) Anaerobic-photosynthetic reclamation of swine waste. Proceedings-Water Reuse Symposium,Washington, DC, II: 924-934.

187. Sukenik A, Shelef G (1984) Algal autoflocculation-verification and proposed mechanism. Biotechnol Bioengg 26: 142-147.

188. Becker EW (1994) Microalgae: Biotechnology and Microbiology. Cambridge University Press, Cambridge, United Kingdom.

189. Uusitalo J (1996) Algal carbon uptake and the difference between alkalinity and high pH ("alkalization"), exemplified with a pH drift experiment. Scientia Marina 60: 129-134.

190. Nurdogen Y, Oswald WJ (1995) Enhances nutrient removal in high-rate ponds. Water Sci Technol 31: 33-43.

191. Nguyen TDP, Frappart M, Jaouen P, Pruvost J, Bourseau P (2014) Harvesting Chlorella vulgaris by natural increase in pH: effect of medium composition. Environ Technol 35: 1378-1388.

192. Smith BT, Davis RH (2012) Sedimentation of algae flocculated suing naturally-available, magnesium-based flocculants. Algal Res 1: 32-39.

193. Ayoub GM, Lee SL, Koopman B (1986) Seawater induced algal flocculation. Water Res 20: 1265-1271.

194. Vandamme D, Foubert I, Fraeye I, Meesschaert B, Muylaert K (2012) Flocculation of Chlorella vulgaris induced by high pH: Role of magnesium and calcium and practical implications. Bioresourc Technol 105: 114-119.

195. Knuckey RM (1998) Australian microalgae and microalgal concentrates for use as aquaculture feeds. Ph.D. Thesis. University of Tasmania, Hobart, Tasmania.

196. D'Souza FML, Knuckey RM, Hohmann S, Pendrey RC (2002) Flocculated microalgae concentrates as diets for larvae of the tiger prawn Penaeus monodon Fabricius. Aquaculture Nutrition 8: 113-120.

197. Ndikubwimana T, Zeng, X, Liu Y, Chang JS, Lu Y (2014) Harvesting of microalgae Desmodesmus sp. F51 by bioflocculation with bacterial bioflocculant. Algal Res 6: 186-193.

198. Kim DG, La HJ, Ahn CY, Park YH, Oh HM (2011) Harvest of Scenedesmus sp. with bioflocculant and reuse of culture medium for subsequent high-density cultures. Bioresourc Technol 102: 3163-3168.

199. Hanifa T, Al-Zuhair S, Al-Marzouqi AH, Haik Y, Farid MM (2011) A review of enzymatic transesterification of microalgal oil-based biodiesel using supercritical technology. Enzyme Res 468292: 1-25.

200. Kim J, Tyu BC, Kim K, Kim BK, Han JI, et al. (2012) Continuous microalgae recovery using electrolysis: effect of different electrode pairs and timing of polarity exchange. Bioresourc Technol 123: 164-170.

201. Zhang J, Hu B (2012) A novel method to harvest microalgae via co-culture of filamentous fungi to form cell pellets. Bioresourc Technol 114: 529-535.

202. Grossart HP, Kiorboe T, Tang KW, Allgaier M, Yam EM, et al. (2006) Interactions between marine snow and heterotrophic bacteria: aggregate formation and microbial dynamics. Aquatic Microbial Ecology 42: 19-26.

203. Salehizadeh H, Shojaosadati S (2001) Extracellular biopolymeric flocculants: recent trends and biotechnological importance. Biotechnol Adv 19: 371-385.

204. Shen Y, Yuan W, Pei ZJ, Wu Q, Mao E (2009) Microalgae mass production methods. Transactions of the ASABE, 52: 1275-1287.

205. Cerqueira AA, Marques MRC (2012) Electrolytic treatment of wastewater in the oil industry. New Technologies in the Oil and Gas Industry, ISBN: 978-953-51-0825-2.

206. Mollah MYA, Morkovsky P, Gomes JGA, Kesmez M, Parga J, Cocke DL (2004) Fundamentals, present and future perspectives of electrocoagulation. J Hazardous Materials 114: 199-210.

207. Uduman N, Lee H, Danquah MK, Hoadley AFA (2011) Electrocoagulation of marine microalgae. Engineering a Better World: Sydney Hilton Hotel, NSW, Australia.

208. Azarian GH, Mesdaghinia AR, Vaezi F (2007) Algae removal by electro-

coagulation process, application for treatment of the effluent from an industrial wastewater treatment plant. Iranian J Public Health 36: 57-64.

209. Ghernaout D, Badis A, Kellil A, Ghernaout B (2008) Application of electrocoagulation in *Escherichia coli* culture and two surface waters. Desalination 219: 118-125.

210. Poelman E, de Pauw N, Jeurissen B (1997) Potential of electrolytic flocculation for recovery of micro-algae. Research, Conservation and Recycling, 19: 1-10.

211. Zenouzi A, Ghobadian B, Hejzai MA, Rahnemoon P (2013) Harvesting of microalgae *Dunaliella salina* using electroflocculation. J Agri Sci Technol 15: 879-888.

212. Santos DMF, Sequeira CAC, Figueiredo JL (2013) Hydrogen production by alkaline water electrolysis. Quimica Nova 36: 176-1193.

213. Alfafara CG, Nakano K, Nomura N, Igarashi T, Matsumura M (2002) Operating and scale-up factors for the electrolytic removal of algae from eutrophied lakewater. J Chem Technol Biotechnol 77: 871-876.

214. De Carvalho Neto RG, da Silva do Nascimento JG, Costa MC, Lopes AC, Neto EFA, et al. (2014) Microalgae harvesting and cell disruption: preliminary evaluation of the technology electroflotation by alternating current. Water Sci Technol 70: 315-320.

215. Ghernaout DC, Benblidia, Khemici F (2014) Microalgae removal from Ghrib Dam (Ain Defla, Algeria) water by electroflotation using stainless steel electrodes. Desalination Water Treat

216. Bektas NH. Akbulut H, Inan, Dimoglo A (2004) Removal of phosphate from aqueous solutions by electro-coagulation. J Hazardous Materials 106: 101-105.

217. Funk WH., Sweeney WJ, Proctor DE (1968) Dissolved-air flotation for harvesting unicellular algae. Water Sewage Works. 115: 343-347.

7

Optimizing Cellulase Production from Municipal Solid Waste (MSW) using Solid State Fermentation (SSF)

Jwan J. Abdullah[1,2*], Darren Greetham[1], Nattha Pensupa[1], Gregory A. Tucker[1] and Chenyu Du[1,3]

[1]Department of Bioenergy, University of Nottingham, UK
[2]Department of Environment, Salahaddin University-Erbil, Iraq
[3]Department of Environment, University of Huddersfield, UK

Abstract

This paper explores the possibility of using an industrially processed municipal solid waste (MSW) for cellulase enzyme production via solid state fermentation (SSF) by *Trichoderma reesei* and *Aspergillus niger*. Both fungi grew well on the MSW substrate and production of cellulase enzymes was optimized for temperature, moisture content, inoculation and period of incubation. The effect of additional minerals, and alternative carbon and nitrogen sources were also examined.

Following optimization a cellulase activity of 26.10 ± 3.09 FPU/g could be produced using *T. reesei* at 30°C with a moisture content of 60% with an inoculums of 0.5 million spores/g and incubation for 168 hours. Addition of extra nitrogen and/or carbon did not improve cellulase accumulation. Acid or alkali pretreatment of MSW led to reduced cellulase production. Crude enzymes produced from MSW by *T. reesei* were evaluated for their ability to release glucose from MSW. A cellulose hydrolysis yield of 24.7% was achieved, which was close to that obtained using a commercial enzyme. Results demonstrated that MSW can be used as an inexpensive lignocellulosic material for the production of cellulase enzymes.

Keywords: Cellulase; Solid State Fermentation (SSF); Municipal Solid Waste (MSW); *Trichoderma reesei; Aspergillus niger;* Composition characterization

Introduction

Municipal solid waste (MSW) management is one of the key topics in environmental protection [1]. In England, around 23.7 million tons of household waste was generated in 2009/10, which equates to 1036 Kg per household [2]. The current technologies for MSW treatment are incineration, landfill, composting and anaerobic digestion. Recently, anaerobic conversion of MSW into biogas has attracted growing interest as a promising method to reduce environmental impact and to generate renewable fuel at the same time. However, some MSW contain over 50% lignocellulosic content [3] and anaerobic digestion may not represent the most efficient process. One alternatively approach is to hydrolyze the lignocellulosic component in the MSW into simple sugars and then ferment these sugars into ethanol. It was estimated that around 152 L of ethanol could be generated from a ton of processed MSW [3].

This enzymatic hydrolysis of lignocellulose to sugars requires a cocktail of enzymes that include mainly cellulase. Cellulase can be produced by various bacteria such as *Bacillus subtilis, Bacillus circulans* [4] (*Bacillus sphaericus-JS1* [5], *Cellulomonas flavigena* [6] and fungi like *Aspergillus sp.* [7] and *Trichoderma reesei* [8]. Among these cellulase producers, *Trichoderma reesei* and *Aspergillus niger* have attracted most attention due to their high cellulase productivity, safe use in industry and the availability of their whole genome sequences. Both submerged fermentation and solid state fermentation (SSF) have been used in cellulase production [9]. In comparison with submerged fermentation, SSF is easy to operate, requires minimum equipment and has the unique advantage of being able to handle insoluble solid substrates, such as lignocellulose. Therefore, it has been widely used in the fermentation of agricultural residues and food processing wastes. Various substrates, such as wheat straw, sugar cane bagasse and oil palm biomass have been examined for cellulase production using *T. reesei* and *A. niger*.

Research has revealed that *T. reesei* can produce 13.4 FPU/g cellulase activity using water hyacinth [8]; 154.58 FPU/g using sugar cane bagasse [10]; 8.2 FPU/g using oil palm empty fruit bunches [11]; and 1.16 FPU/g using rice bran [12] as substrates. Similarly it has been reported that *A. Niger* could produce 24 FPU/g cellulase activity using wheat straw as a substrate [13]. Other substrates such as banana peel, rice straw, corn cob residue, rice husk, banana fruit stalk, and coconut coir pith have all being used for cellulase production [14,15].

Although MSW can contain a high lignocellulosic content, there has been limited research into the use of MSW for cellulase production. Other research demonstrated the possibility of using MSW to produce cellulase via SSF. But by using raw not autoclave MSW [16,17]. In this paper, we report an example of using an industrially processed MSW for cellulase production and then the use of the resulting "cellulase cocktail" for the enzymatic hydrolysis of MSW.

Materials and Methods

Municipal solid waste

Municipal solid waste (MSW) was kindly supplied by Wilson Steam Storage Ltd (UK). The MSW had been subjected to commercial steam

*Corresponding author: Jwan J. Abdullah, Department of Bioenergy, University of Nottingham, School of Biosciences, Sutton Bonington Campus, Loughborough, LE12 5RD, UK, E-mail: jwan4@outlook.com

autoclaving at 165°C to generate an organic rich fraction. The MSW as received was sieved using a 2 mm sieve, and the fraction that passed through the sieve was collected and dried in an oven at 70°C overnight.

MSW characterization

Cellulose and hemicellulose: Cellulose and hemicellulose were analyzed using acid hydrolysis according to the methods described by the National Renewable Energy Laboratory (NREL/TP-500-42618). A 30 mg sample of MSW was added to 1 mL of 12 M H_2SO_4 incubated at 37°C for 1 hour then 11 mL of distilled water was added and the samples was placed in a water bath at 98°C for 2 hours [18]. Then the hydrolyzed simple sugars were determined using a HPLC (Dionex). Monosaccharides (arabinose, galactose, glucose and xylose) were analyzed using Dionex ICS-3000 Reagent-FreeTM Ion Chromatography equipped with Dionex ICS-3000 system, electrochemical detection using ED 40 and computer controller. A CarboPacTM PA 20 column (3 x 150 mm) was used and the mobile phase was 10 mM NaOH with a flow rate of 0.5 mL/min. The injection volume was 10 μL and the column temperature was 30°C.

Lignin: Lignin was determined using the acetyl bromide method as described by Sluiter et al. [19]. A 100 mg sample of MSW was added to 4 mL 25% acetyl bromide in glacial acetic acid and incubated in a water bath (50°C) for 2 hours. The tubes were cooled and a further 12 mL of glacial acetic acid added. Tubes were then centrifuged at 3000 rpm (1609 g) for 5 min. 0.5 mL supernatant was transferred to a new falcon tube and 2.5 mL of glacial acetic acid, 1.5 mL of 0.3 M NaOH, 0.5 mL (0.5 M) hydroxylamine hydrochloride added. Finally glacial acetic acid was added to make a final volume of 10 mL, absorbance was measured at 280 nm and the concentration of lignin calculated by comparison with standards containing 0.4, 0.6, 0.8, 1.0 and 1.2 mg lignin (Sigma Aldrich) [19].

Total nitrogen and protein: Total nitrogen and protein were determined using a Nitrogen/Protein Analyzer (Thermo Scientific Flash EA1112) with L-aspartic acid as the standard and the method as described by Campbell et al. [20].

Total lipid analysis: The lipid content was measured using the Folch method as described by Cequier-Sánchez et al. [21]. MSW sample (400 mg) sample of was added to 12 mL of dichloromethane/methanol (2: 1, v/v) and incubated for 2 hours at room temperature with occasional mixing. Following incubation samples were centrifuged at 1000 rpm (178 g) for 5 mins and the upper organic phase was carefully removed using a glass syringe and transferred into a clean 50 mL glass centrifuge tube. 2.5 mL KCl (0.88%, w/v) was added and the sample was centrifuged at 1000 rpm (178 g) for 5 mins and the lower organic phase was carefully transferred into a pre-weighed glass tube. The lower organic phase was dried under nitrogen gas until all liquid had evaporated. The tube was reweighed to give the total lipid content [21].

Trace element analysis: 2 g sample of MSW was added to 15 mL of concentrated HNO_3 and placed on a hot plate until the volume reached 5 mL. Samples were then filtered using filter paper (No. 42) and distilled water was added to make a final volume of 100 mL. The samples and blank (no MSW) were then analyzed using an Inductively Coupled Plasma Mass Spectrometer (ICP-MS). Major elements (Ca, Mg, K, and Na) were analyzed at the ppm (mg/L) level with the detector operating in analogue mode only. While for the minor elements data were expressed as mg/Kg [22].

Acid and alkali pre-treatments: MSW (2 g) was added to 100 mL of either 1% H_2SO_4 or 2% NaOH and autoclaved at 121°C for 30 min or

15 min, respectively. After autoclaving, samples were neutralized to pH 7.0 using either 1 M NaOH or 1 M H_2SO_4 as appropriate. The samples were then centrifuged at 5000 rpm (4472 g) for 10 min. The solid fraction (pellet) was rinsed three times by adding distilled water to the pellet and re-centrifuged. The solid fraction was then dried overnight at room temperature.

Microorganisms: Two filamentous fungi, *Trichoderma reesei* QM6a and *Aspergillus niger* N402 were kindly donated by Professor David Archer (University of Nottingham, UK). Procedures for storing and cultivating *T. reesei and A. niger* were as previously described by Ries et al. [23].

Solid State Fermentation (SSF): A 6 g (dry weight) of MSW was placed into a 250 mL Duran bottle. Sufficient distilled water was then added to the MSW to give a final moisture content of 60, 70 or 80%. The Duran bottles were autoclaved at 121°C for 15 min. After sterilization the substrate was cooled down to room temperature. Spore suspensions of *T. reesei* or *A. niger* (0.5, 1 or 2 × 10^6 spores/g of dry MSW) were then added. The mash was mixed using a sterilized spatula and approximately 2.0 g of inoculated MSW mash was distributed to Petri dishes. The Petri dishes were then incubated at 25, 28 or 30°C using a static incubator for up to 168 hours.

Addition of mineral solutions, carbon source, nitrogen source and clay: With the aim of improving cellulose production, the addition of mineral solutions, carbon source, nitrogen source and clay was carried out. Two stock solutions of trace metals were prepared; mineral solution 1 contains 0.5 g KH_2PO_4, 0.5 g K_2HPO_4, 1.0 g $(NH_4)_2SO_4$, 0.2 g $MgSO_4$ in 1000 mL distilled water and mineral solution 2 contains 26 g KCl, 26 g $MgSO_4$, 76 g KH_2PO_4 in 1000 mL of distilled water. In the experiments of using a mineral solution, the mineral solution was used to adjust moisture content instead of distilled water. In the experiments investigating the impact of various nutrient/chemical reagent addition, 0.2% or 1% (w/w) cellulose powder, peptone, clay or combined cellulose powder and peptone was added into MSW, as the carbon source, nitrogen source, reagent to adsorb toxic chemicals and combined carbon and nitrogen source, respectively. The addition was mixed with MSW before autoclave.

Extraction and assay of cellulase activity: The method used to extract cellulase activity was as described previously by Pensupa et al. [13]. Fermented solid samples from Petri dishes containing 2.0 g of original dry substrate were mixed with 16 mL of 0.05 M sodium citrate buffer (pH 4.8) and blended for 10 seconds using a food processing blender. The suspension was then poured into a beaker and stirred at 300 rpm for 30 min at 4°C. Samples were centrifuged at 5000 rpm (4472 g) for 10 mins. The supernatant, containing crude enzymes, was retained and used for enzymatic activity analysis.

Cellulase activity was measured as filter paper units (FPU) following the protocol of Adney et al. [24]. Crude extract (0.5 mL) was mixed with 0.5 mL of 50 mM citrate buffer (pH 4.8). A strip (1 × 6 cm, approximately 50 mg) of Whatman No. 1 paper (cellulose substrate) was placed into the tube, and the tube was incubated at 50°C for 1hour. After that, 3 mL of dinitrosalicylic acid (DNS) reagent (prepared according to Adney et al. [24] was added to each tube and the mixture was boiled for 5 mins. After cooling on ice for 5 mins, samples were diluted (0.2 mL of sample mixed with 2.5 mL of distilled water) and absorption was read at 540 nm. One unit (FPU) of enzyme activity was defined as the amount of enzyme required to liberate 1 mmol of glucose per min.

Enzymatic hydrolysis: Fungal extract (8.3 mL) was mixed with 91.7 mL sodium citrate buffer solution (50 mM, pH 4.8) to achieve an enzyme loading rate of around 30 FPU per g dry weight of MSW. A commercial cellulase Ctec2 (Novozyme) at the same enzyme loading was used as a control.

The hydrolysis was started by adding 0.5 g (dry weight) of substrate into the fungal extract or control enzyme solutions, and then the samples were shaken at 150 rpm in a shaking incubator at 50°C for 72 hours. Samples were taken at 0, 2, 4, 12, 24, 48 and 72 hours. Samples were centrifuged at 5000 rpm (4472 g) for 10 mins and the supernatant was analyzed for glucose concentration using a Dionex HPLC.

Transcriptome analysis of *T. reesei* during SSF on MSW: *T. reesei* was cultured on MSW incubated at 30°C under optimal conditions and fungal samples collected at three time points 96,120 and 168 hours and frozen until required. The mycelia were frozen using liquid N_2 and ground using a pestle and mortar. The ground mycelium was then gradually added to 1 mL of Trizol (invitrogen) in a 2 mL tube to give a final volume of around 1.5 mL. The samples were, left at room temperature for 10min. Then 200 µl of chloroform was added to each sample, which were then vortexed and left for 2-3 min. Samples were then vortex again, then centrifuged at 13000 rpm (37788.4 xg) for 10 min. 750 µl of the upper aqueous phase was removed and placed into a 1.5 mL tube. 750 µl of isopropanol was added and samples left for 20 min on ice to precipitate the RNA, Samples were centrifuged at 13000 rpm (37788.4 xg) for 10 min this resulted in a gel like pellet being formed on the side and bottom of the tube. The supernatant was removed and pellet washed with 700 µl of 70% ethanol, centrifuged at 13000 rpm (37788.4 xg) for 10 min. The ethanol was removed and the pellet left to dry in a laminar airflow cabinet for 10-15 mins. The pellet was resuspend in 100 µl of DEPC (Diethylpyrocarbonate) treated water. RNA clean up kit (Qiagen) was used to remove DNA following the manufacturers "on column DNase digest" protocol.

RNA samples were sent for quality control and RNAseq to the service centre. The University of Nottingham. Annotation of the reads and the initial global analyses was carried out by Dr. Martyn Blythe. Three biological replicates were prepared for each time point. Unfortunately, one of the three replicates from day 5 failed the internal RNAseq quality control and as such was not used in the subsequent analysis.

A Filtering Pipeline was used to filter reads with low sequencing score and reads aligned to adaptor sequences. Initially raw reads were trimmed for adapter sequences using Sythe. This was followed by quality trimming using Sickle. Reads that passed the filter were then aligned to the reference and reads were removed which mapped to the rRNA and tRNA genes (as annotated in the genome reference). Then reads that passed the rRNA and tRNA filter were mapped onto the reference genome in the context of known gene exon coordinates by tophat mapping tool. The reference genome used was *T. reesei* Version2 as provided by the Joint Genome Institute (JGI) through their WEB site. It was found however, that the reference genome had poor annotation.

Read counts for each gene were calculated using 'htseq-count'. This program determines the number of uniquely aligned reads per gene. MAPQ30 (Unique) was then used to correctly mapped reads and to generate counts per gene. Gene expression was expressed as Reads Per Kilobase of transcript per Million mapped reads (RPKM) values. The RPKM is simply a normalized read count (stranded/sense reads) for a given gene as defined by the University of California Santa Cruz (UCSC) ref Gene database. The read count of the exon-space of a gene is normalised against the total number of mapped reads (Uniquely and correctly mapped reads with rRNA excluded) in that particular

alignment file, and against the total length of the gene's exon-space.

Statistical analysis

All experiments were carried out in triplicate. Microsoft Excel was used to calculate data means and standard deviations. ANOVA was performed using either Design Expert or SPSS.

Results and Discussion

Composition of MSW

The chemical composition of the MSW was assessed as described in the methods (Table 1). The results showed that the MSW contained significant levels of cellulose, hemicellulose and lignin. The total hydrocarbon content was over 60%, indicating that it was a carbon rich waste stream, which could be used as a potential substrate for biofuel production. The cellulose content of the MSW (27.8 g/100 g) was similar to that reported by Barlaz et al. and Jones et al. [25,26], which were 28.8 and 25.6 g/100 g, respectively. Some researchers however, have reported much higher cellulose in MSW derived principally from the paper, wood and milling industries [27,28].

The hemicellulose content (15.45 g/100 g) was again similar to the 11.9 g/100 g reported by Jones et al. [26]. However, these are higher than several other values for example 5.14, 5.8 and 6.6 g/100 g as reported by Ham et al., Price et al. and Barlaz [29-31], respectively. Lignin content of our MSW (17.7 g/100 g) was close to the values reported by Barlaz et al. and Ham et al. [28,29] these being 12.67; 15.7 and 15.2 g/100 g, respectively. However, some authors reported higher values of lignin for example 25.1 g/100 g [30].

The amount of lipid (11.2 g/100 g) of this MSW was similar to that reported for a MSW sample collected from a Mechanical-Biological Treatment (MBT) plant (Barcelona, Spain), which was 11.52 g/100 g [32]. Low lipid content (4.9 g/100 g) had been reported from a MSW sample derived from food wastes emanating from fruit and vegetable markets, households, hotels and juice production centers [33]. Generally, MSW that contains waste food materials have a higher lipid content than those which are derived from wood and paper [34,35]. The protein and total nitrogen concentrations of this MSW sample were similar to those reported by Ponsá et al. and Rao et al. [32,33]. A high proportion of the nitrogen present in the MSW was in the form of protein (5.9 g/100 g), indicating that the MSW could be a good nitrogen resource for microorganism fermentation in addition to carbon resource.

This variation in composition between the MSW used in this work and that reported by others is to be expected given the variable nature of how the MSW is collected and processed. However, it does demonstrate that he MSW used in this paper could be representative of a much wider range of waste material.

The trace element composition of the MSW is shown in Table 2. Of the major elements (Na, Mg, P, S, K, and Ca), Ca had the highest concentration at 25064.74 ± 1550 mg/Kg. Presence of high levels of calcium was probably due to the lime spray treatment during the sterilization process. Sulphur recorded as the second highest major element (5392.4 mg/Kg). Other elements present at more than 1000 mg/Kg were K, Mg, Na, Al, Fe, Cu and P. The only other elements found in significant quantities were Zn and Mn (Table 2).

Trace elements can be classified into three classes based on their biological function and effects: (1) the essential metals (Na, K, Mg, Ca, V, Mn, Fe, Co, Ni, Cu, Zn, Mo and W); (2) toxic metals (Ag, Cd, Sn, Au, Hg, Ti, Pb, Al, Ge, As, Sb, and Se and metalloids) and (3) non-essential,

Structural component	g/100g
Cellulose	27.8 ± 0.1
Hemicellulose	15.45 ± 0.07
Lignin	17.7 ± 0.05
Lipid	11.2 ± 0.1
Protein	5.9 ± 0.2

Table 1: Chemical composition of municipal solid waste. Mean ± SD (n = 3).

Metals	mg/Kg	Metals	mg/Kg
Na	2509.27 ± 145	Ni	93.47 ± 28
Mg	2722.43 ± 254	Cu	1844.95 ± 116
P	1113.66 ± 151	Zn	612.68 ± 43
S	5392.43 ± 299	As	2.98 ± 0.4
K	2190.69 ± 120	Se	0.13 ± 0.02
Ca	25064.74 ± 1550	Rb	2.32 ± 0.14
B	41.56 ± 4.8	Sr	63.41 ± 7
Ti	21.88 ± 1	Mo	3.39 ± 0.08
Al	3120.61 ± 119	Ag	3.15 ± 0.38
V	23.88 ± 0.9	Cd	33.92 ± 2
Cr	15.96 ± 0.7	Cs	0.15 ± 0.012
Mn	397.73 ± 3	Ba	291.22 ± 24
Fe	3851.89 ± 200	Pb	161.19 ± 9
Co	11.27 ± 0.6	U	0.23 ± 0.018

Table 2: Elements composition of autoclaved municipal solid waste. Mean ± SD (n = 3).

non-toxic metals (Rb, Cs, Sr and T) [36].

The MSW sample contained high concentrations of several of the essential elements and some of the toxic metals. The main sources of heavy metals in MSW are usually batteries (Ni, Zn, and Cd); due to the poor availability of recycling facilities for hazardous wastes and poor public attitudes to waste management [37,38].

In addition other materials such as paints, electronics, ceramics, plastics and inks/dyes can all contribute to the heavy metal burden of MSW [37,39]. Generally, paper fractions contain the highest concentration of these metals [40].

The levels of some of the major elements, such as Na and K are higher in this present study than in some previous reports [41]. However, some previously published data suggested that the K content in our study was actually lower [42]. This again reflects the wide variations expected in MSW and was possibly due to a higher concentration of salty food wastes and plastic materials in the MSW, however, the current results generally agree with those of Park et al. [42]. Similarly a comparison of the present study with others studies shows that the Ca content was almost identical to two previously published papers [43,44], but lower than other published papers [41,45].

Comparison of SSF of MSW and wheat straw

In a previous study, wheat straw had been successfully used for cellulase production in SSF using wheat straw as a substrate [13]. Considering the similarity of the lignocellulosic composition of this MSW sample to wheat straw (cellulose 31.7%, hemicellulose 17.9% and lignin 20.2%), in the first instance a comparison was made between MSW and wheat straw as the sole substrates, using the optimum conditions identified for wheat straw [13] and comparing *T. reesei* and *A. niger* (Figure 1). Our previous results demonstrated that 3 days of SSF was best for cellulase production using wheat straw [13] and this agrees with the result here where after 5 days cellulase production, especially with. *A. niger*, was reduced. At 3 days *A. niger* performed

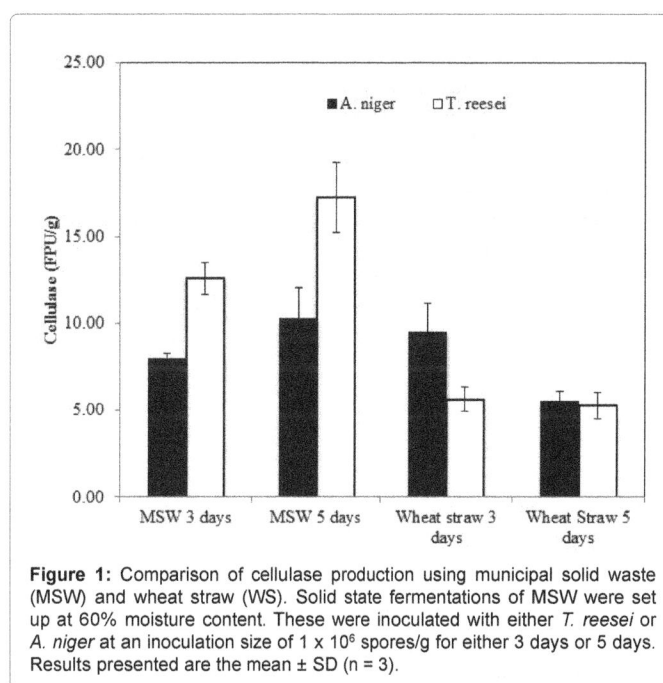

Figure 1: Comparison of cellulase production using municipal solid waste (MSW) and wheat straw (WS). Solid state fermentations of MSW were set up at 60% moisture content. These were inoculated with either *T. reesei* or *A. niger* at an inoculation size of 1×10^6 spores/g for either 3 days or 5 days. Results presented are the mean ± SD (n = 3).

better than *T. reesei* in term of cellulase formation using wheat straw as the substrate. In contrast Figure 1 clearly showed that extending culture time from 3 days to 5 days increased cellulase production in SSF using MSW, possibly indicating that fungi growing on MSW-based substrate required extra time to adapt to the growth environment. The increased fermentation time might be due to the heavy metal content of the MSW. It was also clear that, especially at day 5, *T. reesei* produced higher cellulase activity than *A. niger* when MSW was used as the substrate. A SSF time of 5 days was used as the standard for the following optimization experiments.

Optimization of moisture content and temperature

The combined impact of moisture content and temperature on cellulase production by both *T. reesei* and *A. niger* was investigated. A series of SSF were set up with moisture contents of 60, 70 or 80%, and fermentation temperatures of 25, 28 or 30°C. These were inoculated with either *T. reesei* or *A. niger* spores at a concentration of 1×10^6 spores/g and left to ferment for 120 h (5 days). Cellulase was extracted and assayed at the end of the fermentation and the results are shown in Figure 2. It can be seen that *T. reesei* produced significantly higher amounts of cellulase than *A. niger* under all conditions (p=0.0021). There was a statistically significant effect of moisture content (p=0.003) with 60% giving the highest enzyme activity, at all three temperatures, for both *T. reesei* and *A. niger*. There was also a significant effect of temperature (p=0.0092) on cellulase enzyme production. In this study, the best temperature for *T. reesei* was 30°C, whilst 25°C resulted in maximum production of cellulase by *A. niger*.

Moisture content is generally considered to be a crucial factor that affects oxygen transfer and nutrient accessibility in SSF. High moisture encourages fungal growth, nutrient transportation and enzyme activities, but limits oxygen transfer and facilitates contamination [46,47]. A wide range of moisture contents from 50-89.5% have been used in various studies for cellulase production using *A. niger* [13]. In the case of the MSW used in this study 60% was the minimal moisture content possible due to the water absorbing properties of the substrate.

In addition to the preference of the microorganism, optimum moisture content may also be influenced by properties of the substrate such as porosity and particle size [46,48].

Temperature is another important factor that affects fungal growth and enzyme production. SSF of *T. reesei* and *A. niger* is normally operated within the temperature range of 25 to 30°C [10]. Many studies have been carried out to optimize the incubation temperature for *T. reesei* and *A. niger*. The combined impact of temperature and moisture content has also been reported in similarly studies using spent brewing grains (SBG), rice bran and soybean hulls supplemented with wheat bran [12,49,50].

The highest cellulase activity obtained was 18.98 ± 0.65 FPU/g, at 60% moisture content, 30°C, in SSF using *T. reesei*. As *T. reesei* performed better in SSF of MSW, it was selected for further optimization experiments.

Optimization of inoculation size

Using the moisture and temperature condition determined above the effect of inoculum size on cellulase production by *T. reesei* was examined using spore suspension inoculations of 0.5×10^6, 1×10^6 and 2×10^6 spores/g dry weights MSW. The results are shown in Figure 3. The maximum cellulase activity (19.13 ± 1.5 FPU/g) was obtained in fermentations using 0.5×10^6 spores/g. Increasing the inoculation size from 0.5×10^6 to 1×10^6 and 2×10^6 spores/g, resulted in a significant decrease (p = 0.0045) in recovered cellulase activities to 17.17 ± 2.91 and 15.03 ± 2.81 FPU/g, respectively. Decreased cellulase productivity with an increased inoculation size could be explained by the resultant higher amounts of biomass depleting the nutrient pool and the available oxygen at the early stages of growth, which then affected the cellulase formation [51]. Alternately the decrease in cellulase production even after inoculating the media with higher spore concentrations might be due to the creation of anaerobic conditions or a nutritional imbalance as a result of the more rapid growth of the microorganisms [52,53].

Effect of additions of Minerals and supplements to MSW on cellulase production

Using the optimal conditions determined above the effect of adding additional mineral supplements, additional carbon and nitrogen sources or the inclusion of clay to absorb potentially toxic elements were examined. The addition of mineral solution one did not increase cellulase production, whilst addition of mineral solution two resulted in a significant reduction in enzyme activity (p=0.0044). The results for the addition of the other supplements are shown in Figure 4.

In the case of the other additions, additional carbon (cellulose powder), nitrogen (peptone) or the addition of clay did not result in any significant changes in cellulase production as compared with the control. These results might imply that the MSW has sufficient intrinsic nutrients and that additional supplementation is not required. The only significant differences observed in Figure 4 were between the cellulase production in the presence of 0.2% clay and with some of the fermentations with the addition of carbon and/or nitrogen. The addition of clay may result in the removal of inhibitory heavy metals and thus encourage fungal growth, there was a slight increase in cellulase production following the addition of 0.2% clay but this was not statistically significant. Similarly the addition of alternate carbon or nitrogen sources may act to deflect production of cellulase by the fungus. There was a general decrease in cellulase production observed with the addition of these nutrients but again this was not significant

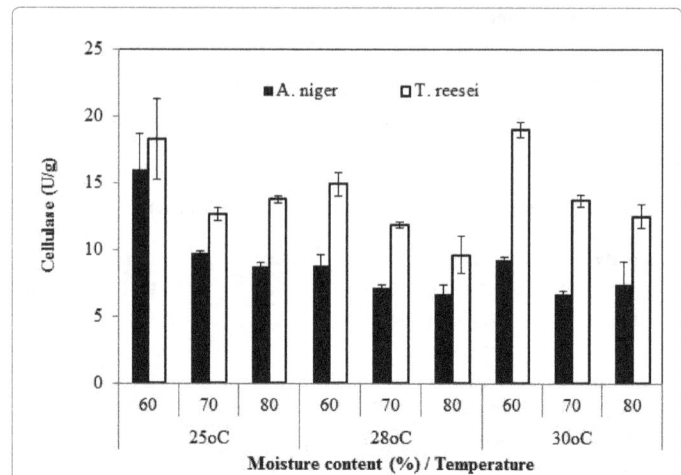

Figure 2: Impact of moisture and temperature on cellulase production. Solid state fermentations of MSW were set up at 60, 70 or 80% moisture content. These were inoculated with either *T. reesei* or *A. niger* at an inoculation size of 1 × 10⁶ spores/g. Plates were then incubated at 25, 28 or 30°C for 120 hours and cellulase production determined. Results presented are the mean ± SD (n = 3).

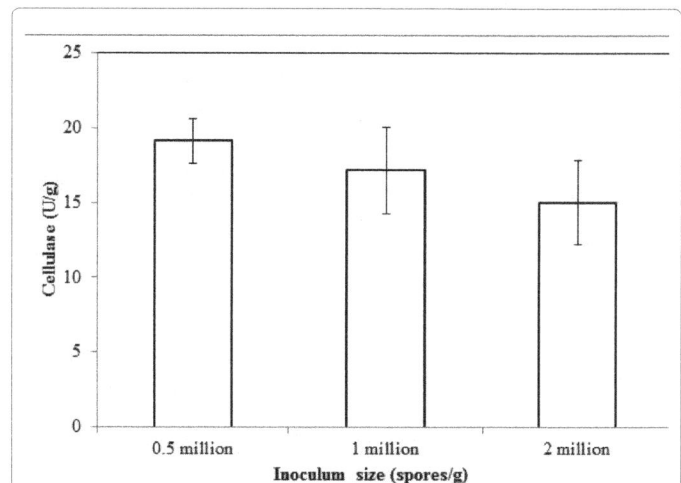

Figure 3: Impact of inoculum dose on cellulose production. SSF of MSW using *T. reesei* at three different inoculation doses were carried out using 60% moisture at 30°C. Crude enzyme was extracted after 120 hours and analyzed for cellulase activity. Results presented are the mean ± SD (n = 3).

when compared to the control.

Effect of pretreating MSW on cellulase production

The impact of pretreatments, designed to increase the accessibility of the cellulose on cellulase production was examined. The MSW was subjected to either an acid or alkali pretreatment prior to SSF. The results showed that in both cases pretreatment actually resulted in reduced cellulase production compared to non-treated MSW. Pretreatment with acid resulted in a significantly lower cellulase production of 47.8% of that obtained in the control fermentations, whilst that with alkali was only slightly higher (65.2%).

Modification of substrate is a common practice to increase the accessibility of the substrate to microorganisms [54,55]. In a previous study, it was found that acid modification of wheat straw significantly enhanced cellulase production in SFF by *A. niger* [13] v. However, a negative impact of pre-treatment on cellulase production has also

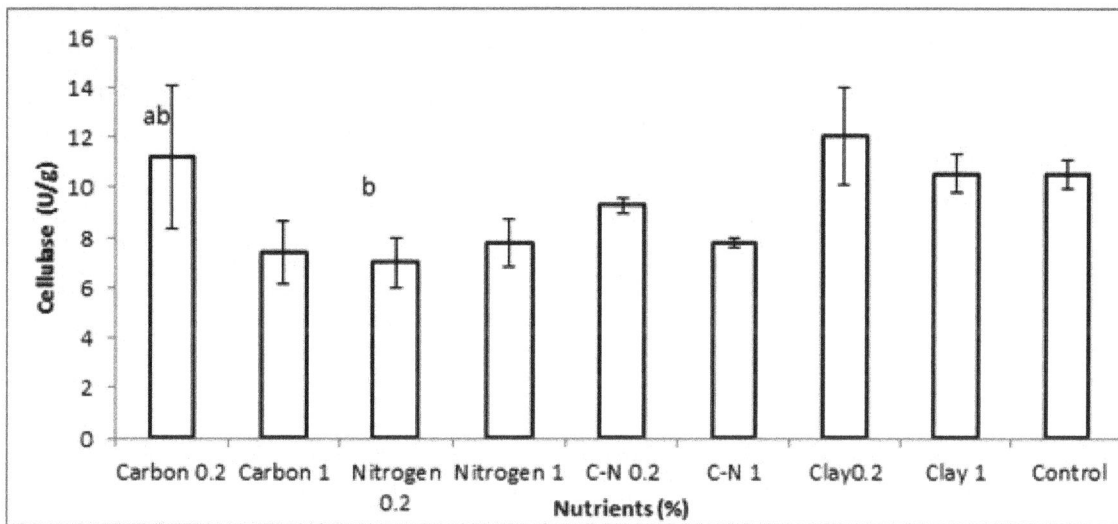

Figure 4: Impact of additional carbon, nitrogen or clay on cellulose production. SSF of MSW using *T. reesei*, at an inoculation size of 1×10^6 spores/g, were carried out using 60% moisture at 30˚C. The effect of adding additional carbon (cellulose powder) or nitrogen (peptone) nutrient sources, either individually or in combination was examined. The addition of clay at either 0.2 or 1 % (w/w) was also tested. All fermentations were compared to a control. Crude enzyme was extracted after 120 hours and analysed for cellulase activity. Results presented are the mean ± SD (n = 3). Bars with the same letters indicate no significant difference (p > 0.05).

been reported by several researchers [56-58]. The decrease in cellulase production could be attributed to (1) structural changes to the lignocellulosic raw materials caused by the alkali pretreatment [56,59]; (2) the generation of inhibitory compounds formed during the acid or alkali pretreatments [56]; (3) the removal of certain nutrients, such as nitrogen by the pretreatment [57,58] and (4) the release of toxic heavy metals that were originally insoluble in the MSW.

Cellulase production under optimal conditions

The optimization of conditions for cellulase production described above utilized a fixed SSF incubation of 120 hours. The effect of extending this culture time from 5 days to 7 days was thus examined. This extended fermentation of cellulase production from 18.53 ± 0.19 to 26.10 ± 3.09 FPU/g. Longer fermentation times were not explored as they were felt to be commercially insignificant. The length of incubation period is a prime concern for the development of a commercial cellulase production process and 7 days may not be viable. Various studies showed that maximum cellulase production from *T. reesei* could be achieved within 72-96 hours, the optimal was at 72 hours using cassava bagasse, wheat bran or rice straw [10]. Also the maximum cellulase produced using apple pomace (2.3 FPU/g) was after 120 hours [60]. The elongated cellulase production period maybe due to the characteristics of this MSW substrate, which contained toxic compounds and substrates from various carbon resources. Compared with homogeneous substrates, fungal cells required a longer time to grow and to express cellulase enzymes.

Hydrolysis of MSW using crude fungal extract

The crude fungal extract obtained from an optimum fermentation was collected. This was then used for the enzymatic hydrolysis of a fresh MSW sample with a cellulase enzyme loading rate of 30 FPU/g dry weights. A commercial cellulase cocktail, Ctec2 (Novozymes), was used as a control at the same enzyme loading rate. Glucose release with time for the hydrolysis is shown in Figure 5. The glucose concentration in the hydrolysis using Ctec2 reached 90% of its final concentration with 24

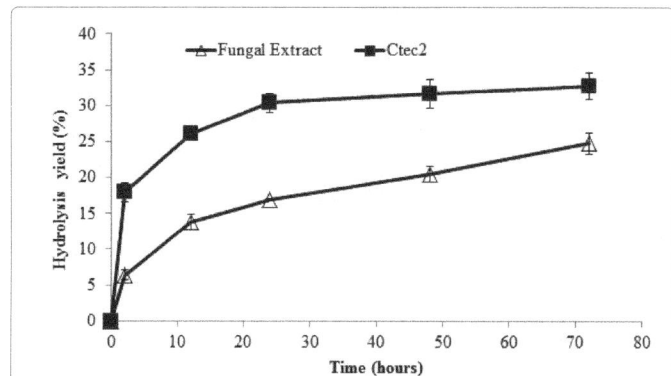

Figure 5: Glucose release during the hydrolysis of MSW using either commercial or MSW derived cellulase cocktails. Untreated MSW was incubated with either a commercial enzyme (Ctec2) or the crude enzyme as obtained from SSF of MSW (MSW). In each case at an enzyme loading rate of 30 FPU/g dry weight. Results presented are the mean ± SD (n = 3).

hours. This represents a 32.8% yield of available glucose in the MSW. In the hydrolysis using the fungal extract from the SSF, the initial glucose release rate was slower compared to that with Ctec2 and the final yield was around 24.7%. Although lower than the commercial enzyme, the hydrolysis achievable with the cellulase extract from *T. reesei* grown on MSW was nonetheless significant.

Transcriptome analysis

Using *T. reesei*, SSF was carried out under optimal conditions and fungal mycelium harvested at 3, 5 and 7 days. RNA was extracted from the mycelium and sent to the service centre at The University of Nottingham for RNAseq analysis. Initial filtering and mapping of the data was carried out by the service centre. Unfortunately one of the three replicates for day 5 failed the quality control and as such has been removed from this analysis. Figure 6 shows the results of the trimming and mapping exercises as carried out by the bioinformatics service. S1-3 represent the triplicate samples from day 3, S4 and 6 the

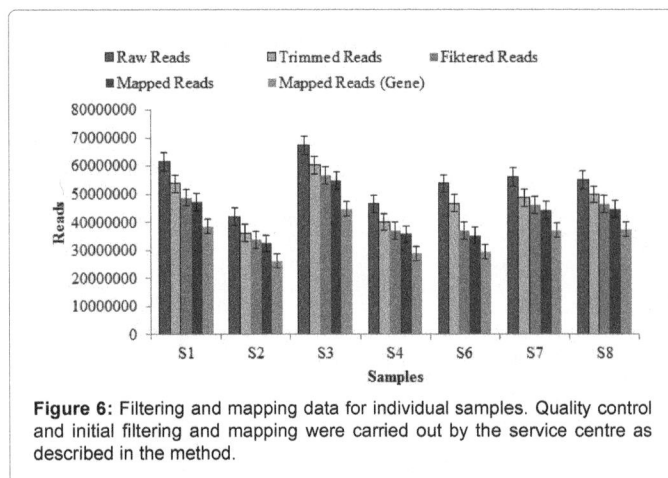

Figure 6: Filtering and mapping data for individual samples. Quality control and initial filtering and mapping were carried out by the service centre as described in the method.

samples from day 5 and S7-9 the triplicate samples from day 7. S5 is missing as this failed the quality control. S9 data was not provided by the bioinformatics centre raw counts ranged from 40 to 68 million and in all cases the filtering and mapping to genes has reduced the number. However, there were still 20 to 38 million reads that could be mapped to genes and a total of 9, 143 individual genes were finally identified and quantified.

PCA analysis of the mapped genes from the individual samples showed three clear clusters of samples 1, 2 and 3; 4 and 6 and 7, 8 and 9. Indicating that there was good replication between the biological replicates and that overall gene expression on the three sampling days was significantly different (data not shown).

The RPKM values for all 8 samples was transferred to an excel spread sheet that has been generated by Dr. Paul Daly (The University of Nottingham) for further analysis.

The spread sheet was designed to specifically analyse those carbohydrate active enzymes that are listed on the CaZy website. Thus the preliminary analysis of gene expression on the three days of sampling was restricted to this class of enzymes and in particular to the glycosyl hydrolases (GH). The analysis was carried out using the nomenclature of Häkkinen et al. [61]. Of the 9,143 individual genes 228 of these mapped onto carbohydrate active enzymes. The number of genes that returned a RPKM of above 0 was 221, 219 and 198 for days 3, 5 and 7, respectively. Of the 228 identified genes 200 were identified as GH. The number of GH genes returning a RPKM value of greater than 0 was 195, 192 and 175 for days 3, 5 and 7, respectively. It is not possible to analyses all of these genes, however a large number were found to be expressed at very low levels with RPKM values below 10. Also for this study we are really only interested in those genes that are expressed very highly since it is assumed that the high level of RNA will correlate with a high protein expression. It is noted that this need not be the case. In that case an arbitrary threshold for the RPKM of 100 was selected and only those genes with a value in excess of this chosen for analysis. It was 197 found that 28, 38 and 10 of the GH genes fell into this category for day 3, 5 and 7, respectively. The highly expressed GH genes fell into 25 classes according to the CAZy database. The number of genes highly expressed in each class at each of the three time points are shown in Table 3 along with their annotated enzyme activities from the CAZy data base.

Expression pattern for selected CAZy genes

The second analysis was to look at the expression pattern of those CAZy genes thought to be most significant. These were those involved in starch metabolism (GH13-amylase and GH15 glucoamylase), hemicellulose metabolism (GH11-xylanase and GH 74 xyloglucanase) and cellulose metabolism (GH6 cellobiohydrolase and GH7 endo glucanase). The data base was searched for all genes in these 6 categories and the expression levels of these at the three sampling points (Figure 7).

In the case of amylase (GH13) the data base returned four independent genes. These along with the other gene involved in starch metabolism (GH15) showed coordinated expression. In all cases there was relatively high expression at day 3 and this remained relatively constant throughout the sampling period. The two genes involved in hemicellulose expression (GH11 and 74) also showed coordinated expression being low but detectable on day 3 and peaking on day 5. For the genes involved in cellulose metabolism the data base returned two independent genes for endo glucanase (GH7) and the expression of both these was coordinated with the other gene (GH6). In this case expression was extremely low on day 3 and peaked on day 5. The overall pattern is consistent with the hypothesis that the fungus is utilising starch, as the most easily degradable substrate, in the early stages of the fermentation but attempts to exploit the hemicelluloses and then cellulose later on. Expression levels at day 7 were all relatively low suggesting that the fungal population is under severe stress at this stage and maybe dying off.

General highly expressed genes

The top 20 most highly expressed genes were identified for each sampling point. These are listed in Table 4. This list is in ascending order of gene ID number, not in order of expression, to allow easier comparisons between the three time points.

Six genes are represented across all three time points. 2 are common between days 3 and 5 and 11 between days 5 and 7. This supports the global analysis of changes in gene expression where there was a larger variation between days 3 and 5 than between days 5 and 7 and again may reflect the acclimatisation into a steady state.

The JGI WEB site was used to obtain putative enzyme identifications for these genes.

Of the 31 genes identified in the search above 18 returned hypothetical proteins of unknown function. The putative identification of the remaining 13 (Table 5).

As might be expected many of the genes encode for house-keeping genes e.g. ribosomal proteins and histones.

The database can also be searched for expression of specific target enzymes with potential commercial value. Of interest would be lipases and proteases. The *T. reesei* database on the JGI WEB site lists 136 gene IDs as having potential lipase activity and 268 with potential protease activity. A manual search for all of these would not be feasible but future work could explore this further. However, a manual search for the lipase genes has identified at least five that are expressed (Table 6) and one of these has an expression level above 100.

Conclusion

In conclusion, results here show that SSF using MSW as a substrate could represent an economical method for the production of cellulase enzyme with low operational costs as MSW is a cheap and abundant

CAZy group and suggested enzyme activity	Number of genes expressed on day 3	Number of genes expressed on day 5	Number of genes expressed on day 7
GH1 (β-glucosidase)	1	1	0
GH2 (β-mannosidase)	1	1	0
GH3 (β-glucosidase)	0	2	0
GH5 (β-1-3-glucosidase)	1	3	0
GH6 (Cellobiohydrolase)	0	1	0
GH7 (Endo-β-1,4 glucanase)	0	1	0
GH11 (β- 1,4 xylanase)	0	1	0
GH12 (β-1,4 glucanase)	0	1	0
GH13 (amylase)	0	1	0
GH15 (Glucoamylase)	0	1	0
GH16 (glucanosyl transferase)	6	3	1
GH17 (β- 1,3 glucosidase)	3	3	2
GH18 (Chitinase)	1	2	1
GH25 (N.O diacylmuramidase)	1	1	0
GH31 (α- glucosidase)	1	1	0
GH37 (α-trehalase)	0	1	0
GH47 (α-1,2 mannosidase)	1	1	0
GH61 (Cu dependent polysaccharide monooxygenase)	3	3	0
GH71 (α- 1.3 glucanase)	1	1	0
GH72 (β-1,3 glucosyl transferase)	3	3	3
GH74 (Xyloglucanase)	0	1	0
GH76 (α-1,6 mannase)	4	5	2
GH92 (α-1.2 mannase)	1	0	0
GH104 (unassigned)	0	0	1

Table 3: Highly expressed GH genes during *T. reesei* fermentation on MSW (RPKM > 100).

Day 3	Day 5	Day 7
3007	44700	44700
53947	45971	45971
56118	49366	49366
64667	53947	61078
65718	61078	65718
68107	65718	66092
68909	68107	66276
70840	70840	70840
72137	72137	72137
73516	73516	73516
74060	81136	81136
81136	82374	103498
82510	106516	106516
105533	106591	106591
111890	109296	109296
119989	111362	111362
121605	121163	121163
123029	121439	121439
123650	121653	121653
124210	123650	12350

Table 4: Gene ID numbers for the 20 most highly expressed genes at each time point. Green highlights are those identified at all three times. Red those common to days 3 and 5; yellow highlights those common between days 5 and 7.

Gene ID	Putative function
49366	Protein turnover
56118	Acetyl C0 binding protein
68107	Ribosomal protein
68909	Ribosomal protein
73516	Glucose repressible gene protein-related protein
81136	Membrane bound protein of unknown function
82510	Histone H4
106516	Glucose/ribitol dehydrogenase
111890	Phosphate transporter
119989	Hydrophobin 2
121605	Actin regulatory protein
123029	Cu^{2+}/ Zn^{2+} superoxide dismutase
124210	Histone H3

Table 5: Putative functions for the gene products.

Gene ID	EC number	RPKM Day 3	RPKM Day 5	RPKM Day 7
32364	3.1.-.-	11.77651	17.56082	0.417827
66324	3.1.1.23	131.8513	56.24179	22.14281
75989	3.1.-.-	10.71609	13.56596	3.480861
119742	3.1.1.3	36.70678	35.01845	20.04622
56427		10.79305	12.6074	1.494388

Table 6: Expression of selected putative lipase genes in T. reesei during SSF fermentation on MSW.

substrate. *T. reesei* recorded the highest production of cellulase enzyme at 30°C with a 168 hours incubation period using 60% moisture content. Crude enzymes derived from this SSF of MSW were able to release sugars from MSW at a rate similar to that of a commercial enzyme preparation.

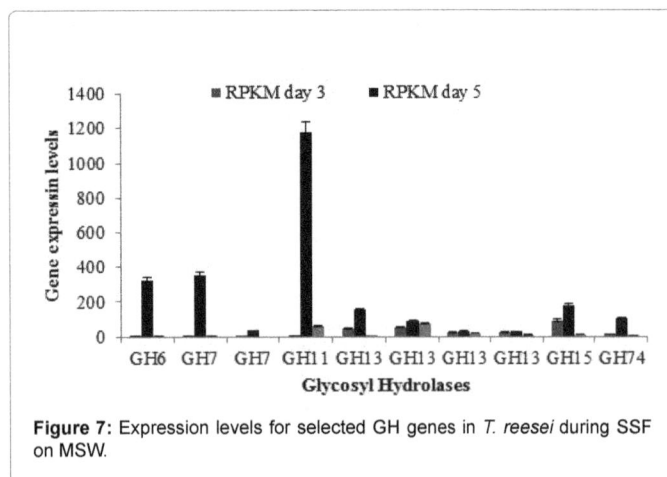

Figure 7: Expression levels for selected GH genes in *T. reesei* during SSF on MSW.

Acknowledgments

The authors gratefully acknowledge the financial support by the Salahaddin University Iraq-Kurdistan region for providing Jwan J. Abdullah's PhD Scholarship and funding this research. We also thank the Biotechnology and Biological Sciences Research Council (BBSRC, BB/G01616X/1) for supporting this research.

References

1. Vergara SE, Tchobanoglous G (2012) Municipal solid waste and the environment: a global perspective. Annual Review of Environment and Resources. 37: 277-309.

2. Keeling C (2011) Canterbury Region Waste Data Report 2009/2010. Environment Canterbury.

3. Li S, Zhang X, Andresen JM (2012) Production of fermentable sugars from enzymatic hydrolysis of pretreated municipal solid waste after autoclave process. Fuel 92: 84-88.

4. Ray AK, Bairagi A, Ghosh SK, Sen K (2007) Optimization of fermentation conditions for cellulase production by Bacillus subtilis CY5 and Bacillus circulans TP3 isolated from fish gut. Acta Ichthyologica et Piscatoria 37: 47-53.

5. Singh J, Batra N, Sobti RC (2004) Purification and characterisation of alkaline cellulase produced by a novel isolate, Bacillus sphaericus JS1. J Ind Microbiol Biotechnol 31: 51-56.

6. Rajoka MI (2004) Influence of various fermentation variables on exo-glucanase production in Cellulomonas flavigena. Electr J Biotechnol 7: 07-08.

7. Guruchandran V, Sasikumar C (2010) Cellulase production by Aspergillus niger fermented in saw dust and Bagasse. J Cell Tissue Res 10: 2115.

8. Zhao SH, Liang XH, Hua DL, Ma TS, Zhang HB (2013) High-yield cellulase production in solid-state fermentation by Trichoderma reesei SEMCC-3.217 using water hyacinth (Eichhornia crassipes). Afr J Biotechnol 10: 10178-10187.

9. Ncube T, Howard RL, Abotsi EK, Jansen van Rensburg EL, Ncube I (2012) Jatropha curcas seed cake as substrate for production of xylanase and cellulase by Aspergillus niger FGSCA733 in solid-state fermentation. Industr Crops Products 37: 118-123.

10. Singhania RR, Sukumaran RK, Pillai A, Prema P, Szakacs G, et al. (2006) Solid-state fermentation of lignocellulosic substrates for cellulase production by Trichoderma reesei NRRL 11460. Indian J Biotechnol 5: 332-336.

11. Alam MZ, Mamun AA, Qudsieh IY, Muyibi SA, Salleh HM, et al. (2009) Solid state bioconversion of oil palm empty fruit bunches for cellulase enzyme production using a rotary drum bioreactor. Biochem Engineer J 46: 61-64.

12. Latifian M, Hamidi-Esfahani Z, Barzegar M (2007) Evaluation of culture conditions for cellulase production by two Trichoderma reesei mutants under solid-state fermentation conditions. Bioresour Technol 98: 3634-3637.

13. Pensupa N, Jin M, Kokolski M, Archer DB, Du C (2013) A solid state fungal fermentation-based strategy for the hydrolysis of wheat straw. Bioresour Technol 149: 261-267.

14. Couto SR, Sanromán MA (2005) Application of solid-state fermentation to ligninolytic enzyme production. Biochem Engineer J 22: 211-219.

15. Klein-Marcuschamer D, Oleskowicz-Popiel P, Simmons BA, Blanch HW (2012) The challenge of enzyme cost in the production of lignocellulosic biofuels. Biotechnol Bioeng 109: 1083-1087.

16. Stutzenberger FJ (1971) Cellulase production by Thermomonospora curvata isolated from municipal solid waste compost. Appl Microbiol 22: 147-152.

17. Gautam SP, Bundela PS, Pandey AK, Khan J, Awasthi MK, et al. (2011) Optimization for the production of cellulase enzyme from municipal solid waste residue by two novel cellulolytic fungi. Biotechnol Res Int 2011: 8.

18. Sluiter A, Hames B, Ruiz R, Scarlata C, Sluiter J, et al. (2006) Determination of sugars, byproducts, and degradation products in liquid fraction process samples. National Renewable Energy Laboratory.

19. Sluiter A, Hames B, Ruiz R, Scarlata C, Sluiter J, et al. (2008) Determination of structural carbohydrates and lignin in biomass. National Renewable Energy Laboratory.

20. Campbell CR (1992) Determination of total nitrogen in plant tissue by combustion. Plant Anal Ref Proc for S US Southern Coop Ser Bull 368: 20-22.

21. Cequier-Sánchez E, Rodríguez C, Ravelo AG, Zárate R (2008) Dichloromethane as a solvent for lipid extraction and assessment of lipid classes and fatty acids from samples of different natures. J Agric Food Chem 56: 4297-4303.

22. Hokura A, Matsuura H, Katsuki F, Haraguchi H, et al. (2000) Multielement determination of major-to-ultratrace elements in plant reference materials by ICP-AES/ICP-MS and evaluation of their enrichment factors. Analytical Sciences 16: 1161-1168.

23. Ries L, Pullan ST, Delmas S, Malla S, Blythe MJ, et al. (2013) Genome-wide transcriptional response of Trichoderma reesei to lignocellulose using RNA sequencing and comparison with Aspergillus niger. BMC Genomics 14: 541.

24. Adney B, Baker J (1996) Measurement of cellulase activities. National Renewable Energy Laboratory 1996: 1-6.

25. Barlaz MA, Eleazer WE, Odle WS, Qian X, Wang YS (1997) Biodegradative analysis of municipal solid waste in laboratory-scale landfills. Environmental Protection Agency.

26. Jones KL, Rees JF, Grainger JM (1983) Methane generation and microbial activity in a domestic refuse landfill site. Eur J Appl Microbiol Biotechnol 18: 242-245.

27. Wang YS, Byrd CS, Barlaz MA (1994) Barlaz, Anaerobic biodegradability of cellulose and hemicellulose in excavated refuse samples using a biochemical methane potential assay. J Ind Microbiol 13: 147-153.

28. Barlaz MA, Schaefer DM, Ham RK (1989) Bacterial population development and chemical characteristics of refuse decomposition in a simulated sanitary landfill. Appl Environ Microbiol 55: 55-65.

29. Ham RK, Norman MR, Fritschel PR (1993) Chemical characterization of fresh kills landfill refuse and extracts. J Environ Engineer 119: 1176-1195.

30. Price GA1, Barlaz MA, Hater GR (2003) Nitrogen management in bioreactor landfills. Waste Manag 23: 675-688.

31. Barlaz MA (2006) Forest products decomposition in municipal solid waste landfills. Waste Manag 26: 321-333.

32. Ponsá S, Gea T, Sánchez A (2011) Anaerobic co-digestion of the organic fraction of municipal solid waste with several pure organic co-substrates. Biosystems Engineering 108: 352-360.

33. Rao MS, Singh SP (2004) Bioenergy conversion studies of organic fraction of MSW: kinetic studies and gas yield-organic loading relationships for process optimisation. Bioresour Technol 95: 173-185.

34. Gallert C, Winter J (2005) Bacterial metabolism in wastewater treatment systems.

35. Hartmann H, Ahring BK (2006) Strategies for the anaerobic digestion of the organic fraction of municipal solid waste: an overview. Water Sci Technol 53: 7-22.

36. Roane TM, Rensing C, Pepper IL, Maier RM (2000) Microorganisms and metal pollutants. Environ Microbiol pp: 403-423.

37. Richard TL, Woodbury PB (1992) The impact of separation on heavy metal contaminants in municipal solid waste composts. Biomass and Bioenergy 3: 195-211.

38. Slack RJ, Bonin M, Gronow JR, Van Santen A, Voulvoulis N (2007) Household hazardous waste data for the UK by direct sampling. Environ Sci Technol 41: 2566-2571.

39. Déportes I, Benoit-Guyod JL, Zmirou D (1995) Hazard to man and the environment posed by the use of urban waste compost: a review. Sci Total Environ 172: 197-122.

40. Hasselriis F, Licata A (1996) Analysis of heavy metal emission data from municipal waste combustion. J Hazardous Materials 47: 77-102.

41. Eighmy TT, Eusden JD, Krzanowski JE, Domingo DS, Staempfli D, et al. (1995) Comprehensive approach toward understanding element speciation and leaching behavior in municipal solid waste incineration electrostatic precipitator ash. Environ Sci Technol 29: 629-646.

42. Park YJ, Heo J (2002) Vitrification of fly ash from municipal solid waste incinerator. J Hazard Mater 91: 83-93.

43. Rodella N, Bosio A, Dalipi R, Zacco A, Borgese L, et al. (2014) Waste silica sources as heavy metal stabilizers for municipal solid waste incineration fly ash. Arab J Chem.

44. Wu HY, Ting YP (2006) Metal extraction from municipal solid waste (MSW) incinerator fly ash-Chemical leaching and fungal bioleaching. Enzyme Microbial Technol 38: 839-847.

45. Xu TJ, Ting YP (2004) Optimisation on bioleaching of incinerator fly ash by Aspergillus niger–use of central composite design. Enzyme Microbial Technol 35: 444-454.

46. Mekala NK, Singhania RR, Sukumaran RK, Pandey A (2008) Cellulase production under solid-state fermentation by Trichoderma reesei RUT C30: statistical optimization of process parameters. Appl Biochem Biotechnol 151: 122-131.

47. Mrudula S, Murugammal R (2011) Production of cellulase by Aspergillus niger under submerged and solid state fermentation using coir waste as a substrate. Braz J Microbiol 42: 1119-1127.

48. Maurya DP, Singh D, Pratap D, Maurya JP (2012) Optimization of solid state fermentation conditions for the production of cellulase by Trichoderma reesei. J Environ Biol 33: 5-8.

49. Francis F, Sabu A, Nampoothiri KM, Ramachandran S, Ghosh S, et al. (2003) Use of response surface methodology for optimizing process parameters for the production of a-amylase by Aspergillus oryzae. Biochem Eng J 15: 107-115.

50. Brijwani K, Oberoi HS, Vadlani PV (2010) Production of a cellulolytic enzyme system in mixed-culture solid-state fermentation of soybean hulls supplemented with wheat bran. Proc Biochem 45: 120-128.

51. Bansal N, Tewari R, Soni R, Soni SK (2012) Production of cellulases from Aspergillus niger NS-2 in solid state fermentation on agricultural and kitchen waste residues. Waste Manag 32: 1341-1346.

52. Haq I, Iqbal S, Qadeen M (1993) Production of xylanase and CMC cellulase by mold culture. Pak J Biotechnol 4: 403-409.

53. Nasir Iqbal HM, Asgher M, Ahmed I, Hussain S, et al. (2010) Media optimization for hyper-production of carboxymethyl cellulase using proximally analyzed agroindustrial residue with Trichoderma harzianum under SSF. IJAVMS 4: 47-55.

54. Krishna C (2005) Solid-state fermentation systems-an overview. Crit Rev Biotechnol 25: 1-30.

55. Koppram R, Tomás-Pejó E, Xiros C, Olsson L (2014) Lignocellulosic ethanol production at high-gravity: challenges and perspectives. Trends Biotechnol 32: 46-53.

56. Aiello C, Ferrer A, Ledesma A (1996) Effect of alkaline treatments at various temperatures on cellulase and biomass production using submerged sugarcane bagasse fermentation with Trichoderma reesei QM 9414. Bioresour Technol 57: 13-18.

57. Brijwani K, Vadlani PV (2011) Cellulolytic enzymes production via solid-state fermentation: effect of pretreatment methods on physicochemical characteristics of substrate. Enzyme Res 2011: 10.

58. Kim D, Cho EJ, Kim JW, Lee Y, Chung H, et al. (2014) Production of cellulases by Penicillium sp. in a solid-state fermentation of oil palm empty fruit bunch. African J Biotechnol 13: 145-155.

59. Kannakar M, Ray RR (2010) Extra cellular endoglucanase production by Rhizopus oryzae in solid and liquid state fermentation of agro wastes. Asian J Biotechnol 2: 27-36.

60. Sun H, Ge X, Hao Z, Peng M (2010) Cellulase production by Trichoderma sp. on apple pomace under solid state fermentation. African J Biotechnol 9.

61. Häkkinen M, Arvas M, Oja M, Aro N, Penttilä M, et al. (2012) Re-annotation of the CAZy genes of Trichoderma reesei and transcription in the presence of lignocellulosic substrates. Microb Cell Fact 11: 134.

8

The Effect of Heating Radiation on the Synthesis and Crystallization of Cordierite Composition Glasses

MKh Rumi*, MA Zufarov, EP Mansurova and NA Kulagina

Institute of Material Sciences SPA, Physics – Sun, Academy of Sciences Republic of Uzbekistan, Tashkent, Uzbekistan

Abstract

The results on crystallization of glasses of the cordierite composition, synthesized under the influence of concentrated radiant flux of different densities, are presented. Synthesis was carried out using a solar furnace or a solar simulator, wherein Xenon lamps of 10 kW power serve as a heat source. We studied glasses of the following stoichiometric composition $2MgO: 2Al_2O_3: 5SiO_2$ without a catalyst and with TiO_2 as a catalyst. The initial raw materials were MgO, Al_2O_3 and quartz-kaolinite-pyrophyllite rock as a main source of SiO_2. The natures of phase transitions in the samples obtained are studied using the X-ray analysis (DRON-UM-1) and the differential-thermal method (Derivatograph Q-1500 D). The absorption spectra are obtained on spectrophotometer SF-56. A comparison of the phase composition of the crystallized samples shows that the crystallization of μ-cordierite and the transition of μ-cordierite to α-cordierite in glasses, synthesized using a Xenon lamp, occurs at lower temperatures than those synthesized using solar radiation, provided the same conditions of synthesis and annealing. Besides of this, in glasses containing TiO_2, the content of Ti^{3+} increases, and a decay of the concomitant phase, magnesium-aluminum-titanate, is activated at annealing temperatures above 1200°C. The differences in the character of the phase formation affect the activity of glass powders to sintering.

It is found that peculiarities of the spectral composition of a Xenon lamp and the Sun affect the nature of the glass crystallization process. A presence of a significant proportion of extreme ultraviolet radiation initiates the crystallization process by the photo-activation mechanism and has the same effect as a rise of the glass crystallization temperature or an increase of the catalyst concentration.

Keywords: Cordierite; Glass; Crystallization; Radiation

Introduction

Peculiarities of crystallization of glasses in the system $MgO: Al_2O_3: SiO_2$ have been studied in many researches. Particular attention has been given to the crystallization of glasses of the cordierite composition ($2MgO: 2Al_2O_3: 5SiO_2$), since various glass-ceramic materials with an optimum combination of dielectric and thermo-mechanical properties are produced on its basis. The most common method of production of cordierite glass-ceramic materials is glass melting in electric or induction furnaces followed by subsequent crystallization. The crystallization of cordierite occurs through the formation of a sequence of intermediate phases. The temperature intervals of the formation and stability of the phases are changed depending on the conditions of synthesis and crystallization of glasses, as well as the type and concentration of catalysts. The effect of such factors, as temperature, time, gas atmosphere and the chemical composition of raw materials, on the properties of these materials has been studied in details [1-6].

In our study, the synthesis of glass is carried out under the influence of concentrated radiant flux from different types of sources. These conditions differ from traditional synthesis methods and can have a significant impact on the crystallization of glasses and the properties of subsequent glass-ceramic materials. However, studying the effect of short-wave radiation on the course of the crystallization of cordierite glasses has much less attention in the literature. The authors of [7,8] using a solar furnace for sintering the powder mixtures to obtain cordierite.

At the same time, in work [9], it is shown that the largest acceleration of the crystallization process in production of lithium-silica-alumina glasses is achieved with use of Xenon arc lamps of different power. The authors believe that since at high temperatures the radiant heat transfer is realized in the glass, the presence of the radiant component of heat transfer initiates the crystallization process by means of the photo-activation mechanism. In paper [10], the effect of ultraviolet radiation on a formation of glass-ceramics in photosensitive glasses is studied. In our conditions, radiation exposure occurs during glass melting, which may also have some influence on the process of the crystallization and phase formation during subsequent heat treatment.

The present work is devoted to the study of the crystallization of glasses of the cordierite composition, synthesized under the action of concentrated radiant flux, and some properties of glass-ceramics, obtained on their basis.

Materials and Methods

The initial components of the glass batch is magnesium oxide MgO (puriss), alumina Al_2O_3 (puriss), as well as natural minerals as a source of SiO_2. These minerals have different compositions, and they contain in addition to basic oxides of silicon, aluminum and magnesium on the average of 4.54 wt% of impurities (Fe_2O_3, Na_2O, K_2O). In this study, a quartz-pyrophyllite-kaolinite rock is used as a silica-containing material (Table 1). Table 2 shows the batch composition for synthesis of a glass of the following stoichiometric composition ($2MgO: 2Al_2O_3: 5SiO_2$) wt %.

---

*Corresponding author: Rumi MKh, Institute of Material Sciences SPA, Physics – Sun, Academy of Sciences Republic of Uzbekistan, Tashkent, Uzbekistan, E-mail: marina@uzsci.net

Chemical composition, wt.%	SiO$_2$	Al$_2$O$_3$	Fe$_2$O$_3$	TiO$_2$	CaO	MgO	K$_2$O	Na$_2$O
	77.96	20.82	0.36	0.10	0.31	0.10	0.20	0.15
Mineralogical composition, wt.%	Silica (SiO$_2$) - 25, Kaolinite (Al$_4$[Si$_4$O$_{10}$] [OH]$_8$) - 25, Pyrophyllite (Al$_2$[Si$_4$O$_{10}$] (OH)$_2$) -50							

Table 1: Chemical and mineralogical composition of the quartz-kaolinite-pyrophyllite rock.

Components	Component Content, wt.%
Magnesium oxide	14.51
Aluminium oxide	22.29
Quartz-kaolinite-pyrophyllite	63.2

Table 2: The batch composition for production of the glass of the composition 13, 78 MgO: 34.86 Al$_2$O$_3$: 51.36 SiO$_2$ (wt.%).

Siliceous raw materials (quartz-kaolinite-pyrophyllite rock) are pulverized preliminarily to particles of the size less than 100 microns. Aluminium oxide is burned annealed at 1450-1500°C for the transition into a stable α-form. The initial components are weighed according to the calculations. Then the mixture is stirred with simultaneous additional grinding in a planetary ball mill to achieve a grind fineness of 5-10 microns. The resulting powders are molded for subsequent glass melting.

The glass synthesis is performed on a solar furnace (The Big Solar Furnace, power 1 MW, Parkent town, Uzbekistan), as well as on a solar simulator, in which focused radiation of two Xenon lamps of 10 kW power is served as a source of heating. Synthesis of glasses is carried out at two flux densities of focused radiation: 1) the flux density of 250 W/cm^2 and 2) the flux density of 600 W/cm^2. Cooling the molten glass is carried out by quenching in water. This method of melt quenching enables to obtain fragile granules, which could subsequently be readily milled for further preparation of glass-ceramics using the ceramic technology. Crystallization of glasses is carried out in the one-stage mode in an electric furnace in the temperature range 850-1100°C. Samples are either fragments of 2-3 mm granules or powders of fineness of 40-80 microns. To determine the ability of powders to sintering, a powder is compressed into tablets with a diameter of 20 mm and a thickness of 5 mm, and then tablets are annealed at temperatures of 900-1350°C. The water absorption of the samples is selected as a characteristic of the degree of sintering. Water absorption was calculated using the following relationship:

W (%) = [(m$_2$ – m$_1$) / m$_1$] * 100

where,

m$_1$ = mass of the dry sample, in g

m$_2$ = mass of the water saturation sample, in g

We investigate glasses of the stoichiometric composition, as well as those supplemented with 12 wt% titanium dioxide (over 100%) as a crystallization catalyst.

The X-ray analysis, the differential-thermal method and optical methods are used to analyze the synthesized materials. The X-ray analysis is performed on the diffractometer DRON-UM 1 (Cu K$_α$ radiation, and Ni-filter).

Derivatograph Q-1500 D is used for the differential thermal analysis (DTA). The heating rate used is 15°C/min. Measurements are carried out up to a temperature of 1200°C. Absorption spectra are studied with SF-56 spectrophotometer in the wavelength range of 100–1000 nm. The resolution of the device is 0.3 nm.

Results and Discussions

Previously, our research [11,12] has shown that during the synthesis of glasses of the cordierite composition under the influence of concentrated radiant flux, the crystallization of α-cordierite occurs through the formation of solid solutions with the structure of high-temperature quartz (quartz-O) and μ-cordierite. On the DTA curves, obtained on the crushed granular samples, the presence of diffuse exothermic peaks in the temperature ranges 700-900°C and 900-1000°C and most pronounced at 1040-1050°C are observed (Figure 1). According to the X-ray analysis, the peaks correspond to the crystallization of the aforementioned phases. In addition, as can be seen from derivatograms, the temperature of exoeffects and their profiles are changed depending on the density of radiant flux and the type of the radiation source.

The X-ray phase analysis of the coarse-grained samples shows (Figure 2) that during the synthesis of glass of the cordierite composition in the absence of catalysts with the use of a Xenon lamp, the crystallization process of μ-cordierite in coarse-grained samples starts at 880°C (density of the radiant flux is 250 W/cm^2).

With a significant overheating of the melt, which corresponds to the flux density of 600 W/cm^2, the amount of crystallizing μ-cordierite increased sharply, while at 980°C the only crystalline phase is α-cordierite, regardless of the flux density. For traditional methods of synthesis of glasses of the cordierite composition, the rate of crystallization of μ-cordierite is extremely small. At a temperature of 850°C, it takes at least 150 hours for μ-cordierite to begin to crystallize [13]. This process is activated under the action of radiant flux.

During the synthesis of glass in the solar furnace, the crystallization begins at higher (by ~ 20°C) temperatures. At the annealing temperature of 880-900°C (flux density of 250 W/cm^2), there are no crystalline phases on the difractograms and at the annealing temperature of 980-990°C, μ-cordierite is the main phase component. For glasses synthesized under high radiant flux density (600 W/cm^2) during annealing at 880–900°C, the crystallization just begins and μ-cordierite remains at the annealing temperature 990°C. To complete the transition μ → α, it is necessary to increase the annealing temperature above 1000°C regardless of the degree of overheating of the melt. A sharper rise of the branch of exothermic effects, a narrow profile and an increase of the peak area, as well as the decrease of corresponding temperature in the DTA curves indicate an increase of the rate of the crystallization process with increasing of flux density, while using the Xenon lamp.

By utilizing powder samples, the phase composition is not changed, but the ratio of the crystallizing phases is changed, showing an increase of μ-cordierite amount. In the DTA curves, the high-temperature exothermic peaks have more distinct forms indicating the surface crystallization of the glass in the absence of catalysts (Figure 3) [14].

A similar effect of radiation was observed for glasses with additions of titanium dioxide as a catalyst. It should be noted that the color of titanium-containing glasses changes from black to yellow. Black glasses are obtained in synthesis in radiation of a Xenon lamp, while yellow glasses are obtained in synthesis in the solar furnace.

On diffractograms of crystallized glasses, in addition to the μ- and

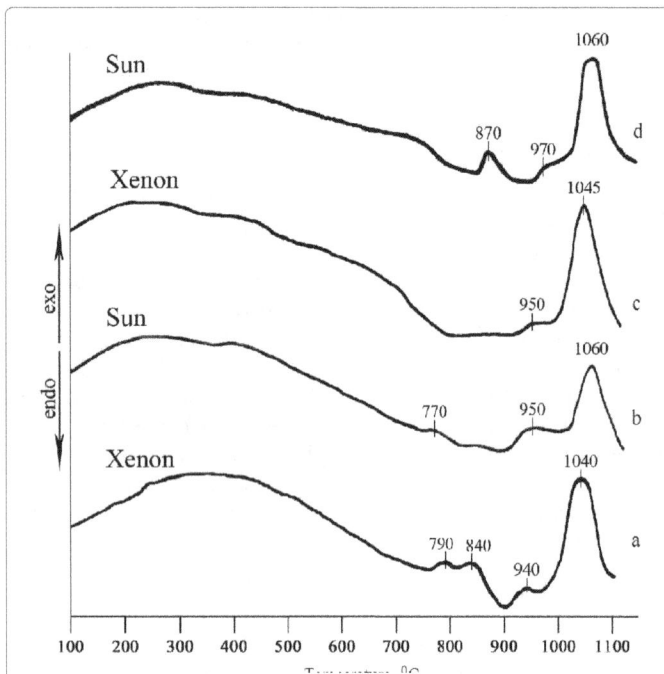

Figure 1: DTA curve fragments of 2-3 mm granules; a,b) flux density 250 W/cm², c,d) flux density 600 W/cm².

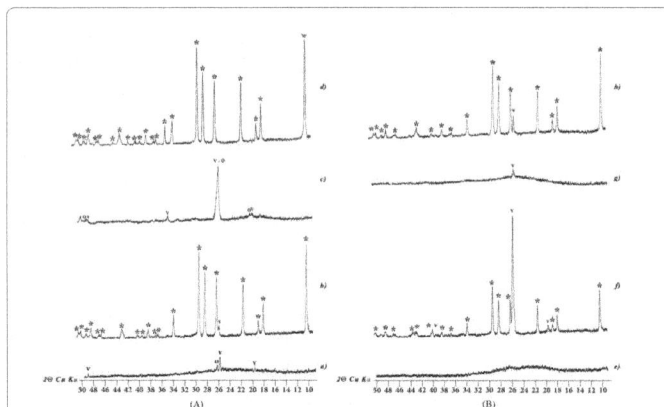

Figure 2: X-Ray diffraction patterns of cordierite composition glass samples: A) heating sources is Xenon lamp, B) heating sources is Sun; a,e) flux density 250 W/cm2, annealing temperature 880-900°C; b,f) flux density 250 W/cm2, treatment temperature 980-990°C; c,g) flux density 600 W/cm2, annealing temperature 880-900°C; d,h) flux density 600 W/cm2, annealing temperature 980-990°C.

α-cordierites, the presence of magnesium-aluminum-titanate (a solid solution $nAl_2TiO_5 \cdot mMgTi_2O_5$) is observed, and its amount increases with increase of content of titanium dioxide [12].

It is found that the synthesis of titanium-containing glasses under the action of radiation from a Xenon lamp in comparison with the solar radiation leads to a decrease of temperature of the corresponding phase transformations during subsequent crystallization from a μ-cordierite to α-cordierite, and at higher temperatures to accelerating of disintegration of magnesium-aluminum-titanate and formation of rutile. It is necessary to note that the phase composition of crystallized glasses with 12 wt % TiO_2, synthesized using the solar furnace, corresponds to the glass composition with 5-10 wt % TiO_2, crystallized at the same temperatures, but synthesized using a Xenon lamp.

Effect of radiation on the phase composition of the crystallized glasses is revealed in the ability of the glass powders, containing 12 wt % TiO_2, to sintering. A comparison of these results with those obtained from the X-ray analysis shows that the degree of sintering depends on the phase composition of the material. An appreciable reduction of water absorption is found only for the samples in which a transition from a metastable modification of cordierite to a stable α-form has been completed. Further intensification of the sintering process occurs with the decomposition of magnesium-aluminum-titanate. It is found that the glass powders, synthesized by using a Xenon lamp, are sintered better than those synthesized in the solar furnace (Figure 4), having the same conditions of synthesis and heat treatment.

Analysis of the absorption spectra shows peculiarities due to the influence of different contents of titanium dioxide, as well as the nature of ionizing radiation.

According to the previous works [15], the absorption band at 400-600 nm corresponds to Ti^{3+}. As shown in Figure 5, the intensity of the absorption band for glasses, obtained in the synthesis using a Xenon lamp, increases with increasing content of TiO_2, which may indicate an increase of the concentration of Ti^{3+}. In this case, the optical density of the glass is increased as well. This fact, along with a shift of the absorption band edge to longer wavelengths, can mean the system disordering. At the same time, this absorption band is practically absent in the cordierite glass (12% TiO_2), synthesized in the solar furnace. The nature of the curve of the latter spectrum is more like the absorption spectrum of a glass with less TiO_2, synthesized using a Xenon lamp.

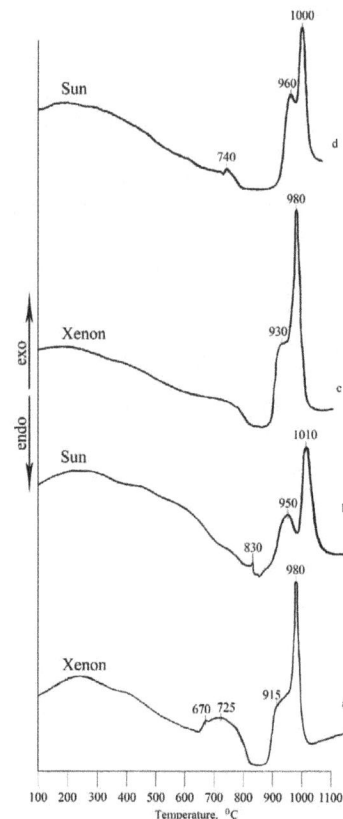

Figure 3: DTA curve or powders of fineness of 40-80 μm; a,b) flux density 250 W/cm², c,d) flux density 600 W/cm².

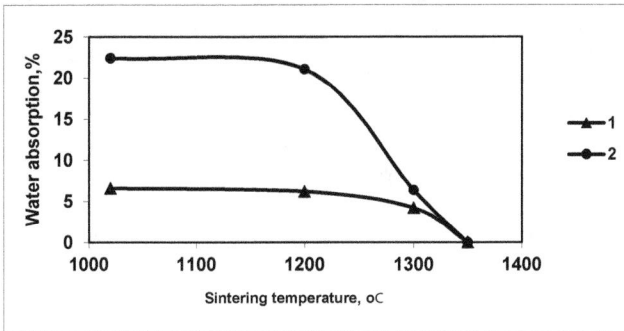

Figure 4: The dependence of the degree of sintering of cordierite glass composition with the addition of 12 wt % TiO_2 on synthesis conditions, the melt holding time is 5 minutes: 1) of a Xenon lamp; 2) the solar radiation.

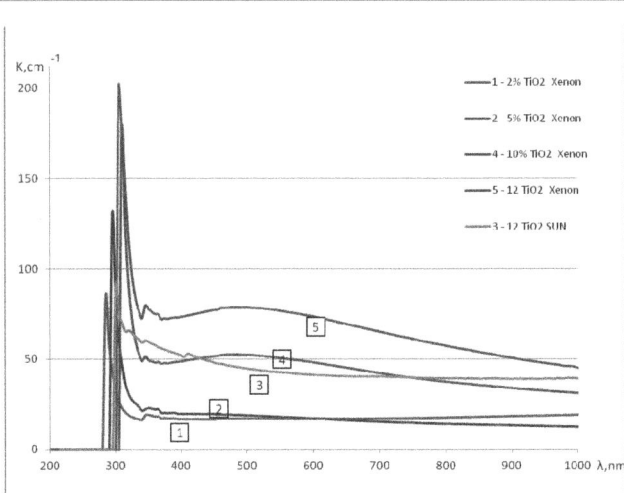

Figure 5: Absorption spectra of cordierite glasses doped by TiO_2 syntheses by irradiation of: Xenon lamp (curves 1,2,4,5), Sun (curve 3).

All absorption spectra have peaks in the region 350-370 nm, and the intensity of peaks remains constant for all of the samples. The origin of the peaks is difficult to identify unambiguously. According to [15], the absorbance at 350-370 nm, along with the absorption band of 400-700 nm may be due to the presence of Ti^{3+}. However, in this case, a correlation of the intensity of these peaks with percentage of TiO_2 should be observed. More likely, the peaks are associated with Fe^{3+} [16] due to the presence of Fe_2O_3 in the initial silica-containing materials. At the same time, there is a possibility of the formation of complex defects of FTi type, where F is the center near the titanium ion that replaces Al^{3+}.

It is known [17-21] that UV radiation has a photo-catalytic influence on a material, which results in a change in the material properties. In our case, this influence occurs at the stage of melting and glass synthesis that leads to the formation of structure imperfections, both on the surface and in the bulk of the melt. One of the effects of such an influence is the formation of anionic vacancies (in this case oxygen vacancies). In the presence of the transition d-elements in a glass, this process results in the possibility of formation of low-valent cations due to the capture of a free electron, for example: $e^- + Ti^{4+} \rightarrow Ti^{3+}$ [22]. In a cordierite glass without a catalyst, a formation of paramagnetic defects is due to the presence of trace contaminants [23], in particular Fe_2O_3, because of

the use of mineral raw materials as a source of SiO_2. Introduction of TiO_2, as a catalyst, results in an increase of the concentration of such defects, including the formation of different complexes based Ti^{3+} (Ti^{3+}-Fe^{3+}, Ti^{3+}-Fe^{2+}, Ti^{3+}-Al^{3+} and so on). Changing the glass structure induces inhomogeneity, promoting segregation. A subsequent thermal treatment of such glasses leads to a bulk fine-grained crystallization with a release of the titanium-containing phase, and, consequently, to improvement of properties of the glass-ceramics.

Thus, the mentioned differences in the rate of crystallization and the nature of phase formation of the glasses studied are quite justified if we take into account the features of the spectral composition of the Xenon lamp and the sun. In particular, the spectrum of the Xenon lamp have a large proportion of extreme ultraviolet radiation in the wavelength region of less than 0.3 micron, which increases the crystallization process due to the photo-activation mechanism [24].

Conclusion

It is found that the crystallization of a glass of the cordierite composition, synthesized by the action of concentrated radiant flux depends on the type of a heat source, along with other known factors. Short-wave radiation activates the crystallization of μ-cordierite, the phase transitions (μ-cordierite to α-cordierite, decay the concomitant phase magnesium-aluminum-titanate in glasses containing TiO_2) and it has the same effect on the processes of crystallization as a rise of the temperature of the glass crystallization or an increase of catalyst concentration.

Acknowledgement

The authors thank Dr. E. Ibragimova for help with the spectrophotometric experiments.

References

1. Jo S, Kang S (2013) TiO_2 effect on crystallization mechanism and physical properties of nano glass-ceramics of MgO-Al_2O_3-SiO_2 glass system. J Nanosci Nanotechnol 13: 3542-3545.

2. Luo XD, Qu DL, Zhang GD, Liu HX, (2011) The Influence of TiO_2 on Synthesizing the Structure of the Cordierite. Adv Materials Res 233-235: 3027-3031

3. Demirci Y, Günay E (2011) Crystallization behavior and properties of cordierite glass-ceramics with added boron oxide. J Ceramic Process Res 12: 352-356.

4. Marikkannan SK, Ayyasamy EP (2013) Synthesis, characterisation and sintering behaviour influencing the mechanical, thermal and physical properties of cordierite-doped TiO_2 J Materials Res Technol 2: 269–275.

5. Shamsudin Z, Hodzic A, Soutis C, Hand RJ, Hayes SA, et al. (2011) Characterisation of thermo-mechanical properties of MgO-Al_2O_3-SiO_2 glass ceramic with different heat treatment temperatures. J Materials Sci 46: 5822-5829.

6. Wang ShM, Kuang FH, Li J (2010) Influence of different Fe_2O_3 content on crystallization of MgO-Al_2O_3-SiO_2-TiO_2 system glass-ceramics. Phase Transitions: A Multinational J 83: 397-403.

7. Oliveira FAC, Shohoji N, Fernandes JC, Rosa LG (2005) Solar sintering of cordierite-based ceramics at low temperatures. Solar Energy 78: 351–361.

8. Xiaohong Xu, Xionghua Ma , Jianfeng Wu , Ling Chen , Tao Xu , et al. (2013) In-Situ Preparation and thermal shock resistance of mullite-cordierite heat tube material for solar thermal power. J Wuhan Univf Technol Mater Sci Ed 28: 407-412.

9. Sirnicky AP, Tikatchinsky ID, Romanovsky MB Issledovanie vozmojnostey intensifikacii processa kristalacii littiyalumosilikatnih stekol. Sb.Kataliziro vannaya kristalizaciya stekla: 109-114

10. Berejnoi AP, Iltchenko LN (1968) Issledovanie natchalnih stady sitalloobrazovaniya v svetotchuvstvitelnih steklah. Neorganitcheskie materiali 4: 584-589.

11. Adylov G, Akbarov R, Singh D, Zufarov M, Voronov G, et al. (2008) Crystallization of μ-and α-cordierite in glass obtained via melting by concentrated radiant flux. Appl Solar Energy 44: 135-138.

12. Adylov G, Akbarov R, Singh D, Zufarov M, Voronov G, et al. (2009) Crystal glass materials based on catalyzed cordierite glass synthesized under exposure to concentrated radiant flux. Glass and Ceramics 66: 120-124.

13. Peter Warwick McMillan (1979) Glass-ceramics. Academic Press.

14. Müller R, Naumann R, Reinsch S (1996) Surface nucleation of μ-cordierite in cordierite glass: thermodynamic aspects. Thermochimica Acta 280- 281: 191-204.

15. Nijankovsky SV, Sidelnikova NS, Baranov VV (2015) Optitcheskoe pogloshchenie i centri okraski v krupnih kristallah Ti : sapfira, virashchennih metodom gorizontalnoy napravlennoy krisstallizacii v vosstanovitelnih usloviyah. Fizika tverdogo tela 57: 763-767.

16. Aseev VA, Nekrasov YA, Homtchenko KV (2010) Obescvetchivanie prirodnih sapfirov. Nautchno-tehnitchesky vestnik Sankt-Peterburgskogo gosudarstvennogo universiteta nautchnih tehnology. mehaniki i optiki 2: 86-89.

17. Laguta VV, Glintchuk MD, Slipenuk AM, Bikov IP (2000) Navedennie svetom sobstvennie defekti v keramike PLZT. Fizika tverdogo tela 42: 2190-2196.

18. Dzwigaj S, Nogier J-Ph, Che M, Saito MT, et al. (2012) Influence of the Ti content on the photocatalytic oxidation of 2-propanol and CO on TiSiBEA zeolites. Catalysis Communications 19: 17-20.

19. http://www.topresearch.org/showinfo-49-186981-0.html

20. Lombard P, Ollier N, Boizot B (2010) EPR study of Ti^{3+} ions formed under ionizing irradiation in oxide glasses. Glass & Optical Materials Division Annual Meeting.

21. Ollier N, Lombard P, Farges F, Boizot B, (2008) Titanium reduction processes in oxide glasses under electronic irradiation. J Non-Crystalline Solids 354: 480-485.

22. http://eknigu.com/lib/P_Physics/PS_Solid%20state/PSa_Applications/

23. Hadakovskaya RY (1978) Himiya titansoderjashchih stekol i sitallov. M: Chimiya.

24. Kozelkin VV (1985) Osnovi infrakrasnoy tekhniki. M Mashinostroenie.

9

Biological Fermentative Methane Production from Brown Sugar Wastewater in a Two-Phase Anaerobic System

Ning Li[2], Jianhui Zhao[2], Rui-na Liu[1], Yong-feng Li[1]* and Nan-qi Ren[2]

[1]*School of Forestry, Northeast Forestry University, Harbin, China*
[2]*School of Municipal and Environmental Engineering, Harbin Institute of Technology, Harbin, China*

Abstract

In this study, a two-phase anaerobic digestion system was established to combine the bioenergy recovery and chemical oxygen demand (COD) removal. The synthetic brown sugar wastewater was used as a substrate. Six system organic loading rates (OLRs) from 12 to 32 kg/(m³·d) were analyzed. Results showed that the highest CH_4 production rate (18.5 L/d) were obtained at OLR= 24 kg/(m³·d). The total energy recovery rate was calculated to assess the overall efficiency of energy recovery capacity. The highest energy recovery rate was 728.67 kJ/d, occurred at OLR=24 kg/(m³·d). Meanwhile, the total COD removal was very high, up to 69.4%. Therefore, the system had a great contribution to energy recovery from brown sugar wastewater.

Keywords: CSTR-UASB; OLR; Hydrogen production; Methane production; Energy recovery

Introduction

With the rapid development of industry and an increase in the standards of living, water pollution is becoming a general phenomenon [1]. Over the years, the widespread use of fossil fuels like coal and oil has already caused serious pollution in the global environment [2], and has even posed a threat to the survival of mankind itself [3]. At the same time, fossil fuels are a non-renewable energy source which can be depleted because of overexploitation [4]. Therefore, the question of how to degrade pollutants quickly, change waste into valuable commodities, and achieve the sustainable development of energy resources has become one of the most urgent problems facing the field of contemporary environmental science [5].

The exploitation and application of new alternative clean energies represent the general trend [6]. As efficient, clean, and environment-friendly sources of energy, hydrogen and methane have aroused people's extensive attention [7]. Continuous flow CSTR-UASB two-phase anaerobic systems have the advantage of high mass transfer efficiency, fast degradation rate of organic compounds and strong ability to produce hydrogen and methane. It can achieve both the removal of pollutants as well as the recycling of new energy which is of great industrial value. Biohydrogen production from wastewater through fermentation is carried out by anaerobic acidogenic bacteria with highly diverse fermentation characteristics [8] and hydrogen production capabilities [9]. After hydrogen production, the effluent contains high content of organic acids. Anaerobic digestion for methane production is an ideal way to utilize metabolites (volatile fatty acids (VFA), and alcohols) from hydrogen production process for additional energy production [10]. The two-phase process separates and enriches acidogens and methanogens in different reactors that may improve the process stability and efficiency compared to traditional one-phase methane production process. Although hydrogen and methane production from waste under lower OLR has been reported [11], performance of the two-phase process under higher OLR was seldom investigated. Moreover, there are few reports on brown sugar wastewater by treatment of a two-phase anaerobic system [12,13]. Investigation on the process performance of two-phase CSTR-UASB under higher OLR may accelerate its application.

In this study, using brown sugar wastewater as the carbon substrate, the performance of continuous H_2 and CH_4 production rates were investigated at different OLRs for CSTR-UASB two-phase anaerobic system.

Materials and Methods

Experimental set-up

This experiment utilized a continuous-flow CSTR-UASB two-phase anaerobic system, with the effective volume of CSTR being 7.0 L and the total volume being 15.8 L. The reactor was equipped with a stirring device which ensured the complete and continuous mixture of microorganisms and water at a stirring speed of 120 r·min⁻¹. Anaerobic conditions in the reactor were ensured through the liquid seal on the shaft; the total volume of UASB was 21.2 L while its effective volume was 9.8 L. There were gas-liquid-solid three-phase separators located in both reactors which had an integrated structure of reaction and settling zone. The reactor walls were wound with resistance wires and a temperature control system maintained a reactor temperature of (35 ± 1)°C in order to ensure high microorganism activity. The continuity of the experiment was maintained by using a peristaltic pump providing water into the reactor at a constant speed. The peristaltic pump could change the influent flow rate and then change hydraulic retention time (HRT) by adjusting its revolution speed. The structure of CSTR-UASB two-phase anaerobic system was as shown in Figure 1. The characteristics of substrate used in this study were shown in Table 1.

Analytical methods

The biogas composition including hydrogen and methane was measured using a gas chromatograph (GC, 6809 N Network GC System, Agilent Technologies, Waldron, Germany) equipped with a thermal conductivity detector (TCD). The column (2 m×5 mm) was

---

*Corresponding author: Yong-feng Li, School of Forestry, Northeast Forestry University, Harbin, China, E-mail: huanjinglining@163.com

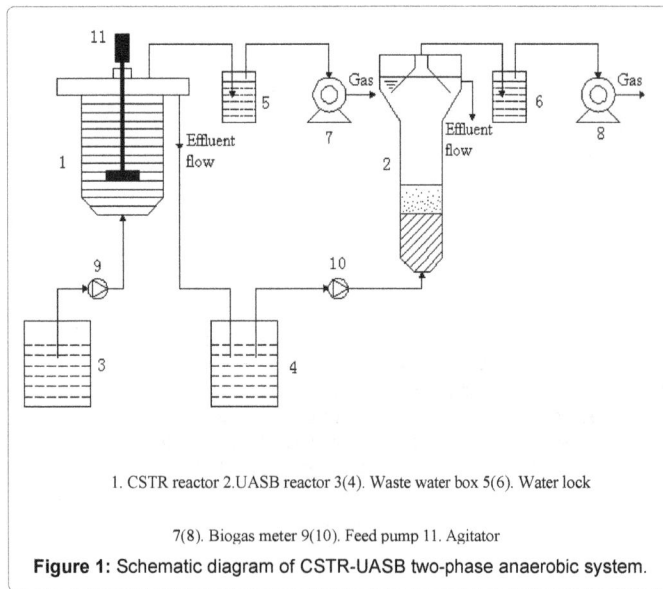

1. CSTR reactor 2.UASB reactor 3(4). Waste water box 5(6). Water lock

7(8). Biogas meter 9(10). Feed pump 11. Agitator

Figure 1: Schematic diagram of CSTR-UASB two-phase anaerobic system.

Parameters	Values	Parameters	Values
Volatile suspended solid (VSS)	1.3 g/L	SO_4^{2-}	1.5 g/L
Total nitrogen (TN)	2.5 g/L	PO_4^{3-}	0.3 g/L
Chemical oxygen demand (COD)	30 g/L	pH	6.3
Total organic carbon (TOC)	10.8 g/L		

Table 1: The characteristics of wastewater used in this study.

filled with porapak Q (50-80 meshes), Nitrogen was used as carrier gas with a flow rate of 40 mL/min.

Volatile fatty acids (VFA) and ethanol in liquid samples were measured by using a gas chromatograph (GC, 6890N Network GC System, Agilent Technologies, Waldbrown, Germany) equipped with a flame ionization detector (FID). The column (Zm) was packed with supporter of GDX-103 (60-80 meshes). The temperatures of the injection port, the oven, and the detector were adjusted to 220°C, 190°C, and 220°C, respectively. The carrier gas was nitrogen at a flow rate of 30 mL/min.

CODs of the samples were measured according to Standard Methods [14]. The pH and ORP were measured by pH meter (PHS-25). A wet gas meter (LML-1) was utilized to measure biogas yield.

The sludge cultivation and operational control parameters

The sludge acclimation and operational control of CSTR reactor: The inoculated sludge of the reactor adopted sludges from a secondary sedimentation tank in a sewage treatment plant in Harbin. It could remove imprities and large particulate matters by precipitation, washing and filtering. Brown sugar water with 10000 mg/L COD was used. The COD: N: P was maintained at a ratio of 1000:5:1 [15] by adding a certain amount of NH_4Cl and KH_2PO_4 in order to supply microorganisms with adequate nitrogen and phosphorus and then cultivated with intermittent aeration for 20 days. During this process, aeration was stopped for 1 h daily so that we could remove the supernatant fluid and add clear water. The mature sludge after domestication was yellow-brown granule with good settlement ability.

The acclimated sludge was then transferred to the CSTR reactors

and started with continuous-flow approach under the conditions of HRT of 6 h, temperature of (35 ± 1) °C, influent pH of 7.00 ± 0.1, OLR of 12 kg/(m^3·d), suspended solid (SS) of 12.81 g/L, volatile suspended solid (VSS) of 8.35 g/L and VSS/SS (biological activity) of 0.65 in inoculated sludge. After about 30 days, the reactor reached a steady state. At this point, the hydrogen production was about 3.5 L/d and the hydrogen content was around 43%. The liquid end products are shown in Table 2, of which the content of ethanol and acetate accounted for 71.5%, mainly for ethanol fermentation.

The sludge acclimation and operational control of UASB reactor: It adopted the same sludge which the CSTR used as the inoculated sludge of the reactor and experienced an identical impurity removal process. Using the effluent of CSTR reactor (liquid fermentation products) as the reaction substrate, the OLR stood at approximately 7.2 kg/ (m^3·d). At the same time, a small amount of NH_4Cl and KH_2PO_4 were added to adjust appropriate nutrition and phosphorus levels in order to maintain COD: N: P in a 200:5:1 proportion. It began to produce methane after about 50 days when the sludge acclimation had completed in the reactor. At that time, SS and VSS in the reactor were 16.28 and 10.36 g/L, respectively. Under the conditions of HRT of 8 h, temperature of (35 ± 1)°C and influent pH of 7.80 ± 0.2, the reactor could operate steadily after another 20 d, producing a methane yield of approximately 5.0 L and a methane content of around 68%.

The control parameters and running status are shown in Table 1 when two reactors reached equilibrium.

After being stabilized, CSTR-UASB two-phase reactors kept HRT a constant. It was from the continuous increase of OLR and the adjustment of the influent pH that the effects of OLR on CSTR-UASB anaerobic system can be observed and studied. The running process in which OLR increased from 12 to 32 kg/(m^3·d) was divided into six stages, each stage increased 4 kg/(m^3·d) and response (run) time was 6 days (Figure 2).

Results and Discussion

Bio-hydrogen and methane production

As can be seen from Figure 3, while OLR increased from 12 to 32 kg/(m^3·d), the bio-hydrogen production in CSTR reactor presented a basic trend of sustained growth while the hydrogen content fluctuated between 30% and 50% whereas the methane production in the UASB reactor first increased and then decreased. On the first day that OLR reached 16 kg/(m^3·d), the methane yield witnessed a sudden rise from

Parameter	CSTR	UASB
HRT(h)	6	8
Temperature (°C)	35 ± 1	35 ± 1
Influent pH	7.1	7.8
Effluent pH	5.0	6.8
Influent COD (mg/L)	3000	1800
Effluent COD (mg/L)	1800	1150
SS (g/L)	13.85	17.61
VSS (g/L)	9.62	11.53
Ethanol (mg/L)	283.5	0
Acetate (mg/L)	64.7	178.6
Propionate (mg/L)	44.6	30.1
Butyrate (mg/L)	87.6	129.5
Valerate (mg/L)	6.3	0

Table 2: Various control parameters and running status after the CSTR and UASB reaching equilibrium.

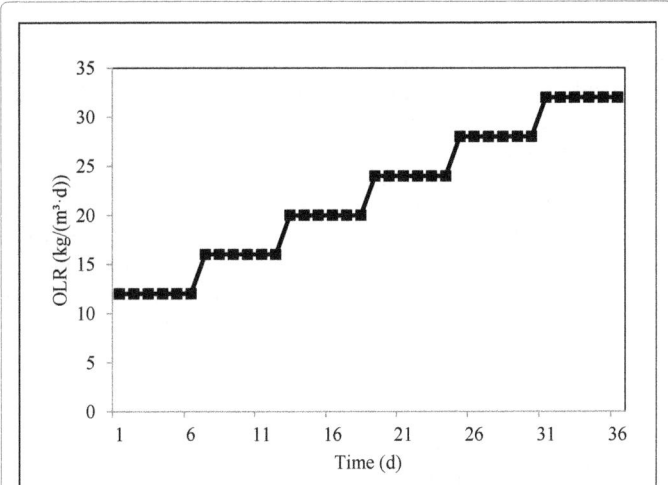

Figure 2: Variation of OLR at different days.

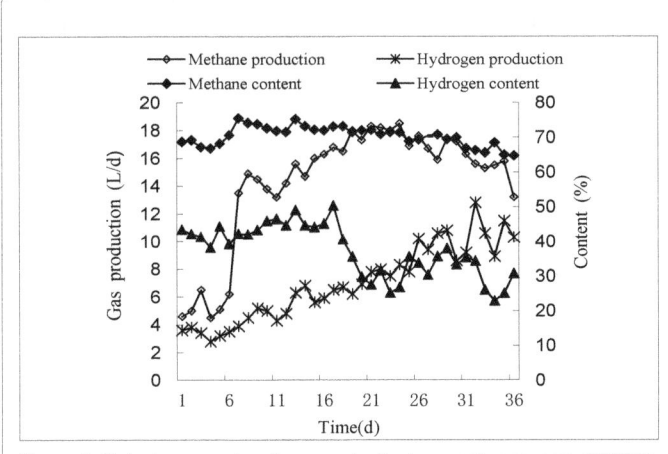

Figure 4: Variation of liquid fermentation products in operation process of CSTR reactor.

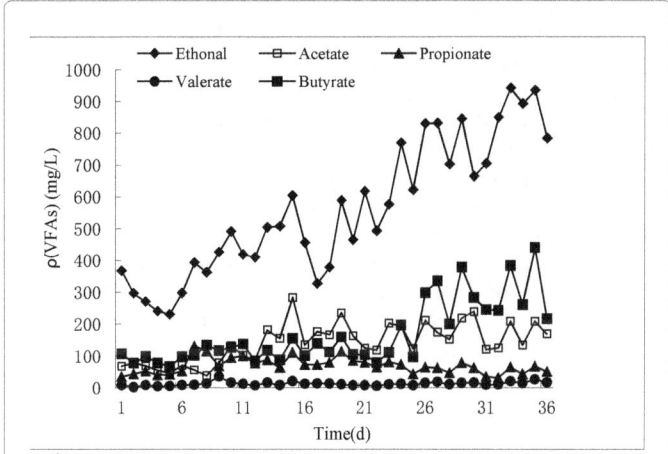

Figure 3: Biohydrogen and methane production in operation process of CSTR-UASB two-phase anaerobic reactor.

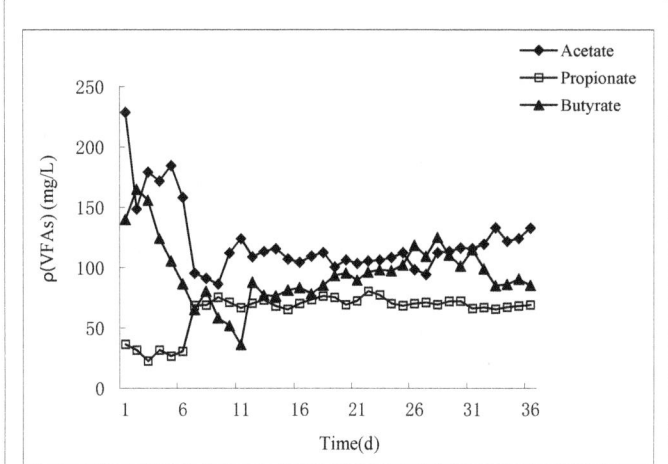

Figure 5: Variation of liquid fermentation products in operation process of UASB reactor.

6.2 to 13.5 L/d. At this point, ethanol and acetate contents in acidogenic phase were 392.7 and 56.1 mg/L, respectively. However, their contents changed into 0 and 95.4 mg/L in methanogenic phase. An explanation for this is that there might have still been a certain amount of acidogenic fermentation bacteria in methanogenic phase, which could take advantage of large amounts of ethanol of acidogenic phase, and ethanol was then oxidized to acetate rapidly. At the same time, methanogens had also begun to continuously use acetate for fermentation in order to produce methane. In the process of OLR gradually increasing from 16 kg/(m³·d) to 24 kg/(m³·d), the methane production yield sustained stable growth. When OLR continued to increase to 32 kg/(m³·d), methane production began to decline and had a relatively large fluctuation. This phenomenon might be caused by methane-producing bacteria, which had a certain impact-resistance to the effluent OLR in acidogenic phase [16]. The maximum tolerance level was 24 kg/(m³·d). Once this limit was exceeded, the activity of methanogenic bacteria for degrading organic matter would be inhibited and methane production would gradually decline.

Variation in liquid fermentation products

Figures 4 and 5 show the variation of liquid fermentation products

in the acidogenic and methanogenic phases. As the influent OLR continued to increase, the total content of liquid fermentation products in acidogenic phase also increased, from 486.4 mg/L at the initial OLR of 12 kg/(m³·d) to 1376.4 mg/L at an OLR 32 kg/(m³·d). When compared to other liquid end products, the upward trend of ethanol was more evident, while other volatile acids fluctuated. The ORP of the CSTR reactor changed between -430 and -320 mV and UASB varied from -460 to -380 mV. In the entire process of operation, the total content of liquid fermentation products in various stages in CSTR represented 486.4′730.5′860.4′975.5′1274.7′1376.4 mg/L while the content of ethanol and acetate were 348.2′499.4′646.1′757.7′935.4 ′1012.5 mg/L, which accounted for 71.6%′68.4%′75.1%′77.7%′73.4% ′73.6% of the total, respectively. It can be suggested that the acidogenic phase was mainly maintaining ethanol-type fermentation.

Compared with the acidogenic phase, there was no ethanol or valerate in the liquid fermentation products of the UASB. When the initial OLR was 12 kg/(m³·d), the acetate content of the methanogenic phase was quite high and the average level was 178.6 mg/L. When the OLR increased to 16 kg/(m³·d), acetate content rapidly decreased to 103 mg/L, subsequently stabilizing at around 110 mg/L. This change might be due to the existence of a large amount of active microorganisms in methanogenic phase [17], which could make full use of ethanol for fermentation in acidogenic phase, resulting in the immediate formation of acetate. With the OLR increasing gradually, microorganisms in methanogenic phase continuously used this acetate for further fermentation. Thus, acetate content would not be too high. The butyrate in methanogenic phase showed a downward trend during the influent OLR of 12 and 16 kg/(m³·d). On the 11th day, the butyrate content reduced to the minimum value of 36.3 mg/L and then leveled out at 100 mg/L with slight fluctuation. This process might be the result of the combined effects of hydrogen-producing acetogens and methanogenus [18]. In addition, since valerate can easily convert to propionate and the conversion rate of propionate is slow, propionate would accumulate to some extent [19]. As a result, the timely adjustment of pH was necessary in order to ensure a higher rate of two-phase anaerobic fermentation of microorganisms.

Variations in COD removal

Figure 6 shows the variation of COD removal for two-phase anaerobic system in operation process. As the figure shows, the maximum COD removal in acidogenic and methanogenic phases was 49.7% and 54.9%, respectively. The total COD removal varied from 60.8% to 72%. When the OLR reached 28 kg/ (m³·d), the total removal was at a maximum, up to 72%. At that time, the energy recovery rate was 712.82 kJ/d (Table 3). An explanation for this is that the anaerobic active micro-organisms had better resistance of high OLR on brown sugar wastewater after being specially domesticated and had better removal effects on COD. In a CSTR-UASB two-phase anaerobic system, the COD removal mainly existed in two areas: Initial organic matter was converted into intermediate products by hydrogen-producing acetogens; intermediate products were further converted into methane by methanogenus [20]. After two stages of degradation, COD removal improved significantly.

Energy recovery rate

Since our two-phase anaerobic system produced a large amount of H_2 and CH_4, the process performance in terms of energy recovery derived from the combination of the two biofuels was calculated according to their combustion heat values. As shown in Table 2, the energy recovery rate tended to increase as OLR increased from 12

to 24 kg/ (m³·d), which is quite obvious because both H_2 and CH_4 production rates increased with increasing OLR. The system achieved the maximum energy recovery rate of 728.67 kJ/d at OLR of 24 kg/ (m³·d), this difference could be attributed to the variation in bacterial population and structure [21]. The optimum rates for energy recovery in the comparable process differed significantly among previous studies (Table 4). The substrate used in this study was more complex compared to other studies. Thus, the CSTR-UASB two-phase anaerobic system may be effective bioreators when applied to energy recovery from brown sugar wastewater. However, further research would be needed to find optimal conditions for higher energy recovery rate. In the future it could be possible to make full use of energy from organic wastewater through CSTR-UASB two-phase anaerobic system.

Conclusions

This experiment used a CSTR-UASB two-phase anaerobic system with artificial brown sugar water as a fermentation substrate to combine the bioenergy recovery and COD removal. As we studied the process of the influent OLR increasing from 12 to 32 kg/(m³·d), results showed that the two-phase anaerobic system operated at OLR=24 kg/ (m³·d) exhibited the best energy recovery rate of 712.82 kJ/d. Meanwhile, the COD removal was up to 69.4%, which meant that the system had good effects on the degradation of brown sugar wastewater as well as energy recovery capacity.

OLR (kg/m³·d)	COD (mg/L)	H₂ production rate (mol/d)	CH₄ production rate (mol/d)	Energy recovery rate (kJ/d)
12	3000	0.15	0.24	228.54
16	4000	0.21	0.63	504.63
20	5000	0.28	0.71	636.47
24	6000	0.33	0.81	728.67
28	7000	0.43	0.76	712.82
32	8000	0.47	0.68	658.42

(A energy recovery rate= H_2 production rate (mol/d) ×242kJ/mol H_2 + CH_4 production rate (mol/d) ×801kJ/mol CH_4)
Table 3: Performance of H_2 and CH_4 production rate as well as energy recovery rate with OLR increasing.

Substrate	Reactor	Culture condition			Maximum energy recovery rate	Reference
	HR	Temperature	pH	COD		
	MR					
Molasses	PBR	35°C	5.5	28g/L	99.62 kJ/L/d	Park et al. [22]
	PBR	35°C	7.0			
Glucose	CSTR	35°C	5.5	3.39g/L	27.77 kJ/L/d	Michael et al. [23]
	CSTR	35°C	7.0			
Cheese whey	CSTR	35°C	N.C	61g/L	201.48 kJ/L/d	Georgia et al. [24]
	PABR	35°C	N.C			
Sucrose	CSTR	35°C	5.2	10~20g/L	189.37 kJ/L/d	Kyazze et al. [25]
	UAF	35°C	N.C			
Olive pulp	CSTR	55°C	N.C	——	9.29 kJ/L/d	Gavala et al. [26]
	CSTR	55°C	N.C			
Food waste	LBR	37°C	N.C	13.1g/L	103.79 kJ/L/d	Han et al. [27]
	UASB	37°C	N.C			
Brown Sugar	CSTR	35°C	5.0	3~8g/L	77.6 kJ/L/d	This study
	UASB	35°C	6.8			

HR: Hydrogenic reactor; MR: Methanogenic reactor; UAF: Up-flow anaerobic filter; LBR: Leaching-bed reactor; PABR: Periodic anaerobic baffled reactor; PBR: Packed-bed reactor; N.C. means Non-control.
Table 4: Comparison of previous performance of two-stage system for energy recovery capacity.

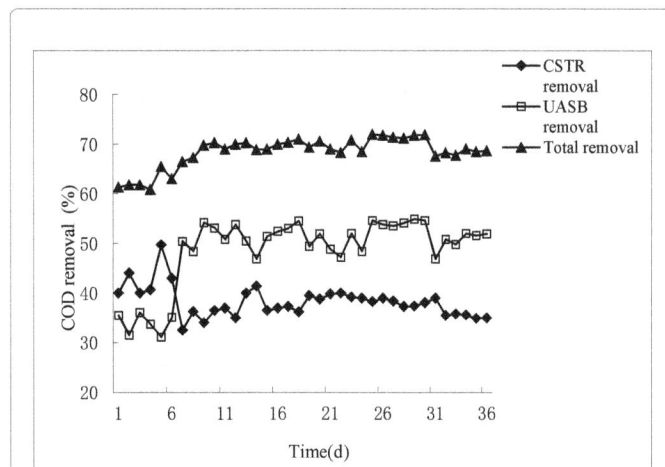

Figure 6: Variation of COD removal in operation process of CSTR-UASB two-phase anaerobic reactor.

Acknowledgements

Financial support from the National Hi-Tech R&D Program (863 Program), Ministry of Science & Technology, China (Grant No.2006AA05Z109) and Shanghai Science and Technology Bureau (Grant No.071605122) and Shanghai Education Committee (Grant No. 07ZZ156) and the Central Special Unversity Funding of Basic Scientific Research (DL09AB06) and GRAP09, Northeast Forestry University are gratefully acknowledged.

References

1. Singh L, Wahid ZA (2015) Enhancement of hydrogen production from palm oil mill effluent via cell immobilisation technique. Int J Energy Res 39: 215-222.

2. Singh L, Wahid ZA (2015) Methods for enhancing bio-hydrogen production from biological process: A review. J Ind Engg Chem 21: 70-80.

3. Pakarinen O, Kaparaju P, Rintala J (2011) The effect of organic loading rate and retention time on hydrogen production from a methanogenic CSTR. Bioresour Technol 102: 8952–8957.

4. Chin HL, Chen ZS, Chou CP (2003) Fedbatch operation using Clostridium acetobutylicum suspension culture as biocatalyst for enhancing hydrogen production. Biotechnol Prog 19: 383–388.

5. Okamoto M, Miyahara T, Mizuno O, Noike T (2000) Biological hydrogen potential of materials characteristic of the organic fraction of municipal solid wastes. Water Sci Technol 41: 25–32.

6. Ren NQ, Wang DY, Yang CP, Wang L, Xu JL, et al. (2010) Selection and isolation of hydrogen-producing fermentative bacteria with high yield and rate and its bioaugmentation process. Int J Hydrogen Energy 35: 2877–2882.

7. Zhu GF, Wu P, Wei QS (2010) Biohydrogen production from purified terephthalic acid (PTA) processing wastewater by anaerobic fermentation using mixed microbial communities. Int J Hydrogen Energy 35: 8350–8356.

8. Hwang MH, Jang N, Hyun SH (2004) Anaerobic bio-hydrogen production from ethanol fermentation: the role of pH. J Biotechnol 111: 297–309.

9. Wang X, Zhao YC (2009) A bench scale study of fermentative hydrogen and methane production from food waste in integrated two-stage process. Int J Hydrogen Energy 34: 245–254.

10. Ueno Y, Tatara M, Fukui H, Makiuchi T, Goto M (2007) Production of hydrogen and methane from organic solid wastes by phase-separation of anaerobic process. Bioresour Technol 98: 1861–1865.

11. Chandra T, Kirschner K, Thuret JY, Pope BD, Ryba T, et al. (2012) Independence of Repressive Histone Marks and Chromatin Compaction during Senescent Heterochromatic Layer Formation. Mol Cell 47: 203-214.

12. Panichnumsin P, Nopharatana A, Ahring B, Chaiprasert P (2012) Enhanced Biomethanation in Co-Digestion of Cassava Pulp and Pig Manure Using A Two-Phase Anaerobic System. J Sustain Energy Environ 3: 73-79.

13. Wang B, Li Y, Wang D, Liu R, Wei Z, et al. (2013) Simultaneous coproduction of hydrogen and methane from sugary wastewater by an "ACSTR H–UASB Met" system. Int J Hydrogen Energy 38: 7774-7779.

14. APHA (1995) Standard Methods for the Examination of Water and Wastewater. (19th edn.) American Public Health Association, Washington, DC.

15. Ren NQ, Chua H, Chan SY (2007) Assessing optimal fermentation type for biohydrogen production in continuous-flow acidogenic reactors. Bioresour Technol 98: 1774–1780.

16. Jin B, Van Leeuwen HJ, Patel B (1998) Utilization of starch processing wastewater for production of microbial biomass protein and fungal a-amylase by Aspergillus oryzae. Bioresour Technol 66: 201–206.

17. Giordano A, Cantù C, Spagni A (2011) Monitoring the biochemical hydrogen and methane potential of the two-stage dark-fermentative process. Bioresourc Technol 102: 4474–4479.

18. Yu HQ, Mu Y, Fang HHP (2004) Thermodynamic analysis on formation of alcohols and high-molecular-weight VFA in acidogenesis of lactose. Biotechnol Bioeng 87: 813–822.

19. Fenton M, Ross RP, McAuliffe O, O'Mahony J, Coffey A (2011) Characterization of the staphylococcal bacteriophage lysine CHAP (K). J Appl Microbiol. 111: 1025–1035.

20. Smith DP, McCarty PL (1989) Reduced product formation following perturbation of ethanol- and propionate-fed methanogenic CSTRs. Biotechnol Bioeng 34: 885–895.

21. Han W, Chen H, Jiao A, Wang Z, Li Y, et al. (2012) Biological fermentative hydrogen and ethanol production using continuous stirred tank reactor. Int J Hydrogen Energy 37: 843–847.

22. Park MJ, Jo LH, Park D, Lee DS, Park JM (2010) Comprehensive study on a two-stage anaerobic digestion process for the sequential production of hydrogen and methane from cost-effective molasses. Int J Hydrogen Energy 35: 6194–6220.

23. Michael C, Nathan M, Christopher C, John B (2007) Two-phase anaerobic digestion for production of hydrogen methane mixtures. Bioresource Technol 98: 2641–2651.

24. Georgia A, Katerina S, Nikolaos V, Michael K, Gerasimos L (2008) Biohydrogen and methane production from cheese whey in a two-stage anaerobic process. Ind Eng Chem Res 47: 5227–5233.

25. Kyazze G, Dinsdale R, Guwy AJ, Hawkes FR, Premier GC, et al. (2007) Performance characteristics of a two-stage dark fermentative system producing hydrogen and methane continuously. Biotechnol Bioeng 97: 759–770.

26. Gavala HN, Skiadas IV, Ahring BK, Lyberatos G (2005) Potential for biohydrogen and methane production from olive pulp. Water Sci Technol 52: 209–215.

27. Han SK, Kim SH, Kim HW, Shin HS (2005) Pilot-scale two-stage process: a combination of acidogenic hydrogenesis and methanogenesis. Water Sci Technol 52: 131–138.

10

Recent Developments in Heat Transfer Fluids Used for Solar Thermal Energy Applications

Umish Srivastva[1*], RK Malhotra[2] and SC Kaushik[3]

[1]*Indian Oil Corporation Limited, RandD Centre, Faridabad, Haryana, India*
[2]*MREI, Faridabad, Haryana, India*
[3]*Indian Institute of Technology Delhi, New Delhi, India*

Abstract

Solar thermal collectors are emerging as a prime mode of harnessing the solar radiations for generation of alternate energy. Heat transfer fluids (HTFs) are employed for transferring and utilizing the solar heat collected via solar thermal energy collectors. Solar thermal collectors are commonly categorized into low temperature collectors, medium temperature collectors and high temperature collectors. Low temperature solar collectors use phase changing refrigerants and water as heat transfer fluids. Degrading water quality in certain geographic locations and high freezing point is hampering its suitability and hence use of water-glycol mixtures as well as water-based nano fluids are gaining momentum in low temperature solar collector applications. Hydrocarbons like propane, pentane and butane are also used as refrigerants in many cases. HTFs used in medium temperature solar collectors include water, water-glycol mixtures – the emerging "green glycol" i.e., trimethylene glycol and also a whole range of naturally occurring hydrocarbon oils in various compositions such as aromatic oils, naphthenic oils and paraffinic oils in their increasing order of operating temperatures. In some cases, semi-synthetic heat transfer oils have also been reported to be used. HTFs for high temperature solar collectors are a high priority area and extensive investigations and developments are occurring globally. In this category, wide range of molecules starting from water in direct steam generation, air, synthetic hydrocarbon oils, nanofluid compositions, molten salts, molten metals, dense suspension of solid silicon carbide particles etc., are being explored and employed. Among these, synthetic hydrocarbon oils are used as a fluid of choice in majority of high temperature solar collector applications while other HTFs are being used with varying degree of experimental maturity and commercial viability – for maximizing their benefits and minimizing their disadvantages. Present paper reviews the recent developments taking place in the area of heat transfer fluids for harnessing solar thermal energy.

Keywords: Solar thermal energy; Solar thermal collectors; Heat transfer fluids

Introduction

Solar energy can be harnessed to generate power by way of either solar photo voltaic route or by concentrating the sun's rays to generate high temperatures and high magnitude of heat (concentrated solar power or CSP). The concentrated heat generated by sun's energy can be transported using suitable heat transfer fluids to heat exchanger where this heat is transferred to convert water into steam and steam turbine is then driven to produce electricity. Out of the two options of solar energy, PV and CSP, CSP based thermal route offers distinct advantages in bringing down the least cost of electricity generated by way of providing storage and dispatch ability of solar power during off-sun periods [1,2]. Solar thermal energy is utilized by capturing the heat of the sun in devices, generally known as solar collectors, designed to maximize the heat absorption through their surfaces exposed to the sun. The heat that is absorbed on the surfaces of such solar collectors is then transferred through a heat transfer media, generally liquid in nature, which takes the collected heat to the point of use. In most of the concentrating solar power plants, sun's heat is captured by a receiver, transferred to a thermo fluid – also known as heat transfer fluid; and this heat from the thermo fluid is then used in a heat exchanger to convert steam from water [3-5]. Solar thermal collectors are defined by the USA Energy Information Administration as low-, medium- , or high-temperature collectors based on their temperatures of operation in the following manner [6].

Low temperature collector

They operate in a temperature range from above ambient to about ~80 °C, used primarily in solar water heating applications, solar based space heating and cooling applications, solar ice making etc. For such low temperature solar thermal applications, following are the heat transfer fluids which are being employed:

Refrigerants/phase changing materials: These are low boiling point but high heat capacity substances used in the solar collectors to transfer heat in applications like solar space cooling and heating, refrigerators, air conditioning, etc [7]. These materials absorb heat from the solar collectors, produce work either by expanding in a turbo-generator of a vapor compression cycle or by dissociating the refrigerant from its absorbent in a vapor absorption cycle [8]. In some of the relatively high temperature applications, higher boiling point refrigerants are used as indirect heat transfer fluids, wherein the heat collected from the solar collectors is transferred to another fluid like water from where the refrigerants pick-up heat to do the required work in the turbo-generators [9,10].

Traditionally, for refrigeration purposes, anhydrous ammonia (R-717) or sulphur dioxides (R-764) were used. However, they were found to have harmful effect when used in domestic application owing to their leakage during use and are thus now mostly used in industrial

*****Corresponding author:** Srivastva U, Indian Oil Corporation Limited, RandD Centre, Faridabad, Haryana, India, E-mail: srivastavau@indianoil.in

applications. For domestic applications, they were replaced by non-toxic, non-flammable, stable, noncorrosive, and non-freezing type's chlorofluorocarbon known as CFC refrigerants most common of them being Freon R-12 [11]. However, since the CFCs have a negative effect on the earth's ozone layer which is the cause of global warming, production of CFC is increasingly being banned in many parts of the world since early 1990s [12].

CFCs have been gradually replaced by the less harmful and comparatively low ozone layer depleting refrigerants' class hydro-chloro-fluoro-carbons known as HCFC such as tetra-fluoro-ethane (R-134a) which is still in use for refrigeration application though there are efforts to phase it out for the same reason [13]. Sometimes refrigerants like HCFC (R123) and HFC (R245fa) are chosen as the working fluids for the Organic Rankine Cycle / Vapor Compression Cycle systems. However, use of these refrigerants are now being restricted with increased environmental awareness. Since, hydrocarbons HCs are comparatively known to be more environmentally friendly, non-toxic, chemically stable and highly soluble in conventional mineral oil, they have also been used in several applications [14]. Owing to better flammability characteristics, hydrocarbons such as butane (R600) and isobutene (R600a) are also chosen as the refrigerants/working fluids for the ORC/VCC system for ice making [15].

Similarly, for solar absorption cooling purposes, refrigerant-absorbent pairs of LiBr-water or water-ammonia are used [16]. Infact, several researchers have used absorption/adsorption cycle as the preferred route of solar cooling and refrigeration. A large variety of working fluids such as activated carbon as adsorbent and ammonia were used, activated carbon/methanol, silica gel/water, olive waste as adsorbent with methanol as adsorbate, composite adsorbents of CaCl2 and BaCl2 developed by the matrix of expanded natural graphite have also been reported to be used [17,18].

Apart from the refrigerants, Phase-Changing-Materials (PCM), in conjunction with solar PV as well as thermal energy, have also been employed in several instances of designing active building architecture through devices known as PV-ST-PCM systems; wherein the heat generated in the solar panels is captured by the PCM which then transfers this heat to a suitable transfer medium such as air or water which can then be utilized for thermal applications in the building such as water heating, space heating, drying processes etc [19]. While PCMs have been used mainly with Building Integrated Photo Voltaics (BIPV) in BIPV-PCM systems at large [20], owing to the poor heat transfer

capability, the innovative PV-ST-PCM systems play the dual role of increasing the PV efficiency on one hand and improving the thermal management of the building space on the other [21]. For storing the latent heat part of solar energy, the choice of suitable PCM to act as heat storage material as well as heat transfer fluid is important as by increasing the thermal storage capacity, the temperature variation can be reduced for long periods of time [22].

There are several materials such as Paraffins, Waxes, Salt Hydrates, Eutectic mixtures of Salt Hydrates etc which were traditionally used as PCMs. The major area of research in PCM integrated with solar energy systems is in finding material and engineering solutions for enhancing the heat transfer capabilities of these systems because while PCMs have exhibited good heat storage ability, their heat dissipation characteristics has not been satisfactory. In their efforts for increasing the heat storage capacity as well as heat transfer capacity of PCMs, incorporation of newer generation materials such as carbon fibres, metallic nano-powders, carbon nano tubes, encapsulated nano-materials, etc are now being increasingly experimented with [23,24].

The important properties such as phase change temperature and heat of fusion of some of the PCMs are listed in Table 1 [25].

Water: Water is the preferred fluid majorly used in low temperature solar collectors for applications such as domestic use, swimming pool heat, solar heating and cooling, etc. Water is abundantly available, is inexpensive and by nature nontoxic. It has high specific heat and is easy to pump owing to having very low viscosity [26]. However, it has a big disadvantage of operation in extreme conditions owing to having comparatively low boiling point (100°C) and high freezing point (0°C). Further, it is easily amenable to lose its neutrality in ph value very soon by picking up contamination thereby causing corrosion hazards and at the same time has a tendency to cause mineral deposits on heat transfer surfaces which reduces its heat transfer capability. The mineral deposits also cause blockages in the piping systems through which water moves within the solar thermal energy collection systems [27].

Water that is used in solar heating application is required to be non-contaminated and clean/treated so as to avoid scaling and mineral deposit formation. Generally, water quality with less than 200 ppm of hardness is used directly in solar water heaters [28,29]. When the incoming water quality is hard, it is recommended to either treat the water by provision of a separate water softener placed at the inlet of the cold tank or an intermediate heat transfer fluid, such as water glycol mixtures in an indirect or a closed loop circuit in the solar water heater [30-33].

Name	Type of Product	Melting Temp. (°C)	Heat of Fusion (kJ/kg)	Source
Astorstat HA17 Astorstat HA18	(Paraffins and Waxes)	21.7-22.8 - 27.2 - 28.3	-	Astor Wax by Honey well (PCM Thermal Solution)
RT26	Paraffin	24 - 26	232	Rubitherm GmbH
RT27		28	206	
Climsel C23	Salt Hydrate	23	148	Climator
Climsel C24		24	108	
STL27	Salt Hydrate	27	213	Mitsubishi Chemicals
S27	Salt Hydrate	27	207	Cristopia
TH29	Salt Hydrate	29	188	TEAP
	Mixture of Two Salt Hydrate	22-25	-	ZAE Bayern
E23	Plus ICE (Mixture of Non-Toxic Eutectic Solution)	23	155	Environmental process system (EPS)

Table 1. Phase Change Temperature and Heat of Fusion of Typical Commercial PCMs

	Water	Ethylene Glycol (% Volume)			Propylene Glycol (% Volume)			Trimethylene Glycol (% Volume)		
		40	50	60	40	50	60	40	50	60
Freezing point (°C)	0	-24	-37	-53	-22	-34	-48	-21	-27	-38
Viscosity (centipoises @ 4.4°C)	1.55	4.8	6.5	9	-	-	-	8.4	11.5	16.9
Specific gravity @ 4.4°C	1	1.07	1.09	1.1	1.03	1.04	1.046	1.04	1.05	1.06
Specific heat (Btu/lb.°F)	1.004	0.84	0.79	0.75	0.89	0.85	0.81	0.87	0.83	0.78
Boiling point (°C)	100	104.4	107.2	111.1	103.9	105.6	107.2	-	-	-
Toxicity	Neutral	More			Less			Least		

Table 2: Comparison of properties of heat transfer fluids used in low-medium temperature solar collectors.

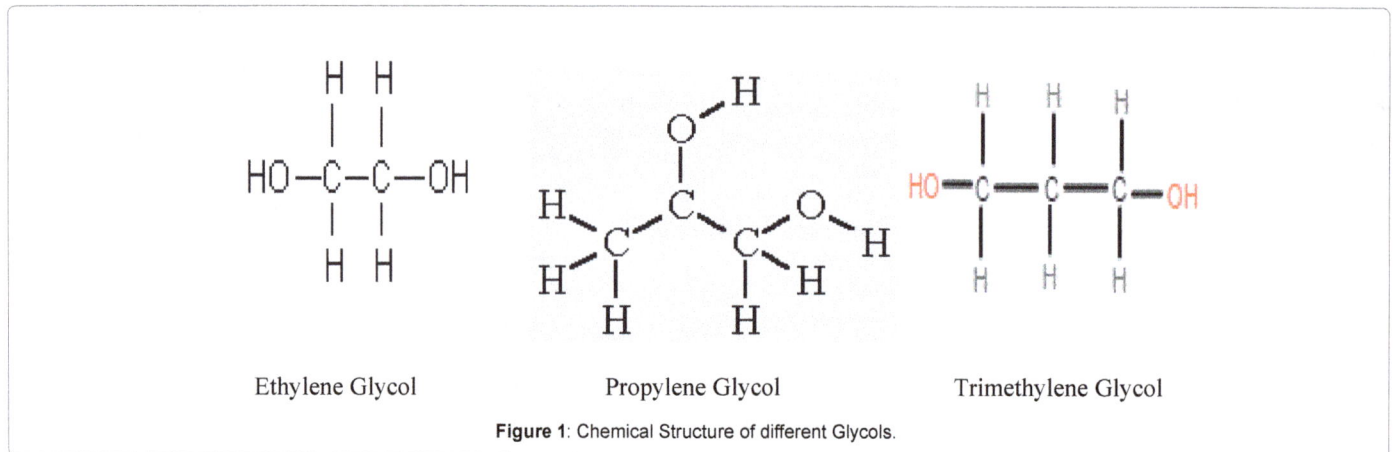

Figure 1: Chemical Structure of different Glycols.

Properties	Aromatics	Naphthenes	Paraffins
Chemical structure	Benzene ring type	Cycloalkane type	Alkane type
Reactivity	Very high	High	Low
Solvency	Very high	High	Low
Oxidation stability	Low	High	Very high
Pour point	Very low	Low	High
Boiling point	Low	High	Very high
Viscosity index	Low	High	Very high
Heat transfer rate	Very high	High	Low
Wax content	Low	High	Very high
Heat transfer operating temperatures	105 – 160°C	150 – 210°C	180 – 280°C

Table 3: Comparison of aromatic, naphthenic and paraffinic heat transfer fluids.

Indirect or closed loop systems use a heat exchanger that separates the potable water from the intermediate fluid, also termed as the "heat-transfer fluid" (HTF), that circulates through the collector. The most common HTF is an antifreeze/water mix that typically uses non-toxic glycol. After being heated in the panels, the HTF travels to the heat exchanger, where its heat is transferred to the potable water. Though slightly more expensive, indirect systems offer freeze protection and typically offer overheat protection as well. Sometimes, water is flown in perpendicular direction through a larger diameter tube where it comes in contact with many small diameter pipes containing heat transfer fluid being heated in the solar heater and indirect heat transfer takes place. This way, periodic operation and maintenance of tube containing harder water becomes easy [34,35,36].

Water-nano fluids: Direct absorption of solar energy in the fluid volume, known as Direct Absorption Solar Collectors (DASC) [37] has been proposed as an effective approach to increase the efficiency of collectors. The concept of DASC was originally proposed by Minardi

andChuang [38] in the 1970s as a simplified design of solar thermal collector and to appreciably enhance the efficiency by absorbing the energy with the fluid volume. They developed a direct collector that absorbed solar radiation by the black fluid (water with 3.0 g/L, Indiaink). However, later on, black fluid like India Ink, as well as other organic and inorganic chromophores have been shown to experience light- and temperature-induced degradation as well as low thermal conductivity [39].

Water-based nanofluids containing carbon nanotubes and stabilized by sodium dodecyl benzene sulfonate (SDBS) as surfactant, have also been experimentally studied. Low nanoparticle volume fraction, ranging from 0.0055% to 0.278%, showed positive effect on density, thermal conductivity and viscosity of nanofluids for temperature range of 20–40°C. Enhancement in density, thermal conductivity and viscosity of nanofluids with volume fraction in nanotubes was observed in comparison to base fluids [40- 42].

The effect of using nanofluids (aluminum/water) was investigated as the working fluid to increase the efficiency of low-temperature DASC [43]. The efficiency enhancement of 10% was obtained using nanofluid-based DASC in comparison with a conventional flat-plate collector. By introducing nanofluids as direct absorber of sunlight in solar collectors, their stability and optical properties as well as their thermal properties improve greatly [44]. However, it has also been reported that the ionic liquids of phosphates are not suitable for applications that require long-term stability at 200°C or more [45,46].

Glycol/water mixtures: Glycol, a synthetic chemical of origin, mixed along with water, often in ratios of 50:50 or 60:40, forms a solution for heat-transfer applications where the temperature in the heat transfer fluid can go below 0°C. Glycols of ethylene or propylene are generally used in heat transfer applications and are generally referred to as "antifreezes". While ethelyne-glycol is a common automotive anti-freeze and is extremely toxic , propylene glycol-water mixture is non-toxic and is used as food-grade anti-freeze [47].

Ethylene glycol-water mixture is traditionally a very commonly used heat transfer fluid in cars and other automobiles, which have to operate in very cold conditions to extremely hot operating temperatures. The properties of glycol-water mixtures varies depending upon the ratios of glycol and water; with the obvious increase in cost component with the increase of glycol content in the mixture [48].

Glycols are generally used in closed loop solar thermal systems using solar collectors that heat the glycol mixture. After getting heated in the collectors, the mixture is pumped in the system and used in a heat exchange to transfer its heat to water kept in the tank. Using glycol-water mixtures helps using the water in low temperature regions for round the clock domestic hot water applications. Glycol-water mixtures are also used for solar water heating applications when the operating temperatures of solar heaters rises beyond the low-temperature-collector region to low-medium-temperature collectors i.e. beyond 100°C [49-51].

Medium temperature collectors

These collectors operate in the temperature range from about 85°C – 270°C, used to supply process heat in various industries such as power, textiles, processing of raw materials etc. Such solar thermal hot water systems — where solar heating is used to provide hot water and low grade steam to process applications — are becoming increasingly common in industrial applications where hot water is at a premium and needed through green initiatives. An important step in optimizing the efficiency of a solar thermal hot water system ion the choice of proper fluid that can assist in transferring heat from the panel mounted on the rooftop down to the hot water heat exchanger.

Water-glycols

At times the temperatures in solar rooftop panels reach above 150°C and so long as the fluid is in circulation, there are no potential dangers at these temperatures. However, sometimes the circulation is interrupted owing to breakdown in power or pumps etc. In such cases the fluid gets stagnated and reaches higher temperatures whereby the glycols can cause a number of operational problems depending upon design of the solar rooftop water heating system. To deal with problems associated with such stagnant fluids, there are certain common designs such as drain back, boil back and allowing the fluid to fill the panel completely [52-56]. In fact, once a glycol based fluid reaches a bulk temperature of above 120°C, its degradation increases drastically making them turn acidic and causing corrosion issues, often needing replacement of the fluids as well as the panel components [57-59].

In order to retard the temperature linked degradation of glycols, corrosion inhibitors are added into them. The corrosion inhibitors are chemicals selected from a group consisting of a water- soluble, alkali metal phosphate, an alkali metal or ammonium hydroxide-neutralized aliphatic monocarboxylic acid having from 8 to 14 carbon atoms and a pH within a range of from 7 to 10; an alkali metal or ammonium hydroxide-neutralized aliphatic dicarboxylic acid having 8 to 14 carbon atoms and a pH within a range of from 7 to 10, an aromatic or substituted aromatic monocarboxylic acid that has from 7 to 14 carbon atoms or its alkali metal salt or ammonium salt, an alkali metal borate, an alkali metal silicate, an alkali metal molybdate, an alkali metal nitrate, and an alkali metal nitrite [60].

The inhibitor – propylene glycol package normally results into a reserve alkalinity of 10 sufficient to protect metals like copper and steel used in solar panel upto a temperature of 120°C. However, when the temperature exceeds 150°C owing to fluid stagnation, the reserve alkalinity is required to be adjusted between 18 to 25 depending upon the glycol being use, so as to minimize the acid formation and corrosion in the system. The following criteria is used to choose the correct glycol-inhibitor package in solar thermal applications:

Thermal stability at temperatures up to 175°C

Non-toxicity

Corrosion protection properties

Optimum reserve alkalinity

Types of glycols: The chemical structures of the three most commonly used glycols for heat transfer applications is given in Figure-1 below:

Ethylene glycol is the most common heat transfer fluid in the glycol family traditionally used automotive applications as well as in solar HVAC industries. Compared to other common types of glycols, it has good thermal and physical properties, is relatively toxic, is not normally used in domestic hot water systems and also degrades much faster at the higher temperature encountered in solar thermal systems [61-63].

Propylene glycol has become the most used antifreeze in some process heat transfer applications and most commercial HVAC systems. It is safe, nontoxic, commonly used in domestic hot water systems, degrades slowly than ethylene glycol but at the same time exhibits lower performance owing to its high viscosity and lower thermal conductivity [64,65].

Another important class of thermally stable glycol, derived from oil or natural gas is trimethylene glycol. Since, it can also be produced from corn sugar, it is known as bio-derived "green" glycol. Compared to propylene glycol manufacturing, there is 30 percent less greenhouse gases in the manufacturing of these green glycols. It has slower thermal degradation rate compared to the conventional glycols, and has superior viscosity at lower temperatures [66].

Table 2 gives a comparison of properties of glycol based heat transfer fluids used in low-medium temperature solar collectors [67].

Propylene glycol has a well-known chemistry and a proven record with process and HVAC systems in the past many years. For solar thermal applications, propylene glycol was most used traditionally but is now being increasingly replaced with the green glycol in developed world countries. Both propylene as well as trimethylene glycols, are

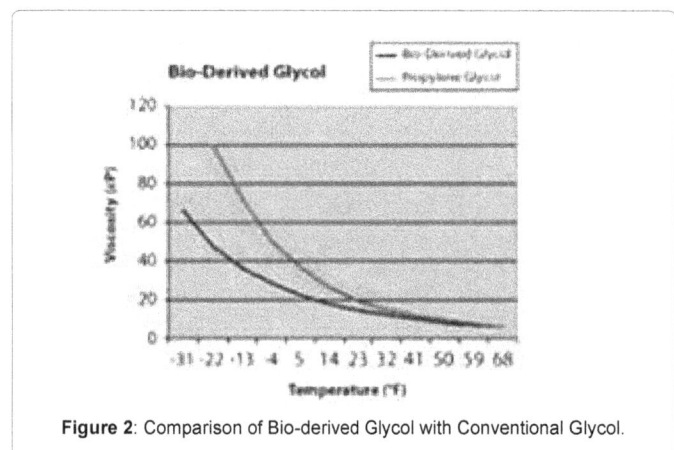

Figure 2: Comparison of Bio-derived Glycol with Conventional Glycol.

	Materials	Thermal conductivity (W/mk)
Metallic Materials	Copper	401
	Silver	429
Non-metallic Materials	Silicon	148
	Alumina (Al2O3)	40
Carbon	Carbon Nano Tubes (CNT)	2000
Base fluids	Water	0.613
	Ethylene glycol (EG)	0.253
	Engine oil (EO)	0.145
Nanofluids (Nanoparticle concentration %)	Water/Al2O3 (1.50)	0.629
	EG/ Al2O3 (3.00)	0.278
	EG-Water/Al2O3 (3.00)	0.382
	Water/TiO2 (0.75)	0.682
	Water/ CuO (1.00)	0.619

Table 4: Thermal conductivity of some materials, base fluids and nanofluids.

safe, non-toxic and are normally inhibited and adjusted for a higher reserve alkalinity to slow degradation. Though, the green glycol provides better thermal and physical properties while being renewable and more environment friendly than propylene glycol, but it has a higher cost. Figure 2 depicts a comparison of bio-derived glycol with the conventional glycol.

Hydrocarbon oils: Hydrocarbon oils can be either of petroleum origin or artificially synthesized in laboratories. As compared to water, the hydrocarbon oils have higher viscosity, lower specific heat and require more energy to pump. These oils are relatively cheap and have a low freezing point.

Most of the heat transfer fluids of petroleum hydrocarbon origin used in industrial applications operate within a temperature range from 120°C to 280°C and their choice for any particular application is governed by the nature of their composition. By this logic, wherever the industrial application uses solar energy for such heat transfers, these hydrocarbons based oils are used as heat transfer fluids. The petroleum hydrocarbon oils can be characterized based on their composition such as aromatic mineral oils, naphthenic oils or paraffinic oils in their respective increasing order of operational temperatures [68].

Though, the aromatic or naphthenic mineral oils are endowed with high heat transfer capacities and lower viscosities, they also have low operating temperatures ranges of say upto 210°C, beyond which these oils degrade very fast. These oils have low pour point and moderate viscosity helping in decreasing the pumping losses and power required for circulation in the system [69].

The paraffinic oils are more stable at low as well as higher temperatures and are thus used for high operating temperatures of about ~280°C despite of not having such good heat transfer properties. All the naturally occurring petroleum based heat transfer fluids are quite toxic by nature and are used in closed loop applications for several years altogether, virtually filled-for-life concept [70].

An indicative comparison of different types of naturally occurring hydrocarbon oils used for heat transfer applications is given in Table 3.

High temperature collectors

High temperature solar collectors are concentrating type high temperatures collectors generating temperatures exceeding 300°C used primarily for industrial processing heat and electricity generation.

These solar collectors use different types of heat transfer fluids ranging from air, water, mineral oils, synthetic oils, molten salts, molten metal's, inorganic minerals, etc and are described below:

Hydrocarbon oils: While there are a few highly refined naturally occurring hydrocarbon oils, mostly paraffinic in nature, which are used in select solar collectors as heat transfer fluids for operating temperatures below 320°C, the high temperature solar collectors majorly uses hydrocarbon oils of synthetic origin. Synthetic fluids are relatively nontoxic and require little maintenance compared to their petroleum counterparts [71].

Synthetic heat transfer fluids are made of glycols, various synthetic fluids such as alkyleated aromatics, terphenyls, and mixtures of bi and diphenyls and their oxides. These are generally of two types - semi-synthetic oils having major constituent as naturally occurring petroleum based minerals oils and some synthetic fatty components added into it as additives to increase it heat transfer abilities or other necessary heat transport properties, - and - fully synthetic heat transfer oils where synthetic components such as eutectic mixtures of biphenyl and diphenyl oxides are most commonly used as heat transfer oils in parabolic trough based concentrated solar power plants round the world [72]. The fully synthetic heat transfer fluids used in parabolic CSP applications have high freezing point and in places where temperature falls below ~12°C. They pose a problem of getting solidified and require additional maintenance to keep them in fluidic conditions [73,74].

Compared to the petroleum based hydrocarbon oils, synthetic oils have excellent heat transfer properties, lower viscosities and high operating temperatures of upto 400°C. Using these synthetic hydrocarbon oils in solar collectors is most convenient in terms of operation and maintenance, but their use is hugely hampered by their maximum operating temperatures of 400°C beyond which these oils deteriorates very fast [75-77].

Researchers are also working to prove the feasibility of utilizing substituted polyaromatic hydrocarbons such as phenylnaphthalenes for solar heat transport application. These hydrocarbon derivatives are expected to retain favorable thermophysical properties and stabilities above 500°C, and can be derived as the byproducts of refining of clean-diesel. Thermophysical properties like low vapor pressure and high resistance to thermal decomposition may make polyaromatic hydrocarbons such as phenylnaphthalenes suitable for heat transfer applications in parabolic solar collectors. Since they are liquids at room temperatures and at the same time can withstand high temperatures of above 500°C, Phenylnaphthalene can potentially replace high-temperature inorganic salts in solar CSP aplications [78].

Nano-fluids: Nano-fluids are a colloidal mixture of base fluid containing nano-sized particles such as oxide ceramics, nitride ceramics, carbide ceramics, metals, semiconductors, carbon nanotubes and composite materials such as alloyed nanoparticles etc and have predominant characteristics because of nanoparticles' small size and high surface area [79]. Research studies on advanced high temperature heat transfer fluids by incorporation of metallic nanoparticles of copper, Al2O3, etc in concentrations of about ~5% by volume into conventional heat transfer fluids have been reported to be significantly improving the thermal transport properties of the HTFs [80,81]. Appreciable increase in thermal conductivity over the base fluids to the tune of about 20% at a 2 vol.% particle loading has been reported. When good dispersion of nanoparticles is obtained, based on the measurement of dynamic viscosity, the nano-fluids behave in a Newtonian manner and the dynamic viscosity increases over the base fluid are minor at

CSP Technology	Operating Pressures (Bar)	Operating Temperatures (°C)	HTF Commonly Used
Trough	15	400	Synthetic hydrocarbon oils
Trough	40	270	Saturated steam
Trough	50-100	400-500	Superheated steam
Tower/Trough	1	500-600	Salts
Tower	1	700-1000	Air
Tower	15	800-900	Air

Table 5: Indicative list of choice of CSP technology versus heat transfer fluids employed.

temperatures of 125°C and above [82-87].

Magnetic nanofluids (MNF) are a special class of nanofluids, exhibiting both magnetic as well as fluid properties. Possibility of using an external magnetic field to control the flow and heat transfer process of the MNF, makes it an interesting choice as a heat transfer medium [88]. Another novel nano-fluid is the carbon nano-tubes mixed nanofluids prepared by proper dispersion and stabilization of the CNT in base fluids (BF) commonly performed using ultrasonication [89]. Significantly higher thermal features of thermal conductivity, convective heat transfer coefficient and boiling critical heat flux compared to the base fluids have been reported n CNT based nano-fluids. Increasing of CNT concentration and temperature further enhances the haet atrsnport properties of CNT based nano-fluids. Increased thermal conductivity of nanofluid in comparison to base fluid by suspending particles is shown in Table 4 below [90]:

Discovered in early 2000's, carbon nano tubes and ionic liquids at room-temperature can be blended to form gels known as "bucky gels of ionic liquids" which has the potential for use in several engineering or chemical processing applications as advanced heat transfer fluids in numerous cooling technologies, heat exchangers, chemical engineering and green energy-based applications such as solar energy. CNT-ionanofluids also exhibit superior thermo-physical and heat transfer properties compared to base ionic liquids. Attractive features of ionanofluids are that they can be designed and fine-tuned through their base ionic liquids so as to obtain desired properties and tasks. These recent research finding on CNT based nanofluids and ionanofluids having ultra-high thermal conductivity, it is being forecasted that these types of fluids are the potential next generation heat transfer fluids [91-93].

Water

Water is being used in some of the high concentrated solar power plants for direct steam generation using linear Fresnel collectors, also in some of the parabolic trough systems as well as in few experimental solar towers. However, its use in CSP application remains in RandD stage and it is mostly used as a preferred heat transfer fluid in low and medium temperature solar collector [94,95].

Direct steam generation (DSG) from water eliminates the need for components like heat transfer fluid, heat exchangers, is non-toxic, have simpler overall plant configuration, allows the solar field to operate at higher temperatures, resulting in higher power cycle efficiencies and lower fluid pumping energies [96,97] Though, DSG still is one of the most promising opportunities for future cost reductions in high concentration solar collectors, it suffers from a number of operational issues such as two-phase flow, higher control requirement, difficult and expensive storage (mostly sensible heat storage), higher temperature gradients, high operating pressures resulting into frequent leakages etc. Compact Linear Fresnel Reflectors (CLFR) were designed to be using only water as its heat transfer fluid in direct steam generation and the trend of DSG is slowly being experimented in Parabolic Solar

Collectors as well as in some of the Solar Towers [98].

Air and compressed gases

Air is one of the most abundantly available substances that can be used as heat transfer fluid. It does not freeze, does not boil and is non-corrosive by nature. However, it has poor heat transport properties in terms of very low heat capacity, poor thermal conductivity and tends to leak out of collectors, ducts and dampers. The system size and engineering requirement for using air as heat transfer fluid is very high and thereby costly in most cases [99].

For very high temperatures such as concentrated solar collectors, there are few experimental projects, where air is being reported to be used directly as heat transfer fluids, especially in concentrated solar towers, where volumetric air receivers are being and the operating temperatures achieved are as high as 550°C. In such high temperature applications, the high cost of engineering systems to handle air as heat transfer fluid can get suitably justified [100].

Simulation methods to comparatively study use of pressurized nitrogen in place of conventional synthetic oil based HTF in parabolic-trough plants using the coordinates of an existing Spanish 50 MWe parabolic-trough plants with 6 h of thermal storage and observed that almost similar net annual electricity productions can happen by replacing the conventional synthetic HTF with pressurized nitrogen which will also be environmentally safe HTF has also been experimented [101].

Another study by theoretical substitution of gas as working fluid in a parabolic trough solar power plant for overcoming flammability and environmental problems associated with conventional synthetic oils was done [102]. The researchers also described a test loop developed at solar research centre of Plataforma Solar de Almería (PSA Spain) for evaluating the effects and technical feasibility of such new concepts in heat transfer oils. They concluded that the high gas pressure can offset pumping power to better acceptable levels but also reported absence of technique to detect the gas leakages from ball joints as a major drawback of using gas in place of oil as HTF.

Molten salts

Salts of sodium or potassium in liquid forms are used as heat transfer fluids in concentrating solar power (CSP) plants as they are able to operate at temperatures well beyond 500°C, improves the power cycle efficiencies to about 40% range, improve the system performance and reduce the Levelized Electricity Cost (LEC). Molten salts are more user friendly and environmentally benign than oils, are easily available in nature and are thus less expensive. However, salts suffer from certain major disadvantages in that they have a high freezing point of about 250°C, which requires lot of auxiliary power during shut-off periods to keep them in molten state and is thus a huge maintenance problem. Further, owing to their corrosive nature, they require expensive operational machineries and system engineering in terms of pipings, pumps, joints etc for their efficient operation [103- 110].

Molten salts have been used as heat transfer fluids in solar towers in a number of concentrated power plants around the world. Their use in solar towers has been quite established though at the same time it makes the electricity generated from such solar towers expensive. Further, since solar plants are operated intermittently, keeping the salts in molten state is a big engineering and economic challenge. However, salts offer an added advantage of being used as a thermal energy storage (TES) medium as well [111-116].

Molten metals

Some of the known metals such as liquid sodium used as heat transfer medium in nuclear industry have also been proposed to be used as a heat transfer fluid for solar application for high temperatures beyond 550°C [117,118]. Liquid sodium has a fairly low melting point, very high boiling point, very high thermal conductivity and fairly high heat capacity. However, sodium as metal as well as liquid sodium, is extremely reactive with water and this makes its use in solar application difficult [119].

Another variant, a eutectic sodium-potassium alloy NaK, has been reported to be used to overcome this problem. Nak has very lower melting point of -12°C which makes it suitable to be used for all practical purpose in CSP applications [120]. Though, NaK has boiling point lower than sodium, lower specific heat capacity and inferior heat transport properties than sodium, owing to the fact that it has subzero melting point, it has been considered as an excellent high temperature heat transfer fluid. Eutectic NaK has a composition of 77.8% potassium by weight and 22.2% sodium by weight and has -12°C melting point. The NaK mixture which has 46% potassium can have a melting point of 20°C and has higher thermal conductivity and specific heat capacity than eutectic NaK. But the inherent danger of reactivity with water remains with NaK as well and thus restricts its use in solar CSP applications. It has been experimentally tried in an experimental solar tower in Spain and there was a huge fire reported causing lot of damage to the system [121]. Another eutectic mixture, Lead-Bismuth (44.5-55.5%) having very high boiling point of above 1500°C, owing to its inherent properties of being non-reactive with water and air, is also being applied in R and D stage in CSP application. Similarly, binary eutectic mixtures of liquid metals like Cadmium-Bismuth, Tin-Bismuth, Bismuth-Zinc, Calcium-Copper etc are other combinations which are being studied under combinatorial material chemistry and synthesis for their usefulness in CSP applications with respect to safety of use, corrosion, pump ability, operation and maintenance issues etc [122].

The choice of heat transfer fluid for high temperature solar applications depend on a number of important factors such as the concentrated solar power technology being employed, the operating temperatures and pressures in the system; all of which governs the cost and performance of the electricity generation. Table 5 below gives an indicative list of the same:

Several authors have carried out comparison studies of some of the in-use as well as promising novel heat transfer fluids including the compatibility studies of these heat transfer fluids with the metallurgies of the system. A comparison of the various properties of some commonly used thermo fluids, solar salts and oil used in high temperature solar collectors is given in Table 6 below [123].

Fluidized solid particles

A very new and novel concept of high temperature resistant HTF in the form of a dense suspension of solid silicon carbide particles in mean diameter of ~ 64 micron range approximating to 30-40% by weight and remaining being air 60-70%; has been recently reported from studies carried out by a European consortium [124]. The dense suspension is circulated upward in a specially designed vertical absorbing tubes located inside a single cavity created in a receiver of alkaline-earth silicate exposed to highly concentrated solar flux reaching a temperature as high as 750°C. The solid particles of silicon carbide has heat capacity as good as the conventional synthetic oil based HTFs but can withstand high temperature without degradation. Heat transfer coefficients as high as 500 W/m²K has been reported to be achieved in this experimental study at very low mean particle velocities in the range of 2.5 cm/s. It has been concluded in this study that the particle volume fraction in the suspension and suspension velocity are the important influencing factors for achieving good heat transfer rates but the same are also some of the engineering challenges to be overcome to make this concept workable on a larger and commercial basis [125].

Conclusions

Solar thermal energy is a technology for harnessing solar energy for thermal applications such as heating-cooling, hot water, industrial heat, power generation etc. Solar thermal collectors are defined by the USA Energy Information Administration as low-, medium-, or high-temperature collectors. The heat from the sun falls on the solar collector which is coated with special solar selective coatings so as to absorb the maximum solar radiations' heat. The heat absorbed on the collector surfaces is then transferred to a heat transfer fluid, generally known as thermo-fluids, diligent choice of which plays a very vital role in determining the overall efficiency of solar energy utilization. The choice of heat transfer fluids for a given solar energy collector depends primarily on the working temperature of the fluid.

In the low temperature solar collectors where traditionally refrigerants like HCFC (R123) and HFC (R245fa) were used rather

	Unit	Sodium salt	Sodium potassium salt	Potassium salt	Salt Hitec XL	Salt Hitec	Salt grade Hitec Solar salt	Solar grade HTF (oil)
Melting point	°C	97.82	-12.6	63.2	120	142	240	15
Boiling point or maximum operating temperature	°C	881.4	785	756.5	500	538	567 (bp 593)	400
Density	Kg/m³	820	749	715	1640	1762	1794	1056
Specific heat capacity	KJ/KG.K	1.256	0.937	0.782	1.9	1.56	1.214	2.5
Viscosity	Pa.s	0.00015	0.00018	0.00017	0.0063	0.003	0.0022	0.0002
Thermal conductivity	W/m.K	119.3	26.2	30.7	Na	0.363	0.536	0.093
Prandtl number		0.0016	0.0063	0.0043		12.89	4.98	5.38

Table 6: Comparison of the various properties of commonly used thermo fluids used in high temperature solar collectors.

extensively, hydrocarbons such as butane (R600) and isobutene (R600a) known to be more environment friendly, non-toxic, chemically stable and highly soluble in conventional mineral oil, have been increasingly used in several applications. For solar cooling and refrigeration purposes, several researchers have reported increased use of absorption/adsorption cycle as the preferred route. Another new research interest has been seen in utilizing PCMs in conjunction with solar energy systems such as PV-ST-PCM wherein the heat of PV panels is absorbed in PCMs which then transfers this heat for solar thermal applications like building space heating, water heating etc. For slightly higher temperature range for water heating applications, direct water, water-ethylene glycol mixtures and water based nano-fluids are increasingly finding place.

The use of medium temperature solar collectors though has extremely large potential as industrial and commercial heat source is rather relatively less explored and applied. For temperatures beyond 100 °C and up to about 150°C, water-glycol mixtures are used wherein propylene-glycols are increasingly replacing traditional ethylene glycol. Another new class of glycol which is more thermally stable and environmentally safe, trimethylene glycol, manufactured by way of bio-processing, is also reportedly being utilized. For the higher side of medium temperature solar collectors, hydrocarbon oils, mostly aromatic or naphthenic in nature are used. Sometimes, semi-synthetic oils, mixture of mineral based hydrocarbons and chemicals are also used for certain high end applications.

High temperature solar collectors are mostly used for power generation and allied uses. These collectors use more stable paraffinic hydrocarbon oils to some extent and eutectic mixture of synthetic compounds, in most cases for upto 400°C. For temperature ranges beyond 400°C, molten salts - mixture of sodium and potassium salts in varying proportions have been used. This category of solar collectors is poised to grow very fast and its growth is majorly hampered by the choice of a suitable thermo-fluid that can withstand higher temperatures for extended period of time. Realizing the important of efficiency increase with every 10° rise of working temperature of the thermo fluids, extensive research is being carried out in synthesizing suitable chemicals that can withstand temperatures beyond 400°C and so far fluid up to 470°C has been reportedly developed in a US lab. To achieve higher system efficiencies, direct steam generation technologies are also being adopted. In few cases, direct heating of air is also being reported to achieve very high temperatures of above 800°C.

For high temperatures solar collectors, another route being adopted by researchers is that of nano-fluids, which are a colloidal mixture of base fluid containing nano-sized particles such as oxide ceramics, nitride ceramics, carbide ceramics, metals, semiconductors, carbon nanotubes and composite materials such as alloyed nanoparticles etc and have predominant characteristics because of nanoparticles' small size and high surface area. Magnetic nanofluids (MNF) as a special class of nanofluids exhibiting both magnetic and fluid properties with a possibility of controlling flow and heat transfer process via an external magnetic field are also being studied. All most all the researchers have reported nano-fluids to exhibit significantly higher thermal features such as thermal conductivity, convective heat transfer coefficient and boiling critical heat flux. Known metals such as liquid sodium or a eutectic mixture of sodium-potassium alloy NaK, that can be used in molten state as thermo fluids, is also reported being used as heat transfer fluid for solar application for high temperatures beyond 550°C. Dense suspension of solid silicon carbide particles also are reported to be used experimentally for high temperatures CSP application such as

solar tower, owing to high temperature stability and heat capacity.

For concentrated solar power applications, a very major need at present is for a workable and user friendly heat transfer fluid that can be operated and maintained easily at low temperatures and is environmentally benign. The available CSP technologies of parabolic troughs are capable of reaching temperatures beyond 400°C but such temperatures are not achieved as the heat transfer fluid, which is synthetic in nature, though very convenient to use becomes prone to thermal degradation beyond 400°C. On the other hand, the molten salts that are used in solar towers can sustain higher temperatures of 650°C or so, becomes very inconvenient to operate and maintain below 250°C which is quite a high freezing point for any working fluid. Hence, any incremental developments of heat transfer fluid with operating temperatures beyond 400°C and ease of maintenance can be a path breaking innovation in CSP technology, per se, and has the potential to place concentrated solar power amongst the most bankable and reliable renewable energy option.

With the available state-of-the-art technologies in heat transfer fluids for concentrated solar power (CSP) plants, which is the solar energy technology of immense importance and need in today's world for generating reliable and despatchable renewable energy, an ideal heat transfer fluid is one which operates as a Thermally Stable, Easily Pump able, having Negligible Vapour Pressure, Fully Compatible, Non-Corrosive, Single Phase Liquid at all operating temperatures having some of the important physical properties as given in Table 6 below.

Acknowledgement

The author would like to acknowledge with thanks the management of Indian Oil Corporation Limited, Research and Development Centre, Faridabad, India and also authorities at Indian Institute of Technology, Delhi, India, for their kind permission to carry out the above study.

References

1. Robert F, Ghassemi M, Cota A (2009) Solar Energy: renewable energy & the environment. CRC Press, Taylor & Francis Group, New York NY, USA.

2. Price H, Lüpfert E, Kearney D, Zarza E, Cohen G, et al. (2002) Advances in parabolic trough solar power technology. J Solar Energy Engg Transactions of the ASME 124 : 109-125.

3. Reddy VS, Kaushik SC, Ranjan KR, Tyagi RK (2013) State-of-the-art of solar thermal power plants—A review. Renew Sustain Energy Rev 27: 258-273.

4. Mekhilefa S, Saidurb R, Safari A (2011) A review on solar energy use in industries. Renew Sustain Energy Rev 15: 1777–1790.

5. Thirugnanasambandam M, Iniyan S, Goic R (2010) A review of solar thermal technologies. Renew Sustain Energy Rev 14: 312–322.

6. http://www.eia.gov/tools/glossary

7. Ziyan HZA, Ahmed MF, Metwally MN, Abd El-Hameed HM (1997) Solar assisted R22 and R134a heat pump systems for low temperature applications. Appl Thermal Engg 17: 455–469.

8. Afshar O, Saidur R, Hasanuzzaman M, Jameel M (2012) A review of thermodynamics and heat transfer in solar refrigeration system. Renew Sustain Energy Rev 16: 5639-5648.

9. El Fader A, Mimet A, Pe´rez-Garcıa M (2009) Modelling and performance study of a continuous adsorption refrigeration system driven by parabolic trough solar collector. Solar Energy 83: 850–861.

10. Florides GA, Tassou SA, Kalogirou SA, Wrobel LC (2002) Review of solar and low energy cooling technologies for buildings. Renew Sustain Energy Rev 6: 557–572.

11. 11. Hassan HZ, Mohamad AA (2012) A review on solar cold production through absorption technology. Renew Sustain Energy Rev 16: 5331-5348.

12. Esen M (2004) Thermal performance of a solar cooker integrated vacuum-

tube collector with heat pipes containing different refrigerants. Solar Energy 76: 751–757.

13. Bolaji BO, Huan Z (2013) Ozone depletion and global warming:Case for the use of natural refrigerant – a review. Renew Sustain Energy Rev18: 49–54.

14. Bu XB, Li HS, Wang LB (2013) Performance analysis and working fluids selection of solar powered organic Rankine-vapor compression ice maker. Solar Energy 95: 271–278.

15. Kenisarin MM (2014) Thermophysical properties of some organic phase change materials for latent heat storage. A review. Solar Energy 107: 553–575.

16. Ullah KR, Saidur R, Ping HW, Akikur RK, Shuvo NH (2013) A review of solar thermal refrigeration and cooling methods. Renew Sustain Energy Rev 24: 499–513.

17. Siddiqui MU, Said SAM (2015) A review of solar powered absorption systems. Renew Sustain Energy Rev 42: 93–115.

18. Berdja M, Abbada B, Yahia F, Bouzefoura F, Oualia M (2014) Design and realization of a solar adsorption refrigeration machine powered by solar energy. Energy Procedia 48: 1226 – 1235.

19. Ma T, Yang H, Zhang Y, Lu L, Wang X (2015) Using phase change materials in photovoltaic systems for thermal regulation and electrical efficiency improvement: A review and outlook. Renew Sustain Energy Rev 43:1273 – 1284.

20. Zhao X (2011) Research Proposal: Developing a novel BIPV façade module enabling enhanced thermal/electrical generation and supply for buildings by using PCM slurry. BR–EeB.

21. Aelenei L, Pereira R, Gonçalves H (2013) BIPV/T versus BIPV/T-PCM: a numerical investigation of advanced system integrated into Solar XXI building façade. 2nd international sustainable energy storage conference. Dublin, Ireland.

22. Pasupathy A, Athanasius L, Velraj R, Seeniraj RV (2008) Experimental investigation and numerical simulation analysis on the thermal performance of a building roof incorporating phase change material (PCM) for thermal management. Appl Thermal Engg 28: 556–565.

23. Benedict LX, Louie SG, Cohen ML (1996) Heat capacity of carbon nanotubes. Solid State Commun 100: 177–180.

24. Kibria MA, Anisur MR, Mahfuz MH, Saidur R, Metselaar IHSC (2015) A review on thermophysical properties of nanoparticle dispersed phase change materials. Energy Convers Manag 95: 69–89.

25. Pasupathy A, Velraj R (2006) Phase Change Material Based Thermal Storage for Energy Conservation in Building Architecture. Int Energy J 7: 147-159.

26. Fanney AH, Klein SA (1988) Thermal performance comparisons for solar hot water system subjected to various collector and heat exchanger flow rates. Solar Energy 40: 1-11.

27. Shah LJ, Furbo S (1998) Correlation of experimental and theoretical heat transfer in mantle tanks used in low flow SDHW systems. Solar Energy 64: 245–256.

28. Islam MR, Sumathy K, Khan SU (2013) Solar water heating systems and their market trends. Renew Sustain Energy Rev 17: 1–25.

29. Hossain MS, Saidura R, Fayazb H, Rahimb NA, Islama MR, et al. (2011) Review on solar water heater collector and thermal energy performance of circulating pipe. Renew Sustain Energy Rev 15: 3801– 3812.

30. Sadhishkumar S, Balusamy T (2014) Performance improvement in solar water heating systems—A review. Renew Sustain Energy Rev 37: 191-198.

31. Lenel UR, Mudd PR (1984) A review of materials for solar heating systems for domestic hot water. Solar Energy 32: 109-120.

32. Liu M, Bruno F, Saman W (2011) Thermal performance analysis of a flat slab phase change thermal storage unit with liquid-based heat transfer fluid for cooling applications. Solar Energy 85: 3017–3027.

33. Gao Y, Zhang Q, Fan R, Lin X, Yu Y (2013) Effects of thermal mass and flow rate on forced-circulation solar hot-water system: Comparison of water-in-glass and U-pipe evacuated-tube solar collectors. Solar Energy 98: 290–301.

34. Srinivas M (2011) Domestic solar hot water systems: Developments, evaluations and essentials for —viability‖ with a special reference to India. Renew Sustain Energy Rev 15: 3850-3861.

35. Ayompe LM, Duffy A (2013) Thermal performance analysis of a solar water heating system with heat pipe evacuated tube collector using data from a field trial. Solar Energy 90: 17–28.

36. Jaisankar S, Radhakrishnan TK, Sheeba KN (2009) Experimental studies on heat transfer and friction factor characteristics of forced circulation solar water heater system fitted with helical twisted tapes. Solar Energy 83: 1943–1952.

37. Otanicar T, Phelan PE, Prasher RS, Rosengarten G, Taylor RA (2010) Nanofluid-Based Direct Absorption Solar. J Renew Sustain Energy 2: 033102.

38. Minardi JE, Chuang HN (1975) Performance of a black liquid flat-plate solar collector. Solar Energy 17: 179–183.

39. Kameya Y, Hanamura K (2011) Enhancement of solar radiation absorption using nanoparticle suspension. Solar Energy 85: 299–307.

40. Halelfadl S, Maré T, Estellé P (2014) Efficiency of carbon nanotubes water based nanofluids as coolants. Experimental Thermal and Fluid Sci. 53: 104–110.

41. Saeedinia M, Akhavan-Behabadi MA, Razi P (2012) Thermal and rheological characteristics of CuO-base oil nanofluid flow inside a circular tube. Int Commun Heat Mass Trans 39: 152-159.

42. Mohammed HA, Bhaskarana G, Shuaiba NH, Saidur R (2011) Heat transfer and fluid flow characteristics in microchannels heat exchanger using nanofluids: A review. Renew Sustain Energy Rev 15: 1502–1512.

43. Hordy N, Rabilloud D, Meunier JL, Coulombe S (2014) High temperature and long-term stability of carbon nanotube nanofluids for direct absorption solar thermal collectors. Solar Energy 105: 82–90.

44. Karami M, Bahabadi MAA, Delfani S, Ghozatloo A (2014) A new application of carbon nano tubes nano fluid as working fluid of low-temperature direct absorption solar collector. Solar Energy Materials Solar Cells 121: 114–118.

45. Chandrasekar M, Suresh S, Bose AC (2010) Experimental investigations and theoretical determination of thermal conductivity and viscosity of Al2O3/water nanofluid. Experiment Thermal Fluid Sci 34: 210-216.

46. Trisaksri V, Wongwises S (2007) Critical review of heat transfer characteristics of nanofluids. Renew Sustain Energy Rev 11: 512–523.

47. Shojaeizadeh E, Veysi F, Yousefi T, Davodi F (2014) An experimental investigation on the efficiency of a Flat-plate solar collector with binary working fluid: A case study of propylene glycol (PG)–water. Experimental Thermal Fluid Sci 53: 218–226.

48. Shukla R, Sumathy K, Erickson P, Gong G (2013) Recent advances in the solar water heating systems: A review. Renew Sustain Energy Rev 19: 173-190.

49. Hossain MS, Saidur R, Faizul M, Sabri M, Said Z, et al. (2015) Spot light on available optical properties and models of nano fluids: A review. Renew Sustain Energy Rev 43: 750–762.

50. Tyagi VV, Kaushik SC, Tyagi SK (2012) Advancement in solar photovoltaic/ thermal (PV/T) hybrid collector technology. Renew Sustain Energy Rev16: 1383– 1398.

51. Wang Z, Yang W, Qiu F, Zhang X, Zhao (2015) Solar water heating: From theory, application, marketing and research. Renew Sustain Energy Rev 41: 68-84.

52. Streicher W (2000) Minimizing the risk of water hammer and other problems at the beginning of stagnation of solar thermal plants – a theoretical approach. Solar Energy 69: 187–196.

53. Vieira ME, Duarte POO, Buarque HLB (2000) Determination of the void fraction and drift velocity in a two-phase flow with a boiling solar collector. Solar Energy 69: 315–319.

54. Cassard H, Denholm P, Ong S (2011) Technical and economic performance of residential solar water heating in the United States. Renew Sustain Energy Rev 15: 3789–3800.

55. Pierrick H, Christophe M, Leon G, Patrick D (2015) Dynamic numerical model of a high efficiency PV–T collector integrated into a domestic hot water system. Solar Energy 111: 68–81.

56. Hobbi A, Siddiqui K (2009) Optimal design of a forced circulation solar water heating system for a residential unit in cold climate using TRNSYS. Solar Energy 83: 700–714.

57. Fan J, Shah LJ, Furbo S (2007) Flow distribution in a solar collector panel with

horizontally inclined absorber strips. Solar Energy 81: 1501–1511.

58. Kong W, Wang Z, Fan J, Bacher P, Perers B, et al. (2012) An improved dynamic test method for solar collectors. Solar Energy 86: 1838–1848.

59. Lauterbach C, Schmitt B, Vajen K (2014) System analysis of a low-temperature solar process heat system. Solar Energy 101: 117–130.

60. Connor K, Cuthbert J (2010) Corrosion-inhibited propyleneglycol/glycerin compositions. WO 2010008951 A1: Dow Global Technologies Inc.

61. Patil PG (1975) Field performance and operation of a flat-glass solar heat collector. Solar Energy 17: 111-117.

62. Martin RLS (1975) Experimental performance of three solar collectors. Solar Energy 17: 345-349.

63. Vargas JVC, Ordonez JC, Dilay E, Parise JAR (2009) Modeling, simulation and optimization of a solar collector driven water heating and absorption cooling plant. Solar Energy 83: 1232–1244.

64. Norton B, Edmonds JEJ (1991) Aqueous propylene-glycol concentrations for the freeze protection of thermo siphon solar energy water heaters. Solar Energy 47: 375-382.

65. Too YCS, Morrison GL, Behnia M (2009) Performance of solar water heaters with narrow mantle heat exchangers. Solar Energy 83: 350–362.

66. http://www.dynalene.com/Articles.asp?ID=283

67. http://www.engineeringtoolbox.com/ethylene-glycol-d_146.html

68. Singh J (1985) Heat transfer fluids and systems for process and energy applications. Marcel Dekker, Inc, New York, USA.

69. Selvakumar P, Somasundaram P, Thangavel P (2014) Performance study on evacuated tube solar collector using therminol D-12 as heat transfer fluid coupled with parabolic trough. Energy Convers Manag 85: 505–510.

70. Bignon MJ (1980) The Influence of the Heat Transfer Fluid on the Receiver Design. Electric Power Sys Res 3: 99 – 109.

71. Oyekunle LO, Susu AA (2005) Characteristic properties of a locally produced paraffinic oil and its suitability as a heat-transfer fluid. Petroleum Sci Technol 22:1499-1509.

72. Coastal Chemical, Hitec solar salt, Product information.

73. Montes MJ, Aba´nades A, Martı´nez-Val JM, Valde´s M (2009) Solar multiple optimization for a solar-only thermal power plant, using oil as heat transfer fluid in the parabolic trough collectors. Solar Energy 83: 2165–2176.

74. Cau G, Cocco D (2014) Comparison of medium-size concentrating solar power plants based on parabolic trough and linear Fresnel collectors. Energy Procedia 45: 101 – 110.

75. Zhai R, Yang Y, Yan Q, Zhu Y (2013) Modeling and characteristic analysis of a solar parabolic trough system: thermal oil as the heat transfer fluid. J Renew Energy: 389514.

76. Wu B, Reddy RG, Rogers RD (2001) Novel ionic liquid thermal storage for solar thermal electric power systems. Novel ionic liquid thermal storage for solar thermal electric power systems, Washington, DC, USA: 445-451.

77. Dow chemical company Dowtherm A - Synthetic organic heat transfer fluid. Product information

78. McFarlane J, Bell JR, Felde DK, Joseph III RA, Qualls AL, et al. (2013) Phenylnaphthalene as a heat transfer fluid for concentrating solar power: Loop Test & Final Report. Energy Transport Sci Div ORNL/TM-2013/26.

79. Javadi FS, Saidur R, Kamalisarvestani M (2013) Investigating performance improvement of solar collectors by using nanofluids. Renew Sustain Energy Rev 28: 232-245.

80. Paul G, Chopkar M, Manna I, Das PK (2010) Techniques for measuring the thermal conductivity of nanofluids: A review. Renew Sustain Energy Rev 14: 1913-1924.

81. Sarkar J (2011) A critical review on convective heat transfer correlations of nanofluids. Renew Sustain Energy Rev 15: 3271– 3277.

82. Sokhansefat T, Kasaeian AB, Kowsary F (2014) Heat transfer enhancement in parabolic trough collector tube using Al2O3/synthetic oil nanofluid. Renew Sustain Energy Rev 33: 636–644.

83. Singh D, Elena TV, Moravek MR, Cingarapu S, Wenhua Y, et al. (2014) Use of metallic nanoparticles to improve the thermophysical properties of organic heat transfer fluids used in concentrated solar power. Solar Energy 105: 468–478.

84. Blake DML, Hale MJ, Price H, Kearney D, Herrmann U (2002) New heat transfer and storage fluids for parabolic trough solar thermal electric plants. New heat transfer and storage fluids for parabolic trough solar thermal electric plants, Zurich, Switzerland.

85. Kotzé JP, von Backström TW, Erens PJ (2011) A Combined Latent Thermal Energy Storage and Steam Generator Concept Using Metallic Phase Change Materials and Metallic Heat Transfer Fluids for Concentrated Solar Power. SolarPACES, Granada.

86. Bergman TL (2009) Effect of reduced specific heats of nanofluids on single phase, laminar internal forced convection. Int J Heat Mass Trans 52: 1240-1244.

87. Laing D, Bahl C, Bauer T, Lehmann D, Steinmann WD (2011) Thermal energy storage for direct steam generation. Solar Energy; 2011; 85(4); pp 627-633.

88. Nkurikiyimfura I, Wanga Y, Pan Z (2013) Heat transfer enhancement by magnetic nanofluids — A review. Renew Sustain Energy Rev 21: 548–561.

89. Sundar LS, Sharma KV, Naik MT, Singh MK (2013) Empirical and theoretical correlations on viscosity of nanofluids: A review. Renew Sustain Energy Rev 25: 670–686.

90. Gupta HK, Agrawal GD, Mathur J (2012) An overview of Nanofluids: A new media towards green environment. Int J Environ Sci 3: 433-440.

91. Godson L, Raja B, Lal DM, Wongwises S (2010) Enhancement of heat transfer using nanofluids—An overview. Renew Sustain Energy Rev 14: 629-641.

92. Murshed SMS, Nieto de Castro CA (2014) Superior thermal features of carbon nanotubes-based nanofluids – A review. Renew Sustain Energy Rev 37: 155-167.

93. Murshed SMS, Nieto de Castro CA, Lourenco MJV, Lopes MLM, Santos FJV (2011) A review of boiling and convective heat transfer with nanofluids. Renew Sustain Energy Rev 15: 2342–2354.

94. Feldhoff JF, Schmitz K, Eck M, Laumann LS, Laing D, et al. (2012) Comparative system analysis of direct steam generation and synthetic oil parabolic trough power plants with integrated thermal storage. Solar Energy 86: 520–530.

95. Odeh SD, Morrison GL, Behnia M (1998) Modeling of parabolic trough direct steam generation solar collectors. Solar Energy 62: 395–406.

96. Phelan P, Otanicar T, Taylor R, Tyagi H (2013) Trends and opportunities in direct-absorption solar thermal collectors. J Thermal Sci Engg Appl 5: 021003-1- 021003-9.

97. Rovira A, Montes MJ, Varela F, Gil M (2013) Comparison of Heat Transfer Fluid and Direct Steam Generation technologies for Integrated Solar Combined Cycles. Appl Thermal Engg 52: 264-274.

98. Gharbia NE, Derbal H, Bouaichaouia S, Saida N (2011) A comparative study between parabolic trough collector and linear Fresnel reflector technologies. Energy Procedia 6: 565–572.

99. Ho CK, Iverson BD (2014) Review of high-temperature central receiver designs for concentrating solar power. Renew Sustain Energy Rev 29: 835-846.

100. Barlev D, Vidu R, Stroeve P (2011) Innovation in concentrated solar power. Solar Energy Materials Solar Cells 95: 2703–2725.

101. Biencinto M, González L, Zarza E, Díez LE, Antón JM (2014) Performance model and annual yield comparison of parabolic-trough solar thermal power plants with either nitrogen or synthetic oil as heat transfer fluid. Energy Convers Manag 87: 238–249.

102. Anton JM, Biencinto M, Zarza E, Díez LE (2014) Theoretical basis and experimental facility for parabolic trough collectors at high temperature using gas as heat transfer fluid. Appl Energy 135: 373–381.

103. Fernández AI, Solé A, Paloma JG, Martínez M, Hadjieva M, et al. (2014) Unconventional experimental technologies used for phase change materials (PCM) characterization: part 2 – morphological and structural characterization, physico-chemical stability and mechanical properties. Renew Sustain Energy Rev.

104. Boerema N, Morrison G, Taylor R, Rosengarten G (2012) Liquid sodium versus Hitec as a heat transfer fluid in solar thermal central receiver systems. Solar Energy 86: 2293–2305.

105. Ren N, Wu YT, Ma CF, Sang LX (2014) Preparation and thermal properties of quaternary mixed nitrate with low melting point. Solar Energy Materials Solar Cells 127: 6–13.

106. Grena R, Tarquini P (2011) Solar linear Fresnel collector using molten nitrates as heat transfer fluid. Energy 36: 1048-1056.

107. Bradshaw RW, Cordaro JG, Siegel NP (2009) Molten nitrate salt development for thermal energy storage in parabolic trough solar power systems. Molten nitrate salt development for thermal energy storage in parabolic trough solar power systems, San Francisco, CA, USA 2: 615-624.

108. Valkenburg MEV, Vaughn RL, Williams M, Wilkes JS (2005) Thermochemistry of ionic liquid heat-transfer fluids. Thermochimica Acta 425: 181-188.

109. Chandrasekar M, Suresh S, Senthilkumar T (2012) Mechanisms proposed through experimental investigations on thermophysical properties and forced convective heat transfer characteristics of various nanofluids – A review. Renew Sustain Energy Rev16: 3917-3938.

110. Baharoon DA, Rahman HA, Omar WZW, Fadhl SO (2015) Historical development of concentrating solar power technologies to generate clean electricity efficiently – A review. Renew Sustain Energy Rev 41: 996-1027.

111. Kosmulski M, Gustafsson J, Rosenholm JB (2004) Thermal stability of low temperature ionic liquids revisited. Thermochimica Acta 412: 47-53.

112. Solé A, Miró L, Barreneche C, Martorell I, Cabeza LF (2013) Review of the T-history method to determine thermophysical properties of phase change materials (PCM). Renew Sustain Energy Rev 26: 425-436.

113. Yang X, Yang X, Ding J, Shao Y, Fan H (2012) Numerical simulation study on the heat transfer characteristics of the tube receiver of the solar thermal power tower. Appl Energy 90: 142–147.

114. Kearney D, Herrmann U, Nava P, Kelly B, Mahoney R, et al. (2003) Assessment of a molten salt heat transfer fluid in a parabolic trough solar field. J Solar Energy Engg 125: 170-176.

115. Kearney D, Kelly B, Herrmann U, Cable R, Pacheco J, et al. (2004) Engineering aspects of a molten salt heat transfer fluid in a trough solar field. Energy 29: 861-870.

116. Frazera D, Stergarb E, Cioneaa C, Hosemanna P (2014) Liquid metal as a heat transport fluid for thermal solar power applications. Energy Procedia 49: 627 – 636.

117. Pacio j, Wetzel TH (2013) Assessment of liquid metal technology status and research paths for their use as efficient heat transfer fluids in solar central receiver systems. Solar Energy 93: 11–22.

118. Kotzé JP, von Backström TW, Erens PJ (2011) A Combined Latent Thermal Energy Storage and Steam Generator Concept Using Metallic Phase Change Materials and Metallic Heat Transfer Fluids for Concentrated Solar Power. SolarPACES

119. Skumanich A (2010) CSP: Developments in heat transfer and storage materials. Renew Energy Focus 11: 40-43.

120. http://blogs.sun.ac.za/sterg/files/2012/10/Kotze-HTF.pdf

121. Vignarooban K, Xu X, Arvay A, Hsu K, Kannan AM (2015) Heat Transfer Fluids for Concentrated Solar Power Systems – A Review. Appl Energy 146: 383-396.

122. Becker M (1980) Comparison of heat transfer fluids for use in solar thermal power stations. Electric Power Sys Res 3: 139 – 150.

123. Flamant G, Gauthier D, Benoit H, Sans JL, Garcia R (2013) Dense suspension of solid particles as a new heat transfer fluid for concentrated solar thermal plants: On-sun proof of concept. Chem Engg Sci 102: 567–576.

124. Flamant G, Gauthier D, Benoit H, Sans JL, Boissière B, et al. (2014) A new heat transfer fluid for concentrating solar systems: Particle flow in tubes. Energy Procedia 49: 617 – 626.

11

Effect of Process Parameters on Yield and Conversion of *Jatropha* Biodiesel in a Batch Reactor

Nassereldeen Ahmed Kabbashi*, Nurudeen Ishola Mohammed, Md Zahangir Alam and Mohammed Elwathig S Mirghani

Bioenvironmental Engineering Research Centre (BERC), Department of Biotechnology Engineering, Faculty of Engineering, International Islamic University Malaysia

Abstract

In a quest for environmental friendly energy source with least pollutants emission due to issues of global warming coupled with dwindling reserve of the fossil fuel, researchers have intensified study on renewable fuels. Among these renewable energy sources, biodiesel stands prominent. Biodiesel production is largely by transesterification of transglycerides of fatty acids almost always in a batch reactor. Of importance in the yield generation and fatty acid methyl esters conversion is the feedstock purity, control of reagents use in production and operation parameters alteration. This is geared towards achieving optimum resource conservation while also minimizing cost and materials wastage. In this study biodiesel was produced from hydrolysate (free fatty acids from hydrolyzed *Jatropha curcas* oil) using calcinated niobic acid catalyst at controlled rates of process parameters. Yield and conversion up to 97.7% and 100% respectively of the alkyl esters produced. This informs the influence of process parameters significantly on the throughput of the final product.

Keywords: Biodiesel; Hydrolysate.

Introduction

Biodiesel is a fuel of organic origin consisting of long chain fatty acids. The application of this fuel in diesel engine offers environmental benefits when compared with fossil fuel. Feed stocks of vegetable oil (Virgin and waste) and animal fats have been explored for production of biodiesel [1]. Other renewable resources from which biodiesel had been produced is algae and yellow lard [2]. Biodiesel is an environmentally friendly alternative liquid fuel that can be used in any diesel engine with little or no engine restructuring. Interest in organic oils consideration for biodiesel production has been kindled on account of its less polluting nature and its renewable source when compared with fossil diesel fuel [3]. Biodegradable fuels like biodiesels have an expanding range of potential applications as they are less environmental polluting. Therefore, there is growing interest in degradable diesel fuels that degrade more rapidly than petroleum fuels [4]. Biodiesel beckons increasing consideration from institutions and individual desire home brew biodiesel production.

The production of biodiesel from oil origin has utilized feed stocks from available biomass in the world over. For instance, in USA, the combined vegetable oil and animal fat production amounts to about 35.3 billion pounds per year [5]. Similarly, Brazilian production has witnessed predominant increase with about 2 million tons/year in 2009 utilizing about 60 plant oils [6]. Moreover, as at 2002, of the world production of biodiesel about 84% was from rapeseed oil, 13% from sunflower while palm oil and soybean biodiesel both record 1% [7] subject to the availability of these seed oil and the energy needs in the various locations of the world.

Transesterification primarily is the main catalyzed chemical reaction between triglycerides and alcohol to afford mono-esters [8]. The long and branched chain triglyceride molecules are transformed to monoesters and glycerin [9]. In transesterification, triglycerides are converted in sequential order to diglycerides and monoglycerides in reversible reactions. However, this approach is feasible if the FFA content is within the allowable limit. High FFA from hydrolyzed oil requires a prior esterification in biodiesel production.

The main vegetable oil materials of no food usage for biodiesel production are plant species such as *Jatropha* or *ratanjyote* or *seemaikattamankku* (*Jatropha curcas*), karanja or honge (*Pongamia pinnata*), nagchampa (*Calophyllum inophyllum*), rubber seed tree (*Hevca brasiliensis*), neem (*Azadirachta indica*), mahua (*Madhuca indica and Madhuca longifolia*), silk cotton tree (*Ceiba pentandra*), jojoba (*Simmondsia chinensis*), babassu tree, *Euphorbia tirucalli*, microalgae, *etc*. Their accessibility and easy growth benefits in various location and climates of the world as well as their relative affordability place them an edge over majority of oil food [10].

The non-food oils materials such as *Jatropha*, microalgae, neem, karanja, rubber seed, mahua, silk cotton tree, *etc*. are easily available in developing countries and support the economy of the developing nations commensurately when compared to oils used for food materials [11]. The oils from neem (*Azardirachta indica*) and rubber (*Hevea brasiliensis*) have high free fatty acid (FFA) content. The presence of FFAs in any feedstock materials put the material to danger of soap formation when an alkaline catalyst is use in the process. Moreover, there is also constraint of biodiesel-glycerol separation in the downstream process.

Jatropha curcas oil is a potential feed stock for biodiesel production. Apart from palm oil and algae no other biodiesel feedstock is capable of producing high yield of biodiesel as *Jatropha curcas* [12]. And as such, the food-energy feud of palm oil and the complexity involve in algae-oil generation among other constraints has intensify the improved interest in *Jatropha curcas* for consideration as feedstock material in biodiesel production.

The seed oil of *Jatropha* was utilized as a diesel fuel alternative

---

*Corresponding author: Kabbashi NA, Bioenvironmental Engineering Research Centre, Department of Biotechnology Engineering, Faculty of Engineering, International Islamic University Malaysia, Malaysia, E-mail: nasreldin@iium.edu.my

during World War II and can as well serve as blends with conventional diesel [13,14]. Hence, the use of *Jatropha curcas* and *Pongamia pinnata (Karanja)* indicate a more suitable feedstock for synthesis of renewable fuel such as biodiesel [15,16]. The oil content of *Jatropha* and Karanja are very high reaching up to 30-50% [12,13].

Biodiesel is economically feasible in majority of oilseed-producing regions of the world [17]. Biodiesel is a technologically feasible substitute to petro-diesel, but biodiesel selling price doubled that of fossil diesel in most advance countries which is in part dependent on control measures of the reagent used in the production [17]. Although, biodiesel is still currently produced in relatively small scale compared to fossil fuel, present market price is not competitive. Hence, biodiesel at the present economic situation does not satisfactorily rivaled petro-diesel [17].

Commercially, biodiesel production is essentially same depending on whichever route is adopted in production. The process involves reaction of oil and fat with alcohol in the presence of a catalyst. The catalysts utilized over time have ranged from alkaline based catalyst (NaOH and KOH) which were used for commercial production. Others are heterogeneous base catalyst, acid based homogenous and heterogeneous catalyst as well enzyme catalyst [11].

Apart from the feedstock constituting the highest cost of biodiesel synthesis amounting to about 80% of the operating cost [18], the amount of other materials used in the process which depends largely on the control of process parameters such as temperature, catalyst dose, reaction time and molar ratio of alcohol used [19]. It is thus imperative that the amount or concentration of these parameters were determined in order to avoid wastage while also ensuring sufficient amount for optimum conversion and yield generation. This paper presents the finding of effect of process parameters on FAME yield and conversion using fluffy niobic acid calcined at 150°C.

Materials and Method

Calcination and characterization of powdered acid catalyst

The powdered niobic acid ($Nb_2O_5.H_2O$) catalyst utilized in this study was calcined in an indigenous "Iso Temp-220" furnace for duration of 4 hrs at 150°C. The white fluffy niobic acid powder was subsequently stored in the desiccator until needed for use. Characterization of the calcined catalyst was done using Fourier Transformed Infrared Spectroscopic (FT-IR) analysis.

Batch esterification reaction

The batch esterification reaction was carried out in 250ml screwed-cap shake flask and the content of the flask was made to react by monitoring the operating parameters in an incubator shaker (INFORs AG CH-4103 BOTTMINGEN). 16gm *Jatropha curcas* hydrolysate (FFAs produced by enzymatic hydrolysis of crude *Jatropha curcas* oil) was put in the flask with a catalyst loading varied between (1.0-5.0 wt% relative to the FFA) while variation of the methanol to oil molar ratio was between 3:1 to 7:1. The basis of methanol to oil ratio selection was relative to the molar weight of 819 g/mol of the oil. The reaction temperature was varied between 45-65°C and agitation rate ranges between 100-500 rpm to investigate the influence of the various tested parameters. The reaction was also monitored starting from 3-7 hrs to test the effect of time on the yield and conversion of the FFA.

After completion of every experimental run, the effluent was centrifuge in a (Rotina 38 Zentrifugen D-78532 Tuttlingen). The oil and unreacted methanol phase were decanted into a separating funnel and was left overnight to separate the oil and excess methanol. The final biodiesel product was incubated in an oven for 3-4 hrs to eliminate the moisture generated during the process. Each experiment was carried out in successive triplicate and estimation of biodiesel yield was based on the weight of biodiesel produced and weight of the FFA used in the reaction while the FFA conversion was due to the amount of initial FFA and the residual FFA as measure of the acid value [3] respectively.

The standard error of ±1-2% was obtained from the estimation of both the yield and conversion for the set of the experimental triplicates.

$$Yield\ (\%) = \frac{weight\ of\ biodiesel}{weight\ of\ FFA} \times 100 \qquad (1)$$

While the % conversion was determined by estimating the acid value after the reaction and was calculated by the equation below

$$C = \left(1 - \frac{Av_t}{Av_0}\right) \qquad (2)$$

Where C is the FFA conversion, Av_t=final acid value and Av_0= initial acid value before esterification reaction.

Results and Discussion

Fourier Transform Infra-Red Spectroscopy (FT-IR) is use for elucidation of structures of absorption, emission, photoconductivity or Raman scattering of substance [20]. It shows the absorption peaks of sample which corresponds to frequencies of variations between the bonds of the atoms of the material's constituents. The FT-IR of the powdered niobic acid catalyst after calcination is presented in Figure 1 below. The spectra of the calcined solid acid has broad band at 3500 cm^{-1} due to the O-H stretching mode of hexagonal groups [21]. Moreover, peak observed between 2850 cm^{-1} and 2900 cm^{-1} is attributed to asymmetric stretching of CH_3 group. The small peak observed close to 2400 cm^{-1} is attributed to C=O stretching vibrations of ketone, aldehyde, lactone and carbonyl group [22]. Peak observed at 1800 cm^{-1} also informs a C=O stretching vibrations of ketone and aldehyde groups relative to attached NH_2 groups.

Batch esterification reaction

Temperature effect: Study on the effect of temperature variation on the esterification of *Jatropha curcas* hydrolysate is as shown in Figure 2a. At the commencement of the reaction, higher rate was predominant relative to concentration of free fatty acids. Almarales et al. [23] reported a FAME conversion at 30 minutes reaction time at high temperature of 200°C. In this study, optimum FAME conversion and yield was favoured up to 60°C at 15% relative to the initial yield in the reaction. Higher temperature above 60°C lowers FAME yield in comparison with result of FAME conversion at 60°C. Comparable result of study carried out by Marchetti et al. [24] and Patil and Deng [25] show temperature effect to have positive contribution on fatty acids esters conversion up till 60°C which is in agreement with finding from the study. Thus, suffice to conclude that temperature above 60°C does not favour conversion of the FFA as boiling point of methanol is gradually reached.

Effect of oil-alcohol molar ratio: The effect of molar ratio of methanol to oil on FFA conversion and yield was investigated from 3:1 to 7:1. Figure 2b shows the influence of methanol on conversion of FFA and yield. Methanol-oil ratio of 5:1 offered the best combination for optimum conversion and yield. This finding is comparable with that reported by Almarales et al. [23] in hydroesterification of *Nannochloropsis oculata* microalgae' biomass to biodiesel using Al_2O_3 supported Nb_2O_5 catalyst. Increasing the methanol ratio further resulted in reduced conversion. Ramadhas et al. [8] reported that acid

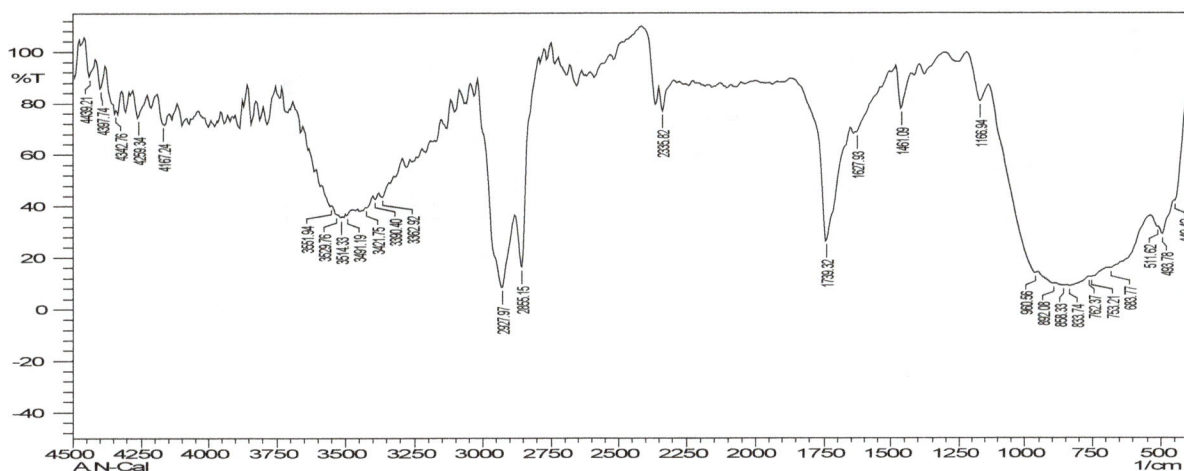

Figure 1: Ft-IR analysis of calcined powdered catalyst.

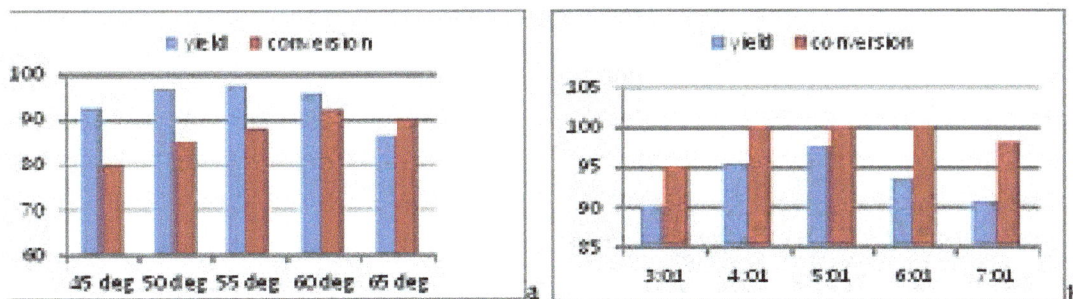

Figure 2a and 2b: Temperature and molar ratio effects on yield and conversion of FAME from FFA in %.

esterification reduces high FFA oil with high amount of methanol while also occurring at longer reaction time. Contrary to finding in this study lower methanol t oil ratio produced higher yield and conversion reaching 100% at 4:1 methanol to FFA ratio.

Effect of mixing rate: Agitation rate of 400rpm produced the effective conversion; nevertheless less magnitude was recorded at further increase on the agitation rate for the yield of the biodiesel and its FAME conversion (Figure 3a). This may be attributed to shift of reaction towards the reverse reaction. Agitation significantly affects the reaction rate; insufficient mixing could inhibit the reaction rate, which could lower the product formation.

Conversely, higher agitation impacts negatively on the reaction and higher energy is also expended while lower throughput is recorded. Thus the yield of biodiesel above 400rpm agitation rate was observed to be lowered as the evaporation of methanol causes insufficient contact with the substrate.

Effect of catalyst dosage: The powdered niobic acid catalyst dose was used to investigate FAME conversion and yield. The considered variation (1-5 wt%), optimum biodiesel yield was recorded at 4 wt% of the catalyst while the conversion continue to increase with 5 wt% of catalyst above 90% (Figure 3b). Carvalho et al. [26] reported that catalyst with acidic sites such as niobium oxide produce good conversion in esterification process. The catalyst is thermally stable and diffusion problem is reduced. Almarales et al. [23] obtained 92.24% conversion using Nb_2O_5 (Al_2O_3) and 87.43% using pure Nb_2O_5 at higher catalyst

loading of 20% each. In this study a higher yield and conversion was achieved with lower catalyst dose. Comparable result was achieved by Umdu and Erol [27] in transesterification of *Nannochloropsis oculata* microalgae lipid to FAME with Al_2O_3 supported on CaO and MgO Catalyst with result such as that recorded in this study.

Reaction time effect: In this reaction step, conversion of the *Jatropha curcas* hydrolysate from to methyl esters was investigated over 7 hrs periods. The yield was proportional with the time increment while FAME conversion attained optimum value of 96% at 6 hrs reaction time. Figure 4 presents the effect of reaction time of FAME conversion and yield.

Srilatha et al. [20] studied the effect of reaction time on esterification of palmitic acid and sunflower fatty acid using heteropoly tungstate supported on niobia catalyst. Similar to the findings in this study FFA conversion was observed to increase with time. However lower reaction time (1-2 hrs) was achieved compared to 6 hrs optimum recorded in this study. This may be due to pure niobic acid used in the study compared to the heteropoly tungstate supported on niobia.

Conclusion

In the esterification reaction study to convert *Jatropha curcas* hydrolysate from hydrolyzed *Jatropha curcas* oil to biodiesel, we showed that parameters influence on yield and conversion rate were estimated. Optimum yield and conversion was determined to be 96% and 100% respectively at 5:1 methanol to FFA ratio, agitation of

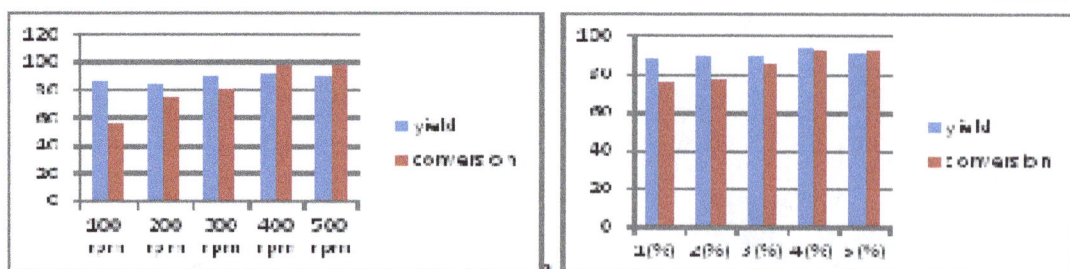

Figure 3a and 3b: Effect of mixing rate and catalyt dosage on FAME conversion and yield from FFA in %.

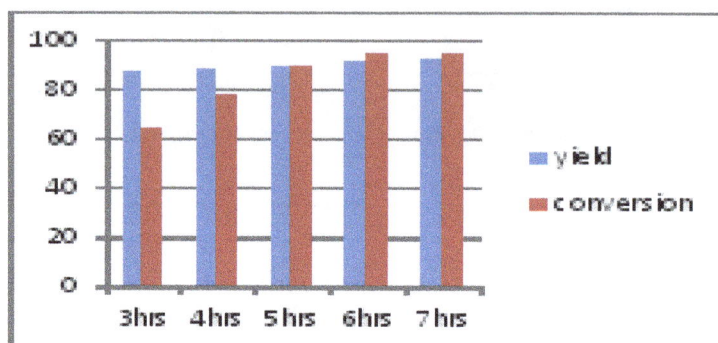

Figure 4: Effect of reaction time of FAME conversion and yield from FFA in %.

400rpm, catalyst loading of 4 wt%, temperature of 60°C and reaction time of 6 hrs. All the control parameters significantly affect FAME conversion while yield generation was less affected in comparison with the lowest control parameters from commencement of the reactions. It is thus sufficient to conclude that the control of the parameter best affect the conversion more than the yield as the FAME Conversion was observed to continually increase rapidly while the yield recorded lower or insignificant increase as the reactions proceeded.

References

1. Hess MA, Hass MJ, Foglia TA (2007) Attempts to reduce NO exhaust emissions by using reformulated biodiesel. Fuel Process Technol 88: 693-699.

2. Moser BR (2009) Biodiesel production, properties and feedstocks. Invitro cell Dev Biol-Plant 45: 229-266.

3. Mohammed NI, Kabbashi NA, Alam Md Z, Mirghani ME (2014) *Jatropha curcas* oil characterization and its significance for feedstock selection in biodiesel production. IPCBEE 65: 57-62.

4. Ma F, Hanna MA (1999) Biodiesel production: A review. Bioresour Technol 70: 1-15.

5. Pearl GG (2002) Animal fat potential for bioenergy use. Bioenergy The Biennial Bioenergy Conference, Boise, USA.

6. Aranda DAG, da Silva CCC, Detoni C (2009) Current processes in Brazilian biodiesel production. Int rev chem Engg 1: 603-608.

7. Korbitz W (2002) New trends in developing biodiesel world-wide. Asia Bio-Fuels: Evaluating and Exploiting the Commercial Uses of Ethanol. Fuel Alcohol and Biodiesel, Singapore.

8. Ramadhas AS, Jayaraj S, Muraleedharan C (2005) Biodiesel Production from high FFA rubber seed oil. Fuel 84: 335-340.

9. Ikwagwu OE, Onorugbu IC, Njoku OU (2000) Production of Biodiesel using rubber [*Hevea brasiliensis*]. Seed oil. Ind Crops Prod 12: 57-62.

10. Karmee SK, Chadha A (2005) Preparation of biodiesel from crude oil of Pongamia pinnata. Bioresour Technol 96: 1425-1429.

11. Demirbas A (2008) Biodiesel: 'A realistic Fuel Alternative for Diesel Engine. Springer Publisher, UK.

12. Azam MM, Waris A, Nahar NM (2005) Prospects and potential of fatty acid methyl esters of some non-traditional seed oils for use as biodiesel in India. Biomass Bioenergy. 29: 293-302.

13. Foidl N, Foidl G, Sanchez M, Mittelbach M, Hackel S (1996) *Jatropha curcas* L. As a source for the production of biofuel in nicaragua. Bioresour Technol 58: 77-82.

14. Gubitz GM, Mittelbach M, Trabi M (1999) Exploitation of the tropical seed plant *Jatropha Curcas* L. Bioresour Technol 67: 73-82.

15. Meher LC, Saga DV, Naik SN (2006a) Technical aspects of biodiesel production by transesterification-a review. Renew Sust Engerg Rev 10: 248-268.

16. Meher LC, Dharmagadda VSS, Naik SN (2006b) Optimization of alkali-catalyzed transesterification of Pongamia pinnata oil for production of biodiesel. Bioresour Technol 97:1392-1397.

17. Pahola TBenavides Urmila Diwekar (2012) Optimal control of biodiesel production in a batch reactor Part I: Deterministic control. Fuel 94: 211-217.

18. Demirbas A (2002a) Biodiesel from vegetable oils via transesterification in supercritical methanol. Energy Convers Manag 43: 2349-2356.

19. Srilatha K, Lingaiah N, Prabhavathi Devi BLA, Prasad RBN, Venkateswar S, et al. (2009) Esterification of free fatty acids for biodiesel production over heteropoly tungstate supported on niobia catalysts. Appl Catalysis 365: 28-33.

20. Ramani K, Boopathy R, Mandal AB, Sekaran G (2012) Preparation of acidic lipase immobilized surface-modified mesoporous activated carbon catalyst and thereof for hydrolysis of lipids. Catalysis commun 14: 82-88.

21. 22. Kennedy LJ, Mohan das K, Sekaran G (2004) Integrated biological and catalytic oxidation of organics/inorganics in tannery wastewater by rice husk based mesoporous   activated carbon-Bacillus sp. Carbon 42: 2399-2407.

22. Almarales A, Chenard G, Abdala R, Aranda DGA, Reyes Y, Tapanes NO (2012) Hydroesterification of *Nannochloropsisoculata* microalga's biomass to biodiesel on Al$_2$O$_3$ supported Nb$_2$O$_5$ catalyst. Natural Sci 4: 204-210.

23. Marchetti JM, Miguel VU, Errazu AF (2007) Heterogeneous esterification of oil with high amount of free fatty acids. Fuel 86: 906-910.

24. Patil PD, Deng S (2009) Optimization of biodiesel production from edible and non-edible vegetable oils. Fuel 88: 1302-1306.

25. Carvalho L, Britto P, Matovanelli R, Camacho L, Antunes OA (2005) Esterification of the fatty acid of palm by heterogeneous catalysis. 13th Brazilian Congress of Catalysis and 3rd Mercosur Congress on Catalysis, 4: 1-4.

26. Umdu MT, Erol S (2009) Transesterification of *Nannochloropsis oculata* microalga's lipid to biodiesel on Al_2O_3 supported CaO and MgO catalysts. Bioresourc Technol 10: 2828-2831.

12

Characterization of Aluminum Doped Zinc Oxide (Azo) Thin Films Prepared by Reactive Thermal Evaporation for Solar Cell Applications

Mugwang'a FK[1]*, Karimi PK[2], Njoroge WK[2] and Omayio O[2]

[1]*Department of Physics, Pawni University, Kenya*
[2]*Department of Physics, Kenyatta University, Kenya*

Abstract

Aluminium doped Zinc Oxide (AZO) thin films have been deposited using reactive thermal evaporation technique using an Edward Auto 306 Magnetron Sputtering System. Transmittance and reflectance data in the range 300 nm-2500 nm were obtained using UV-VIS NIR Spectrophotometer Solid State 3700 DUV for all the thin films samples that were prepared. Transmittance values of above 70% were observed. The optical measurements were simulated using SCOUT 98 software to determine optical constants and optical bad gap of the thin film. The optical properties in these films were varied by varying Aluminums doping percentages. It was observed that the transmission over the visible range decreased as the concentration of Aluminum increased. This is due to free carriers coupling to the electric field hence increasing the reflection. Optical band gap for various samples of Aluminum doped thin films show a direct allowed transition and a shift in the optical absorption edge as the Aluminums concentration increased. These results show values of band gap ranging between 3.2 eV and 3.5 eV. Between 0% - 3% the optical band gap reduces. This is followed with widening of the band gap for doping between 4%- 6%. Urbach energy gradually increased with increasing band gap. The band gap reduced due to formation of localized states near the conduction band corresponding to increase in Urbach energy.

Keywords: Aluminums doped Zinc Oxide (AZO); Thin films; Optical properties; Reactive thermal evaporation technique; Solar cell applications

Introduction

Zinc Oxide (ZnO) films have become technologically important due to their range of electrical and optical properties, together with their high chemical and mechanical stabilities, which make them suitable for a variety of applications such as flat panel display electrodes and gas sensors. Moreover, these films can be used as surface acoustic wave devices, because of their large piezoelectric constant, and also as solar cells, since their optical band gap (3.3 eV) is wide enough to transmit most of the useful solar radiation [1]. ZnO is an n-type semiconductor and its conductivity can be controlled by thermal treatment or by adequate doping [2]. The doping of ZnO films with the group III elements can increase the conductivity of the films. In comparison with other elements, Aluminum and Gallium are the best dopants because their ionic radii are similar to that of Zn^{2+} [3,4].

Many techniques have been used for fabricating ZnO films, such as chemical vapour deposition, pulsed laser deposition, dc reactive sputtering, spray pyrolysis and the sol–gel process [3]. The structural, optical, and electrical properties of ZnO and ZnO:Al films prepared by thermally evaporating zinc acetate and $AlCl_3$ in vacuum have been investigated in detail, together with the effects of heat treatment in air and vacuum. The properties of the deposited ZnO and ZnO:Al films depend on the deposition parameters such as substrate temperature, evaporation rate of zinc acetate and Aluminum concentration [5].

The Al-doped ZnO film exhibits remarkable electrical conductivity, together with high charge carrier density and mobility [6]. The ZnO doped with Al^{3+} is used extensively for photo-electronic devices [6], Theoretically, spatially organized ZnO doped with Al^{3+} could result in improving electrical properties. For this reason, only few studies of the conduction mechanism in heavily Al-doped ZnO films have been reported. Slightly doping was explained by a limited incorporation of Al into the ZnO lattice, and Aluminum acts as a donor [6].

Selmi et al., [1] studied deposition time and its effects on the properties of ZnO:Al films. It is shown that films grow with the hexagonal c-axis perpendicular to the substrate surface. The morphological characteristics show a granular and homogenous surface and the cristallinity of the films are enhanced with increased deposition time. The deposited films show good optical transmittance (80%–90%) in the visible and near infrared spectrum.

Transparent Conductive Oxides (TCO) films are degenerate wide band gap semiconductors with low resistance and high transparency in the visible range. For these reasons these materials, are widely used in optoelectronic applications such as flat panel displays, solar cells and electro chromatic devices. Usually, the TCO films are n-type semiconductors such as Indium Tin Oxide (ITO), Tin Oxide (SnO_2) and Zinc Oxide (ZnO), whereas ITO film is the one most used in these devices up to now. Recently, Al doped ZnO (ZnO:Al) film is one of the materials which could replace the ITO films. The direct optical band gap of ITO films is generally greater than 3.75 eV although a range of values from 3.5 to 4.06 eV have also been reported in the literature [7].

This research study uses ZnO:Al. It is widely used because the films have electrical and optical properties similar to those of ITO, and because it is stable in a hydrogen atmosphere. Thin ZnO films should be doped by aluminum, since it has been remarked that extrinsic donors due to the dopant atom are more stable than intrinsic donors due to the native defects. The electrical conductivity in ZnO:Al film is higher due to the Al^{3+} ions in substitutional sites of the Zn^{2+} ions and the Aluminum interstitial atoms, in addition to Oxygen vacancies and Zinc interstitials [4]. The Bond enthalpy of Zn-O is 159 ± 4 kJ/mol while that of Al-O is 511 ± 3 kJ/mol [4].

*Corresponding author: Mugwanga KF, Albert-Ludwigs-Universitat Freiburg, Kenya, E-mail: mugwanga.mugwanga@gmail.com

Experimental Techniques

Deposition of Aluminum doped zinc oxide (ZnO:Al)

Zinc (99.9% purity) and Aluminum (99.99% purity) were mixed at varying doping percentages of Aluminum (0-6%) and then heated in closed glass tube until they melted to form a compound. Glass substrates were cleaned to remove stains on them by boiling in dilute chromic acid to remove surface contaminants and rinsing thoroughly with distilled water and ethanol and allowed to dry completely. The substrate was then mounted on a rotating substrate holder and the compound was then placed in a Molybdenum boat. The chamber was covered tightly and pumped down to 5.0×10^{-6} mbars. A current of 4.0 A was supplied to the heater to evaporate the materials at a temperature of about 800 K. The shutter was removed to permit deposition on glass substrate in the presence of oxygen which was let into the chamber. Since Zinc is more reactive than Aluminum, ZnO:Al thin films were formed. The Bond enthalpy of Zn-O is 159 ± 4 kJ/mol while that of Al-O is 511 ± 3 kJ/mol [4].

Thin film thickness measurements

Thin film thickness was estimated using Tencor alpha step surface profilometry (resolution of 5 Å) equipment with a diamond stylus of radius 12.5 μm. During measurement, the stylus was moved across the film surface while keeping the sample and the sample stage stationary. The step created during the deposition process enabled the film's thickness to be read directly as the step height. SCOUT 98 software was also used to simulate the film thickness. This was used to validate the measurements obtained by Tencor alpha step surface profilometry equipment with comparisons to thickness with Quartz crystal monitor.

Optical measurements

Optical measurements (reflectance and transmittance) in the spectral range from 300 nm - 2500 nm were carried out using UV/VIS/NIR 3700 double beam Shimanzu spectrophotometer. Photons of selected wavelengths and beam intensity I_o (photons/cm²-s) were directed at the film of thickness (t) and their relative transmissions observed. Wavelengths of photon are selected by the spectrophotometer. Photons with energies greater than band gap (E_g) are absorbed while those with energies less than E_g are transmitted. The spectrophotometer had two radiation sources; a deuterium lamp for UV range and a halogen lamp for visible (VIS) and near infrared (NIR) range. The radiation source changed automatically to access the wavelength range during measurements. During transmission measurements, samples were placed in front of the integration sphere and behind it during reflection measurements. SCOUT 98 software was used to simulate transmittance data to get the optical constants like absorption coefficient among others. Drude, OJL, Tauch Lourntz, Extended Drude and Harmonic Oscillator models were used to simulate the data. These models are inbuilt in the SCOUT 98 software [8]. The models simulate refractive index, dielectric function, absorption coefficient real and imaginary parts and energy loss parameters.

Sheet resistivity measurements

The four point probe technique (Figure 1) was used to measure the sheet resistivity of the Aluminum doped ZnO semiconductor thin film samples. With a symmmetrical square geometry adopted, the four leads from the probe head were connected to Keithley Source Meter via relay switching circuit as per the Van der Pauw set-up for Voltage and Current measurements [9-11].

Discussion of Experimental Results

Optical characterization of AZO thin films

In this research study, optical constants from near normal reflectance and transmittance data for ZnO:Al are studied. SCOUT 98 software [8] was used to simulate the transmittance data to generate the corresponding optical constants. Drude and OJL models are essential in simulations of transmittance data.

Optical characterization of ZnO:Al thin films: The optical transmittance spectra of ZnO:Al films as a function of wavelength in the range (300 - 2500 nm) were plotted in Figure 2. The high transmission ($\geq 70\%$) is understood because ZnO is a semiconductor with wide direct band gap of 3.3 eV [1,3]. Due to the high transmission, these films have good optical properties for solar cells window applications. It was observed that the transmission over the visible range decreases as the concentration of Aluminum increases. This is due to free carriers coupling to the electric field hence increasing the reflection. This agrees very well with Elmin et al. [12] who reported significantly reduced transmission when ZnO was doped with higher percentages of Aluminum.

From the transmittance graph on Figure 2 the absorption edge

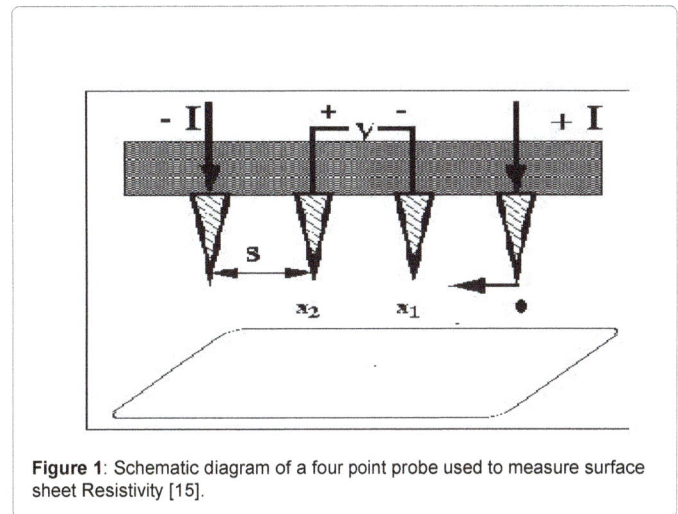

Figure 1: Schematic diagram of a four point probe used to measure surface sheet Resistivity [15].

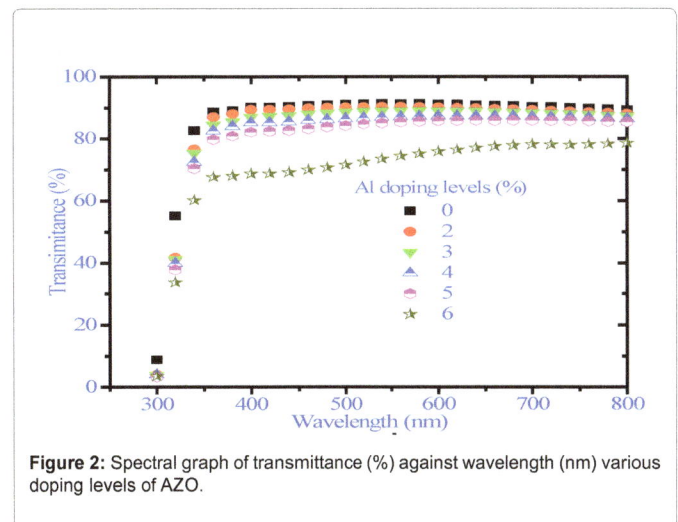

Figure 2: Spectral graph of transmittance (%) against wavelength (nm) various doping levels of AZO.

shifted toward higher wavelength as the doping levels increased. All films exhibited sharp fundamental absorption edge. Reflectance data (Figure 3) show that the average reflectance is below 45% within the visible range. All films exhibit low absorption in the visible spectrum range. The film had thickness range between 95 nm to 130 nm.

Simulated and Experimental graphs for AZO transmittance data: The experimental and simulated spectral for transmittance data were plotted against wavelength for the different samples. Using the SCOUT 98 software, the simulated curves fitted perfectly in to the experimental curves as shown in Figure 4 (b) - (e) below. From the graphs, the optical

band gap energies and thickness of the films were simulated.

Optical band gap and urbach energy for AZO thin films: Optical band gap for various samples of Aluminum doped thin films as simulated from SCOUT 98 software are shown in Table 1. The reduction in transmittance also led to variation in band gap. This may be due to the Oxygen vacancies and the behaviour of free carrier's concentration with increasing the ZnO doping. This compares very well with studies conducted by Wang Wang [13] and Shadia [14] who both got values of band gap ranging between 3.2 eV and 3.5 eV. Between 0 - to 3% the optical band gap reduces. This is followed with widening of the band gap for doping between 4% - 6%.

From Table 1, Urbach energy gradually increased with increasing band gap. This is consistent with the variation of the conductivity of the film with doping concentration. When band gap was reducing due to formation of localized states near the conduction band, it corresponded to increase in Urbach energy. Increased impurity formed more localized states within the band gap despite that beyond the 3% doping there was increased scattering. Increased doping at 4% to 6% did not enhance conductivity despite Urbach energy in from Figure 5, there was a general drop in the band gap up to a doping level of 3%. Decrease in optical band gap energy can be attributed to creation of new donor levels in the forbidden zone; and a shift in the fermi level causing a change in the band structure of the films. At room temperature aluminum atoms occupy the zinc sites in the ZnO lattice. They are singly ionized donors giving one extra electron. There after the band gap starts to widen, this

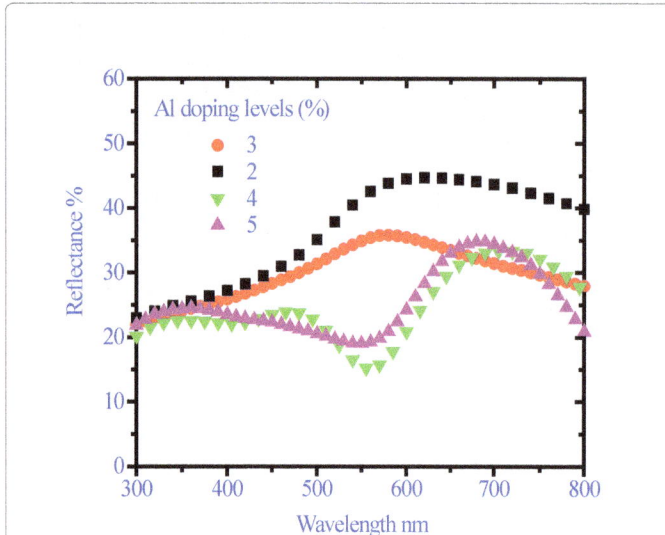

Figure 3: Spectral graph of reflectance % in the visible, IR, and UV regions against wavelength (nm) for different doping levels of AZO.

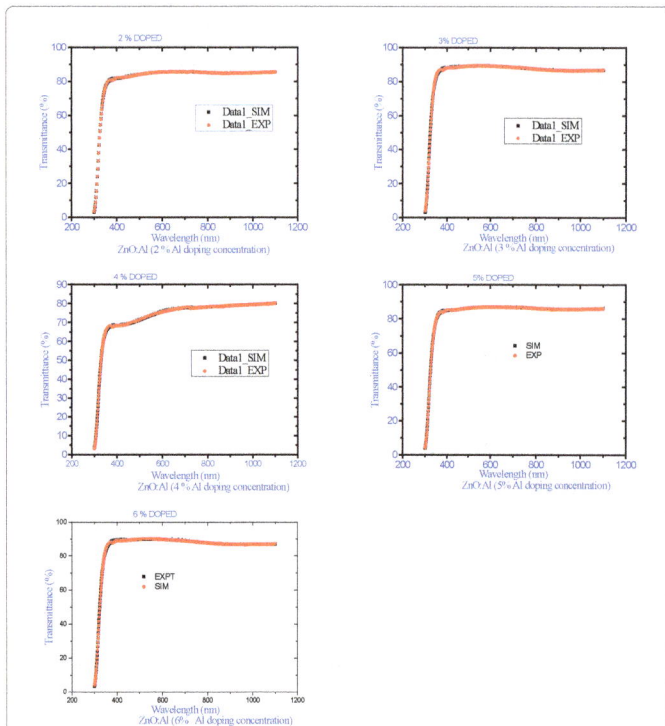

Figure 4: Doping Concentration of Al: ZNO and transmittance spectra for ZnO doped with Al at various Al doping levels.

Doping % of ZnO With Al	Optical bandgap ±0.2 eV	Thickness (nm) ± 5nm	Urbach energy ± 0.01 (x10⁻⁴) eV
0	3.34	113	2.02
2	3.28	115	2.04
3	3.18	112	2.07
4	3.21	108	2.09
5	3.32	101	2.12
6	3.42	98	2.18

Table 1: Values of the optical band gap at different doping levels for AZO.

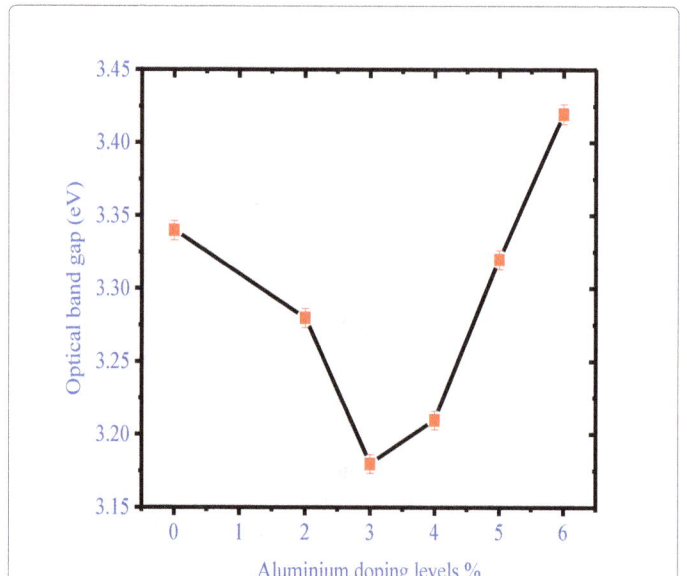

Figure 5: Variation of optical band gap for ZnO:Al with various doping levels of Aluminum (%).

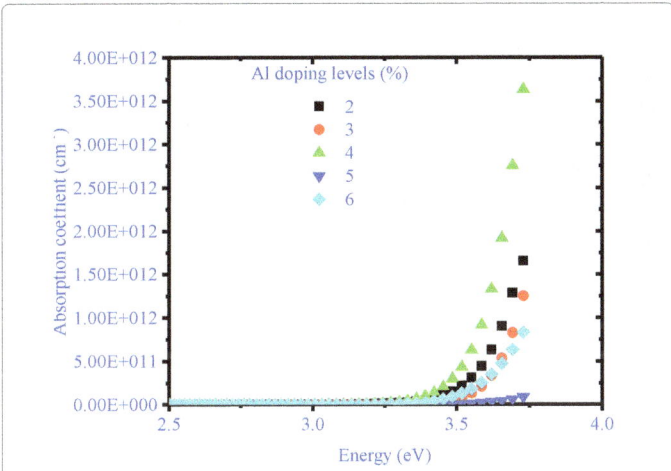

Figure 6: Graph of variation of absorption coefficient for various Al doping levels against photon energy.

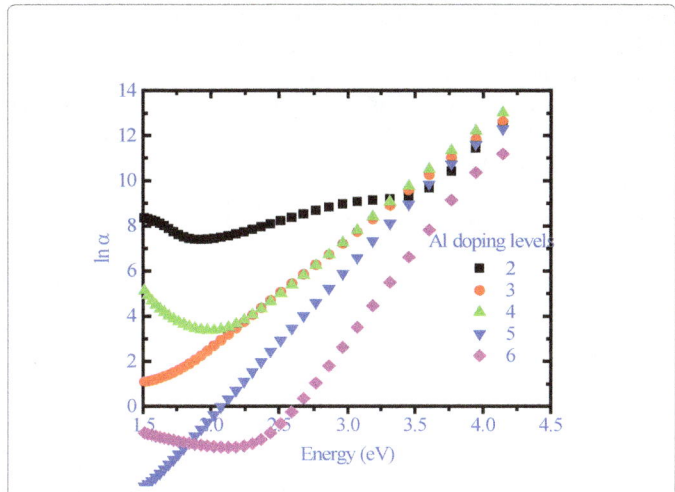

Figure 7: Graph of lnα against energy for various Al doping levels.

can be explained by Burstein Moss effect [15]. Increased concentration of donor atoms causes more electrons to occupy states at the bottom of the conduction band causing it to be filled up with donor electrons which results in widening of band gap. It is clear that the variation of α (cm⁻¹) versus photon energy (eV), Figure 6 which is near straight line, indicate the presence of direct optical transitions. Excess doping also destroys the structure stoichometry hence reducing the conductivity of the ZnO films [16].

The Urbach energy which is interpreted as the width of tails of localized states in the gap region were calculated from the following relationship, $Eu = (d\ln\alpha/d\hbar v)^{-1}$. It is obvious when ($In\alpha$) is plotted against ($\hbar\omega$), the inverse of slope will be the value of tail localized state [17]. The values of Urbach energy Eu for all composition are tabulated in Table 1. This is extracted from spectral graph in Figure 7.

It is observed that, the Urbach energy increased gradually with increasing the Aluminum doping levels (Table 2). This may be due to the fact that, ZnO doping causes a shift in the optical absorption edge therefore change in the band structure of the films. Variation in the optical absorption edge of the films with increasing the ZnO ratio indicate that, the dopant ratio is responsible for the width of localized states in the optical band of the films and causes an increase in the energy width of localized states thereby affecting the optical energy gap.

The band tailing (Urbach energy) is the effect of impurity or disorder and any other defects. In the exponential-edge region (Figure 6), the absorption coefficient is expressed by the Urbach relationship. The Urbach's absorption edge is formed in the region of photon energies below the forbidden band gap. The interaction between lattice vibrations and localized states in the tail of the band gap of the compound has a significant effect on the optical properties of the thin film. The increase in aluminum doping creates crystal strain in ZnO crystal since Al-O has a higher bond enthalpy of 511 kJ/mol compared to 159 kJ/mol of Zn-O. This increases structural defects and thereby increasing Urbach energy.

Electrical characterization of ZnO:Al thin films

Using a four point probe [11] the following results for sheet resistivity of AZO thin films were realized as in Table 3.

Surface sheet resistivity at room temperature was as in Table 3. The

Doping % of ZnO With Al	Urbach energy ± 0.01 (x10⁻⁴) eV
0	2.02
2	2.04
3	2.07
4	2.09
5	2.12
6	2.18

Table 2: Values of Urbarch energy at different doping levels of ZnO with Al.

Aluminium doped Zinc Oxide	
Doping %	Sheet resistivity ±2 Ω-cm
0	48.23
2	44.26
3	36.24
4	44.71
5	51.72
6	55.21

Table 3: A summary of electrical surface sheet resistivity for AZO thin films.

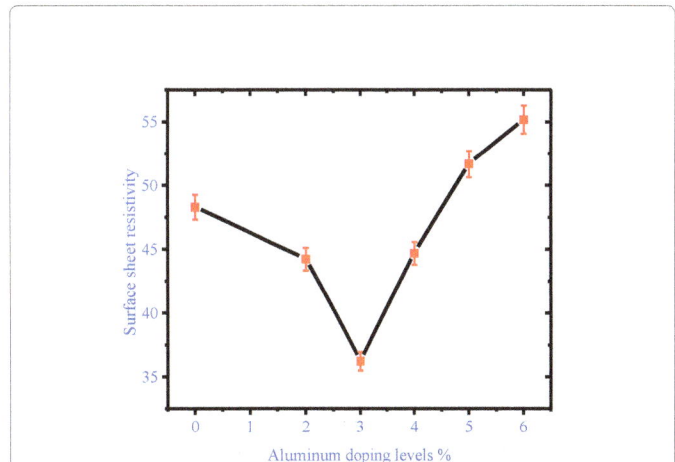

Figure 8: Graph of Aluminum doping levels with sheet resistivity.

electrical properties greatly depend on deposition parameters and film thickness. During the deposition, the substrate temperature was kept at 780 K. Figure 8 shows the variation of doping concentration with sheet resistivity. The mobility of the ZnO:Al reduce with the doping concentration, Such behaviour was expected as a result of substitution doping of Al^{3+} at the Zn^{2+} site creating one extra free carrier in the process. As the doping level is increased, more dopent atoms occupy lattice sites of Zinc atoms resulting in more charge carriers. This leads to a higher polarization of the electron by the addition of Al and thus increasing the electron phonon coupling. Strong scattering implies short carrier lifetimes and thus lower mobility.

Lower electron mobility causes reduction in conductivity hence high sheet resistivity. Thus, the resistivity increases with the doping concentration. However, after a certain level of doping, the do-pant atoms in the crystal grain and grain boundaries tend to saturation. In this case, high doping concentration will lead to a large quantity of ionized impurity. This ionized impurity provides strong scattering centre's for charge carriers. According to the Conwell-Weisskoft theory [5], when degenerate charge carriers are scattered by impurity ions, the energy dependence of ionized impurity scattering mobility due to a high doping concentration will lead to a larger quantity of ionized impurity, resulting in a decrease in the mobilities of the ZnO:Al films [5]. This result compares well with other studies [18,19] who reported sheet resistivity of 40– 60 Ω – cm.

Composition and structural characterization of ZnO:Al thin films

X-ray diffraction of ZnO:Al thin film (as-deposited): XRD measurements show structural make-up and size of crystalline structures for AZO. The X-ray diffraction (XRD) spectra of all the AZO films show the presence of a sharp peak indicating that the films are highly oriented as shown in Figure 9. The XRD spectra of the films reveal that they have crystalline structure with the main diffraction peak located at angle $2\theta = 36^0$. Neither metallic Zinc or Aluminum peaks nor Zinc Oxide peaks were observed from the XRD patterns. This implies that Aluminum atoms replace zinc in the hexagonal crystal lattice. Peak intensity of the XRD diffraction reflections is determined by the crystalline grain, size and structure and axis orientation. These results are in agreement with those of Mujdat, [20]. XRD analysis reveals that the film exhibit only the (002) peak, indicating that they have preferred orientation, implying a c-axis growth perpendicular to the substrate surface. The dominant (002) peak becomes sharper, indicating the well-established c-axis orientation of ZnO:Al thin films. This shows the crystallinity of the AZO thin film [21].

Elemental composition of ZnO:Al thin film (as-deposited): Table 4 shows the elemental percentage concentration of the thin films as obtained by the X-ray florescence (XRF) MiniPal 2 machine. The analysis was carried out to ascertain the elemental composition of the thin films. The peak-based analysis technique is used where elemental intensities of thin films are calculated and respective spectral background obtained. Figure 10 shows the peak based analysis of the elemental composition of thin film samples.

Conclusion

Deposition of thin films of ZnO:Al was done by reactive thermal evaporation techniques. The reflectance and transmittance data of the films were measured. Aluminum-doped ZnO films have high transmittance above 70% and the corresponding reflectance of below 45% within the visible range. AZO observed optical band gap ranging

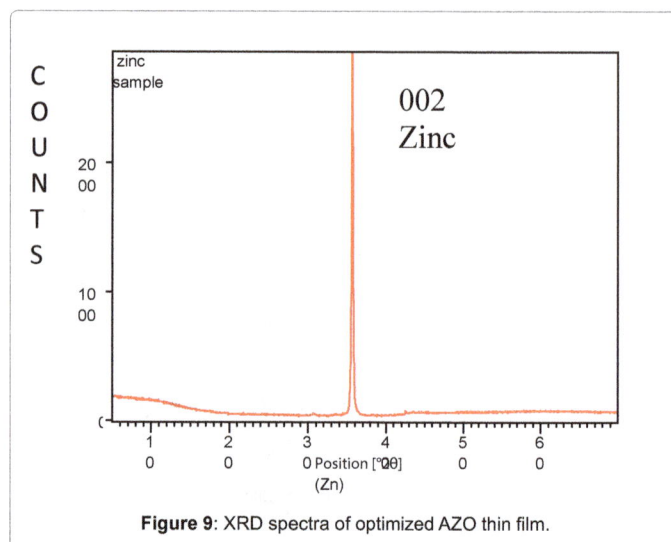

Figure 9: XRD spectra of optimized AZO thin film.

Compound/Elemental	Percentage Levels ± 0.05 %
Al	1.23
Si	59.0
Zn	39.77
SiO_2	53
ZnO	46.8
Al_2O_2	0.2

Table 4: XRF elemental percentage composition of optimized Al doped ZnO thin films.

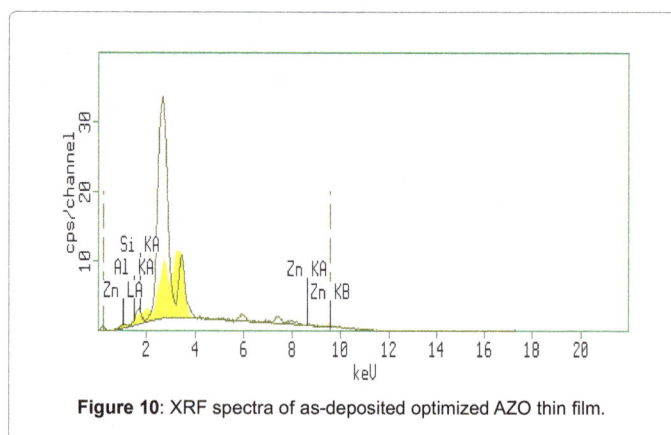

Figure 10: XRF spectra of as-deposited optimized AZO thin film.

between 3.18 eV and 3.42 eV. Conductivity increased with doping levels for AZO from 3.34 eV for undoped ZnO to minimum of 3.18 eV at 4% Aluminum doping. The XRD spectra indicate that the films were crystalline in nature for AZO.

Acknowledgments

We express our gratitude to Dr C Migwi, the chairman physics department, Kenyatta university for his personal encouragement during this research; Dr Kaduki chairman physics department, university of Nairobi for his support during the research; Mr Simon Njuguna (KU), Mr Boniface Muthoka (UON) and Mr Omucheni (UON) for their technical support during laboratory work; optoelectronics group members Kirwa, Muga, Agumba, Tuwei, masinde among others for their moral support.

References

1. Selmi M, Chaabouni F, Abaab M, Rezig B (2008) Superlattices and

Microstructures Studies on the properties of sputter-deposited Al-doped ZnO films. Superlattices and Microstructures 44: 268– 275.

2. Jang K, Park H, Jung S, Duy NV, Kim Y, et al. (2010) Optical and electrical properties of 2 wt.% Al_2O_3-doped ZnO films and characteristicsof Al-doped ZnO thin-film transistors with ultra-thin gate insulators. Thin Solid Films 518: 2808–2811.

3. Silva RF, Maria ED, Zaniquelli (2004) Aluminum-doped zinc oxide films prepared by an inorganic sol.gel route. Thin Solid Films 449: 86-93.

4. Tanusevskia A, Georgieva V (2010) Optical and electrical properties of nanocrystal zinc oxide films prepared by dc magnetron sputtering at different sputtering pressures. Appl Surface Sci: 19964-19965.

5. Jin Ma, Feng Ji, Hong-lei Ma and Shu-ying Li. (2000). Preparation and properties of transparent conducting zinc oxide and Aluminum-doped zinc oxide films prepared by evaporating method. Solar energy materials & Solar Cells 60: 341-348.

6. Lina SS, Huanga JL, Sajgalik P (2005) Effects of substrate temperature on the properties of heavily Al-doped ZnO films by simultaneous r.f. and d.c. magnetron sputtering Surface & Coatings Technology 190: 39– 47

7. Gupta N, Alapatt GF, Podila R, Singh R, Poole KF (2009) Prospects of Nanostructure-Based Solar Cells for Manufacturing Future Generations of Photovoltaic Modules. Int J Photoenergy 10: 1155.

8. Theiss W (2000) Scout thin films analysis software handbook, edited by Theiss M Hand and Software Aachen German.

9. Brown M, Jakeman F (1996) Theory of four point probe technique as applied to film layers on conducting substrates. Brit J appl Phys 17: 1146-1149.

10. Chapin DM, Fuller CS, Pearson GL (1954) A New Silicon p-n Junction Photocell for Converting Solar Radiation into Electrical Power J Appl Phys 25: 676.

11. Agumba JO (2010) Design and fabrication of a simple four point probe system for electrical characterization of thin films. Thesis, Department of Physics. Kenyatta University.

12. Bacaksiz E, Aksu S, Yýlmaz S, Parlak M, Altunba M (2009) Structural, optical and electrical properties of Al-doped ZnO microrods prepared by spray pyrolysis. Thin Solid Films 133: 245–253.

13. Wang ZL (2004) Zinc Oxide Nanostructures: Growth, Properties and Applications. J Phys 16: 829-858.

14. Shadia J, Naseem M, Riyad N (2009) Electrical and Optical properties of ZnO:Al thin films prepared by pyrolysis technique. Faculty of Science; Physics department, University of Jordan; Ammam, 11942, Jordan.

15. Liu Y, Li Q, Shao H (2009) Optical and photoluminescent properties of Al-doped zinc oxide thin films by pulsed laser deposition. J Alloys Compound Thin solid phys 485: 529-531.

16. Ogwu AA, Darma TT, Bourquerel E (2007) Electrical resistivity of copper oxide thin films prepared by reactive magnetron sputtering. J Achieve Mat Manufactur Engg 24: 172.

17. Rnjdar RM, Ali AB (2007) Optical Properties of Thin Film. Sulaimani University College of Science Physics Department 24.

18. Zhao Y, Xinhua G, and Wang W (2002) R&D Activities of Silicon-based thin film solarcel in China. Thin Film Device Technol 203: 714-720.

19. Yoo J, Lee J, Kim S, Yoon K, Park J, et al. (2005) High transmittance and low resistive ZnO:Al films for thin film solar cells. Thin Solid Films 480: 213– 217.

20. Mudjat C, Saliha I, Yasemin C, Fahrettin Y (2007) The effects of Aluminium doping on the optical constants of ZnO thin films. J Material Sci 19: 704-708.

21. Ogwu AA, Bouquerel E, Ademosu O, Moh S, Crossan E, et al. (2005) An investigation of the surface energy and optical transmittance of copper oxide thin. lms prepared by reactive magnetron sputtering Thin Film Centre. Electronic Engg Phys Div Acta Materialia 53: 5151–5159.

13

Design and Construction of a Water Scrubber for the Upgrading of Biogas

Temilola T Olugasa* and Oluwafemi A Oyesile

Department of Mechanical Engineering, University of Ibadan, Ibadan, Nigeria

Abstract

This paper discusses the results of studies conducted on raw biogas produced from a prototypic biogas production plant located at the Teaching and Research Farm, University of Ibadan, Ibadan. This setup consists of a mixing chamber, a biogas digester and a stabilizing unit, locally designed and fabricated. It further discusses preliminary and detailed design coupled with the construction of an effective and efficient technology used in purifying raw biogas generated from the prototypic biogas production plant; this technology is otherwise known as the Water Scrubbing technology. The Scrubbing system consists of the Water scrubber with iron wool packed bed connected to a 500 litre water tank, and two tyre tubes which were used in storing the pre scrubbed (raw) biogas and the scrubbed (purified) biogas. The water scrubber has an inlet for the entry of the raw biogas and a discharge for the exit of the scrubbed biogas. Raw biogas from the plant was stored in a tyre tube and directly fed into the Water scrubber housing the iron wool packed bed, the purified biogas from the exit was also collected into another tyre tube. The samples of the gas mixture were taken before and after scrubbing and analyzed with Pascal Manometric Glass Tube technique. Results indicated that methane content of the scrubbed/ purified biogas was raised from 58% to 82% due to the reduction of Carbon dioxide and Hydrogen Sulphide. CO_2 was reduced from 31% to 14% while H_2S was reduced from 1% to 0.4%.

The corresponding Energy content of the purified biogas was evaluated to be 41MJ/kg which is higher than that of the raw biogas which was evaluated to be 29MJ/kg.

Keywords: Biogas; Water scrubber; Packed bed; Methane; Purify; Upgrade

Introduction

In Nigeria, we are posed with the problem of generating electricity, producing adequate electricity for the entire population of the country has been observed as a bone of contention. There is also a problem of finding and using alternative energy sources. The major source of electricity generation in Nigeria is fossil fuels; these fuels have an adverse effect on the biological system because they facilitate global warming. As a result of this development, researches have shown that renewable fuels are genuinely important in solving these problems. An example of a renewable fuel is biogas which simply means "fuel from biological matter". A more comprehensive definition was given by Olugasa et al. [1] and described biogas as a of the mixture of carbon dioxide, CO_2 and inflammable gas Methane, CH_4 which is produced by bacterial conversion of organic matter under anaerobic (oxygen-free) conditions.

However, biogas has some limitations; these limitations have in one way or the other restricted the commercial use of biogas. The limitations have also reared its ugly head in the awaiting success story of this impressive source of energy. These limitations include low energy content and a challenging difficulty in compressing and storing biogas. Another notable limitation of biogas is the fact that there have been little technological advancements in the production of biogas. The low energy content is majorly caused by unwanted constituents or impurities in the biogas like Carbon dioxide (CO_2), hydrogen sulphide (H_2S), siloxanes, halogens etc. There is therefore an urgent and essential need to purify biogas before it can be used to generate adequate electricity and be used as a vehicle fuel; hence, the need for this study. A lot of processes have been developed in increasing the energy content (Methane content) in biogas. Some of these processes are Polyethylene Glycol scrubbing, chemical absorption, pressure swing adsorption, Bio trickling filter, Cryogenic separation, Iron absorptive media, biological scrubbing and most importantly the Water Scrubbing technology. According to Vijay [2] the water scrubber's advantage over other purification techniques is its simplicity, availability of water; it's suitability for biogas enrichment in rural areas, and characteristic as a universal solvent. In addition to these, the packing material in the scrubbing setup increases the contact time between the biogas and water. This work is therefore aimed at providing an effective and efficient scrubbing technique that would be capable of removing significant amounts of Carbon dioxide and Hydrogen sulphide, resulting to an increase in the energy content of biogas and the recommendation of the commercial use of purified biogas in Nigeria [3-10].

Materials and Methods

Assessment and selection of a biogas plant

The biogas plant at the Teaching and Research Farm, University of Ibadan was selected as the case study. This biogas plant has a bio-digester of 2 m³ capacities, a mixing chamber and a stabilizing unit. 8 kg of Cow dung was transported from the dairy farm, University of Ibadan to the location of the biogas plant at the Teaching and Research Farm. The system was charged with cow dung at the ratio of 1:1 (water to cow dung) and left in an open area with ambient daily average temperature of 31.5°C and monitored for a period of 14 days. After proper mixing in the mixing chamber, the system was left air-tight to ensure an anaerobic environment inside the digester and manually stirred occasionally every day.

A biogas collection setup made up of a ½ inch hose, 50 cm long was

*Corresponding author: Olugasa TT, Bsc, Department of Mechanical Engineering University of Ibadan, Oyo State Nigeria, E-mail: temilola18@ yahoo.co.uk

designed and connected to the biogas plant for the collection of raw biogas to be tested and analyzed, and for the subsequent use of the raw biogas to be scrubbed.

Design of the biogas scrubber

The design of a packed bed water scrubber involves the following steps:

i.   Assumptions of basic data.

ii.   Solubility data generation.

iii. Material balance and determination of water flow rate.

iv.   Selection of packing material.

v.   Determination of column diameter.

vi.   Determination of the height of the packed bed column.

vii. Selection of packing support and water distributor.

Assumptions of basic data

The basic data assumed during the design of the scrubber were:

Inlet pressure of the biogas = 100 kPa

Inlet temperature of biogas = 25 °C

Volume of Biogas to be Scrubbed= 0.050 m³

Percentage of carbon dioxide in biogas = 35%

Partial pressure of CO_2= 0.35 kPa

Solubility data generation: Henry's Law was used to determine the solubility of CO_2 in water. Solubility of CO_2 in water at 1 bar and 298K is given as 2857 Pa.m³/mol

Henry's Law:

$$P_t = K_H C_{max} \qquad (1)$$

C_{max} = Saturation concentration of CO_2 in mol/m³

K_H = Henry's coefficient [Pa.m³/mol] = 2857 Pa.m³/mol

P_i = Partial pressure of CO_2 component in biogas (P_a)

This equation was used in determining the Henry's law constant of Methane; it gives us a guideline as to the solubility of methane in water. It further introduces us to the relationship between the Henry's law constant and the concentration of methane in water, an inverse relationship. The Henry's law constant of methane is very high due to the insolubility of methane, compared to that of Carbon dioxide and Hydrogen sulphide which are both soluble in water.

Material balance and determination of water flow rate

Material/mass balance equation: Rate of Accumulation of species I in the volume element = (Rate of inward flow of species I in the volume element) – (Rate of outward flow of species I from the volume element) + (Rate of species I generation in the element)

$$P_{A1} - P_{A2} = \left(\frac{F_l \pi}{F_g C_T} \right)(C_{A1} - C_{A2}) \qquad (2)$$

$F_l / F_g = 10 \qquad (3)$

Equations 2 and 3 were used in calculating the pressure drop in the scrubber with known parameters like the ratio of molar flow rate of water to biogas, molar density of water, total pressure and difference in concentration of biogas at inlet and outlet.

Ergun   Equation:

$$\Delta P = \frac{150\mu(1-\varepsilon)2V_s L}{\varepsilon^3 D_P^2} + \frac{1.75(1-\varepsilon)\rho V_s^2 L}{\varepsilon^3 D_p} \qquad (4)$$

Equation 4 was used in determining the superficial velocity of carbon dioxide using parameters such as gas density, void fraction, equivalent spherical diameter of packing, dynamic viscosity of the gas, length of the packed bed and pressure drop.

Flow Rate Equation: $Q = V_s A$ \qquad (5)

Equation 5 was used in verifying the Area of the reacting column using the calculated superficial velocity and the molar flow rate of the gas.

Determination of volume of reactor, packed bed height and diameter:

Volume of a cylinder: $V = AH$ \qquad (6)

$$V_\gamma = hA_{cs} \frac{F_g}{\pi K_G \alpha}(P_{A2} - P_{A1}) \qquad (7)$$

With a specified volume and a known area, the height of the packed bed was also established using equation 7

Selection of packing material: A packing material used in enhancing the contact time (interfacial area) between the gas and water. The packing material selected for the scrubber was iron wool with the following specifications:

Equivalent spherical diameter of packing=5×10^{-3}m

Void fraction=0.5

Iron wool was selected because it removes hydrogen sulphide from biogas

Selection of packing support and water distributor: Metal sieves were placed at the top and bottom of the middle section of the scrubber to act as support for the iron wool packing.

Water was supplied to the scrubbing tower by means of a connected overhead 500 litre tank and the water was distributed by means of a water sprayer connected to the top section of the scrubber.

Basic Parameters

P_A Partial pressure

C_A Concentration

P_{A1} Partial Pressure at inlet

P_{A2} Partial Pressure at outlet

F_l Molar flow rate of liquid

F_g Molar flow rate of gas

π = Total Pressure

C_T Molar density of water

C_{A1} Concentration of CO_2 in water at inlet

C_{A2} Concentration of CO_2 in water at outlet

ΔP Pressure drop

μ Dynamic viscosity of gas

ε Void fraction

V_s Superficial Velocity

L Length of bed

ρ Density of gas

D_p Equivalent spherical diameter of packing

Equivalent spherical diameter of packing=5×10^{-3} m

Void fraction=0.5

Gas (methane) density = 0.72 kg/m^3

Gas (Carbon dioxide) density = 1.98 kg/m^3

Gas (methane) viscosity = 7.39×10^{-6} m^2/s at 25^{0}C

Viscosity of carbon dioxide = 0.0855×10^{-6} m^2/s at 25 ^{0}C [3,4]

CAD Models

Detailed designs of a Water Scrubber with iron wool packed bed were produced using CATIA software. These are shown in Figures 1, 2 and 3. Figure 2 shows the 3-dimensional view of the scrubber, while Figure 3 shows the details of the scrubbing process as being counter current with the biogas coming in from the side of the tank and going out from the top of the tower while the water is introduced from the top of the scrubber and the flows out from the bottom of the tower.

The scrubbing setup consists of the water scrubber with an iron wool packed bed connected to a five hundred litre tank containing pure water with the aid of a piping network that consists of a one inch pipe, reducers (2 inches to 1 inch), 2 inches and 1 inch nipples and a 1 inch elbow, Teflon tapes were used to tighten the internal and external threading in the pipe and pipe fittings. The scrubber was also connected

Figure 2: Testing of the Scrubbing System.

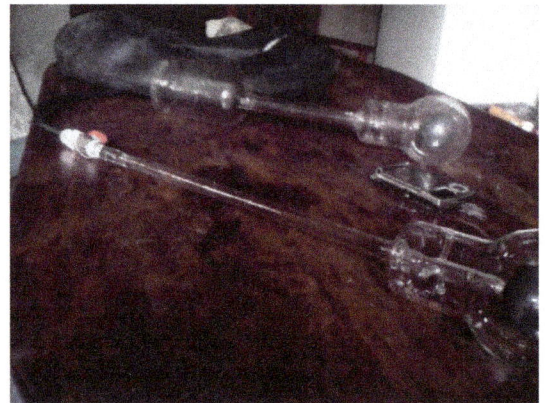

Figure 3: Pascal Manometric Glass tube.

to two tyre tubes for collection of raw and purified biogas respectively at the inlet and the discharge of the water scrubber. The tyre tubes were connected to the water scrubber with the aid of ½ inch hoses, ½ ball valves, ½ inch reducers, ½ inch nipples, top gut gum, and araldite glue. Raw biogas was stored in one of the tyre tubes and pressurized before it was sent into the bottom section of the water scrubber. The iron wool packed bed was used to enhance the contact time (interfacial area) between the biogas and water, and also to react with the hydrogen sulphide in the biogas. Pressurized water was also sprayed from top with the aid of a shower sprayer to absorb the CO_2 and H_2S from pressurized biogas. The movement of the biogas was achieved by means of upward displacement, downward delivery elemental technique. Purified biogas

Figure 1: Isometric view of the Water scrubber with an iron wool packed bed.

was then collected into another tyre tube where it was stored for further analysis [11-14].

Chemical analysis of raw and purified biogas

The chemical analysis used in determining the composition of biogas was done using Pascal Manometric glass tube. 30cm³ of the gas was trapped into the Pascal Manometric glass tube via the gas regulator. The Pascal Manometric glass was filled with known volume of fractionating reagents mixture which consists of 1M magnesium perchlorate, 1M Sodium hydroxide, 1M Barium sulphate and 1M Nitric acid. The fractionation uses the redox principle in which the reduction oxidation process will precipitate the fractions of the gases. The percentage of the gas fractions was got

Percentage of CH4 in biogas $= \dfrac{a \times 16.04}{\text{Volume of biogas used}}$

Percentage of CO_2 in biogas$= \dfrac{b \times 44.01}{\text{Volume of biogas used}}$

Percentage of H_2S in biogas$= \dfrac{c \times 34.06}{\text{Volume of Biogas used}}$

Where a, b and c are the volume of CH_4, CO_2 and H_2S gases trapped respectively [5].

Results and Discussion

Design Specifications

Using the Material/ mass balance equation, the Ergun equation, the flow rate equation, the relationship between volume, area and height of a packed bed; the following design specifications of a water scrubber with iron wool packed bed were obtained:

A superficial velocity of 2.4×10^{-3} m/s was obtained; the packed bed area of 0.1158 m²; a diameter of 38.4 cm and a packed bed height of 60.5cm.

By means of an indispensable design process, coupled with the right design specifications, as well as accurate construction and fabrication, along with precise and proper assembling; Figure 4 shows an enhanced purification technique which is the water scrubber with iron wool

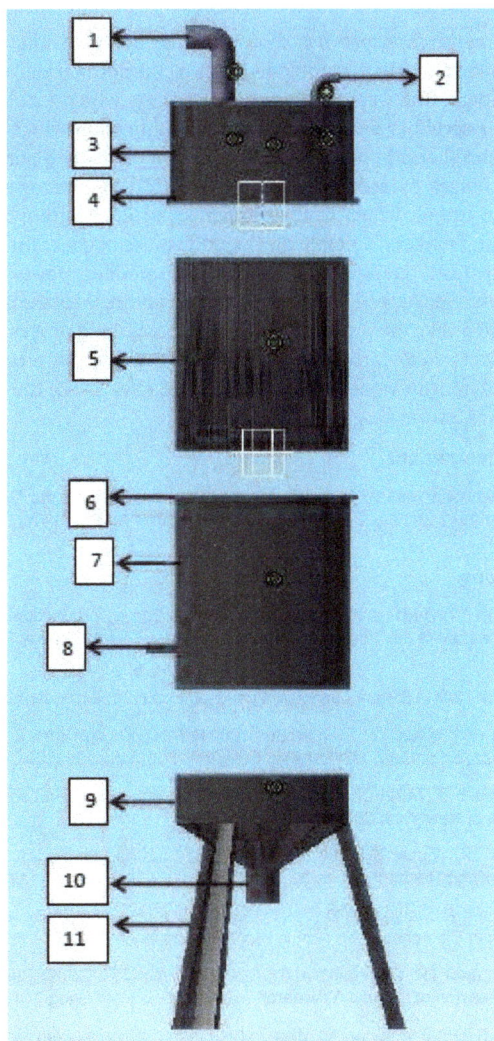

Part Nos.	Description	Specification	Material used
1	Water inlet	1 inch diameter pipe	Polyvinyl chloride pipe
2	Gas discharge	½ inch diameter pipe.	Polyvinyl chloride pipe
3	Head	10 inches diameter, 8 inches long.	Mild steel pipe
4	Flange	10 ¾ inches internal diameter, $13\frac{1}{4}$ inches external diameter, $1\frac{1}{2}$ inches height.	Mild steel circular bar
5	Packed bed	9 inches diameter, 20 inches long	Steel wool
6	Sieve	10 inches diameter	Mild steel circular bar
7	Body	10 inches diameter, 20 inches long	Mild steel pipe
8	Gas inlet	½ inch diameter pipe	Polyvinyl chloride pipe
9	Base	10 inches diameter, 8 inches long.	Mild steel pipe
10	Water discharge	2 inches diameter valve	Stainless steel valve
11	Stand	24 inches long.	Iron angle bar

Figure 4: Exploded view of the Water Scrubber with an iron wool packed bed and the Bill of Materials.

packed bed which has been proffered to improve the qualities and increase the energy content in biogas.

Most scrubbers use raschig rings/ balls as the contactors for the packed bed. This study has however, explored the use of a packed bed in order to investigate if there will be increased absorption of hydrogen sulphide, which is only absorbed in water scrubbing in small

The scrubbing setup shown below in Figure 5 comprises of the water scrubber with iron wool packed bed connected to a five hundred litre tank which contains pure water with the aid of a pipe network. The water scrubber is connected to two tyre tubes for collection and storage of raw and purified biogas respectively at its inlet and discharge.

Chemical Analysis

Figures 6 and 7 show diagrams (pie charts) which represent the composition of raw biogas (pre scrubbed) and composition of purified biogas (scrubbed). The figures show that there is an increase in methane content from 58% to 82%, due to removal of Carbon dioxide and Hydrogen Sulphide. Carbon dioxide was reduced from 31% to 14% while Hydrogen Sulphide was reduced from 1% to 0.4%. This implies that CO2 was reduced by 55% and H2S was reduced by 60%, while CH_4 increased by 41%. The mass of raw biogas was measured to be 1.26 kg while that of purified biogas was measured to be 1.18 kg.

The corresponding Energy content of the purified biogas is

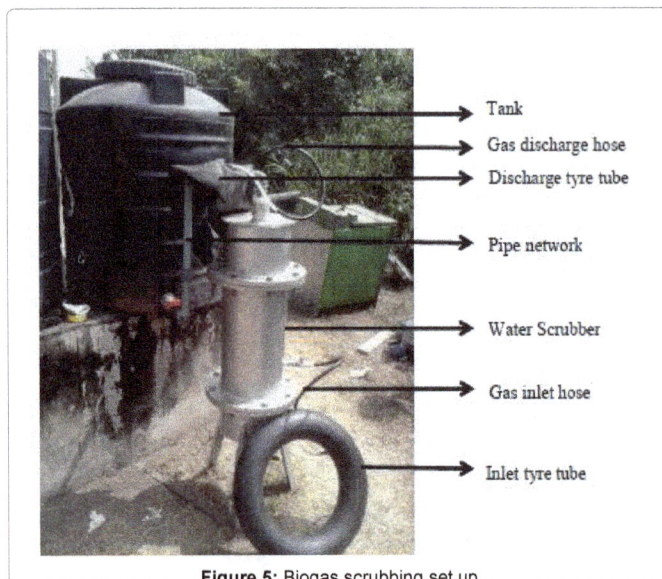

Figure 5: Biogas scrubbing set up.

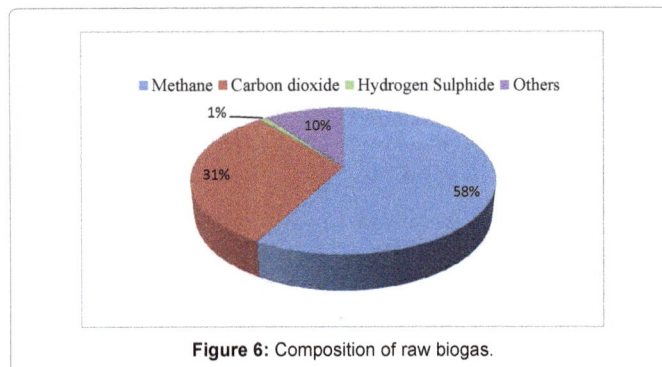

Figure 6: Composition of raw biogas.

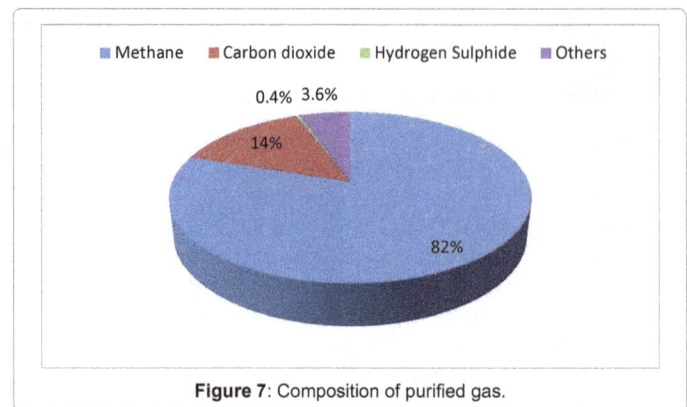

Figure 7: Composition of purified gas.

evaluated to be 16.4 MJ/kg or 27.675 MJ/m³ which is more than that of the raw biogas, 11.6 MJ/kg or 19.575 MJ/m³. [15-17].

Conclusion

Many new and stimulating developments have been discovered in recent years through the study of biogas and its vast technology, though they are of worthy note, there are still more benefits that have not been optimally utilized.

This research paper has made a huge effort in offering a very important solution to the problems affecting the optimal and commercial use of Biogas. The water scrubbing technology which involves the use of a water scrubber with an additional modification which is the iron wool packed bed has been proven to achieve eighty two percent (82%) purified biogas by reducing Carbon dioxide and Hydrogen Sulphide to a large extent, which in turn enhances adequate compression and storage; and makes it suitable for the utmost use of generating adequate electricity, faster cooking and fuel for automobiles, power generators and boilers. The by-product of the water scrubbing process is fairly easy to dispose unlike the products of chemical scrubbing process because it contains very weak carbonic acid. This shows that water scrubbing with the use of an iron wool-packed bed is quite effective in the removal of CO_2 and H_2S from biogas.

Acknowledgement

The Author's wish to acknowledge the grant provided by the Department of Mechanical Engineering, University of Ibadan, Nigeria for this study.

References

1. Olugasa TT, Odesola IF, Oyewola MO (2013) Energy production from biogas: A conceptual review for use in Nigeria. J Renew Sustain Energy Rev 32: 770–776.

2. Vijay VK (2011) Biogas purification using water scrubbing systems.

3. Peters MS, Klaus D. Timmerhaus (2004) Plant Design and Economics for Chemical Engineers. (5th edn.) Mc Graw-Hill Companies Inc: 88.

4. Levenspiel O (1999) Chemical Reaction Engineering. (3rd edn.) John Wiley and Sons Inc, New York: 685.

5. Borilek JK, Winner K (1998) Pascal modified method for biogas fractionation. Chem Engg Technol Series no 15: 345.

6. Borjesson P, Mattiasson B (2007) Biogas as a resource-efficient vehicle fuel. Trends in Biotechnol. 2nd Nordic Biogas Conference, Malmo.

7. Bajracharya TR, Dhungana A, Thapaliya N, Hamal G (2009) Purification and Compression of Biogas: A research experience. J Instit Engg 7: 1-9.

8. Eze JI (2010) Preliminary Studies on Biogas Scrubbing system for Family sized Biogas Digester. Global J Sci Frontier Res 10: 13-17.

9. Jones J, Partington B (2008) Biogas cleaning and Uses, Agriculture and Rural Development, Alberta. AGRI-FACTS, Practical Information For Alberta's Agriculture Industry. Agdex 768-775.

10. Kapdi SS, Vijay VK, Rajesh SK, Guar RR (2003) Feasibility study on purification and compression of biogas for rural areas. International Conference in Energy and Rural Development MNT, Jaipur, India.

11. Karki AB, Shrestha JN, Bajgain S (2005) Biogas: as renewable source of energy in Nepal. SNV Publisher, Nepal.

12. Kossmann W, Pönitz U, Habermehl S, Hoerz T, Krämer P, et al. (2011) Biogas Digest. Biogas Basics, Information and Advisory Service on Appropriate Technology (ISAT) 1.

13. Prasertsan S, Sajjakulnukit B (2006) Biomass and biogas energy in Thailand: potential, opportunity and barriers in Renewable Energy. Renew Energy 31: 599-610.

14. Zhao Q, Leonhardt E, MacConnell C, Frear C, Chen S (2010) Purification Technologies for Biogas Generated by Anaerobic Digestion. CSANR Research Report: 001.

15. Ilyas SZ (2006) A Case Study to Bottle the Biogas in Cylinders as Source of Power for Rural Industries Development in Pakistan. World Appl Sci J 1: 27-130.

16. Tippayawong N, Promwungkwa A, Rerkkriangkrai P (2007) Long-term operation of a small biogas/diesel dual-fuel engine for on-farm electricity generation. Iranian J Sci Technol Transact B: Engg 34: 167-177.

17. Wise DL, Rato B (1981) Analysis of systems for purification of fuel gas. Fuel gas production from biomass. CRC Press 2.

14

Reaction of Hydrothermally Altered Volcanic Rocks in Acid Solutions

Georgina Izquierdo Montalvo[1], Alfonso Aragón Aguilar[1], F. Rafael Gómez Mendoza[2] and Magaly Flores Armienta[3]

[1]*Instituto de Investigaciones Eléctricas, Reforma 113, Col. Palmira, Cuernavaca, Morelos, CP, Mexico*
[2]*Paseo Cuauhnáhuac 8532, Col. Progreso Jiutepec, Morelos, Mexico*
[3]*Comisión Federal de Electricidad, GPG. Morelia, Michoacán, Mexico*

Abstract

Rock matrix stimulation has been used to clean, to recover and to enhance well productivity in oil systems. Recently, for the same purposes this methodology began to be applied in geothermal systems. In order to investigate the solubility of altered volcanic rocks in acid solution used in rock matrix stimulation; experiments were carried out on samples of igneous hydrothermal altered rocks from the Los Humeros geothermal reservoir. Industrially, the common acid solutions used during acid well stimulation are HCl 10% and a mixture of HCl 10% and HF 5%. In this work, experiments were conducted in the laboratory using the referred acid solutions at atmospheric pressure and temperature of 110 ± 5°C.

The chemistry, the mineralogy and the permeability of selected rocks from Los Humeros geothermal field were determined before and after the reaction with each acid solution. Mineral dissolution is selective and depends on the permeability of the rocks, the type and the intensity of hydrothermal alteration.

As it is expected, Calcite readily reacts with acids leaving empty cavities, veins and micro fractures (worm holes). Calc-silicates are resistant to acid solutions. If Calcite is absent dissolution of minerals is observed in the external surfaces of the specimen in contact with the acid solution giving rise to a rough texture and leaving the rock matrix unreacted

Keywords: Acid stimulation; Matrix acidizing; Acid dissolution of rocks; Productivity enhancement; Permeability enhancement

Introduction

Rock matrix stimulation has been a methodology used for years to clean, to recover and to enhance well productivity in oil systems. Some years ago this methodology began to be applied in some geothermal systems in Philippines, Indonesia and the United States. Not always being successful especially in volcanic reservoir rocks.

As originally designed, matrix acidizing has been applied successfully in both carbonate and sandstone formations; the main purpose in carbonate formations is to form conductive channels called wormholes, through the formation rock [1]. The acid solution penetrates beyond the near wellbore region extending and forming smaller channels branching off the main wormhole. In sandstone formation, matrix acidizing treatments usually are designed primarily to dissolve acid soluble material deposited in pore network near the wellbore.

In carbonate rocks, the acids commonly used are: Hydrochloric, Acetic and Formic. In sandstone formations, the acids commonly used are: Hydrochloric, Acetic, Formic and Hydrofluoric. Where a siliceous carbonate formation is treated, HF is used in combination with HCl.

To minimize or to eliminate the effects of scale deposition as well as restore or improve permeability, several methodologies have been used in geothermal fields. Among others: matrix acidizing, hydraulic fracturing, thermal fracturing and chemical stimulation.

Hydraulic fracturing is commonly used although not many successful cases are known; it is considered as an option in geothermal fields to improve wells with poor reservoir connectivity [2,3].

Thermal fracturing produces thermal shock by injection of cool water. It is a well-documented method but it is not suitable to eliminate scales.

As it was mentioned, matrix stimulation is an old methodology used to enhance and recover well productivity in oil systems. Nowadays has been extended to the geothermal industry in wells that have shown reduced productivity either by clogged pores and fractures or scale formation. Diluted Hydrochloric acid is used widely; it is known that easily dissolves scales such as calcium carbonate and is used extensively in oil field operations throughout the world. On the other hand, is extremely reactive with sulfur scales formed by pyrite, chalcopyrite, galena, among others forming secondary products.

At low concentrations, HCl and mixtures of HF and HCl have been used in acidizing operations. HCl is selected to treat limestone and calcite in veins, pores and scales. A mixture of HF and HCl is used to dissolve silicates and silica. A mixture of 12% HCl–3% HF (called regular mud acid) is commonly used [4].

Chemical stimulation using chelating agents such as ethylenediaminetetracetic acid (EDTA) or nitrilotriacetic acid (NTA) have been proposed as an alternative treatment. Such agents have the ability to chelate, or bond, metals such as calcium. This procedure has been studied in the laboratory as a method to dissolve calcite in geothermal reservoirs [5]. They found that the rate of calcite dissolution is not as fast as using strong mineral acids.

Before an acid treatment will be designed, some important features have to be taken into account, such as the type of formation damage, reservoir geology, mineralogy, reservoir fluids and scale deposition in production wells and in the formation. An understanding of the

*Corresponding author: Aguilar AA, Instituto de Investigaciones Eléctricas, Reforma 113, Col. Palmira, Cuernavaca, Morelos, CP, Mexico, E-mail: gim@iie.org.mx

formation type and other properties such as porosity, permeability, intensity and type of hydrothermal alteration are also critical. In geothermal systems rock matrix stimulation must be designed especially for each system and for each well.

In hydrothermal systems secondary mineralogy is the result of water-rock interaction and depends on the physical and chemical conditions in the reservoir: temperature, pressure, liquid and rock composition, fluid pH, steam-water ratio etc. Mineral deposition in pores and fractures in the reservoir rocks may reduce porosity and in consequence productivity may decrease. A common mineral assemblage in volcanic high temperature geothermal systems, where the rocks interact with neutral to basic pH fluids, is represented by chlorites, quartz, epidote and other minerals such as calcite, chain and ring silicates etc. From these minerals calcite readily reacts in diluted HCl solutions; chlorites partly react in diluted HCl-HF solutions.

Another common problem in geothermal wells is the scale formation inside the well casing and on the wall rocks; which also contributes to reduce productivity. Deposition of inorganic scale may occur during well production. Depending on the well conditions, on the reservoir rocks and on the nature of the fluids different scale types may form. Common scales include calcite, calcium sulfate, silica and iron sulfides.

Acid stimulation techniques have been successfully used in some geothermal systems. During the past decade has been applied in fields such as Salton Sea, in Philippines and Indonesia. In Mexican geothermal fields acid stimulation of wells has been carried out in Los Azufres, Mich. and Cerro Prieto B. C.

In this work a selection of reservoir rocks of the Los Humeros geothermal field (LHGF) were used to perform solubility experiments using the common acid solutions used in acid stimulation of wells. Sets of rocks were provided by the Comisión Federal de Electricidad in order to carry out laboratory experimentation.

The best results were obtained in samples where calcite was present in pores or in fine or wide veins.

Mineralogical changes are not noticeable, except in samples containing calcite. The bulk mineralogy remains the same.

The objective of the experimental work was to evaluate changes in the igneous rocks mainly on the mineralogical composition and on the petrophysical and mechanical properties of rocks.

The Los Humeros Geothermal Field: The Los Humeros is one of the four geothermal fields currently operating in Mexico; it has an installed capacity of 93 MW. The field is inside the Los Humeros volcanic caldera, which lies at the eastern end of the Mexican Volcanic Belt, Figure 1.

After more than 20 years of production, as most geothermal fields, the field has undergone to several processes affecting production. Among others, normal decline of production, self-sealing by scale deposition in the formation and the pipelines, other processes like silicification of reservoir rocks, etc.

In LHGF two important features must be taken into account: one the complex lithology due to the geological events that formed the large caldera and the occurrence of aggressive fluids before and after production started.

Recently [6], mineralogical evidences on the effects of fluid acids on the reservoir rocks of the LHGF have been presented. The low pH fluids react with the rock changing its chemical and physical properties.

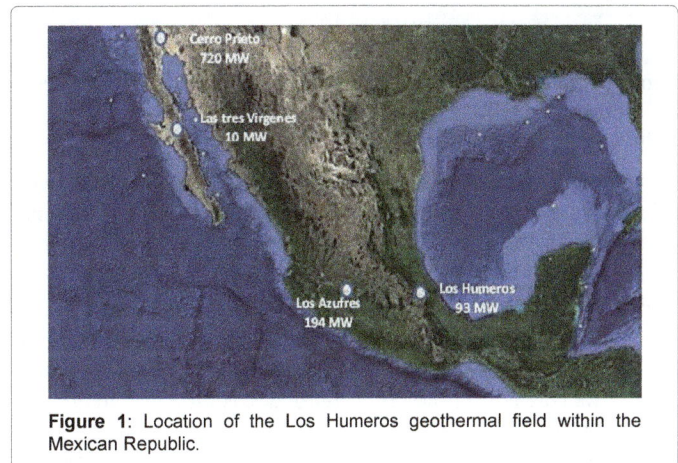

Figure 1: Location of the Los Humeros geothermal field within the Mexican Republic.

If calcite was present, dissolution occurs, total or partial sealing of vugs and fracture increasing and decreasing permeability. Due to the complexity of the processes that have affected the reservoir rocks, an acid stimulation work must be considered individually for each well and a specific acid treatment must be designed.

In relation to the Los Humeros caldera, a series of geologic events have occurred [7]. The complex geology has been summarized in four Units [8]:

Unit 1. Post-caldera volcanism. Is composed of andesites, basalts, dacites, rhyolites, flow and ash tuffs, pumices, ashes and materials from phreatic eruptions. The unit contains shallow aquifers.

Unit 2. Caldera volcanism. This unit is mainly composed of lithic and vitreous ignimbrites from the two collapses (Los Humeros and Los Potreros). It also includes rhyolites, pumices, tuffs and some andesitic lava flows, as well as the peripheral rhyolitic domes. This unit acts as an aquitard.

Unit 3. Pre-caldera volcanism. It is composed of thick andesitic lava flows, with some intercalations of horizons of tuffs. The characteristic accessory mineral of the upper andesites is augite and the lower andesites is hornblende. Both packages include minor and local flows of basalts, dacites and eventually rhyolites. This unit contains the geothermal fluids.

Unit 4. Basement. This basement unit is composed of limestones and subordinated shales and flint. This unit includes also intrusive rocks (granite, granodiorite and tonalite) and metamorphic (marble, skarn, hornfels), and eventually some more recent diabasic to andesitic dikes.

Materials and Methods

Cores of a suitable size, coming mainly from Unit 3, were selected to carry out the matrix acidicing experimental work. As mentioned, Unit 3 is where the geothermal fluids are contained and where the highest hydrothermal alteration of rocks is found.

The mineralogy of several cores from different wells was determined by X-ray diffraction and petrography.

Core A, from a well located at the south of the LHGF, from a depth of 1200-1203 m; has a gray color with greenish shades. It is classified as andesite, its characteristic accessory mineral is augite; it shows medium intensity of hydrothermal alteration. Minerals identified are: Plagioclase, quartz, augite altered to chlorite, calcite, epidote and traces of hematite.

Core B, from a well located almost at the central part of the LHG, from a depth of 1500-1503 m; has a gray color with greenish shades. It is classified as andesite, compact, fine grain, with aphanitic to porphyric texture. It shows small irregular fractures sealed by hematite.

Core C, from a well located near the central part of the LHGF, from a depth of 1300 - 1303 m has a light gray color, is classified as andesite whose dominant accessory mineral is hornblende. It shows high intensity of hydrothermal alteration represented by epidote, chlorite, quartz, calcite and mica (illite).

Core D, from a well located at the central part of the LHGF, from a depth of 2000 and 2004 m. This core is rather different from the others. This core possibly was an altered andesite; which apparently was affected by aggressive fluids leaving only a frame of micro crystalline quartz, well preserved plagioclases and traces of chlorite and pyrite.

Experiments were carried out in acid and temperature resistant vessels. Pre-weigh core fragments were placed consecutively in each acid solution (HCl 10%; HCl 10% + HF 5%; HCl 10%).

Experiments were conducted at atmospheric pressure in a controlled temperature oil bath at $110 \pm 5°C$ for 1 hour using the same specimen for each solution. After each treatment, samples were recovered from the acid solution, immersed in distilled water and were leaved to dry at room temperature. Always the weight of dry specimens was registered. Weight lost is relative, because samples loose particles during the reaction and during the handling.

At the end of the third reaction, small chips of the cores were finely powdered for mineral identification by X-ray diffraction. An Ital Structure diffractometer with filtered CuKά radiation was used. Chemical analysis of reacted rocks was carried out by ICP OS (Thermo scientific, iCAP 6300).

Klinkerberg permeability was determined in the same briquette before and after the acid treatment by measuring the absolute permeability by the stable state technique at room temperature and constant pressure using nitrogen as work fluid (Contreras and García, pers. comm.).

Results

In core A, calcite is present filling holes and fine veins, after the reaction with HCl, empty holes and veins were observed indicating complete calcite dissolution; also it was noted that acid penetrates the rock matrix. A fragment of the same core without calcite, after acid treatments show only bleaching of the external surface of the specimen; also after the reaction with the binary solution the external surface of the specimens turned bleached and rough. Figure 2, at the left shows a fragment of core A before any treatments, below of it a cylinder after the reaction with HCl 10%. At the right the same fragment after the third treatment in HCl 10% solution.

Chemical analyses of treated samples show lower concentration of major elements compared to the original and untreated specimen. Chemical, physical and mineralogical changes are related to the original rock composition, time of reaction and the type of acid solution. In core A, calcium concentration decreases due to calcite dissolution. Calc-silicates react with the mixture of HCl-HF decreasing in Si, the concentration of most of the major elements show slight decrease. The major weight loss is observed in Core A due to the important dissolution of calcite.

The fragment of core B used in the experiments is a fine grain andesite, with no apparent alteration; except by the presence of hematite, which appears disperse in fine veins and fractures. After the treatments no important weight loss was registered; also changes in permeability are small. Again the surface of the core fragment shows roughness resembling an etched glass with superimposed hematite veins (Figure 3). That means that acid solution react on the surface rock forming minerals without entering in to the matrix of the core. At the conditions of our experimentation hematite is insoluble in acids.

Next example is core C. It corresponds to an andesite with high intensity of hydrothermal alteration represented by epidote, chlorite, quartz, calcite and small amounts of mica. The specimen selected for the experiment has a fracture filled by quartz and epidote. As the sample lacks in calcite the reaction with acids is at the contact surfaces. Figure 4 shows, a slice of the core C before any treatment showing a vein sealed with quartz and epidote. At the end of the acid treatments, the specimen is bleached and no apparent dissolution of quartz and epidote is observed. Small broken chips of the sample show that fluids do not penetrate to the rock matrix. As expected, when calcite is absent, negligible differences in loss weight and in permeability data are recorded. That means that the mixture of HCl + HF is not able to dissolve quartz and epidote.

Core D, this core is a particular one; it has been studied extensively to understand other processes that locally affected the deep reservoir rocks [6]. In Figure 5 two specimens from different well are compared. At the left, a fragment of core B which is an andesite; after the three acid treatments, the specimen looks bleached by the action of acids showing small patches of hematite. At the right, core D before acid treatments.

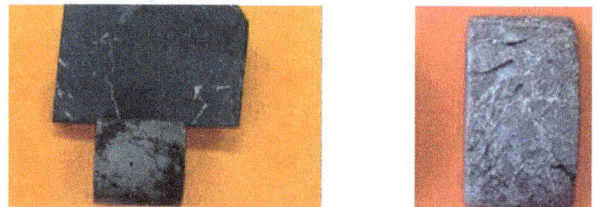

Figure 2: Core A. At the left, on top, the original fragment of the core A, below it after HCl treatment; at the right the same fragment after reaction in HCl+ HF solution. Open conduits and rough surface are observed. Dimension of specimens are 3.5 cm in diameter and 3- 4 cm high.

Figure 3: At the left a piece of the untreated core B (15 cm x 8 cm). At the right an image of a fragment of core B after reaction with the binary acid mixture, (16X).

Figure 4: A fragment of core C, before acid treatment, with a vein sealed by quartz and epidote. After reaction in acid solutions the cylinder shows a bleached surface and no dissolution of quartz and epidote. The size of the cylinder is 3.5 cm of diameter and 4 cm height.

Figure 5: At the left a fragment of core B after acid treatments in the laboratory; at the right a fragment of core D as it is, without any acid treatment in the laboratory.

It is remarkable that their physical appearance is similar; both look bleached. The chemical, mineralogical, petrophysical and mechanical data of core D, suggest that the rock from 2000 m deep wazs exposed naturally to the action of an acid fluid.

Assuming that the rock from core D was in contact with a low pH fluid in the reservoir, the action of such fluid was to bleach and to leach components leaving a bleached silicified mass. Probably the residence time of the fluid in contact with the original rock was not enough to form new minerals typical from acid environments; instead, the fluid reacted in the same way as we observe in the laboratory. The product in the natural system was a bleached, leached and silicified mass. The petrographic and X-ray diffraction analysis of core D indicate that quartz is the main component, plagioclase, and traces of chlorite and pyrite also are present. In our laboratory experiments using core D showed no changes when reacting with HCl 10%, mineralogy, chemical composition and permeability were exactly the same after interaction of the rock with the hot acid solution. That is because in HCl the main components of this core are insoluble. However, after the reaction with the binary acid mixture important changes were observed in the chemistry and in the permeability of the rock sample. The mineralogical composition was the same.

Table 1 presents chemical composition data, for major oxides in weight %, of a fragment of core D before any treatment. Also includes chemical data of fragments of core D after acid treatments. Samples were analyzed in a Thermo ICP-OS by the author Izquierdo G. As can be noted, after the interaction with HCl the sample shows slight change in concentration of major oxides. After the interaction with the binary solution, the chemistry changed particularly in SiO_2.

As part of the characterization of core D, tensile strength was determined before reaction in acid solutions. This value was compared to the value obtained for core samples from other wells of the same field before acid treatment. Table 2 includes data for core D compared to core B; also core E is included even it was not part of this work. From data can be seen that tensile strength for core D is almost half of the value for the other two cores, even they are hydrothermally altered. This result has been considered as an evidence of the interaction of deep rocks of the reservoir with acid fluids [6]. As tensile strength is a destructive test it was not possible to determine in reacted specimens; for sure will be affected by the reaction with acids.

As far as permeability of core D is concerned, the initial value is the same as the measured after the reaction with HCl. That means that there were no minerals to dissolve so the difference in weight lost is negligible. When core D was reacted in the binary solution the permeability increases and the weight lost was considerable.

Table 3 is a summary of Klinkenberg permeability data before and after acid treatments. The permeability increased in samples where calcite was present and because of the light leaching of the contact surface between the sample and the solution. The permeability increases even more after the reaction with the binary mixture of acids.

Data for core D supports the assumption that the original rock was depleted in primary and secondary minerals and was a silica rich mass (microcrystalline quartz). After the reaction with HCl 10% as mentioned the permeability has the same value. With the binary mixture the permeability increases considerably because bonds of the micro crystalline quartz are easily to break down as for silicates and calc-silicates.

All solutions where the rocks reacted were preserved for two weeks at room conditions in order to have evidence of the formation of new

Sample	SiO_2	Fe_2O_3	MgO	CaO	Na_2O	K_2O
Core D before treatment	65.940	1.019	0.644	1.951	4.111	8.980
After HCl 10 %	65.475	0.919	0.643	1.876	4.046	7.284
After HCl 10%:HF 5%	59.511	0.752	0.879	2.578	2.991	6.772

Table 1: Chemical composition of a fragment of core D before and after acid treatments.

Sample	Length (cm)	Diameter (cm)	Tensile strength (MPa – Bar)
Core D	2.53	5.14	2.21 – 22.1
Core B	2.53	5.14	4.89 -48.9
Core E	2.53	5.14	4.53 -45.3

Table 2: Tensile strength data of core D, which has been altered by acids fluid in the natural system and data for cores hydrothermally altered

Sample	Klinkenberg permeability Before(mD)	Klinkenberg permeability After (mD)
Core D HCl 10%	2.18	2.18
Core D HCl 10% + HF 5%	2.18	57.6
Core B HCl 10%	0.096	0.11
Core B HCl 10% + HF 5%	0.11	5.92
Core E HCl 10%	0.073	0.284
Core E HCl 10% + HF 5%	0.284	1.47

Table 3: Permeability values before and after acid treatments.

phases from them. Under this condition no new phases were formed from these acid solutions. Solid material at the bottom of vessels were separated, dried and analyzed by XRD, the identified phases are the rock forming minerals.

Results indicate that working with igneous rocks, before any attempt to acid stimulation of wells; it is very important to known the type and intensity of hydrothermal alteration, the nature of the rocks we are dealing with and the processes that have been occurred in the reservoir.

Conclusions

Even the results were obtained at micro scale in the laboratory, compared to the size of a geothermal reservoir; give an approach of what may be expected in an industrial acid stimulation work.

The results showed that the effectiveness of matrix stimulation of igneous rocks will depend on the type and on the intensity of the hydrothermal alteration.

Calcite reacts rapidly with both acid solutions leaving open pores and veins (wormholes); while calc silicates react only superficially leaving much of the rock matrix unreacted.

Other minerals like chlorites in veins or cavities will be partially dissolved. Ring and chain silicates are hard to dissolve; as well as quartz, hematite and epidote. The binary mixture will penetrate in empty veins and cavities or just will act on the surface of the rock leaving a rough surface enough to promote fluid circulation.

X-ray diffraction of the chemically treated samples shows the same mineralogy as the original rock samples. That means that acid react by dissolving first contact minerals leaving much of the rock matrix unaffected, except when calcite is present.

When calcite is absent, the third treatment after the binary acid mixture does not show any difference in the physical and chemical characteristics of rocks.

No new phases are formed by the interaction between the rocks and acid solutions.

The Klinkerberg permeability increases in samples where calcite was present. Values are relative to the distribution and the amount of calcite in each specimen.

Acknowledgements

The authors want to express their gratitude to the authorities of the Gerencia de Proyectos Geotermoelectricos from the Comisión Federal de Electricidad of Mexico (CFE) by their cooperation in freely supplying samples. Some data were taken from the final report of the contract No 9400046929 between CFE and IIE.

References

1. Kalfayan L (2008) Production enhancement with acid stimulation: Penn Well Corporation.

2. Flores M, Davies D, Couples G, Palsson B (2005) Stimulation of geothermal wells, can afford it? Proceedings of the World Geothermal Congress. Antalya, Turkey.

3. Sandrine P, Francois-David V, Patrick BNS, André G (2009) Chemical stimulation techniques for geothermal wells: experiments on the three-well EGS system at Soultz-sous-Forêts, France. Geothermics 38: 349-359.

4. Malate RCE, Austri, JJC, Sarmiento ZF, Di Lullo G., Sookprason PA, et al. (1998) Matrix stimulation treatment of geothermal wells using sandstone acid. Proceedings. 32th Workshop on Geothermal Reservoir Engineering Stanford University. Stanford, CA.

5. Mella M, Kovac K, Xu T, Rose P, McCulloch J (2006) Calcite dissolution in geothermal reservoirs using chelants. Geothermal Resources Council Transactions 30: 347-352.

6. Izquierdo G, Aragón A, Gómez FR, López S (2014) Evidencia Mineralógica del efecto de fluidos ácidos sobre las rocas del yacimiento geotérmico de Los Humeros, Puebla. Geotermia Revista Mexicana de Geoenergía 27.

7. Gutiérrez-Negrín LCA (1982) Litología y zoneamiento hidrotermal de los pozos H-1 y H-2 del campo geotérmico de Los Humeros, Pue. CFE, Internal report number 23-82.

8. Viggiano JC, Robles J (1988) Mineralogía hidrotermal en el campo geotérmico de Los Humeros, Pue. I: Sus usos como indicadora de temperatura y del régimen hidrológico. Geotermia Revista Mexicana de Geoenergía 4: 15-28.

15

Dynamic Energy Modelling for Data Centres: Experimental and Numerical Analysis

Eduard Oró[1*], Alvaro Vergara[2,3] and Jaume Salom[1]

[1]Catalonia Institute for Energy Research, IREC, Spain
[2]University of Freiburg, Friedrichstr 39, 79098, Freiburg, Germany
[3]Pontificia Universidad Católica de Chile, Santiago, Chile

Abstract

The total energy demand of data centres has experienced an important increase in the last years. Due to their unique nature, data centres demand enormous amounts of energy. Therefore, they are ideal candidates for implementing actions to reduce the energy consumption and thus improve their ecological footprint while at the same time reduce their operational costs. The aim of this work is to develop and to validate with experimental data a dynamic energy model of a real data centre located in Barcelona. The dynamic energy model is then used first to characterize the energy consumption and the energy efficiency of the infrastructure and second to see the benefits of the implementation of different energy efficiency strategies into the data centre cooling system portfolio. The results show an average Power Effective Usage (PUE) of 1,74 while some of the proposed strategies can achieve important energy reductions, up to 21%, in the cooling energy consumption. Therefore, the combination of them can achieve important reductions in the overall data centre energy consumption. The validated energy model can then be used to study the benefit of the implementation of different energy efficiency strategies in other data centres.

Keywords: Data centre; Dynamic energy model; Experimental analysis; Energy efficiency; Energy consumption

Introduction

Data centres are continuously growing in size, complexity and energy demand due to the increasing demand for storage, networking and computing. These unique infrastructures run 24 h a day, the 365 days of the year and they are up to 100 times more energy intensive than conventional office buildings. Nowadays, 40% of the total energy consumption is attributed to cooling and therefore the development of effective and efficient strategies to reduce cooling demand is required. Recently the data centre industry [1,2] has taken consciousness of the need of the implementation of energy efficiency strategies and the use of renewable energy sources in data centre not only to show their environmental commitment, but also to reduce the operational cost. In that sense, Oró et al. [3] presented a literature review on the implementation of energy efficiency strategies and the integration of renewables into data centres portfolio.

In conventional data centre the cooling infrastructure is divided into cooling production mainly air cooled chillers which produce chilled water and cooling distribution which distributes chilled water to Computer Room Air Handling (CRAH) units. Therefore the evaluation of the cooling system performance in data centres should be focused on cooling production and air management which has the objective to keep Information Technology (IT) equipment intake conditions within the recommended ranges with the minimum energy consumption. Due to the recent increase of data centre industry many researchers have been focusing on the implementation of energy efficiency strategies into the cooling system. In that sense, a number of case studies reveal that the fan energy savings in 70-90% range and chiller energy savings in 15-25% range are achievable with effective air management [4]. Lu et al. [5] evaluated the air management and energy performance of the cooling system of a data centre in Finland. They investigated for that specific facility the possibilities of energy savings (mainly the reduction of the fan speed) and heat reuses for space heating and hot water. Similarly, Lajevardi et al. [6] analysed a small data centre located in the Gresham City Hall (United States) over

a period of six weeks proposing different energy efficiency and thermal management issues. Choo et al. [7] evaluated experimentally and numerically the energy efficiency performance of a medium size data centre at the campus of the University of Maryland. They also assessed energy conservation measures such as eliminating unnecessary CRAH units, increasing the return set point temperature, using of cold aisle containment and implementing free cooling. To evaluate data centre air management performance, Computational Fluid Dynamics (CFD) modelling of IT equipment and indoor temperature distribution is required regards the air distribution system and IT server's operation requirements. Even though CFD analysis provides valuable inputs for data centre air management system, they are useless for real time data centre cooling management infrastructures and for dynamic energy systems evaluation. Therefore, other modelling techniques are needed to overcome these problems, reaching a compromise between time and cost required by the simulation tool and reliable information. The use of dynamic energy model using Transient System Simulation program (TRNSYS) [8] can overcome this problem. Kim et al. [9] studied the feasibility of the integration of a hot water cooling system with a desiccant-assisted evaporative cooling system for data centre air conditioning using TRNSYS. Recently, Depoorter et al. [10] developed a dynamic energy model using TRNSYS to assess the potential of direct air free cooling in the data centre portfolio around Europe.

*Corresponding author: Eduard Oró, Catalonia Institute for Energy Research, IREC. Jardins de les Dones de Negre 1, 08930, Sant Adrià de Besòs (Barcelona), Spain, E-mail: eoro@irec.cat

The aim of this paper is to develop a dynamic energy model for data centre facilities in order to evaluate the benefits of the implementation of different energy efficiency strategies in the cooling system. The dynamic model is validated with experimental data from a real data centre located in Barcelona (Spain) and is also used to characterize the energy consumption of the data centre.

Methodology

Operational requirements

The American Society of Heating, Refrigerating, and Air-Conditioning Engineers (ASHRAE) thermal guidelines [11] define recommended and allowable temperature and humidity ranges for four environmental classes, two of which are applicable to data centres. The recommended envelope (Table 1) defines the limits under which IT equipment would most reliably operate while still achieving reasonably energy efficient data centre operation. However, it is acceptable to operate outside the recommended envelope for short periods of time without risk of affecting the overall IT equipment reliability.

Data centre characteristics

The data centre is an operative facility with an IT capacity of 115 kW which is used to provide computing and information services for the Polytechnic University of Catalonia (Spain). The data centre was constructed in 2006 and started operations in 2007. It has an IT room area of 285 m² and the whitespace is located on a second basement surrounded by other refrigerated areas. Figure 1 shows the scheme of the data centre and the equipment distribution. The facility is composed by 70 racks of data and 12 racks of communication equipment. The theoretical maximal power consumption is 4 kW per rack but at the present moment the real average power consumption is between 1.5 and 2 kW per rack. Some racks are distributed in cold and hot aisle containment and some other racks are placed with no containment in the whitespace. Hot and cold aisle containment is an effective strategy to ensure a properly air management inside an aisle of racks. However, if proper installation is not done air inefficiencies can also be present. The racks which are not enclosed are directly mounted over the perforated tiles of the raised floor or they have a frontal air tide. This situation reduces the airflow velocity and also forces some of the air to directly bypass the rack without exchange heat with it.The infrastructure is connected to the main grid but there is also an emergency generator in case of energy supply failure. Figure 2 shows schematically the

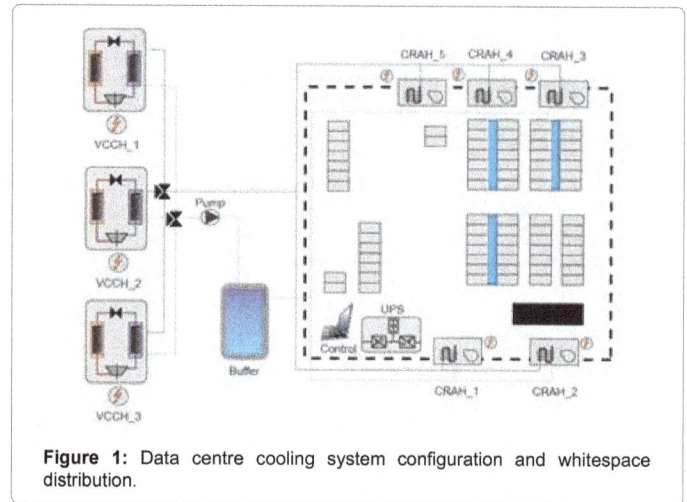

Figure 1: Data centre cooling system configuration and whitespace distribution.

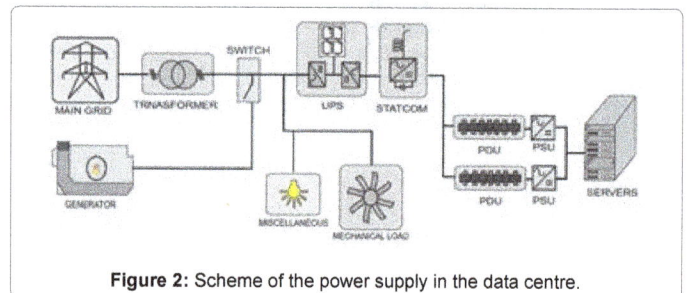

Figure 2: Scheme of the power supply in the data centre.

main electrical and mechanical components of the data centre. The installation has N+1 redundancy in the chillers, and 2N in the power distribution units (PDU) and the power supply units (PSU). Notice that all the electrical components are located inside the whitespace and thus contribute to the IT room thermal load.

The power consumed by the IT equipment and the electrical losses of the equipment is converted into heat [1] and therefore reliable thermal management is essential to provide an adequate environment for IT devices [5]. The refrigeration system is composed of five CRAH units which use chilled water from three water-air chillers. A raised floor is used to distribute the chilled air to the bottom of the racks and the exhaust warm air leaves the room direct to the CRAHs as it is shown in Figure 1. The 3-way valve of the CRAH controls the cooling exchange between the chilled water and the exhaust air. This valve actuates (opening or closing) in function of the return air temperature from the whitespace. Each unit keeps a constant air flow rate to the IT room of 8500 m³/h while the total chilled water flow rate is 39.9 m³/h. There is also a 1 m³ buffer tank after the chillers to prevent water temperature fluctuations. The main characteristics of the cooling equipment are listed in Table 2.

Dynamic energy model

To evaluate the effect of airflow efficiency improvement in the data centre, a dynamic energy model using TRNSYS has been developed. The model is based on a component-by-component approach. Information from the equipment manufacturers, data centre operators and data collected directly in the facility was used to build the energy model. This paper is focused on analysing the cooling system and therefore the whitespace characterization is assumed as a black box. This black box calculates the return air temperature in function of the current IT load, miscellaneous loads, and the air inlet conditions following Equation 1.

Equipment Environment Specifications				
Class	Product Operation		Product Power Off	
	Dry-Bulb Temp. range	Humidity range	Dry-Bulb Temp. range	Humidity range
		Recommended		
A1-A4	18 to 27 °C	5.5°C DP to 60% RH and 15°C DP		
		Allowable		
A1	15-32	20% to 80% RH	5-45	8% to 80% RH
A2	10-35	20% to 80% RH −12°C DP and	5-45	8% to 80% RH
A3	5-40	8% RH to 85% RH	5-45	8% to 80% RH
A4	5-45	8% RH to 90% RH	5-45	8% to 80% RH

Table 1: ASHRAE environmental classes for data centres [11].

Vapour Compression Chiller	
Model	STULZ 822
Cooling capacity [kW]	77.7
Rated EER	2.6
Water flow rate [m³/h]	13.3
CRAH	
Model	Uniflair TDCR 1200A
Cooling capacity [kW]	37
Energy consumption [kW]	2.5
Air flow rate [m³/h]	8500
Water Pumps	
Energy consumption [kW]	3
Water flow rate [m³/h]	42

Table 2: Specifications of main equipment.

$$\dot{Q}cooling = \dot{m}air \cdot Cp,air \cdot (Tair,return - Tair,supply) \qquad Eq.1$$

Where $\dot{m}air$ is the air mass flow rate, cp, air is the air specific heat capacity, Tair, supply is the air inlet temperature to the whitespace and Tsupply, return is the air return temperature. Notice that the total cooling demand ($\dot{Q}cooling$) is defined as the IT load ($\dot{Q}IT$) plus the additional loads ($\dot{Q}add$):

$$\dot{Q}cooling = \dot{Q}IT + \dot{Q}add \qquad Eq.2$$

$$\dot{Q}add = \dot{Q}CRAC,e + \dot{Q}electrical + \dot{Q}miscellaneous - \dot{Q}loss \qquad Eq.3$$

Where $\dot{Q}CRAC,e$ is the electrical loss inside the white space of the CRAH units, $\dot{Q}electrical$ is the electrical loss inside the whitespace (it are) by means of PDU, wiring, UPS, filter, electric board, $\dot{Q}miscellaneous$ is the miscellaneous losses inside the whitespace such as the lighting, ventilation, working people, etc. and $\dot{Q}loss$ is the heat loss through the wall to the environment.

The whitespace is cooled by CRAHs which were modelled with a cooling coil using a bypass approach (type 508). In order to control the chilled water flow rate to the CRAH and the humidification ration a proportional controller (type 1669) acts on a flow diverter (type 11) and an evaporative cooling device (type 507). The water is cooled by three chillers which are simulated using the well-known type 655. The energy model also consists of other main system components including single speed fan (type 112), pipes (type 31), etc. which are available in the TRNSYS library. Those components were connected according to the system configuration already described and shown in Figure 3. The typical meteorological year (TMY2) data was used to obtain the weather conditions for Barcelona. The TMY2 data sets are the typical values of meteorological elements for a one-year period from Meteonorm [12].

Results and Discussion

Model validation

For operational data, the water return temperature ($T_{w.return}$) was used as input of the model. The model was validated by comparing the simulation results with the operational data during 12 hours. In the validation, the supply chilled water ($T_{w.supply}$), the return water temperature ($T_{w.return}$), the cooling consumption ($P_{cooling}$) and the total consumption (P_{total}) were selected to validate the model. The actual chiller set points are different between them. Chiller #1 starts when

the return water temperature is above 9.5°C, when it is higher than 12.5°C then chiller #2 also starts and chiller #3 does not start till the return water is above 13.5°C. Moreover, the outlet water set point is set at 6°C. For the CRAH units, the return air temperature set point is set at 25.5°C. Figure 4 shows the modelled and the real temperature and power values over a period of 12 hours. According to these results, the predicted values by the model make good agreement with the actual operational data and therefore the consistency of the dynamic model is demonstrated.

Data centre energy characterization

Once the dynamic energy model proposed has been validated with experimental data from the data centre, it is used to evaluate the energy characterization of the installation over an entire year. Figure 5 shows the disaggregate energy consumption of the data centre after one year of operation. As expected the main consumers are the IT equipment and

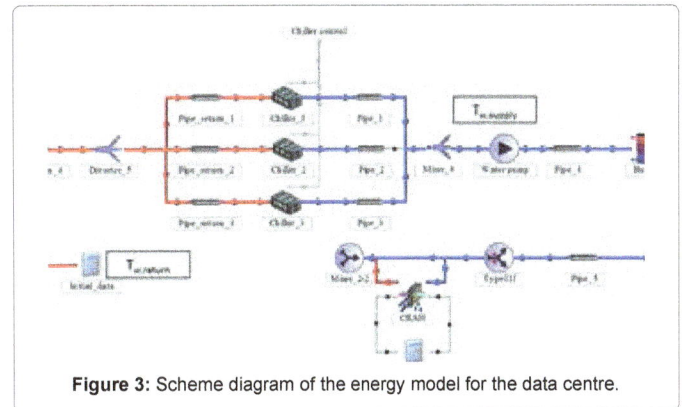

Figure 3: Scheme diagram of the energy model for the data centre.

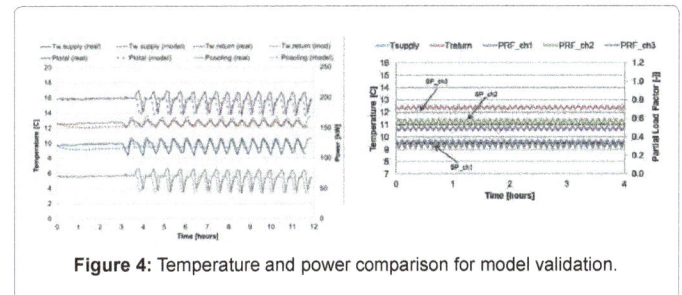

Figure 4: Temperature and power comparison for model validation.

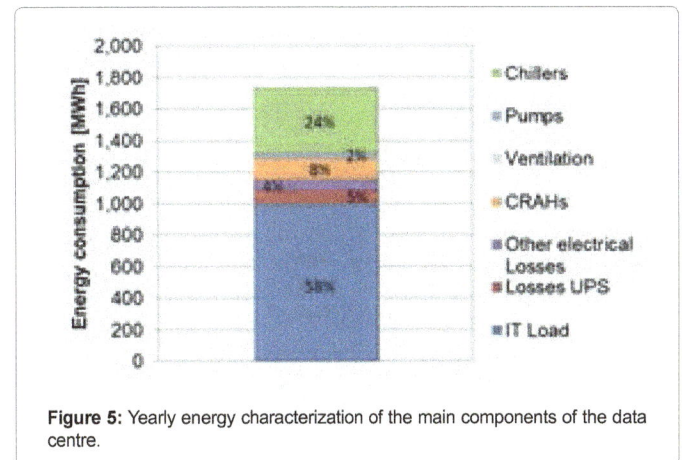

Figure 5: Yearly energy characterization of the main components of the data centre.

the chillers, being 82% of the total energy consumption. The cooling system which takes into account the consumption of the chillers, water pumps, CRAHs and ventilation represents up to 34% of the total energy consumption. Therefore, a potential of energy saving due to the implementation of better operational strategies is possible. The Power Usage effectiveness (PUE) metric is the most common metric used in the data centre industry. It measures the energy efficiency of the installation by dividing the total energy consumed by the facility with the IT energy consumed. In the present data centre the average PUE value is 1,74 which is in line with the self-reported PUE's values from the Uptime Institute 2013 data centre industry survey [13].

Cooling management strategies

The dynamic energy model allows the estimation of the data centre energy consumption under different energy management strategies. Different scenarios have been analysed:

- IT room air inlet temperature increased.

- Modification in the chillers working sequence.

- Conditioning of the chiller air.

- Increasing water chilled temperature.

IT room air inlet temperature rise: Increasing the IT room supply temperature has been suggested as the easiest and most direct way to save energy in data centres. However, as Patterson [14] noted, just implementing a higher inlet air temperature while still relying solely on mechanical cooling, may not improve the efficiency of the cooling system. That conclusion is confirmed by the results of the simulation, where an increase in the air inlet temperature has negligible results in the cooling energy savings of the infrastructure. Increasing the inlet air temperature from 18°C to 27°C and maintaining constant the chilled water temperature, a cooling energy reduction of only 2% was observed. Since the total heat that has to be removed from the IT room is the same and the temperature difference between air inlet and outlet is constant since the air volume flow is also constant, thus the fans does not experience less consumption. On the other hand, the chiller water pump operation is highly affected for this measure reducing its energy consumption drastically but it does not affect at the overall picture since its consumption is not significant.

Modification in the chillers working sequence: It is well known that most chillers operate more efficiently in partial load using variable speed compressors and pumps, resulting in a reduced water and refrigerant flow. Therefore, when using chillers at partial load, the Energy Efficiency Ratio (EER) will increase. However, an optimum exists at a certain partial load, depending on the characteristics of the chiller and below this the energy efficiency starts to decrease again. Moreover, chiller sequencing control is an essential function for multiple-chiller plants that switches on and off chillers in terms of data centre instantaneous cooling load. It significantly affects both inlet air temperature control and data centre energy consumption [15,16]. In the present installation a chiller sequencing control is implemented which switch on or off chillers according to a direct indication of the system, the return water temperature. Therefore, firstly chiller #1 is activated when cooling is needed; if more cooling is needed then chiller #2 is activated and just in the case that a high level of cooling is needed chiller #3 is switched on. Table 3 shows the initial and the modified water set point temperatures for each of the chillers. Notice that in the reference configuration when the return water temperature was between 9.5 and 12.5°C only chiller #1 was activated (Figure 6). In the

		Chiller #1	Chiller #2	Chiller #3
$T_{water.IN}$ (Return water temperature)	Reference	9.5	12.5	15
$T_{water.IN}$ (Return water temperature)	New version	9.5	11	12

Table 3: Return water temperatures to activate the chillers.

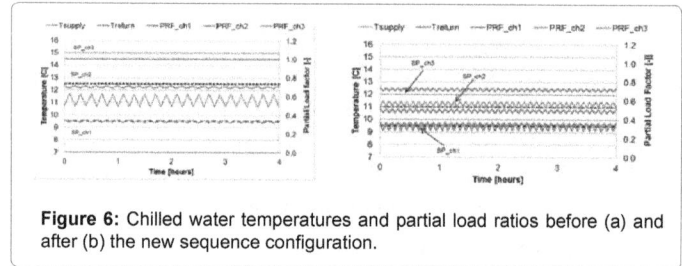

Figure 6: Chilled water temperatures and partial load ratios before (a) and after (b) the new sequence configuration.

new configuration the activation temperature for the first compressor of chiller #2 and #3 has been reduced while the set point temperature for the second compressor has been enhanced. Notice that in the reference situation a small increase in the inlet water temperature activated the full load of each chiller. This modification in the current configuration allows the chillers working at partial load while maintaining the supply water temperature.

This new sequence configuration was implemented in the control strategy of the data centre model and was evaluated over a period of one year. Figure 7 shows the energy consumption of each of the chillers of the installation and the energy reduction. The total energy reduction was around 12%, representing more than 4% in the overall data centre energy consumption. Therefore an annual energy savings of 50.4MWh are expected. Notice that chiller #3 increased its energy consumption since in the reference case it is normally switched off.

Conditioning of the chiller air: It is well known that lower dry bulb temperature of outdoor air enhance the EER air-cooled chiller systems [17]. This outdoor air cooling can be achieved using adiabatic cooling process which reduces heat through a change in air humidity rate; this phenomenon happens due to in the process of absorbing water the air uses its enthalpy reducing then its temperature. Figure 8 shows the ambient air temperature during 7 days of summer at Barcelona and the air temperature once it is conditioned by adiabatic cooling. Notice that the conditioned air is plotted for total saturation at 100% relative humidity and at 90% relative humidity. In peak hours the temperature decrease can be up to 5°C. The model has adopted total air saturation during the entire year. The results show that for Barcelona an energy reduction of 4% in the cooling system can be achieved. Therefore an annual energy savings of 17.62 MWh are expected. However, the implementation of this process besides the adiabatic cooling system also needs to account for water consumption. With no implementation of optimization process in the system; so the system always adds water to the air in order to reach total saturation even though the temperature reduction potential is really low i.e. during nights the temperature reduction can be less than 1°C, the yearly water consumption is 330 m³. This water consumption obviously reduces the overall economical savings of the implementation of this strategy. Moreover, an additional water pump must be installed but its energy consumption will not affect the overall energy consumption of the system.

However, the overall potential of this strategy increases when it is applied to locations with higher dry bulb temperature and low humidity thanks to its ability to reduce of reducing temperature by absorbing water. This happens for instance in continental climate zones like Madrid. The model has been used to see the results of this implementation in Madrid showing a potential energy reduction of the cooling system of 6%.

Increasing chilled water temperature: In this scenario the authors wanted to study the effect on the energy consumption of the infrastructure when the chilled water temperature is raised from 6 (reference case) to 11°C while the inlet air temperature is kept constant at its maximum. In this scenario the water temperature increases is selected following ASHRAE and chiller manufacturer recommendations and while the maximum air inlet temperature is set to 27°C based on the ASHRAE recommendations. When increasing

the supply air temperature, the IT room environment will obviously be changed. This strategy needs to be studied in detail in order to avoid hot spots due to air management inefficiency. Figure 9 shows the energy consumption of the chillers under different scenarios (supply air temperature at 27°C while enhancing chilled water temperature). With proper system management conditions annual cooling energy savings up to 21% can be achieved. Therefore an annual energy savings of 88.20 MWh are expected. At a certain point, even though the water chilled temperature is increased the energy consumption is almost constant and it is expected to decrease. This phenomenon occurs since the heat transfer between the water and the air decreases (lower temperature difference between them) and thus the water flow rate has to be increased in order to supply the same amount of cold to the IT room.

However, the dynamic nature of IT equipment cooling fans and semiconductor leakage current in the CPU due to higher temperatures may diminish or even negate the cooling system gains due to an enhancement of the air inlet temperatures. Moss and Bean [18] highlighted the importance of undertaking a holistic analysis of the data centre energy consumption taking into account not only cooling energy reduction by the implementation of energy efficiency strategies such as free cooling or higher inlet air temperatures but IT equipment performance due to higher inlet air temperatures. Figure 10 shows the increase of energy consumption (percentage in function of the energy consumption at 18°C) of 3 different servers in function of the air inlet temperature. While the air inlet temperature does not play a significant role in the energy consumption of some architecture (server X) in other servers can increase the overall energy consumption up to 25%. Therefore, further research should be done in order to optimize the management system minimizing the overall data centre energy consumption and taking into account particular IT equipment consumption.

Conclusions

In the last years, the increasing energy demand of the data centre industry has put the sector under pressure to limit its environmental impact and reduce energy costs. Industry and researchers have made a lot of effort to overcome this situation and many energy efficiency measures have already been investigated and implemented. Efficient cooling production and air management in data centres plays an important role reducing the overall cooling energy consumption. This paper aimed to develop a dynamic energy model of a real 115 kW IT data centre located in Barcelona (Spain). The proposed model is then validated with experimental data collected during 2 weeks

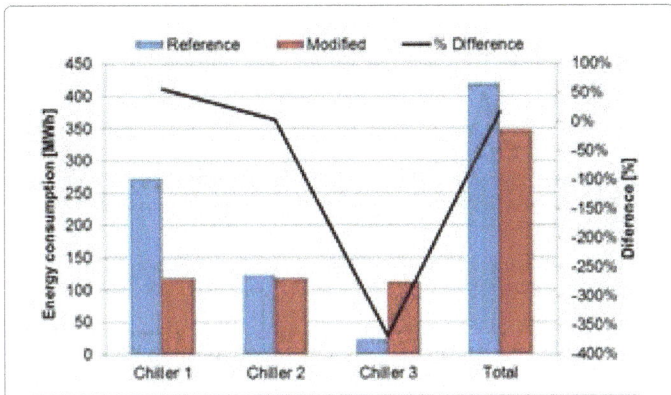

Figure 7: Energy consumption of the chillers before and after the new sequence configuration.

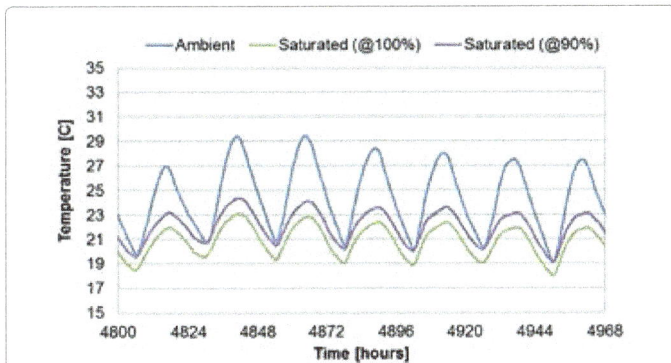

Figure 8: Ambient air temperature at different conditions during summer.

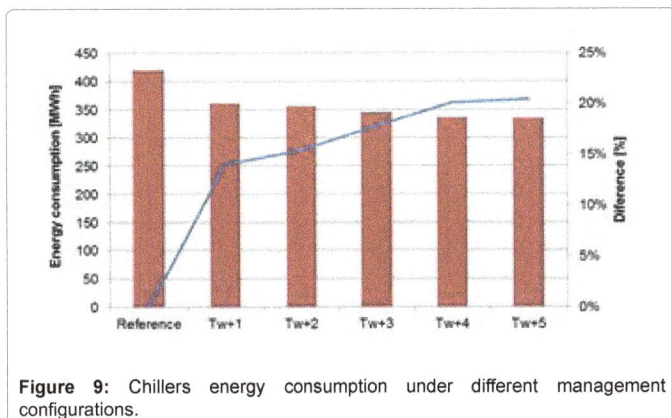

Figure 9: Chillers energy consumption under different management configurations.

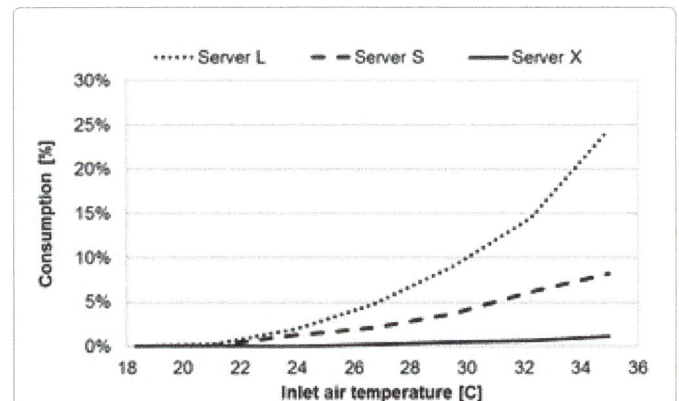

Figure 10: IT equipment consumption in function of inlet air temperature.

under normal operation. First, data centre energy characterization was studied being the cooling system energy consumption up to 34% of the total energy consumption of the facility. Moreover the PUE value was also calculated being 1.74. Second, the already validated dynamic model was used to study the benefit of the implementation of different energy efficiency strategies. The strategies proposed and studied were an increase of the inlet air temperature, a modification of the chiller working sequence, an analysis of the improved energy performance of the air-cooled chiller with cooled air and an increase of the water chilled temperature. The results show that the only increase of the IT room supply temperature has no significant decrease in the energy savings. This strategy should be implemented in parallel with others such as the use of air free cooling but not when operating with solely mechanical cooling. The actual chiller working sequence (one chiller after the other) can be improved maintaining the IT room temperatures. This strategy has the potential of reducing the cooling energy consumption up to 12%. Adiabatic cooling can help improving the energy performance of air-cooled chiller system. The implementation of this strategy in the studied infrastructure can reduce the cooling consumption up to 4%. However, this strategy aims for an initial investment and operational costs (water consumption and maintenance). Notice that the location of the data centre in Barcelona (high ambient relative humidity) reduces the overall potential of this strategy which will be much more efficient in continental climate zones such as Madrid. Moreover, for the correct implementation of this strategy, a control algorithm to optimize the use of cooling water reducing the annual water consumption while enhancing the benefit of the air saturation. Finally the chilled water temperature increase while the inlet air temperature was kept as its maximum (27°C from ASHRAE recommendations) was also studied. The results show that with proper system management control annual cooling energy savings up to 21% can be achieved. It is worth to highlight that at certain point the cooling energy consumption keeps almost constant and it is expected to increase due to an increase of the water pumps. This strategy should also be studied in detail since due to the dynamic nature of IT equipment (mainly for variable velocity internal fans and semiconductor leakage current in the CPU) can diminish or even negate the cooling system gains due to an enhancement of the air inlet temperatures. Therefore, further research should be done in order to optimize the management system minimizing the overall data centre energy consumption and taking into account particular IT equipment energy consumption.

Acknowledgments

The research leading to these results has received funding from the European Union's Seventh Framework Programme FP7/2007-2013 under Grant Agreement n° 608679 - RenewIT.

References

1. http://www.google.com/green/energy/

2. https://www.apple.com/environment/renewable-energy/

3. Oró E, Depoorter V, Garcia A, Salom J (2015) Energy efficiency and renewable energy integration in data centres. Strategies and modelling review. Renew Sust Energ Rev 42: 429-445.

4. Herrlin MK (2010) Data Center Air Management Research. Application assessment report #0912, ANCIS Incorporated.

5. Lu T, Lü X, Remes M, Viljanen M (2011) Investigation of air management and energy performance in a data center in Finland: Case study. Energy and Buildings 43: 3360-3372.

6. Lajevardi B, Haapala KR, Junker JF (2014) Real-time monitoring and evaluation of energy efficiency and thermal management of data centers. Journal of Manufacturing Systems 37: 511-516.

7. Choo K, Galante RM, Ohadi MM (2014) Energy consumption analysis of a medium-size primary data center in an academic campus. Energy and Buildings 76: 414-421.

8. http://www.trnsys.com

9. Kim MH, Ham SW, Park JS, Jeong JW (2014) Impact of integrated hot water cooling and desiccantassisted evaporative cooling systems on energy savings in a data centre. Energy 78: 384-396.

10. Depoorter V, Oró E, Salom J (2015) The location as an energy efficiency and renewable energy supply measure for data centres in Europe. Applied Energy 140: 338-349.

11. http://ecoinfo.cnrs.fr/IMG/pdf/ashrae_2011_thermal_guidelines_data_center.pdf

12. http://meteonorm.com/

13. (2013) Data Center industry survey. Uptime Insitute.

14. Patterson MK (2010) The Effect of Data Center Temperature on Energy Efficiency. Intel Corporation, USA.

15. Chang YC, Lin FA, Lin CH (2005) Optimal chiller sequencing by branch and bound method for saving energy. Energy Conversion and Management 46: 2158-2172.

16. Liao Y, Huang G, Sun Y, Shang L (2014) Uncertainty analysis for chiller sequencing control. Energy and Buildings 85: 187-198.

17. Yu FW, Chan KT (2011) Improved energy performance of air-cooled chiller system with mist precooling. Appl Ther Eng 31: 537-544.

18. Moss D, Bean JH (2009) Energy impact of increased server inlet temperature. White paper American Power Conversion.

16

Cells Harvesting of Tropic Ocean Oleaginous Microalga Strain *Desmodesmus* Sp. WC08

Sen Zhang[1,3], Ping-huai Liu2*, Jiang-wei Wu[2] and Qing Wang[2]

[1]*Science Island Branch of Graduate School, University of Science and Technology of China, China*
[2]*College of Materials and Chemical Engineering, Hainan University, China*
[3]*China National Tobacco Quality Supervision and Test Center, Zhengzhou, China*

Abstract

For biomass recovery of tropic ocean oleaginous microalgae strain *Desmodesmus* sp. WC08, eight flocculation methods (pH adjustment, $Al_2(SO4)_3$, polyacrylamide, $AlCl_3$, $Ca(OH)_2$, $FeCl_3$, alum and chitosan) were evaluated and optimized. The results indicated that ferric chloride, aluminum sulfate, aluminum chloride and chitosan exhibited high flocculation ability and their flocculation efficiency were all beyond 94% at the optimal dosage (0.15, 0.4 and 0.03 gL^{-1}, respectively). Chitosan displayed the most tremendous potential for biomass recovery from culture broth based on its feasibility and safety. Acetic acid and hydrous chloride, used to dissolving chitosan, had no significant difference in the flocculation efficiency. And when the pH of culture broth was set 5 or 6, the flocculation efficiency of chitosan was higher and the required flocculation time was less. More dose chitosan was needed to harvest the biomass following the cell growth stage. Overall, the optimal flocculation reagent for harvesting biomass of *Desmodesmus* sp. WC08 is chitosan, and its optimal flocculation conditions are: just when the pH of the culture broth was set to 6 and the dosage of chitosan was 0.03 g/L at the end of microalgae cultivation, more than 110g of algal biomass can be recovered just by per 1 g of chitosan. Meanwhile, there is little residual chitosan found in the final flocculation supernatant which can be reused to some extent.

Keywords: Microalgae *Desmodesmus sp*; Flocculation; Cells harvesting; Chitosan

Introduction

In recent decades, microalgae have become more and more attractive because of a wide range of applications. For example, algal biomass can be used as food source both for human and animal, fertilizer, biofuel, fine chemicals, pharmaceuticals and water treatment [1]. Especially for the purpose of biofuel, microalgae have been attracting much attention from the scientific community to government [2]. However, microalgal cells harvesting is a bottleneck technology all the time for all kinds of miroalgal applications based on algal biomass, due to their small size (5-20 μm), low culture concentration (0.5-5 gL^{-1}) in culture medium, and high cost (about 20-30% of total production cost) [3,4].

To date, there are several available techniques for microalgae harvesting, such as centrifugation, filtration, air floatation, gravity sedimentation, flocculation and so on. Centrifugation and filtration are two kinds of common solid-liquid separation technology, which are most usually used to collect microalgal slud. But centrifugation is an energy-intensive process and is unfit to be used in large-scale [5]. While filtration is ease to operate, the membranes can be rapidly fouled by extracellular organic matter and minority of microalgae which are vimineous or have big size may be more suitable [6]. Gravity sedimentation is convenient and cheap, but that is just for those microalgae which have good self-sedimentation nature. Air floatation is also confirmed to be efficacious for microalgae harvesting. However, in contrast, flocculation is most popular among these methods. Because it is effective, convenient and low-cost for harvesting microalgae from large quantities of culture broth [7]. Besides, it can largely improve harvesting efficiency of other separation technology by pretreatment.

Flocculation is a process in which freely-suspended cells are aggregated together to form large particles and cells can then be easily harvested by sedimentation [8]. Flocculation is usually induced by various factors, mainly including chemicals which are called flocculants, for example, aluminum and ferric salts, as well as cationic polymers.

Different flocculants have different performance and effects on different microalgae strains. Many chemical flocculants are inexpensive and effective for variety of microalgae strains, but most of them are toxic and have some adverse effects on downstream process. Chitosan, a natural biopolymer, is a linear poly-amino-saccharide and has marked advantages over commonly used chemical flocculants [9]. It has high flocculation ability, and is non-toxic and biodegradable. However, the flocculation ability usually changes with the relative conditions to some extent, such as flocculant dosage, pH of culture medium, settling time and so on.

In our previous study, tropic ocean oleaginous microalgae strain *Desmodesmus* sp. WC08 was identified as one possible candidate for biodiesel production based on its biomass productivity, lipid productivity and fatty acid profile [10]. However, fewer studies were performed to investigate cells harvesting of tropic ocean microalgal species of the *Desmodesmus* genus. So in this work we have explored the harvesting potential of tropic ocean microalgae strain *Desmodesmus* sp. WC08 with different flocculants. In addition, the effects of dosage, sedimentation time, acid solution, pH of culture medium and growth stage on flocculation efficiency are also discussed. Finally, the effects of recycled water from harvesting process on the growth and lipid accumulation of microalgae *Desmodesmus* sp. WC08 are further investigated.

Corresponding author: Ping-huai Liu, College of Materials and Chemical Engineering, Hainan University, Haikou, 570228, P.R. China, E-mail: zhangsn2013@sina.com

Materials and Methods

Microalgae strain and growth conditions

Microalgae strain *Desmodesmus* sp. WC08 was obtained from Hainan University at Haikou, China. Microalgae were grown photo-autographically in BG-11 medium added 20 gL^{-1} sea salt. The cells were grown in 250 cm^3 culture medium in a sterilized air-bubbled (from the bottom) transparent glass-tube (600 mm×30 mm i.d.). Mass cultivation was carried out in 1 L culture medium in a sterilized air-bubbled (from the bottom) transparent glass-tube (600 mm×60 mm i.d.). All cultures were incubated at room temperature and the light intensity was approximately 160 μmol m^{-2} s^{-1} with continuous illumination.

Effects of different flocculants with different dosages

Seven flocculants Ca(OH)$_2$, Al$_2$(SO4)$_3$, FeCl$_3$, AlCl$_3$·6H$_2$O, AlK(SO$_4$)$_2$·12H$_2$O, polyacrylamide, chitosan) and pH regulation, were used for collecting cells of microalgae *Desmodesmus* sp. WC08 from culture medium. The same dosages (0.10, 0.20, 0.30, 0.40, 0.50 and 0.60 gL^{-1}) were used for the flocculants Ca (OH)$_2$, Al$_2$ (SO4)$_3$, AlCl$_3$·6H$_2$O and AlK(SO$_4$)$_2$·12H$_2$O. The dosages of FeCl$_3$ was 0.03, 0.06, 0.09, 0.12, 0.15 and 0.18 gL^{-1}, respectively. The dosage of polyacrylamide was 0.05, 0.10, 0.15, 0.20, 0.25 and 0.30 g L^{-1}, respectively. And the dosage of chitosan was 0.01, 0.02, 0.03, 0.04, 0.06 and 0.08 gL^{-1}, respectively. The pH was set to 5, 6, 7, 8, 9, 10, 11 and 12, respectively. After adding the required amounts of flocculants on the 45 cm^3 microalgal culture medium in a 50 cm^3 test jar, the 0.5-1 cm^3 supernatant samples were taken and measured according to 2.5.1. The detection time of flocculation efficiency were set at 0, 15, 30, 60, 90 and 120 min, respectively, for all flocculation experiments. And every experiment was carried out triplicate.

Effects of flocculation conditions with using chitosan as the flocculant

Effects of pH of culture broth: When the pH of culture media was set 5, 6 and 7, respectively, with hydrochloric acid at different dosages (0.01, 0.02, 0.03, 0.04, 0.06 and 0.08 gL^{-1}, respectively), flocculation efficiency (E$_f$) of chitosan in different pH was measured and evaluated at 30 and 120 min, respectively.

Effects of acid solvents dissolving chitosan: Two kinds of acid solvents, acetic acid and hydrochloric acid, were used to dissolve chitosan for evaluating their each effects on pronation of chitosan. In this experiment, the dosages of chitosan were set to 0.02, 0.03 and 0.04 gL^{-1}, and the pH of culture broth was set to 6 for measuring E$_f$ at 30 and 120 min, respectively.

Effects of culture time of cells: The effects of culture time on flocculation efficiency of chitosan were further investigated at 4th, 8th, 12th, 16th and 20th d from the culture beginning. In this test, the dosage of chitosan was 0.01, 0.02, 0.03, 0.04, 0.06 and 0.08 gL^{-1}, respectively. And chitosan was dissolved by hydrochloric acid and the final pH of culture broth was set to 6. All E$_f$s was measured at 30 min once the chitosan was added into the culture broth.

Supernatant reuse assay

Growth, biomass and lipid content were further investigated in the supernatant of culture broth flocculated by chitosan. Therefore, three culture medium (fresh brine BG11, culture broth flocculated by chitosan (FS), and FS + nutrient salt of BG11 medium) were compared under the same culture conditions of microalgae strain *Desmodesmus* sp. WC08. Cultivation was carried out in 250 cm^3 culture medium in a sterilized air-bubbled (from the bottom) transparent glass-tube (600 mm×30 mm i.d.). All cultures were incubated at room temperature and the light intensity was approximately 160 μmol m^{-2}s^{-1} with continuous illumination. Cell growth was measured by the absorbance at 680 nm with a UV spectrophotometer. And the microalgae biomass were all collected by centrifugation (6000 g, 10 min) and dried by a freeze-dryer. Every experiment was carried out triplicate.

To quantify lipid contents, lipids were extracted using the solvent system of chloroform-methanol (2:1, v/v) following a slightly modified method of Floch [11], which was briefly described as follows. After freeze-drying algal cells, algal dry powder of 0.1 g was mixed with 2 cm^3 of chloroform-methanol (2:1, v/v) solvent and stirred for 30 min at room temperature. Supernatant was collected by centrifugation at 6000 g for 5 min, and the residual algal slud went through the same solvent extraction procedure twice to ensure that most lipid was extracted. All crude oil extract was combined together and 0.1% NaCl was added to give a final solvent ratio of chloroform: methanol: water of 8:4:3 (v/v/v) and mixed for 3 min, then centrifuged at 6000 g for another 10 min. The lower organic phase was collected and evaporated at room temperature on previously weighted glass capsules until constant weight. Then the weight of the crude lipid obtained from each sample was measured gravimetrically. Experiments were performed in triplicate.

Analytical methods

Determination of flocculation efficiency: 45 cm^3 of algal culture broth was placed in a 50 cm^3 test jar. Flocculant was added at the designated concentration and mixed rapidly for 30 s followed by slowly mixing for 3 min. After mixing, the algal cells were allowed to settle down. An aliquot of supernatant was taken from the test jar height of one-thirds to measure the optical density (OD$_{680}$) at different designed time. Fresh brine BG11 medium was used as a blank reference. Each assay was carried out in triplicate. E$_f$ was calculated as follows:

$$E_f(\%) = (1 - OD_t / OD_o) \times 100$$

Where OD$_o$ is the optical density of samples taken at original time and OD$_t$ is the optical density at time t.

Determination of chitosan concentration: The quantitative determination of chitosan in the final supernatant of culture broth flocculated by chitosan was carried out in routine analysis method reported by Wischke and Muzzarelli [12,13].

Result and Discussion

Effects of different flocculants on flocculation efficiency

Flocculant dosage has been recognized as a crucial parameter in flocculation processes because it can influence both the rate and the extent of flocculation sedimentation [14]. And different flocculants have different flocculation ability on the cells harvesting for different microalgae species. Tropic ocean oleaginous microalgae strain *Desmodesmus* sp. WC08 is identified as a possible candidate for biodiesel production in previous study in our lab. Cells harvesting has been studied a lot for several microalgae species in large-scale cultivation. However, rare study was reported for this genus ocean microalgae strains. Cationic flocculants and pH regulation were investigated for the biomass harvesting of tropic ocean microalgae strain *Desmodesmus* sp. WC08.

The results were showed in Figure 1 for the flocculation efficiency of different flocculants. As Figure 1 showed, it displayed a worse flocculation ability found for using pH regulation, polyacrylamide, calcium hydroxide and alum to precipitate cells of microalgae

ability.

In contrast, the other four flocculants has exhibited the quite high E_f. Changes of these four flocculants were quite sharp within 15 min and then it either turned slowly or reached the balance. For aluminum sulfate (Figure 1B here), the highest flocculation efficiency was [94.08 ± 0.33%] obtained by 0.6 gL^{-1} of aluminum sulfate at 90 min. And it has significant difference ($p<0.05$) with other concentrations at different sedimentation time. For aluminum chloride (Figure 1D), after 15 min, the changes of E_f went up slowly for 0.1 -0.3 gL^{-1} of aluminum chloride, and it reached the balance and was overlapped for 0.4-0.6 g L^{-1} of aluminum chloride. So 0.4 gL^{-1} was a saturated concentration for aluminum chloride to harvest biomass of microalgae Desmodesmus sp. WC08. The highest flocculation efficiency [(98.13 ± 0.03)%] was obtained with such concentration at 120 min, which had no significant difference ($p>0.05$) with the other sedimentation time except for 15 min. Namely, when the dosage reached 0.4 gL^{-1}, over 98% of biomass can be recovered after 15 min. For ferric chloride (Figure 1F here), the optimal flocculation efficiency [(94.38 ± 0.24)%] was gained by 0.15 gL^{-1} of ferric chloride at 120 min. And there was no significant difference ($p>0.05$) among 0.15, 0.12 and 0.18 gL^{-1}. Meanwhile, the E_f of 120 min was significantly different ($p<0.05$) with others sedimentation time using 0.15 gL^{-1} of ferric chloride. For chitosan (Fig.1-H here), there were few differences among 0.02-0.08 gL^{-1}. After 120 min, the highest flocculation efficiency [(97.49 ± 0.22)%] was got when the concentration was 0.03 gL^{-1} and it was not significantly different ($p>0.05$) with 0.02 gL^{-1} of chitosan. Dosage which was beyond 0.03 gL^{-1} led to the lower flocculation efficiency. This phenomenon was also observed by other researchers and the reason might be that excess amino group of chitosan resulted in supersaturated positive charge in culture broth [15].

As Figure 1 and the analysis above showed, $AlCl_3$, $FeCl_3$, $Al_2(SO_4)_3$ and chitosan exhibited a high flocculation efficiency over 94% at a relatively short time, while the other four flocculation methods, especially for pH regulation, showed lower flocculation efficiency. The flocculation efficiency of $Al_2(SO_4)_3$, $FeCl_3$, $AlCl_3$ and chitosan sharply increased from beginning to 15 min of sedimentation with less dosage, especially for chitosan, which exhibited a similar flocculation ability with $AlCl_3$, while the dosage was just less than 1/10 of $AlCl_3$. It reported that flocculation by metal salts may be unacceptable if biomass was used in certain aquaculture (or other) applications [16]. Therefore, flocculants based on natural biopolymers were a safer alternative, example for chitosan, which is non-toxic, biodegradable, renewable and ecologically acceptable [17,18]. So it was more suitable for chitosan to flocculate the cells of tropic ocean microalgae Desmodesmus sp. WC08 based on its safety and environment friendly. However, if there are no strict demands or special needs for algal biomass in downstream processing, ferric chloride, aluminum sulfate and aluminum chloride would also be a good available choice.

Effects of flocculation conditions using chitosan

Effects of pH of culture broth: When the sedimentation time was at 30 min, the E_fs of different pH in culture broth using chitosan as the flocculant were showed in Figure 2A. The E_f was displayed as pH6 > pH5 > pH7, when the concentration was under 0.03 gL^{-1}. And both for pH 5 and pH 6, the highest flocculation efficiency was obtained when using 0.03 gL^{-1} of chitosan. Then the flocculation efficiency gradually decreased with the increase of chitosan concentration. For pH 7, the flocculation efficiency gradually increased with the increase of chitosan concentration. And when the dosage was beyond 0.04 gL^{-1}, the flocculation efficiency turned stable. Afterward, when the concentration was higher than 0.05 gL^{-1}, the flocculation efficiency was

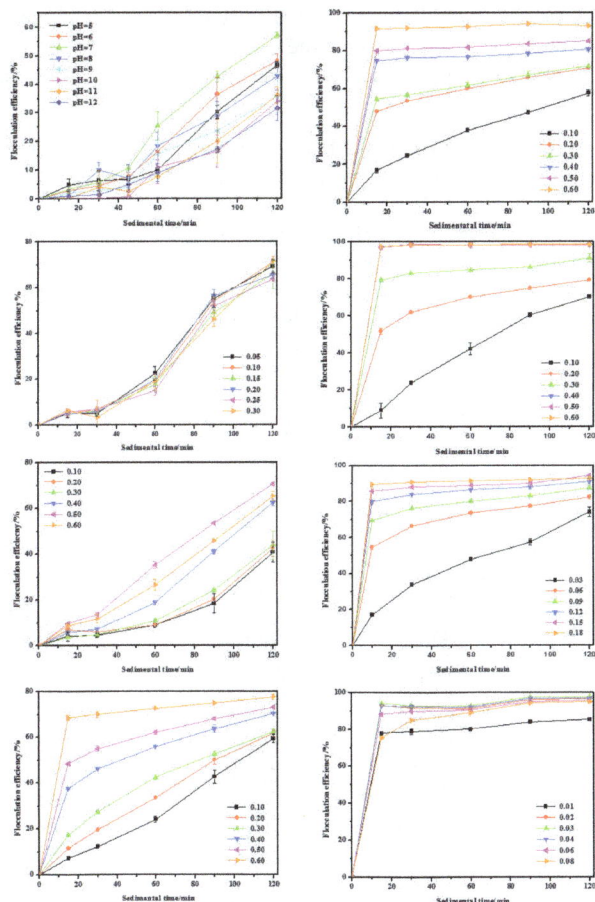

Figure 1: Effect of the different dosages (g/L) of eight flocculants, (A) pH, (B) $Al_2(SO4)_3$, (C) polyacrylamide, (D) $AlCl_3·6H_2O$, (E) $Ca(OH)_2$, (F) $FeCl_3$, (G) $AlK(SO_4)_2·12H_2O$ and (H) chitosan, at different sedimentation time on the flocculation efficiency of tropic ocean oleaginous microalgae Desmodesmus sp. WC08 at the end of cultivation.

Desmodesmus sp. WC08. For pH regulation (Figure 1A and 1B), the highest E_f [(57.20 ± 1.15)%] was got by pH 7 at 120 min. And when the pH was between 5 and 7, the flocculation efficiency at 120 min rose with the increase of pH and the situation was opposite when the pH was 8-12. So pH regulation was less effective for cells recovery of microalgae Desmodesmus sp. WC08. For polyacrylamide (Figure 1C), the flocculation efficiency increased with the sedimentation time extension. And there was no significant change ($p>0.05$) of flocculation efficiency among different concentrations of polyacrylamide. When the dosage reached 0.3 gL^{-1}, the highest E_f was just [(71.50 ± 0.90)%] at 120 min. So that is not a good choice for the cells harvesting of microalgae Desmodesmus sp. WC08. For calcium hydroxide (Figure 1D and 1E), when the dosage reached 0.5 gL^{-1}, the flocculation efficiency [(70.51 ± 0.70)%] was highest at 120 min, which had significant different differences ($p<0.05$) with 0.4 and 0.6 gL^{-1}. Alum, (Figure 1G), the flocculation efficiency rose with the increase of the dosage. The highest flocculation efficiency (77.75 ± 1.17)% was got by 0.6 gL^{-1} at 120 min, which had significant differences ($p<0.05$) with other concentrations. The flocculation efficiency of these four tested flocculation methods were all less than 80%, so they were unfit for cells harvesting of ocean microalgae Desmodesmus sp. WC08 because of their bad flocculation exhibited as pH7 > pH6 > pH 5.

When the sedimentation time was at 120 min (Figure 2B), several changes were different with 30 min of sedimentation. When the concentration was less than 0.05 gL^{-1}, the E$_f$ was exhibited as pH5 > pH6 > pH7. And then they tended to be no difference.

The flocculation mechanism of chitosan can be understood as neutralization and polymer bridging [19]. Because of faster reaction performance, neutralization may play a major role when the pH was lower (pH=5 or 6) in the culture broth. Hence, E$_f$ of chitosan quickly reached the top. When the pH was higher (pH=7), the protonation of chitosan was not enough and the effect of neutralization was weak. So the other flocculation mechanism -bridging,may play a main role in the culture broth with higher pH. In this flocculation mechanism, for more flocculation, more dosage flocculants are needed for the action of bridging. So it needed more chitosan for obtaining the highest flocculation efficiency in higher pH of culture broth.

It is more suitable to harvest the microalgal cells when the pH of culture broth and the sedimentation time are 6 and 30 min, respectively. Although when the pH was 5, the E$_f$ was also satisfying. However, it meant more acid needed to be added into culture media for lower pH regulation. And there were some serious challenges about equipment corrosion and environment pollution with using more acid. And less sedimentation time meant higher harvesting efficiency with the precondition of nearly same E$_f$. So the optimal pH and flocculation time for harvesting microalgae *Desmodesmus* sp. WC08 using chitosan are 6 and 30 min, respectively.

Effects of dissolving solvent of chitosan: It was reported that the structural behavior and solubility of chitosan in different acid solutions may be different and this differences may be caused by the nature of acid solutions and degree of deacetylation and amino group of chitosan [15,20]. The dilute organic acid was the common solution to dissolve chitosan and acetic acid was more often. Hydrochloric acid was one of most common acid used in most industrial production, which was more advantageous in price than acetic acid. And it was discovered that chitosan displayed high solubility in hydrochloric acid [21]. So it is imperative to investigate the flocculation performance of chitosan dissolved in hydrochloride acid solution for harvesting microalgae *Desmodesmus* sp. WC08.

Figure 3 showed the difference for flocculating microalgae *Desmodesmus* sp. WC08 at 30 min of sedimentation time with acetic acid and hydrochloride acid as the dissolving solvent of chitosan. There was no significant difference ($p > 0.05$) for acetic acid and hydrochloride acid in different concentrations. The same results were also obtained for 120 min of sedimentation time (the data was not given). Although the flocculation efficiency was slightly higher with acetic acid as chitosan

solvent than hydrochloride acid, it had no significant difference ($p > 0.05$) for harvesting cells of microalgae *Desmodesmus* sp. WC08. It is a desired result, because it meant that hydrochloride acid can take place of acetic acid to be used as the dissolving solvent of chitosan to collect cells when microalgae *Desmodesmus* sp. WC08 is cultivated in large-scale.

Effects of cell growth time: Tropic ocean microalgae *Desmodesmus* sp. WC08 were cultured in artificial seawater BG11 medium for 4, 8, 12, 16 and 20 d in order to test the influence of the growth stage on flocculation efficiency using chitosan at pH 6. It is recognized medium composition, cell concentration and cell properties were changing with culture time [18]. So it is believed that cell growth stage may be an important factor impacting on flocculation of microalgal cells. So this test has been done to understand the flocculation performance of microalgae *Desmodesmus* sp. WC08 in different growth stage.

As shown in Figure 4, the dosage of chitosan increased with the culture time. The highest flocculation efficiency reached 91% by 0.01 gL^{-1} chitosan for the culture time of fourth and eighth day. And for the twelfth and sixteenth day, the highest flocculation efficiency was

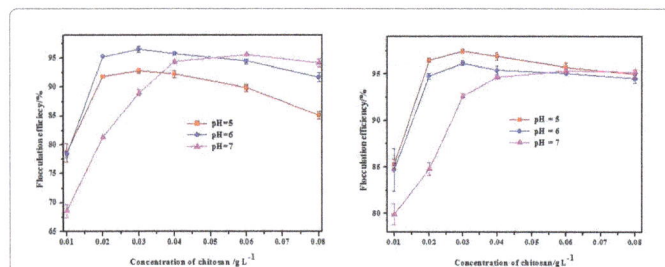

Figure 3: Effects of acid solution on the flocculation efficiency of chitosan for biomass harvesting of tropic ocean oleaginous microalgae *Desmodesmus* sp. WC08.

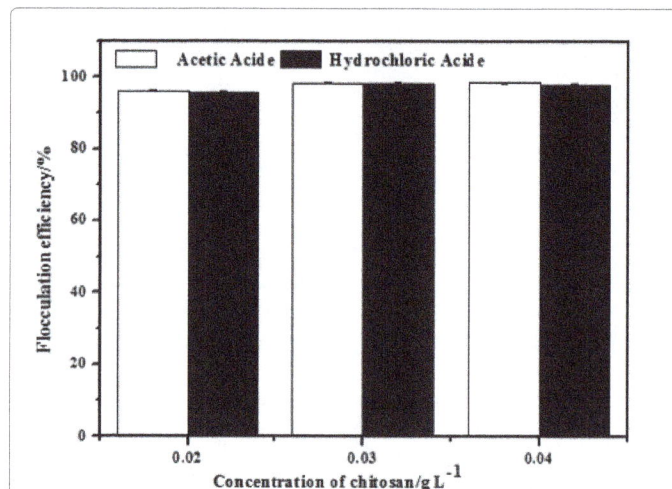

Figure 2: Effects of different pH on the flocculation efficiency of chitosan for biomass harvesting of tropic ocean oleaginous microalgae *Desmodesmus* sp. WC08 at the end of cultivation with different dosages at sedimentation time of 30 min (A) and 120 min (B), respectively.

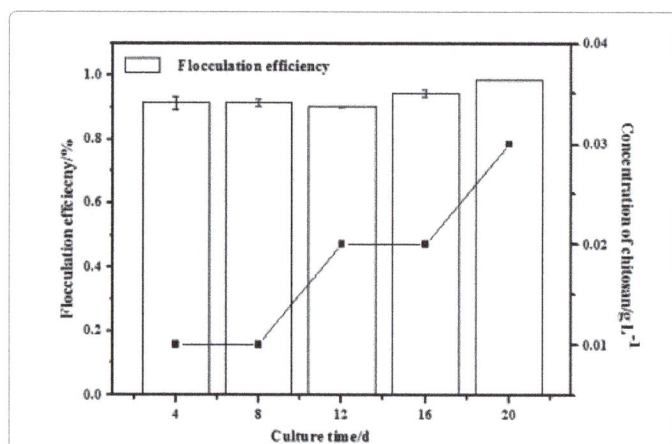

Figure 4: Effects of cells growth stage on the flocculation efficiency of chitosan for biomass harvesting of tropic ocean oleaginous microalgae *Desmodesmus* sp. WC08.

beyond 95% by 0.02 gL^{-1} chitosan. At the end of cultivation, the highest flocculation efficiency was beyond 98% by 0.03 gL^{-1} chitosan. Meanwhile, the biomass was 0.52, 0.74, 1.29, 2.42 and 3.18 gL^{-1} in 4th, 8th, 12th, 16th and 20th d of culture time, respectively. Based on biomass and flocculation efficiency, per 1g chitosan can recovery microalgal biomass of 47, 68, 58, 114 and 104 g in each culture time, respectively. So the optimal cell harvesting of tropic ocean microalgae *Desmodesmus* sp. WC08 is at end of cultivation (16-20d) and the optimal cells harvesting can be obtained beyond 110 g biomass just by per 1 g chitosan.

Supernatant reuse assay

Chitosan residual in the supernatant: After cells harvesting, residual chitosan in the supernatant of culture broth flocculated by chitosan was determined using dye-binding assay described Wischke and Muzzarelli [12,13]. And the relationship between chitosan concentration (mg/L) and (OD_{575}-OD_{750}) is obtained in this experiment as following equation:

$$C_{chitosan} \ (mg \ L^{-1}) = 81.64 \times (OD_{575} - OD_{750}) - 1.87 \ (R^2=0.998)$$

According to this relation equation, the final residual chitosan was calculated just (4.12 ± 1.52) mgL^{-1} in the supernatant at the end of cell cultivation under the optimal flocculation conditions of chitosan (that is, the pH of culture broth was set to 6 and the dissolved solvent of chitosan is hydrochloride acid, and the original chitosan concentration was 0.03 gL^{-1}). It's showed that most chitosan play a role of sedimentation and settle down slowly by binding to microalgal cells. Namely, about 86.27% of chitosan added into the culture broth exists in the final recovered microalgal biomass and 13.73% of chitosan is free in the supernatant and nearly play little positive role in the process of flocculation.

Growth, biomass and lipid content in the recycled culture medium: If the supernatant could be recycled, it would be quite significant for saving water and reducing the cost of cultivation. It's reported that recycled water from its harvesting culture can be reused and has some positive effects on biomass production [22].

In Figure 5, it is showed that although the biomass and lipid yield of FS culture medium is lower than that in the fresh culture medium (BG11) and there is a significant difference ($p<0.05$) existing between them. And the recycled culture medium FS+BG11 can be reused well because of pretty good growth and lipid yield. The nutrition concentration is the main differences among the three culture mediums. Higher nutrition concentration in SF+BG11 leads to higher biomass and lipid yield than FS, but it is lower than fresh brine BG11 culture medium. Maybe it is because the BG11 salts in SF+BG11 is a little higher than fresh culture medium, so more effort must be taken to optimize the required additional nutrition salts in the supernatant. Besides, some metabolites in FS may be an important factor to play some role to lead to such a result, so more detailed research must be further studied for supernatant reusing.

Conclusions

Eight flocculation methods were investigated for harvesting tropic ocean oleaginous microalgae strain *Desmodesmus* sp. WC08. Ferric chloride, aluminum sulfate, aluminum chloride and chitosan exhibited quite high flocculation ability. Among these flocculants, chitosan held the most tremendous potential for efficient and safe biomass recovery from culture broth. However, if there are no strict demands or special needs for algal biomass in downstream processing, ferric chloride, aluminum sulfate and aluminum chloride will be also a good available choice for harvesting microalgae *Desmodesmus* sp. WC08.

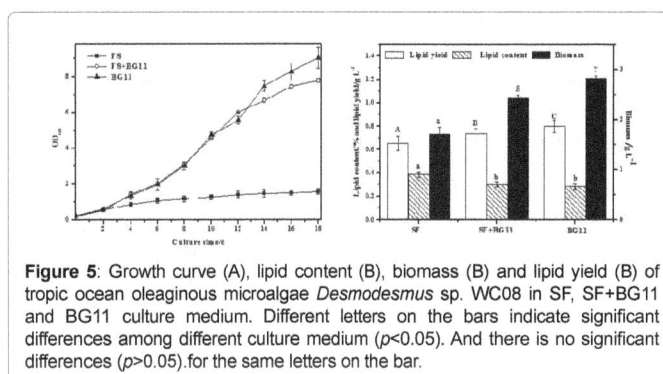

Figure 5: Growth curve (A), lipid content (B), biomass (B) and lipid yield (B) of tropic ocean oleaginous microalgae *Desmodesmus* sp. WC08 in SF, SF+BG11 and BG11 culture medium. Different letters on the bars indicate significant differences among different culture medium ($p<0.05$). And there is no significant differences ($p>0.05$).for the same letters on the bar.

Partial acid culture medium (pH5-6) was better for more efficient biomass harvesting of tropic ocean microalgae *Desmodesmus* sp. WC08 with less dosage of chitosan. Acetic acid and hydrous chloride have no distinct difference on the protonation of chitosan to harvest microalgae *Desmodesmus* sp. WC08. With increasing of microalgae growth time, more dosage of chitosan was gradually needed. And finally, more than 110g biomass of tropic ocean microalgae *Desmodesmus* sp. WC08 can be harvested just by per 1 g chitosan at the end of cultivation. In addition, there is little chitosan residues in the final supernatant of culture broth flocculated by chitosan and the final supernatant has quite great potential for recycling.

Acknowledgements

We acknowledge the financial support from the National Science and Technology Support Program of China (NO.2011BAD14B01), the Provincial Science and Technology Program on Modernization of Traditional Chinese Medicine of Hainan (NO.ZY201327) and Innovation Fund Project for Technology Based Firms (13C26244604892).

References

1. Azma M, Mohamed MS, Mohamad R, Rahim RA, Ariff AB (2011) Improvement of medium composition for heterotrophic cultivation of green microalgae, Tetraselmis suecica, using response surface methodology. Biochem Eng J 53: 187.

2. Karemore A, Pal R, Sen R (2013) Strategic enhancement of algal biomass and lipid in Chlorococcum infusionum as bioenergy feedstock. Algal Res 2: 113.

3. Molina Grima E, Belarbi EH, Acién Fernández F, Robles Medina A, Chisti Y (2003) Recovery of microalgal biomass and metabolites: process options and economics. Biotechnol Adv 20: 491.

4. Gudin C, Thepenier C (1986) Bioconversion of solar energy into organic chemicals by microalgae. Adv Biotechnol Process 6: 73.

5. Heasman M, Diemar J, O'connor W, Sushames T, Foulkes L (2000) Development of extended shelf-life microalgae concentrate diets harvested by centrifugation for bivalve molluscs–a summary. Aquac Res 31: 637.

6. Babel S, Takizawa S (2010) Microfiltration membrane fouling and cake behavior during algal filtration. Desalination 261: 46.

7. Wu Z, Zhu Y, Huang W, Zhang C, Li T, et al. (2012) Evaluation of flocculation induced by pH increase for harvesting microalgae and reuse of flocculated medium. Bioresour Technol 110: 496.

8. Guo SL, Zhao XQ, Wan C, Huang ZY, Yang YL, et al. (2013) Characterization of flocculating agent from the self-flocculating microalga Scenedesmus obliquus AS-6-1 for efficient biomass harvest. Bioresour Technol 145: 285.

9. Vandamme D, Foubert I, Meesschaert B, Muylaert K (2010) Flocculation of microalgae using cationic starch. J Appl Phycol 22: 525.

10. Zhang S, Liu PH, Yang X, Hao ZD, Zhang L, et al. (2014) Isolation and identification by 18S rDNA sequence of high lipid potential microalgal species for fuel production in Hainan Dao. Biomass Bioenerg 66: 197-203.

11. Folch J, Lees M, Sloane-Stanley G (1957) A simple method for the isolation and purification of total lipids from animal tissues. J Biol Chem 226: 497-509.

12. Wischke C, Borchert HH (2006) Increased sensitivity of chitosan determination by a dye binding method. Carbohyd Res 341: 2978.

13. Muzzarelli RA (1998) Colorimetric determination of chitosan. Anal Biochem 260: 255.

14. Chen L, Wang C, Wang W, Wei J (2013) Optimal conditions of different flocculation methods for harvesting Scenedesmus sp. cultivated in an open-pond system. Bioresour Technol 133: 9.

15. Rashid N, Rehman MSU, Han JI (2013) Use of chitosan acid solutions to improve separation efficiency for harvesting of the microalga Chlorella vulgaris. Chem Eng J 226: 238.

16. Granados M, Acien F, Gomez C, Fernández-Sevilla J, Molina Grima E (2012) Evaluation of flocculants for the recovery of freshwater microalgae. Bioresour Technol 118: 102.

17. Vandamme D, Foubert I, Muylaert K (2013) Flocculation as a low-cost method for harvesting microalgae for bulk biomass production. Trends Biotechnol 31: 233.

18. Xu Y, Purton S, Baganz F (2013) Chitosan flocculation to aid the harvesting of the microalga Chlorella sorokiniana. Bioresour Technol 129: 296.

19. Uduman N, Qi Y, Danquah MK, Hoadley AF (2010) Marine microalgae flocculation and focused beam reflectance measurement. Chem Eng J 162: 935.

20. Papazi A, Makridis P, Divanach P (2010) Harvesting Chlorella minutissima using cell coagulants. J Appl Phycol 22: 349.

21. Zhao QS, Cheng XJ, Ji QX, Kang CZ, Chen XG (2009) Effect of organic and inorganic acids on chitosan/glycerophosphate thermosensitive hydrogel. J Sol-gel Sci Technol 50: 111.

22. Farid MS, Shariati A, Badakhshan A, Anvaripour B (2013) Using nano-chitosan for harvesting microalga Nannochloropsis sp. Bioresour Technol 131: 555.

17

De-carbonization of Electricity Generation in an Oil and Gas Producing Country: A Sensitivity Analysis over the Power Sector in Egypt

Arash Farnoosh* and Fendric Lantz

IFP Énergies Nouvelles, IFP School, 228-232 Avenue Napoléon Bonaparte, F-92852 Rueil-Malmaison, France

Abstract

Fossil fuels are used in power generation in oil and gas producing countries due to the resource availability. However, the growing electricity demand, the potential exports revenues associated to hydrocarbons as well as the environmental policies have to be taken into account for the definition of the electricity generation mix. Thus, the development of the power generation capacities according to the resource availability and the economic factors (demand and costs) is investigated through a modeling approach. Over the past ten years, Egypt has become an important gas producer and a strategic gas supplier for Europe. Moreover, natural gas represents around eighty percent of the Egyptian power sector mix. However, this extensive share of natural gas in power generation mix could not be sustainable in long-term due to the limited hydrocarbons' resources of Egypt. In this study, the current and future power generation situation of the country is analyzed through a dynamic linear programming model. Finally, a power generation strategy based on a gradual integration of nuclear and renewable is suggested.

Keywords: Energy sector; Egypt; Power generation mix

Introduction

The increasing trend of the electricity demand is mainly associated to both economic development and demographic evolution in most of the countries. To meet this need, the main sources of electricity production used in the world are still the power plants using fossil fuels (coal, gas and oil to a lesser extent) that provide 67% of electricity production in the early 2010s, followed by hydro plants (16%) and nuclear (13%). Other renewable types of power plant (wind, solar, geothermal, biomass and etc.) provide the rest of that production. In Europe if we are taking example, renewable account for nearly 20% of electricity production. As part of the European "20-20-20" initiative, the development of electricity from renewable is supported by incentive policies based on guaranteed purchase prices or bids for the construction of power generation units. For example Germany, under its Energiewende plan adopted in 2011, is accelerating its energy transition to almost total abolition of non-renewable power units in its electricity generation mix in long term. MENA[1] region is not an exception in this global electricity-generation-decarburization trend and is following the same path. Expansion of non-fossil energy in MENA countries is driven by a number of key factors: energy security enhancement, major energy demand growth, urbanisation, water scarcity and of course environmental concerns. With high fossil fuel prices resulting in both huge bills for net oil-importing countries and opportunity costs for net oil-exporting countries, non-fossil resources have become an increasingly attractive alternative to domestic oil and gas consumption. From 2008 to 2011, non-hydro renewable resources for power generation more than doubled to reach almost 3TWh[2] and grew at faster rate than their conventional counterparts [1].

The most appropriated power generation mix is implemented to reach the electricity demand according to the resource and the technologies availability. Thus, in a large number of oil and gas exporting countries, fossil fuels are used for power generation to provide electricity to a growing population with a low cost. However, the growing electricity needs of the population, the environmental concerns and the potential value of oil and gas resources on foreign markets make that the optimal electricity generation mix has to be designed according to these constraints or targets. Egypt is a typical case for such problem with a growing population and gas reserves which could be used for exports and power generation. Over the past decade, Egypt had solid economic growth due to its rising exports and investment and also its strong national consumption. Energy sector has been highly interconnected with economic activity of the country. Most of the energy demand growth came from growing industrial production and robust population expansion. Energy demand growth has also been promoted by the governmental subsidies coming from exports revenue (mainly hydrocarbon resources). Unfortunately this subsidization policy contributed a lot to fiscal deficit of the country. Recently government has announced several times the suppression of these subsidies. No action has been taken place regarding this issue until now and it seems that nothing will be realized (at least in the short-term future) due to social events and uncertainties that the country is currently facing with following the Arab Spring and recent socio-political movements.

Egypt's highest export revenue comes from natural gas. However, its production is slowing down largely because of the lack of foreign investments (notably from International oil Companies). This production decline will also impact the petrochemical industry fed with natural gas as row material. Natural gas is the key fuel in Egypt, especially in industry and power sector which is the largest energy consumer sector of the country.

In this paper, electricity generation mix of the country is explored through a linear programming model. A sensitivity analysis is

---

[1] Middle East and North Africa

[2] Tera-Watt-Hours

*Corresponding author: Farnoosh A, IFP Énergies Nouvelles, IFP School, 228-232 Avenue Napoléon Bonaparte, F-92852 Rueil-Malmaison, France
E-mail: arash.farnoosh@ifpen.fr

performed on several economic factors to point out the crucial role of the discount rate as well as the carbon price on investment decisions. For this purpose, economic context of Egypt is presented in section 2 and section 3 specifically dedicated to the power sector. The methodology is developed in section 4 and the results are analysed section 5.

Energy and Environmental Policy and the Power Generation Mix Development

Oil and gas resources availability aims to define the best strategy to use them for the people's welfare over a long period. Furthermore, the decision process for energy policy has to deal with a lot of uncertainties concerning both the potential amount of resources and the economic activity along several decades. The undesirable effects of oil revenues on the long term economic activity are clearly analyzed in the so-called Dutch Disease case [2]. In this context, the definition of the electricity mix to reach a growing energy demand has to take into account the revenues from the resource exports, the power generation costs and the environmental policy.

Egyptian Ministry of Planning defined the energy strategy of the country by issuing its 6th Five Year Plan (2007-2012). The plan mostly included the investment plans for electric power, oil and natural gas industry. Energy efficiency improvements, security of supply and willingness to adopt nuclear technology were also considered as chief strategic targets. Oil and gas sector promotion consist mainly efforts targeting the expansion and intensification of the exploration activities and completion of the 20-year 10 billion dollar Petrochemicals Master Plan (lunched in 2002 for constructing 24 petrochemical units across the country by the end of 2022). And the strategy for the power sector aims to improve efficiency, promote renewable energies and security of supply for all sectors, encourage the development of grid in rural regions and facilitate more interconnection with neighbouring nations (Ministry of Planning 2007). This 5-year plan has been revised and discussed in 2011 again but no official strategy has been yet released. However, most probably increasing focus on export maximizing, upstream investment incentives and ensuring demand satisfaction will be the key components.

Egypt was first Arab nation signed the Kyoto protocol in 1999. From then Egypt seeking to diversify its current energy mix by increasing usage of renewable energy sources such as hydro, wind and solar. The Renewable Energy Expansion Plan, adopted in 2008, sets target for renewable sources to reach 20% of total domestic energy supply by the year 2020. 12% will be provided by wind and hydro [3]. However, at the moment there is no solid support scheme (stable feed-in tariffs for example) in place for the promotion of renewable sources. The total energy related CO_2 emissions of the country since 1990s is shown in Figure 1.

The Egyptian Environmental Affairs Agency defines the country's environmental policies. The entity established in 1982 and thereafter the Ministry of State for the Environmental Affairs was created. Environmental policy of the country (National Environmental Action Plan) addresses environmental issues and strategies for encouraging effective use of energy in different oil sector activities, expansion of gas network and use of natural gas.

Egyptian government provides subsidies for various types of fuel such as natural gas, kerosene, butane, diesel, gasoline and fuel oil. Gas

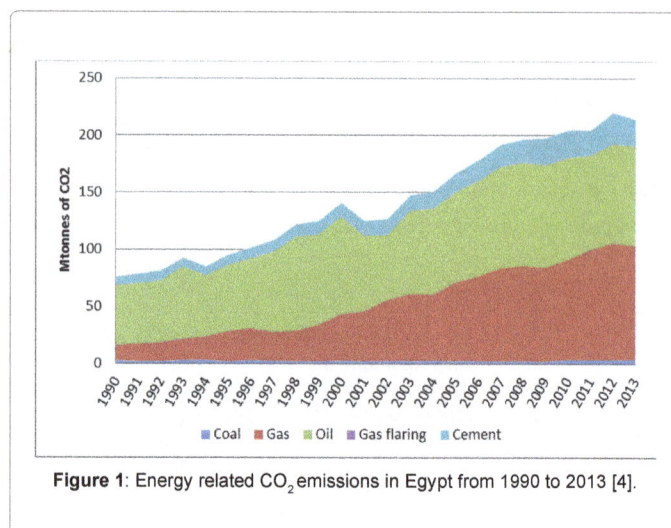

Figure 1: Energy related CO_2 emissions in Egypt from 1990 to 2013 [4].

prices heavily subsidized for industrial usage and power generation to bring more incentives to both sectors for switching from oil and oil products to gas and thereby letting more oil for export. Global fuel price rising in the international markets resulted in more restricted government budget. Moreover, cheap gas prices compare to global prices boosted domestic gas demand. Following the national demand increase and no reaction concerning these subsidies, Egypt became a net importer of oil in 2010 [4,5]. This trend will most probably continue given the intensive depletion observed in the Egyptian oil fields in addition to the national demand increase. Several announcements have been made by the government to decrease energy subsidies. For instance, in 2007, the Egyptian government announced its intention to phase out subsidies for natural gas for both energy intensive and non-intensive industries with different time horizons, respectively in 2009 and by the end of 2013. However, following economic crisis, the government fixed natural gas and electricity prices for all industries. Egypt spent around 20.3 billion dollars for energy subsidies in 2010, equivalent to almost 13% of the country's GDP [6]. Nevertheless, subsidy reforms (particularly in residential and commercial sectors) seem to be very unlikely to be occurred, especially in power sector, under current peculiar socio-political situation of the country.

Power Sector Overview

Organization, market and regulation

Egyptian power sector went through some restructuring and unbundling reforms in 2001. The existing vertically integrated monopolistic system was unbundled into six generation, one transmission and nine distribution companies. Under the supervision of the Ministry of Electricity and Energy, the Egyptian Electricity Holding Company still owns 90% of generation and distribution sectors and 100% of the transmission company. The Egyptian Electricity Holding Company (EEHC) is the only entity empowered to approve and construct any generation capacity or to buy power from international private developers of electricity. Even though the 2001 unbundling reforms aimed to eventually privatize the sector, but Electricity Holding Company remained 100% public and it is very unlikely to see any privatization process in the near future.

Egyptian Electricity Holding Company (EEHC) consists of totally sixteen electricity companies separated according to the region in which they operate and also the type fuel they use. Cairo, East Delta, Middle

Delta, West Delta and Upper Egypt are the thermal power companies while Hydro Plans Company is in charge of all hydro generation across the country. Several privately own power units have also finance and built under BOOT (Build, Operate and Transfer) financing scheme put in place in late 2002 by the Egyptian government. Port Said East Power Company, the Sidi Krir Generation Company and the Suez Gulf Company are examples of these private operators. There are currently three International Private Producers operating in Egypt. The first international operator was US-based InterGen, a joint venture of Bechtel Enterprises and Shell Generating Limited, along with some local partners to operate Sidi Krir BOOT project.

At the moment power market in Egypt is organized in the "Single Buyer"[3] structure. Egyptian Electricity Transmission Corporation sells power from the generation entities (including private independents) to the 9 regional distribution companies. Approximately 10% of the Egypt's distribution grid is owned by 6 small private companies who manage the sale of mid and low voltage power to final consumers. These companies are as following: Global Energy Company, the Alexandria Carbon Black Company, the Om El Goreifat Company, the National Electricity Technology Company and finally the Mirage Company. For the purpose of controlling and regulating all the issues related to generation, transmission, distribution and consumption, the Egyptian Electric utility Organization and Consumer Protection agency was created in 1997 by the government. Many other specialized regulatory authorities have also been established to regulate the various areas of the power sector, such as Nuclear Power Plants Authority, New and Renewable Energy Authority, Hydro Power Projects Execution Authority and etc.

Electricity supply and power plants

Egypt has increased its generation capacity from 15.5 GW in 2000 to almost 27 GW in 2010. Power output has also been doubled from 78.1 TWh in 2000 up to 148 TWh by 2010 [7]. EEHC had to deal with some outages in 2010 during peak hours because of the growing usage of air-conditioners during hot days. Egyptian government announced ambitious goals for increasing capacities to satisfy the growing domestic demand. EEHC is currently applying the 6th Five Year Plan targets capacity additions of 7 GW over the 2007-2012 periods [8]. The plan includes 3 GW of Combined Cycle and 4 GW of Steam Turbine capacities. Recently, EEHC has also proposed the 7th Five Year Plan for 2012-2017 periods, including an additional 5.25 MW of Combined Cycle plus 7.15 MW of Steam Turbines [8].

Concerning renewables, in 2007, the Renewable Energy Expansion Plan adopted for renewable penetration of 20% in to the network by 2020, where hydro power represents 5.8%, wind 12% and 2.2% from other renewable energy sources, especially solar [8].

Combined cycle and steam units (both using natural gas as fuel) accounted for 62% of the total capacity in 2010. These technologies have been considerably promoted by the Egyptian government since 2000 as gas production increased and subsidies over natural gas encouraged the investment in this technology. In 2011, Al Damietta and Al Shabab power plants with total capacity of 1.7 GW were added to the network.

Egypt started producing hydro power in 1960's after the construction of the Aswan High Dam station. Since then, no new major project has been realized. In 2010, total capacity of hydro was 2.8 GW accounting for 9.5% of total generation [7].

Oil-firing power plants account for 18% of power generation of the country it has not historically been encouraged by the government

because of its expensive price leading to very high subsidies for the government. Oil has been mainly used in the peak summer months for meeting air-conditioning demand. Share of fossil fuels (oil and natural gas) in the total power generation of the country accounts for almost 90% of fuel types used for national electricity demand's satisfaction (Figure 2).

Nuclear power has also been proposed several times by the Egyptian government. Plan to develop this technology were put in place in the 1980's. 1000 MW nuclear capacity were proposed at El Dabaa on the Mediterranean coasts. Project was halted due to the huge costs and safety reasons following the Chernobyl accident. In 2006, following an increase in international oil and gas prices and rising domestic demand of power, the nuclear program revised by the government. Finally, in 2010 Egypt launched a tender for 1.2 GW El Dabaa Plant with forecasted cost of 1.5 billion dollars and commissioning date of 2019 [9,10]. Figure 3 illustrates the entire electricity infrastructure and power plant stations of Egypt.

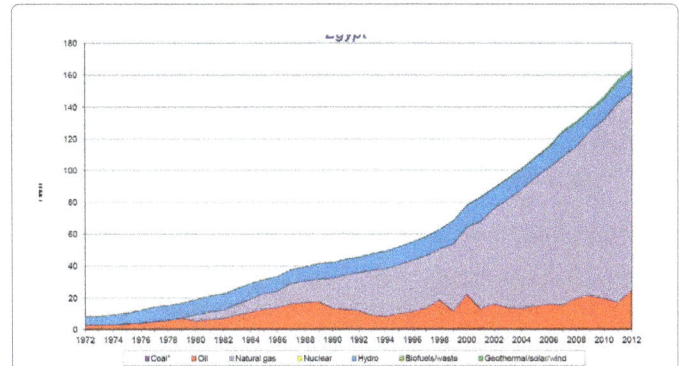

Figure 2: Electricity generation by source in Egypt in 2012 [9].

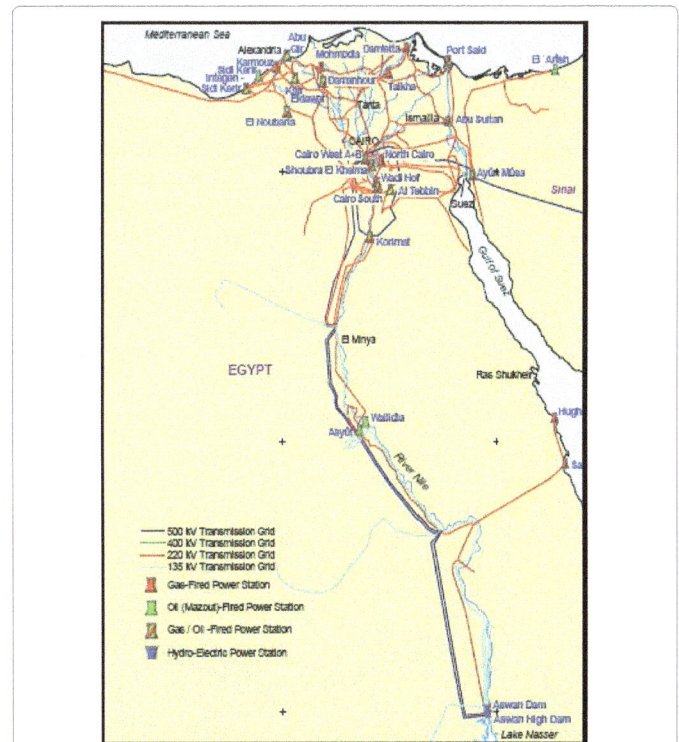

Figure 3: Electricity generation and transmission infrastructure in Egypt [11].

---

[3] For more information regarding this market model please refer to appendix A.

Methodology

A linear programming optimization framework was used to assess the costs and savings of expanding the role of non-fossil fuel based power sources in electricity supply. LP (linear program) cost minimizing is an approach that systematically evaluates potential power supply to satisfy the demand at the best societal cost. This method analyses what would be the incremental cost if each source of power generation were to integrate the electricity supply of the country. In pursuit of this objective, a review of relevant non-fossil and fossil based power unit choices on the basis of resource potential, cost and economic benefits is provided. Several choices of technologies that are or are expected to be technically and economically feasible over the next two decades have been identified and incorporated into the modelling effort.

Electricity generation should be provided by a large set of power plants which are characterized by different technologies associated to a very large spectrum of fixed and variable costs. Consequently, this leads to an optimal usage and investments so as to satisfy the current and future demand. Optimizing the overall electricity cost of production by the different types of plants enables us to rank various production means. Indeed, when electricity demand increases and the available power (in the lowest cost category of generation means) is not enough, the system must switch to the generation-mean whose cost category is just one step above the previous one. In other words, the utilisation of power plants are ranked according to their growing running cost (so-called "merit-order" process).

The main contribution of this study is to analyse the optimality of the Egyptian power generation mix via LP models (based on the above-mentioned structure) and to reveal the most optimal decisions for the next 20 years of the national electric system under different proposed investment scenarios through the dynamic model. Afterwards, the sensitivity analysis is realised to measure the competitiveness of non-fossil power sources with fuel-based ones under various discount rate and carbon price scenarios.

During the past decades, a huge body of literature related to the application of sophisticated energy optimization and simulation scenarios have been carried out for optimal planning of the future national energy systems [12-16]. Grouping existing literature, there are several studies seem to be related to the optimization of the use of non-fossil sources and the assessment of existing tools and optimal penetration rates of these technologies in the power systems [17-21]. A study for Algeria, Morocco and Tunisia has been done by Brand and Zingerle so as to analyse the impact of renewables and non-fossil technologies' integration into their electricity systems. For instance, Mazhari et al. used system dynamics and agent based modelling approach in order to find the most optimal and economical mixture of storage capacities and solar plants.

Various types of linear programming models have also been used for future optimal generation mix simulations. Xydis and Koroneos [22], stated the role of solid wastes in future energy systems, while Chang and Li [23], pointed out the role of all the renewable energies options for the future generation mix of ASEAN countries.

Although numerous studies have been conducted on the optimization and simulation of future energy systems with various rates of pure renewables penetration, limited papers have appeared on the optimization of power systems with both nuclear penetration and renewables imposition which is the main focus of this study.

Total electricity generation cost minimization, is one of the main

modelling approaches in power generation modelling. Examples of such models include POLES[4] [24], MARKAL[5] and TIMES[6] [25]. Many other examples have also been developed by consultants and utilities themselves and are not therefore published. The basic idea of these models is to explain electricity prices from the marginal generation cost. In this case, assumption over the future electricity prices does not have to be made. Focusing on minimum generation cost implies minimizing the cost to be transferred to the final consumers, irrespective of the electricity price. The main advantage of this method is to analyse the producer behaviour facing with a mix of different types of constraints such as economic, technical and environmental ones. Our approach is similar, in the way that a linear dynamic model is developed where the total costs are to be minimized under certain constraints developed in the next section.

Optimizing the overall production cost of electricity via various types of power plants enables to "prioritize and rank" the different means of production. Indeed, when electricity demand increases and the power available in the category of lowest cost is not enough, then it should implement the generation mean whose cost category is immediately above. This leads to a prioritizing of different equipment based on their operating (variable) costs which allows defining a dispatching of different equipment on the annual load curve. Generation Mix management, made by the cost minimization objective corresponds to an economic optimum: at each time step, the marginal cost (the cost to satisfy a request for additional MWh) is equal to the operating cost of production with the marginal equipment. All equipment with lower production cost will be used and in theory, no more expensive equipment will operate.

In medium and long-term decision-making process, optimization techniques can become very helpful, particularly if we take into account the investment decisions and costs associated with each additional capacity. Model proposed in this study, is solved using dynamic linear programming so as to consider those investment trends to satisfy the growing electricity demand of the country.

Power generation mix structure of the country is modelled under GAMS 24.0.2 (General Algebraic Modelling System) software within CPLEX as a solver. This cost minimization model contains the objective cost function that must be minimized and the demand constrains that have to be satisfied. For the current power generation mix of Egypt (in our model, the year 2010) the production capacities must be respected and in the case of long-term optimization, investments are allowed.

The constraints of the model are the demand equations, the capacity constraints and the investment equations. In the demand equations for each season, the sum of the power generated by the power plants is greater than the demand. On the supply side, the power loaded from each unit is lower than the power capacities times the seasonal availability coefficients. Finally, the installed capacities are equal to the sum of the existing units and investments.

The model is developed based on a long time period. This period is split in several sub-periods associated to the time index t with n(t) years. In each sub-period, we consider a representative year denoted by a(t). Thus there are b(t) years before period t defined as follows:

$$b(t) = \sum_{k}^{t-1} n(k)$$

[4] Prospective Outlook on Long term Energy Systems

[5] MARKet ALlocation

[6] The Integrated MARKAL-EFOM System

The model basic structure is as following:

$$\text{Min} \sum_t \left[\sum_i \sum_s \sum_m \left(\gamma_t \times E_{ia(t)} \times H_s \times R_m \right) P_{isma(t)} + \sum_i \left(\varphi_t \times I_{ia(t)} \right) C_{i(t)} \right]$$

With,

$P_{isma(t)}$: is the Power loaded (called) on the grid by each equipment of type i, for the season s in the representative year a(t) with demand randomness factor of m (MW)

H_s: Length of the season s (hours)

$E_{ia(t)}$: Variable cost of production of each equipment i at the representative year a(t) ($/MWh)

t : the time period (step)

a(t) : representative year of the period t

R_m : probability of having randomness factor of m

$I_{ia(t)}$: investment in the unit i at the representative year a(t) ($/kW)

$C_{i(t)}$: capacity to build for unit i at the period t (MW)

γ_t is the discount factor applied to the annual costs of each period. We assume that the costs are the same for all the year of a given planning period, thus it is defined as:

$$\gamma_t = \frac{1}{(1+r)^{b(t)}} \sum_{k=1}^{n(t)} \frac{1}{(1+r)^k}$$

And φ_t is the discount factor applied to investments:

$$\phi_t = \frac{1}{(1+r)^{b(t)}}$$

where r is the discount rate.

Hence, the total discounted cost of different installed units is minimized according to the electricity demand and available capacity. Different discount factors were applied for the variable and investment costs. As a matter of fact, in this model the variable cost could be different in each year (according to the yearly utilization rate of each power plant) during the life-time of the power plant and the discounting operation has to be adapted accordingly. Instead, the discounting operation corresponding to future investments is less complex since, by convention, the investment occurs in year 0 (initial investment or the so-called "overnight cost") and it can be modeled as repayment of annuities (yearly fixed costs) throughout the life time of each power plant.

For each period, supply (capacity) and demand sides' constraints are as following:

Capacity constraint:

$$\frac{1}{\tau_{is}} P_{isma(t)} \leq \sum_i \alpha_{it} C_{ia(t)}$$

With,

α_{it}: availability coefficient of the capacity of equipment i activated in year t. It measures the capacity reductions that occur after the construction of a plant.

τ_{is}: coefficient of availability in each season for each equipment i

And the evolution of production capacity (new additional investment) during the modelled time horizon is satisfied by the following dynamic power-unit-fleet relation:

$$C_{i,t} = C_{i,t-1} + U_{i,t} \quad \text{with} \quad U_{i,t} \geq 0$$

In which, $C_{i,t}$ and $C_{i,t-1}$ represent the capacity of equipment i during two consecutive years, and $U_{i,t}$ is equal to the capacity evolution of unit i in year t.

Demand constraint:

All the equipment must provide the seasonal power required for the satisfaction of the consumers' demand and this must be done for each random event m.

$$\sum_i P_{isma(t)} \quad \geq \quad D_{sma(t)}$$

D_{st}: loaded power on the grid for the season s (MW)

Empirical Analysis

Parameters of the model

Hereby, the demand and costs structures are presented in addition to the techno-economic data used for each power unit in the optimization model. Figure 4 shows a typical daily electricity demand curve (load-curve) of Egypt.

Therefore, three demand fractions are considered: H1, H2 and H3. H1 represents the base-load and H2 and H3 represent respectively the semi-base and peak daily demands. Thereafter, this 3-fractionned structure of the daily demand is spread to two different seasons: S1 and S2. S1 represents summer season in which we generally observe the peak demand periods (caused by the air-conditioning effect) and S2 goes for winter season. These demand-compositions for the fractioning hours and seasons hypothesis, are shown in Figure 5.

Demand randomness factors (m_x) and their associated probabilities (R_m) introduced in the model assume 10% variability of the registered demand in both negative and positive directions.

Demand increase forecasts for 2020 and 2030 are expected to be respectively equal to 35 and 17 per cents [26]. The forecasted electricity demand used in this model is summarized in Table 1.

As the amount of hydroelectricity remains constant, identical to that of 2010 which is equal to 14 TWh [9], during modelled time horizon (owing to the already saturated potential of hydroelectricity in Egypt), we subtracted the hydro share directly from the demanded electricity. This process has been also applied for the case of other renewable resources, solar and wind. In other words, the amount of renewable production (based on the Egyptian government target for 20% of renewable share as described in section 2) has been imposed on the loaded power as must-run production units, of course in consistent with their associated availability factors. Hence, whatever the cost of production, these renewables would be always placed at the top of the merit order (generation mix ranking curve) in the model. That's why the generation costs of these units do not impact the decision making process of the model and the competition (in terms of generation cost) would be between nuclear, gas and fuel power plants.

So as to cover the risk related to the intermittent production of solar and wind power plants, we have introduced in the model a necessary investment in the fossil-fuel power plants that play the back-up role in case of insufficient load factor which generally happens during peak

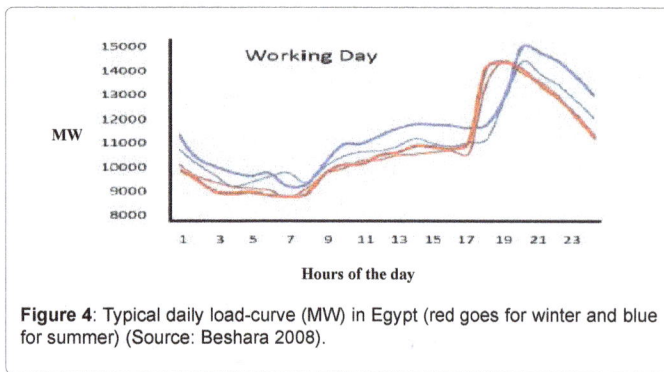

Figure 4: Typical daily load-curve (MW) in Egypt (red goes for winter and blue for summer) (Source: Beshara 2008).

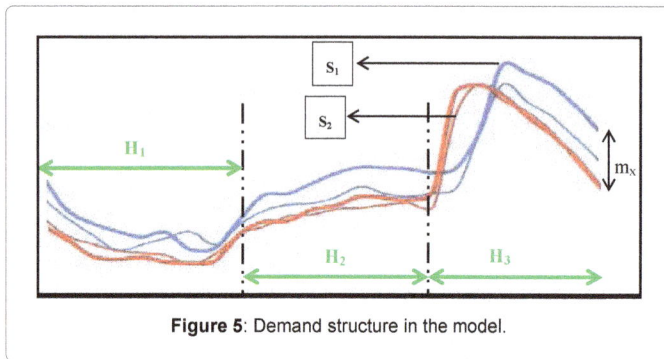

Figure 5: Demand structure in the model.

Total electricity demand in Egypt (TWh/y)				
2000	2005	2010	2020	2030
78	109	**148**	**200**	**236**

Table 1: Egyptian power demand. [8,26].

consumption, especially in summer. In most of the regions around the world, lowest values of capacity factor for the intermittent technologies are observed during peak demand periods. This is also the case of Egypt with hot and not necessarily very windy summers.

In this model the absence of production from intermittent means is compensated by the least expensive (in terms of total cost) thermal power units which have around 100% of availability (load factor equals to 1) except for the ex-ante planned maintenance.

Wind speed can widely fluctuate in a rather short-time period. These fluctuations cause the need to rapidly compensate for large amounts of increased or decreased production with other power plants in the system. The most reliable way to answer these variations is to use pumped storage and hydro storage facilities which have very quick ramp (start-up) possibilities with relatively large power volume capacities. Unfortunately there is not enough potential for these technologies in Egypt due to its climatic situation. However, gas and fuel power plants can also quickly start and make up for the losses in production. Even though the existing and already operational flexible power plants could be used to provide the needed flexible back up for renewable, but this works only in very short-term. In longer-terms, with the aging of existing power plants and integration of more renewable in the system (up to 20%), construction of conventional back-up power plants would be vital for the stability of the Egyptian power system.

It is also worth to mention that nuclear power can also play a flexible back-up role in power systems. Contrary to what is commonly believed, nuclear power plants have (on average) very responsive load gradients

(about 5% of load per minute) even though their start-up time is very long from both warm and cold conditions. For the time being this flexibility potential exist only in very experienced countries in realm of nuclear industry such as France and Germany for example. Therefore, flexibility analysis of nuclear plants is out of the scope of this study due to the fact that Egypt will be a newcomer in the nuclear sector (if the country adopt for the installation of before-mentioned power plants in the time horizon of this study). Under the assumption of 20% renewable integration (for both years 2020 and 2030), at least 4GW and 6GW of flexible back-up facilities would be needed respectively for the years 2020 and 2030. These added capacities do not include the replacement of retired old-age existing power units during the studied period. The necessary replacement capacity is calculated by the model without any flexibility concern for the future power plants. Therefore less flexible plants (such as nuclear in our case) have also been considered. This is not the case of our additional cost calculation for back-up units in the model.

Fuel costs are calculated per MWh on the basis of price information available for gas, oil and uranium [6,7]. In the case of gas price, the minimum average price of large gas producing countries like Canada, US, Australia and Russia (6 $/MMBtu) is considered, where domestic prices of natural gas can decouple from international market prices. This averaged price could be a good representative of international gas price for Egyptian power sector, although the real (strongly subsidized) domestic gas price is much lower for the Egyptian power producers. And for oil, Dubai dated average price over the last 4 years has been considered (80 $/bbl), even if sometimes oil products are used in power generation which are more or less expensive than the crude itself. Despite the fact that this study is done under the assumption of stable fuel prices for the matter of simplicity; this should not be considered or interpreted as any sort of prediction of stable energy markets.

In the case of uranium the task is entirely different because the price of U_3O_8 (so-called "yellow cake") only counts for about 5% of the total cost of power production and therefore any volatility in the price has very small impact on the total cost of electricity generation. Spot-market plays a very limited role for the nuclear fuel (at different stages) and most of the activities are carried out under long term contracts. In the model it is assumed that the nuclear fuel price is equal to 7 $/MWh until fuel fabrication process, plus 2.5 $/MWh more for transport, storage and eventually reprocessing and final disposal [9,27].

Apart from fuel costs, which have already been described, the other variable and fixed costs of each type of power plant are also essential for the decision making process of the model. Plants' life-time and efficiency should also be incorporated in the model so as to be able to evaluate the potential amount of electricity (from technical point of view) that each power plant could produce. Table 2 provides the techno-economic properties of various thermal power plants used in the model. As a matter of fact, year 2010 has been used as the base case for our modelling purpose due the accurate access to complete and detailed techno-economic data (load-duration, costs, efficiencies and…) for that year and moreover as a result of the political issues that happened recently in the country, not many changes have been taken place in terms of investment and costs in renewable energy sources. The almost constant trend of investments in renewable installed capacities in Egypt between 2010 and 2013 is shown in Figure 6.

Simulation results and economic analysis

Model has been run for over the period 2010-2040. Investments are allowed in the model during all of the periods and time steps so

Techno-economic data for each type of power plant

Plant type	Nuclear Plant	CCGT Plant	Fuel Plant
Efficiency (%)	33	57	38
Investment cost ($/Kwe)	2050	534	364
Life cycle (years)	60	30	30
Fix O&M cost ($/Kwe)	46	8	8
Variable O&M cost ($/MWh)	0.8	1	0.3

Table 2: Techno-economic data for each type of power plant [27].

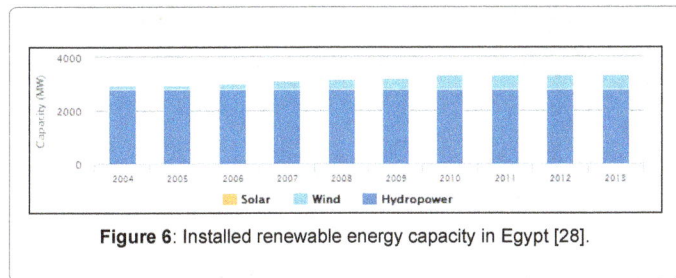

Figure 6: Installed renewable energy capacity in Egypt [28].

as to reach the final electricity demand increase. Seasonal and daily demands have been coupled with the randomness factors already described in the modelling frame-work section of this paper. However, sensitivity analyses on the model's parameters (electricity demand, power generation cost) point out that the both primal and dual results have significant changes when the parameters are modified. Thus, it is decided to take into account the uncertainties on the model's parameters through various discount factor assumptions are run sensitivity analyses on the discount rate.

The major impact of discount rates is on the value of total cost generation cost per MWh which itself includes investment, OandM and fuel costs. In this scenario carbon cost is equal to zero and therefore direct emissions resulting from fossil fuel power plants usage have been neglected.

For discount rates below 5%, total demand increase is satisfied with nuclear energy which is considered as the most viable and economic way of generating electricity. Almost 10% of the total investment takes place in the base year 2010. This is almost tripled in the final year 2030. Nonetheless most of the investment occurs in the middle periods between 2010 and 2030. For example in 2020 around 60% of the total investment decision has been realized and the model recommends 9.5 GW of investment in total installed capacity of the country (Figure 7).

For discount rates above 5% other fossil resources, particularly CCGT (combined cycle gas turbines) power plants, become more economic. For instance at 8% discount rate, the model suggests about 1.8 GW of investment in total capacity with CCGT power plants (consuming only natural gas as a fuel) from the beginning of our base (reference) year of 2010. In 2020 (middle period) model suggests not only CCGT technologies but also fuel power plants. Total amount of suggested investment in fuel power plants reaches almost 35% of total additional capacity in 2020. The remaining capacity investment is still in CCGT technologies. The model considered 100% fossil-based generation mix (as the most optimal one) up to at least 2025. From then on, nuclear technology becomes again the most optimal solution to answer the further increase of electricity demand. The fact that technologies within huge initial investment costs (so-called overnight costs) and long construction times become more economic only at the end of the period, could be explained by their notable sensitivity to large discount rates. Moreover, as we have assumed in our model that

the last periods' demand will remain constant for a very long period of time (an assumption used for increasing the reliability, stationary and rationality of the dynamic model for investment decision making), nuclear power becomes less risky and optimal solution for long-term demand satisfaction. Economic viability of this long-term decision-making strategy turns out to be less rational for discount factors higher than 8% and even fully disappears for discount factors rates above 10%.

By looking at the results in Figure 7 it is also noticeable that for the discount rate values above 10%, investments in fuel power plants turns out to be optimal from the beginning and becomes even the only optimal choice after 12%. Short construction time (compare to the other technologies) and rapid return on investment are the main reasons behind this expensive 100% fuel-based plants investment. Prompt satisfaction of accelerating electricity demand with least costs, is also another reason. However, by moving further in time and giving more time to the investor(s), more capital intensive technologies such as CCGT come into action once more.

It should not be forgotten that the above conclusions obtained under the zero carbon emission price assumption and they can be totally altered by setting a certain amount of CO_2 price in the model. Henceforth, CO_2 costs are introduced in the model. Carbon emissions' amounts were integrated as physical property of each fossil fuel type by taking into account the thermal efficiency of each fossil power plant. Initially, the CO_2 price of 10€ per tonne was designated and then the model was run again. Investment results under this assumption for the same discount rate intervals are shown in Figure 8.

For the discount rates up to 5%, nuclear power remains again the most optimal choice and other technologies are not competitive at all (except as a back-up plant to compensate renewable intermittencies). Significant modification compare to the pervious case (without emissions) can be noted in the discount range of 8% to 10%. In this range, nuclear energy is still present as an economical source of power; for instance around 8% of discount rate, nuclear energy could provide up to 70% of total electricity sector investment of Egypt as a most optimal power unit. However fossil plants start to occupy a bigger share in the power generation mix of the country in 10% discount rate case.

Uncertainty about climate policy is one of the greatest risk factors that investors in power sectors are dealt with at the moment. Climate policy may have a weighty impact on power generation costs with different options. If ambitious carbon reductions are to be achieved

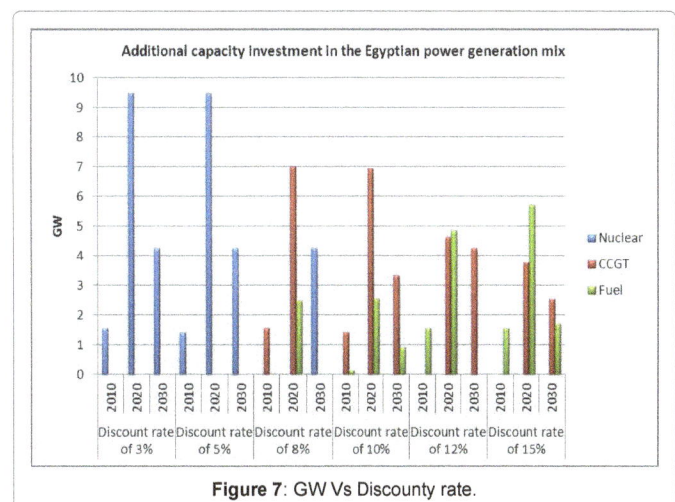

Figure 7: GW Vs Discounty rate.

globally, the power sector may need to be rapidly decarbonized in many regions. However, the decarbonisation trend observed in non-OECDs in much slower than that of OECDs. Uncertainty about future climate policy (hereby integrated via various CO_2 prices) thereby creates significant insecurity about generation costs of different technologies.

Hence, a sensitivity analysis designed for different CO_2 prices so as to better demonstrate the impact of carbon price increase on the power generation structure of Egypt and obviously the promotion of non and less CO_2-emitting technologies, respectively nuclear and CCGT, compare to fossil fuel based ones. Egyptian optimal generation capacity additions proposed by the model under different CO_2 price scenarios are shown in Figure 9.

Finally, it is important to mention that Egypt became a net importer of oil in 2010 (our reference year). In our model we assume that fossil fuel prices (oil and gas) are equal to that of international markets. Hence, if Egypt continues to provide natural gas to power producers under subsidies (with final price lower than that of international markets), all the suggested investments in fuel power plants should be replaced by gas units. This could become also applicable for nuclear units after certain level of subsidies. And on the contrary, under total subsidy-suppression scenario in addition to less uncertain investment and political environment (leading to smaller discount rates) nuclear power choice could be the most economic and optimal solution. Not only it will provide cheaper power but also help to free certain share of domestic gas production for export into international markets. Nevertheless, we should not forget that certain amount of power (almost 20% according to our model) must be still afforded by fossil fuel plants, with rapid start-up time, to assure the back-up role for the 20% integration of intermittent renewable in the Egyptian electricity mix. Finally, an attempt to analyse the pass through effect of intensive subsidies in the wholesale and retail power tariffs of Egypt was performed. For this purpose, a static cost-minimization model (without investment) of Egyptian power supply has been constructed for the reference year 2010. In this model demand's variation is based only on the peak/base periods and seasons. Hence neither medium nor long term demand increase scenarios were applied. The shadow values (marginal values) associated with the loaded power (model's output) for each season and each hour corresponds to the marginal values produced by the last power unit (MWh). Observation of those values for our static model (in reference year 2010) indicates that the long-term marginal cost of electricity production is around 72$ per MWh. Actually this value is the average of all the marginal values generated by the model for each season and hour of the day. Due to the fact that the technology does not change during peak hours, it can be used as a

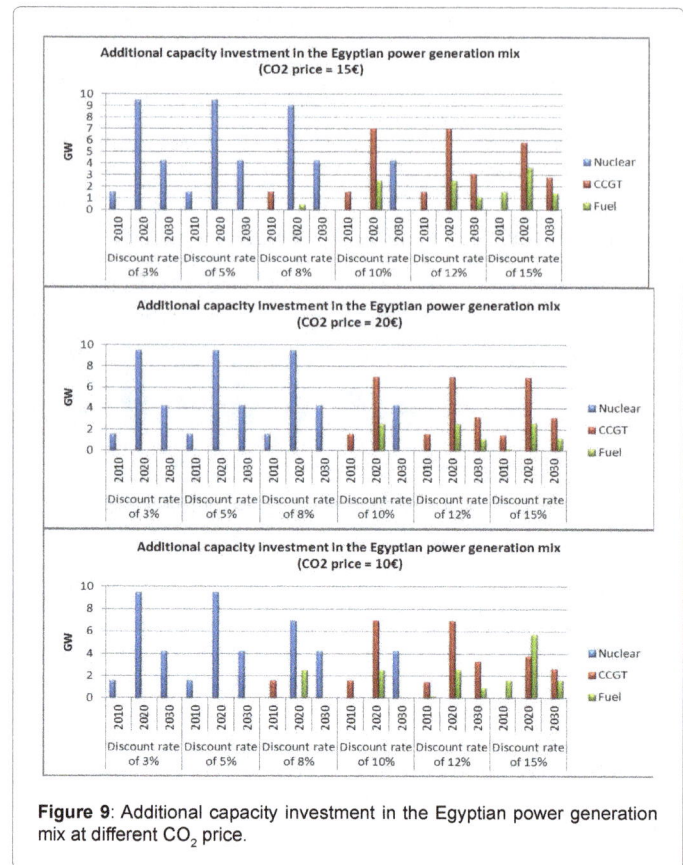

Figure 9: Additional capacity investment in the Egyptian power generation mix at different CO_2 price.

proper indicator of total marginal cost.

The weighted average of Egyptian electricity tariffs (multiplying the share of each consumer by its related tariffs) is equal to approximately 45$ per MWh. Table 3 shows the Egyptian electricity tariffs for each category of demand and consumption. This value is less than 60% of the marginal value given by the model. Hence, if for example the marginal pricing criteria as an optimal way of electricity pricing is considered (in which short-run and long-run marginal costs are equal and future investments are guaranteed), the existing tariffs are far below the optimal level [29]. In other words, the allocated utility of fossil fuels (including subsidies) associated to the power generation is higher than the potential value of these fuels (oil and gas) for a probable export or unsubsidized usages in the power and other energy intensive sectors. This observation confirms the distorted optimality of the current heavily-subsidized power sector of Egypt, in terms of both fuel prices and final tariffs.

Conclusion

The economic analysis of the power generation mix with an optimization model point out that, according to resource availability and the future expected electricity needs, being mainly dependent on national fossil fuel reserves for power generation is not an economic optimum. The gas resources could be exported and more power units could be based on renewable resources or nuclear power plants. Moreover, investment in nuclear power units for the demand satisfaction of the next 20 years (between 2020 and 2030) in addition to 20% integration of renewables in the generation mix can reduce the CO_2 emission of the Egyptian power sector by almost 25 million tons

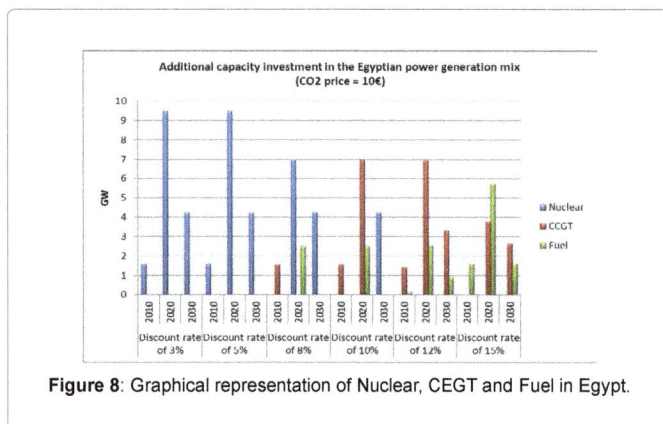

Figure 8: Graphical representation of Nuclear, CEGT and Fuel in Egypt.

Current Electricity Tariff Structure (1 Pt ≈ 0,14 $)	
Sector	*Average Price (Pt/KWh)*
Residential	30
Commercial	40
Agriculture	11
Industry	20

Table 3: Egyptian Electricity Tariffs [8].

per year.

However, these choices are affected by the evolution of costs and demand over twenty years period. Thus, the choices of a low or a high discount rate strongly impact the power generation mix and consequently the CO_2 emission rates.

Efficient utilization of the energy resources concerning the electricity sector requires a considerable promotion of the alternative non-fossil techniques. Even though the renewable sources of power generation can be used efficiently at very decentralized and local scales, yet intermittent nature of these technologies does not permit to provide a large scale continues base-load power. Besides, the need for more fossil-fuel-based back-up power plants would become inevitable to guarantee the national power system equilibrium.

Therefore, a power generation strategy based on a gradual integration of nuclear and renewable is suggested. A power generation mix, based on an optimal choice of fossil, nuclear, hydraulic and other renewable energy, is considered to be the most appropriate way of electricity production in Egypt.

Appendix A: Single Buyer Model

In this restructured electricity market, networks (whether transmission or distribution) remain regulated while generation is exposed to competition. For the networks the incentives for capital investments are function of the regulation imposed by the regulatory authorities. Contrarily in the case of generation no explicit price control applies, nevertheless the regulators may monitor generation adequacy and establish additional market and tariff-based incentives to encourage new investments in the sector.

Under a single buyer model only new capacity development is exposed to competition, while the continued operation of plants with respect to output would be exempt from competition and would rather run under (usually long-term) power purchase agreements. The single buyer is responsible to determine capacity requirements and could also direct the technology decision through suitable conditions included in the call for tender for new capacity.

In this model the revenue that a generator is allowed to receive under its contract with the single buyer is normally contains two main components, availability payments and energy payments. The energy payments are intended, among other things, to recompense the generator for the costs associated with operating the plant, that is fuel and variable OandM costs. The availability payments are anticipated to provide the generator with revenue to cover the cost of capital, including a normal rate of return, and the fixed OandM costs (Figure 1A).

References

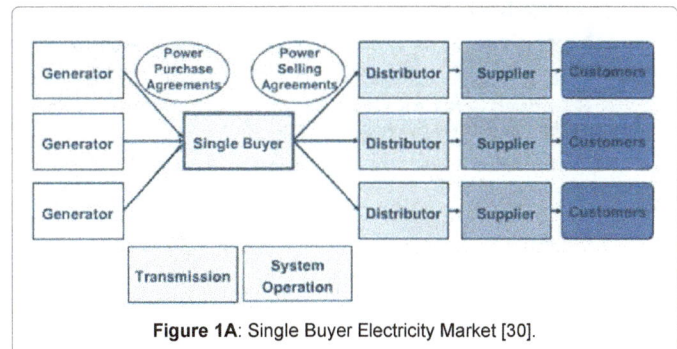

Figure 1A: Single Buyer Electricity Market [30].

Renewables status report: 21.

2. Corden WM, Neary P (1982) Booming sector and de-industrialization in a small open economy. The Econom J 92: 825-848.

3. Ministry of Electricity & Energy (2010/2011) New & Renewable Energy Authority (NREA), Annual Report.

4. http://www.globalcarbonatlas.org

5. BP (2015) Statistical Review of the World Energy: 48.

6. World Bank (2014) World Development report: 362.

7. IEA (2012) World Energy Outlook 2012: 696.

8. EEHC (2012) Egyptian Electricity Holding Company Annual reports 2010/2011/2012.

9. IEA (2014) Electricity Information: 883.

10. Selim TH (2009) On the economic feasibility of nuclear power generation in Egypt. The Egyptian Centre for Economic Studies (ECES).

11. http://www.geni.org

12. Haidar AMA, John PN, Shawal M (2011) Optimal configuration assessment of renewable energy in Malaysia. Renew Energy 36:881–888.

13. Hainoun A, Seif Aldin M, Almoustafa S (2010) Formulating an optimal long-term energy supply strategy for Syria using MESSAGE model. Energy Policy 38:1701–1714.

14. Nielsen SK, Karlsson K (2007) Energy scenarios: a review of methods, uses and suggestions for improvement. Int J Global Energy Issues 27:302–322.

15. Ostergaard PA (2009) Reviewing optimisation criteria for energy systems analyses of renewable energy integration. Energy 34:1236–1245.

16. Sorensen P, Norheim I, Meibom P, Uhlen K. (2008) Simulations of wind power integration with complementary power system planning tools. Electric Power Sys Res 78:1069–1079.

17. Kaldellis JK, Kavadias KA, Filios AE (2009) A new computational algorithm for the calculation of maximum wind energy penetration in autonomous electrical generation systems. Appl Energy 86:1011–1123.

18. Karlsson K, Meibom P (2008) Optimal investment paths for future renewable based energy systems-using the optimisation model Balmorel. Int J Hydrogen Energy 33:1777–1787.

19. Kazagic A, Merzic A, Redzic E, Music M (2014) Power utility generation portfolio optimization as function of specific RES and decarbonisation targets – EPBiH case study. Appl Energy 135: 694-703.

20. Lund H (2009) Renewable energy strategies for sustainable development. Energy 32: 912-919.

21. Segurado R, Krajacic G, Duic N, Alves L (2011) Increasing the penetration of renewable energy resources in S. Vicente, Cape Verde. Appl Energy 88:466–472.

22. Xydis G, Koroneos C (2012) A linear programming approach for the optimal planning of a future energy system. Potential contribution of energy recovery from municipal solid wastes. Renew Sustain Energy Rev 16: 369-378.

23. Chang Y, Li Y (2013) Power generation and cross-border grid planning for the integrated ASEA electricity market: A dynamic linear programming model.

1. REN21 (2013) Renewable Energy Policy Network for the 21st Century: MENA

Energy Strategy Rev 2: 153-160.

24. Criqui P (2001) Prospective Outlook on Long-term Energy Systems, IEPE Report: 9.

25. Loulou R, Goldstein G, Noble K. (2004) Documentation for the MARKAL Family of Models. Energy Technol Sys Analysis Prog: 386.

26. IRENA (2014) Africa Power sector: Planning and Prospect for Renewable Energy: 44.

27. IEA-NEA (2010) Projected Costs of Generating Electricity, OECD Publishing: 216.

28. IRENA (2015) Renewable Energy Capacity Statistics: 44.

29. Boiteux M (1949) La tarification des demandes en pointe: application de la théorie de la vente au coût marginal. Revue Générale de l'Electricité.

30. Petrov K (2010) The Single Buyer Model, KEMA report: 7.

18

Power Line Communication System for Grid Distributed Renewable Energy

Nguyen TV [1,2], Petit P[1], Aillerie M[1], Charles JP [1] and Le QT [2]

[1]Lorraine University, LMOPS-EA 4423, 57070 Metz, France

[2]Quang Ninh University of industry, Quang Ninh, Vietnam

Abstract

The multi-sources nature of renewable energy production can be taken into account thanks to involving the solutions of distributed architecture based on individual DC-DC converters, connected to a direct current (DC) bus. Associated to this architecture, to assume simply the communication between the modules, one solution is the use of the DC bus power bus to support the communication between optimizers and a central controller using a power-line communication approach (PLC). The current work consists, at first, in the analysis of the pertinent information necessary to exchange between DC-DC converters and a central controller and, at second, by the development of a new hardware solution for the PLC from the conception of the communication system to the realization of a prototype. The various possible devices connected on the bus or networks are considered as programmable logic controllers, various sensors, microcontrollers and grid inverters. At minima, the information to exchange between the various devices may include the maximum power point of photovoltaic modules (MPP) and the temperature of the individual sources. At first, the ASCII Modbus protocol was chosen in the present work to assume the PLC communication on the DC bus. The interfacing circuitry between the DC bus and PLC controller is achieved by TRSV04 transceiver and power coupling circuit.

Keywords: Power-line communication (PLC); Direct current bus; Smart dc-dc converter; Optimizer; Renewable energy; MODBUS protocol

Introduction

The suitability of the proposed network for DC system is assessed by its compliance with the requirements set for specific applications in renewable energy generation and conversion system. We present a simple hardware implementation of a distributed architecture based on smart DC-DC optimizers integrating both, the power conversion and interface communication stages. This work is done in the aim of a realization taking into account the efficiency of the global system and its economical approach with a low-cost communication solution PLC_{DC}.

This choice of PLC system eliminates the need of using extra wires or complex wireless interfaces to assume the inter-communication between individual converters.

Power line communications systems operate by impressing a modulated carrier signal on the DC bus. The data is transmitted safely and reliably. Finally, development of such PLC system in distributed architecture based on DC bus configuration amounts to solve the problematic of the plug-in and interface of a small level information signal emitter-receiver with a high voltage power line module.

The described system is based on network architecture for the DC bus system that meets all the requirements presented above and only PLC_{DC} developments will be presented in this contribution; the main energy converter functions, with their MPPT (Maximum Power Point Tracking) algorithms and the self-power stage were presented in previous contributions [1-5]. This means also that the PLC_{DC} systems can be considered in each converter, as an additional stage to the main energy conversion stage, without modification of its basic structure. It is to be noted that a self-power supply stage can also be added in the optimizer. In this paper PLC interface is controlled by a dedicated Peripheral Interface Controller (PIC). Nevertheless, an optimizer integrating both the functions of tracker and PLC_{DC} master-slave control by a PIC microcontroller can easily be used.

In the aim of the integration of a communicant system in a renewable energy generator system, the current work presents the design and the realization, up to a complete laboratory prototype allowing the test and evaluation of all the technical choices and involved concepts, a system able to communicate on DC power lines. In the present work, the corresponding hardware and software elements of the data transmission interfaces for both, the individual converter, i.e. the slave interface, and the controller, i.e. the master interface were developed. The final realization of several prototypes, slaves and master units, and the communication between them, are tested and some improvements are suggested to achieve industrial qualification.

Design of a Master-Slaves PLC System on DC Bus

The work is on the Amplitude Shift Keying (ASK) structure of modulation, which is commonly considered in industrial communication network. ASK works by assigning unique pattern binary digits to different amplitudes by representing digital data as variations in the amplitude of a carrier power signal remaining constant. The advantages of ASK are its simplicity due to the associated low bandwidth requirements, so it is easy to implement transmitter and receiver with a small number of components and the detection is also facilitated. In this structure of modulation, the information carrier is a rectangular-wave signal in the 50 kHz frequency range. In ASK, the On-O Keying (OOK) method is one possible modulating method used for the data transmission [6,7]. This modulated signal is superimposed on the 400V DC voltage. In the present realization, to implement the OOK method, we have used a PIC16f876 microcontroller, realizing a half duplex power-line transceiver. It is to be noted that such kind of microcontroller or equivalent integrating circuit is generally used to drive the individual DC-DC converters as power sources in distributed

*Corresponding author: Vinh Nguyen, Lorraine University, LMOPS-EA 4423, 57070 Metz, France, E-mail: vinhnt@qui.edu.vn

DC bus architecture.

Taking into account the previously presented study and the chosen approach for renewable energy generator, we have developed a PLC system with two different types of circuits for the transmitter-receiver: one, functioning in a slave mode dedicated to the individual optimizers, a second one considered as the master, being the interface with the central management controller. These circuits are presented in Figure 1 with upper and lower parts for the slave and master circuits, respectively. In the chosen solution, the receiver parts of the slave and master circuits are identical and only the transmitter circuits are different. For experimental validation of the technological choices that we have done in the present study, the DC bus voltage is specifically adjusted to 400 V, as determined by the transformation ratio fixed in each individual DC-DC converter. The PIC microcontroller, a PIC16F876 in our application, rep-resented in Figure 1a, is also used for the MPP tracking in the energy conversion part of the step-up DC-DC converter which is installed corresponding to a PLC slave. The PIC microcontroller creates a signal frequency carrier for the controller switch MOSFET M as well as for M1 of the PLC master-slaves circuits. The comparator converts data levels to the reference ones to obtain the two digital levels 0V and 5V.

The main electronic component of the receivers of the PLC master and slave circuits is a transistor mounted in a common emitter mode, which role is to adapt PLC signal from the DC bus to the microcontroller input, pin RC7/RX of the PIC microcontroller. In this circuit, the received signal enters first to the demodulator, which recovers the original data. An interfacing branch is used to isolate the receivers from the 400 V dc environments. It is composed by R1, C2 and combine with the filter circuit L1, C3.

In the proposed PLC transmitter part of the slave circuit, the effect of distortion is minimized, by a careful selection of a constant and stable carrier frequency (fc). Therefore, an oscillator is built using the PIC microcontroller generating modulating signal at a frequency of 50 kHz. The amplification of the signal was designed using a Q2N2226 transistor, dedicated for low-voltage and high-speed applications, especially in inductive circuits. The interfacing circuit consists of a forward-converter transformer where both primary and secondary windings conducting simultaneously with opposing magneto-motive forces along the mutual flux path. The difference of the magneto-motive forces is responsible for maintaining the magnetizing flux in the core. When primary winding current is interrupted by switching o the switch, the dotted ends of the windings develop negative potential to oppose the interruption of current blocking the diode, and thus, interrupting the conduction. To reduce the current delivered by the DC bus in the secondary coil, thus avoiding possible saturation, we added, in Figure 1b, a self, Ls, and a capacitor, C1, in the interfacing circuit of PLC system master-slaves. Each PLC slave circuit connected on the DC power-line corresponds to a node. The current passing through the PLC transmitter part of the slave circuit is equal the output current of DC-DC converter. The purpose of the Ls coil, Figure 1b, is to limit the saturation current in the circuit magnetic of Lp1-Lp2 transformer. Thus this work design is the same as a forward-converter transformer on the DC bus. The current coming from all source-nodes of the DC bus crosses the PLC transmitter part of the master circuit.

Communication and Transmission Protocol Implemented in PLC Modules

The work on implementation of the MODBUS protocol in master like slaves microcontrollers. An exhaustive amount of information about the protocol is given on the website [8,9]. Referring to those documents the ASCII mode protocol for PLC slave and master was developed and tested successfully. Experiments consisted in reading two slaves connected to HVDC bus. Reading frames sent by master were composed of thirteen ASCII characters, including longitudinal redundancy checksum byte (LRC), necessary to exchange dependability. As shown in Figure 2, the response frame from slave was composed of fifteen ASCII characters, with the need to compute LRC byte depending on the value of transmitted data before response frame sending. The ASCII characters are described below:

Reading frame:

- Start delimiter
- Slave adress1
- Reading request
- Momery words address to read
- Carriage return and line feed

Response frame:

- Start delimiter
- Slave adress1
- Reading request
- Number of bytes
- 6 blank bytes
- Carriage return and line feed

In the response frame the blank characters are replaced by the four ASCII characters of the two read data bytes added with the computed LRC.

Obviously, received LRC byte was always compared to compute one's, both for master and the two slaves. So LRC byte allows to checked frame integrity: thanks to all this mean possible serious disturbances on DC bus can be instantaneously detected as it was done in our power line communication design on the DC [10-12]. Available baud rates are limited by the carrier wave frequency (maximum baud rate equal to about 10% of the carrier frequency) to take account of the receiver filter delay group, and DC power line impedance. Concerning the microcontroller, the only limitations are the highest baud rate of the universal synchronous asynchronous receiver transmitter (USART) (115200 bit/s with 20 MHz CPU clock).

In spite of different role, PLC master and PCL slaves consist of common blocks: the same microcontroller, the TRSV04 (described further), the power line interfacing circuit and the power supply. The PLC master microcontroller generates the packets of data to TRSV04 transceiver whose role is to superimpose these packets on the HVDC bus using ASK at the programmed carrier frequency. Considering the slave receiver, the TRSV04 detects the carrier and then decodes the modulation and deliver logic levels to microcontroller. The same process occurs when a slave unit replies to the master. As explained above all the packets are transmitted according to the ASCII Modbus protocol.

In the master module shown in the Figure 3a, see a LCD 16×2 (Liquid Crystal Display - 2 lines with 16 characters per line) for data display. It can operate both in 4-bit or 8-bit mode. All useful data concerning PV (Photo Voltaic) system can be displayed: PV module temperature,

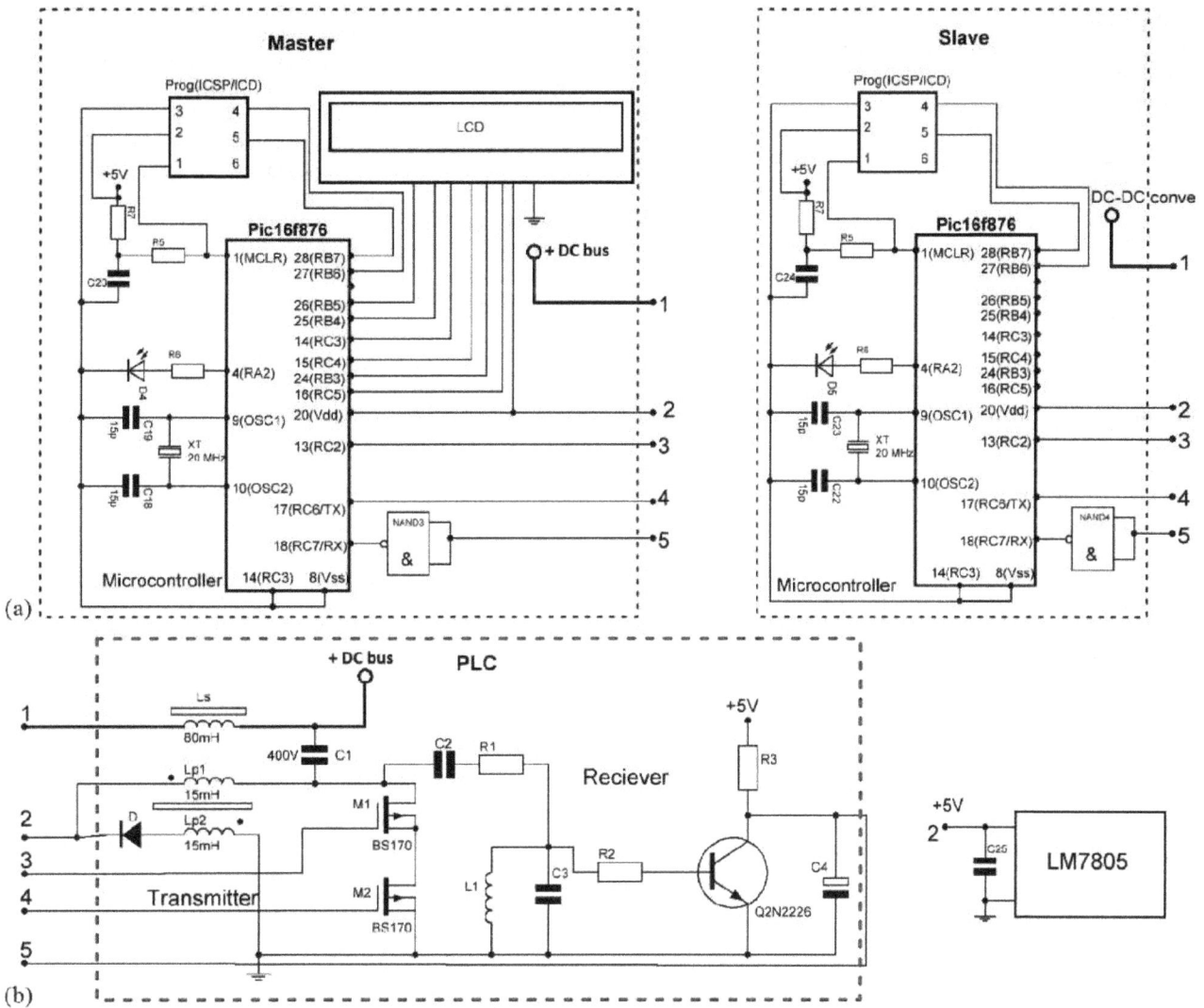

Figure 1: a) The PLC system master-slave; b) Implementation of the microcontroller for transmitter and receiver parts of PLC master and slave.

| : | 0 | 1 | 0 | 3 | 1 | 0 | 0 | 2 | E | A | \r | \n |

Reading frame on slave 1

| : | 0 | 1 | 0 | 3 | 0 | 2 | 0 | 0 | 0 | 5 | F | 9 | \r | \n |

Response frame from slave 1

Figure 2: MODBUS frames in ASCII type.

PV module maximum power point (MPP), which can be detailed by electrical measures like PV module voltage, PV module current and DC bus voltage directly realized by optimizers (as shown in Figure 3b). Tests have consisted in exchanges between the master and two slave units by reading of the slave sensors, the number can extend for the module slaves correspond with the number of the DC-DC converters connect parallel on DC bus [13-17]. A simple potentiometer has been used as an analog sensor and, after ADC conversion, the measurement transferred in a 16 bits format via the DC bus. For the moment the quantification of all the PV system measurement is limited to 10 bits (microcontroller resolution in the case of laboratory optimizers). But measurement devices with higher resolution can be considered like, as evoked above, PV module temperature but also irradiance useful for PV system diagnosis.

In the case of optimizers which are considered as slaves, their principal function is obviously energy conversion with the highest efficiency based on a Maximum Power Point Tracker (MPPT) but we can add other functions like PV module monitoring, with the ability to detect and diagnose local shadowing due to buildings around, trees, leaves or simply dirt on PV modules, and to measure a deformation of the I/V curve. Follow a few modifications it would be easy to imagine a way to implement an intrusion detection system, especially in the case of PV modules set on the ground. When an intrusion occurs, input power of the optimizer changes like in the case of a fugitive shadowing and can be computed from to the input optimizer voltage and the current sensors. The power change detection is diagnosed by optimizer as intrusion information. This status located to one particular PV module is sent to the PLC master. Nevertheless, to avoid false alarm detection, it would be necessary to verify intrusion status on other optimizers before displaying security message on the master LCD. A complete monitoring system should allow parameters transmission to optimizers, like specific operating points (for instance safety voltage for firemen) or independent control of distributed sources.

Implementation

The two modules transmitter and receiver coupled together by the medium used for data transmission, i.e., the DC bus are shown in the Figures 4,5 and 6. Both the master and slaves boards are directly connected to the DC bus. The microcontroller (PIC16F876A) of PLC interface transmitter module is programmed to transmit ASCII characters composed of seven data bits and one even parity bit which are fed into the TRSV04. The TRSV04 transmits the data at the programmed frequency of 50 kHz by binary ASK technique on the DC power line.

The receiver TRSV04 of PLC master detects the carrier amplitude and converts it to a logical value as shown further in Figure 4 towards the microcontroller. These data are treated before displaying on the LCD. The medium chosen for these first tests is an ASI bus cable (Actuator Sensor Interface) which could even be convenient to transmit power, from to sufficient section (higher than 2 mm²). Exchanges between master and slave can be considered as quite dependable in the first experimented case of data transmission without power on the DC. No errors were detected by LRC function of Modbus, in the case of small distance between master and the two slaves. More over we observed no transmission errors with PLC modules in the case of significant distance between master and slaves as described in following lines.

The experimental tests were performed with master and slave modules connected via a DC bus of around 150 meters long as shown in Figure 6, with 100 meters between master and the first slave module

and 50 meters between the two slaves. In Figure 7 and Figure 8 shows oscilloscope traces of the transmitted signals owing on the bus.

In Figure 7 see, first signal (yellow), a reading frame transmitted via the TX pin of the master microcontroller. The fourth wave is the signal visible on the RX pin of master microcontroller (green signal): we can obviously see response frame from slaves, but also master reading frame itself. The second signal (blue) corresponds to the RX pin of a slave module microcontroller: we easily identify master frame but also proper slave transmission which is visible alone on third wave corresponding to slave TX pin (violet signal). Reading frame duration (first signal) is a little higher than theoretical calculus: thirteen characters conformed to USART protocol (one bit start, seven data bits, one parity bit and one stop bit) would have 104 ms with 2400 baud rate. In fact few additional delays have been programmed to insure transmission reliability, so we can estimate real duration around 120 ms for reading frame. In the case of slave response, fifteen characters have a theoretical duration of 120 ms. Real duration is measured around 140 ms. So, a complete exchange has a minimal duration of around 260 ms, implying a DC bus refreshing frequency order of lower 4 Hz. Even RTU mode is implemented with the hope to double refreshing frequency, the relative slowness of this bus induces to plan a maximum of functions directly embedded in distributed optimizers (local energy counting for example). In the case of urgent safety state dedicated to firemen or for maintenance, ability to control all the optimizers in the same time (classical zero addressing) should allow a response in only 120 ms (ASCII Modbus protocol).

Not visible on Figure 8, the PLC information signals present a rectangular shape with amplitude of V normally superimposed on the DC bus. We note a small delay for this signal to establish to a stable level, implying a systematic group delay. The shape at the bottom is the logical level obtained at the output after demodulation (blue signal) of the ASK modulation (green signal) of the DC bus signal. It is easy to understand that the speed limitation is mainly due to the global delays observed here Figure 9 shows at the upper shape the voltage observed on the DC bus when a transmission is send to the slaves (green signal). The second curve corresponds to the signal modulation obtained accorded on the carrier frequency i.e. 50 kHz with the impedance over the power line 1,5 Ohm and influence of the impedances output of the DC-DC converters connect parallel on the DC bus.

Future scope

A specific study must be achieved to evaluate the influence of high scale of impedance variations of the DC bus. Some improvements must be experimented concerning the master unit and especially for the dimensioning of the passive filtering elements which have to drive high level of currents supplied by all the connected generators. The scope appears as limitless since communication requires no additional wiring. At first our study was dedicated to on grid architectures implying multi renewable energy sources (PV modules, wind generators…) and charges like inverters, with the aim to increase the global efficiency in procedure transmission of energy in spite of possible disturbances like important shadowing. DC bus is also appropriate to smart monitoring including other devices like industrial control components (programmable logic controllers, graphic terminals . . .) or simply a computer connected to DC bus via PLC master and naturally to Internet for all the conveniences induced and can install security in the PLC system to supervision the dispersed sources.

Conclusion

This contribution presents new developments of communication

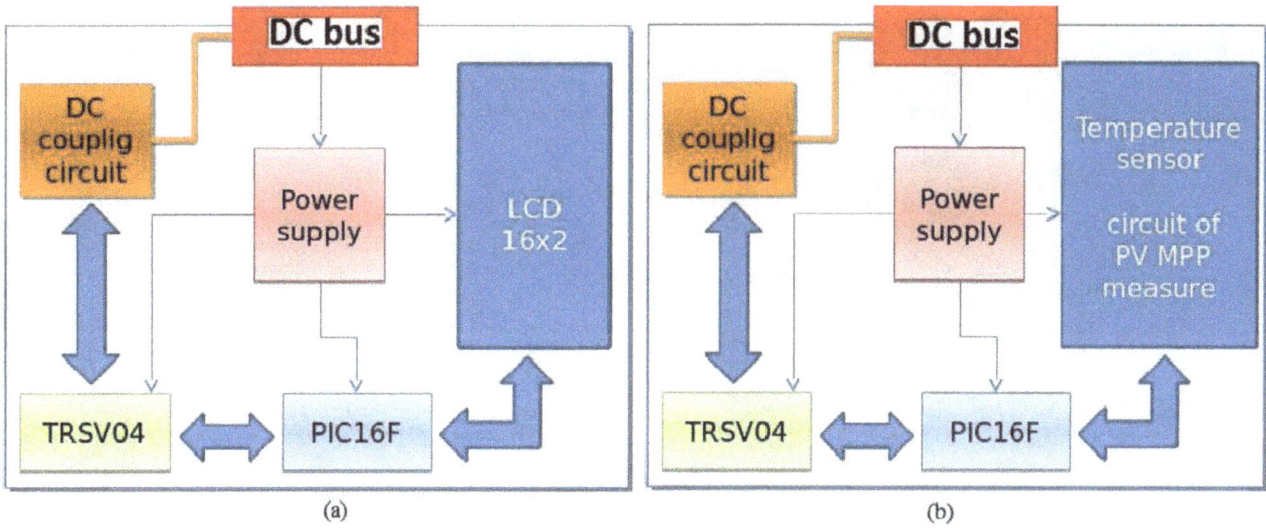

Figure 3: a) PLC master module block diagram; b) PLC slave module block diagram.

Figure 4: The transceiver board of the master TRSV04.

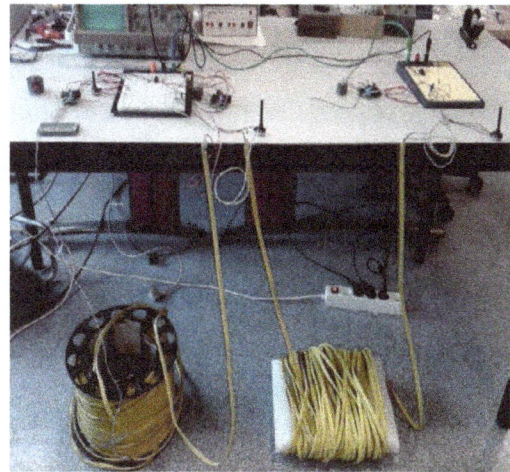

Figure 6: The Transmitter and Receiver Coupled to the DC bus.

Figure 5: The transceiver board of a slave TRSV04.

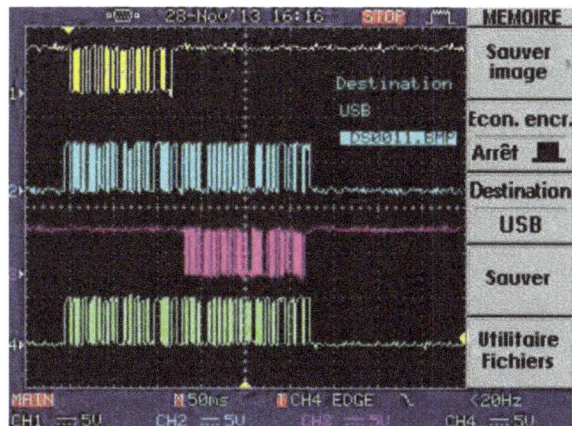

Figure 7: Master and slave frames on oscilloscope: microcontroller USART signals and transceiver PLC data.

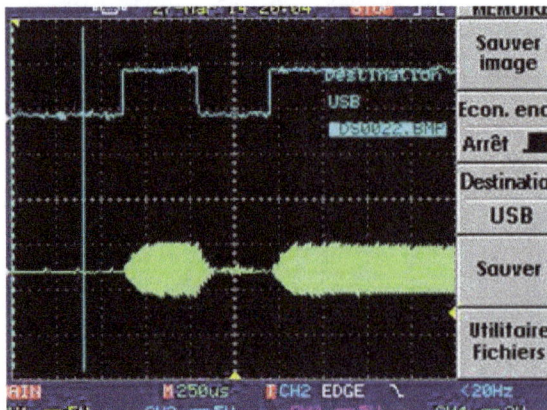

Figure 8: Signal of receiver module of the PLC master-slave.

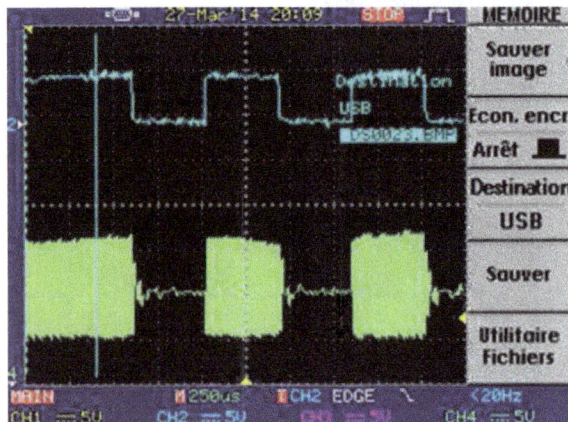

Figure 9: Signal of receiver module of the PLC master-slave.

system dedicated for distributed renewable energy production generators. The basic concepts initially fixed for this original development of communication system are based on a low cost, low frequency carrier, avoiding propagation phenomenon and allowing long distances.

Thus, this work explores, up to the realization of a complete prototype system, the communication between individual devices (DC-DC converters and a master controller) constituting a renewable energy generator and using power lines (a DC bus) as information support based on a PLC technology at a very low frequency carrier. The communication protocol is based on the widely accepted Modbus protocol. The presented system communicates successfully and is able to receive and transmit data without any errors. This system constitutes the base for an implementation in a massively parallel architecture

in the case of stand-alone PV systems, and even self-consumption solutions integrating smart storage. As no extra wires dedicated to communication are required in the proposed solution dedicated for power electricity transport, the reduce of the cost should be noticeable, by reducing the wire section because of high voltage of the DC bus and by the present implemented simple communication system.

References

1. Petit P, Aillerie M (2012) Integration of individual DC/DC converters in a renewable energy distributed architecture. Industrial Technol, 2012 IEEE International Conference , Athens Greece 802-806.

2. Zegaoui A, Aillerie M, Petit P, Sawicki JP, Charles JP, et al. (2011) Dynamic behaviour of PV generator trackers under irradiation and temperature changes. Solar Energy 85: 2953-2964.

3. Zegaoui A, Aillerie M, Petit P, Sawicki JP, Jaafar A, et al. (2011) Comparison of Two Common Maximum Power Point Trackers by Simulating of PV Generators. Energy Procedia 6: 678-687.

4. Nguyen TV, Petit P, Maufay F, Aillerie M, Jafaar A, et al. (2013) Self-powered high efficiency coupled inductor boost converter for photovoltaic energy conversion. Energy Procedia, 36: 650-656.

5. Petit P, Zegaoui A, Sawicki JP, Aillerie M, Charles JP (2011) New architecture for high efficiency DC-DC converter dedicated to photovoltaic conversion. Energy Procedia, 6: 688-694.

6. WA Atia, RS Bondurant (1999) Demonstration of return-to-zero signaling in both OOK and DPSK formats to improve receiver sensitivity in an optically pre amplified receiver. Presented at the LEOS'99, San Francisco, CA, Paper TuM3.

7. Neha, Shrirao, Ajay, Thakare (2013) Design of Digital Modulators: BASK, BPSK and BFSK using VHDL. Int J Adv Res Comp Sci Software Engg 3: 382-386.

8. http://modbus.org/tech.php1.

9. Shabarinath BB, Gaur N (2013) Modbus communication in microcontroller based elevator controller. In Control, Automation, Robotics and Embedded Systems (CARE), 2013 International Conference, Jabalpur, India.

10. Nguyen TV, Petit P, Maufay F, Aillerie M, Charles JP (2013) Power -line communication (PLC) on HVDC bus in a renewable energy system. Energy Procedia 36: 657-666.

11. Nguyen TV, Petit P, Maufay F, Aillerie M, Charles JP (2014) Power-line communication between parallel DC-DC optimizers on High Voltage Direct Current bus. WIT Transactions on Ecology and the Environment. Energy Production and Management in the 21st Century: The Quest for sustainable, WITPRESS, 190: 1297-1308.

12. Nguyen TV, Petit P, Sawicki JP, Aillerie M, Charles JP (2014) DC power-line communication based network architecture for HVDC distribution of a renewable energy system. Energy Procedia 50: 147-154.

13. Pinomaa A, Ahola J, Kosonen A (2011) Power-line Communication-Based Network Architecture for LVDC Distribution System. IEEE International Symposium on Power-line Communications and Its Applications 358-363.

14. www.abb.com.

15. Hrasnica H, Haidine A, Lehnert R (2004) Broadband power-line communications: Network Design. John Wiley & Sons, West Sussex, England.

16. Vanfretti L, Hertem DV, Nordstrom L, Gjerde JO (2011) A Smart Transmission Grid for Europe: Research Challenges in Developing Grid Enabling Technologies. IEEE Transactions on Smart Grid.

17. Roychoudhuri C, Tayahi M (2006) Spectral Super-Resolution by Understanding Superposition Principle and Detecting Processes. IJOMT, 1: 146-153.

19

Investigation on the Exergy Performance of a Central Receiver Power Plant

DI Ramadan Ali Abdiwe and Markus Haider*

Vienna University of Technology, Institute for Exergy Systems and Thermodynamics, Austria

Abstract

The present paper describes the exergy analysis of a Central Receiver System (CRS) power plant. The plant consists of a thousand heliostats with an area of 130 m² each, an external receiver with an area of 59 m² and a height of 70 m, a steam generator, two steam turbines with a reheater in between, two feed water heaters and a condenser. EBSILON®Professional software was used to obtain the exergy efficiency and the irreversibility in each component of the power plant to pinpoint the causes and locations of the thermodynamic imperfection. The model analyzed and tested the effect of two design parameters including the Direct Normal Irradiation (DNI) and the outlet temperature of the Heat Transfer Fluid (HTF) on the exergy performance. The obtained results show at a constant DNI the maximum exergy loss occurs at the Receiver followed by the heliostat field and the power cycle has the lowest exergy loss. The increase of the DNI affects negatively the exergy efficiency of the overall system. The variation of the outlet temperature of the HTF has an impact of the exergy performance of the receiver subsystem as well as the overall system; the increase of the outlet temperature from 450°C to 600°C leads to an increase the exergy efficiency of the receiver to about 5% and an increase the exergy efficiency of the overall system to about 1%.

Keywords: Central receiver system; External receiver; Exergy efficiency; Irreversibility

Nomenclature

η_{II}:	Exergy efficiency(%),
W_{net}	Net output of the overall system (W)
$W_{turbine}$	Output of both turbines (W)
W_{pump}	Work for both pumps (W)
y & z	Mass fractions
m_{st}	Steam mass flow rate (kg/s)
m_{ms}	Molten salt mass flow rate (kg/s)
T_o	Atmospheric temperature (°C)
T_{wi}	Inlet water at the condenser (°C)
$T_{(rec,sur)}$	Receiver surface temperature (°C)
I_{total}	Overall system irreversibility (W)
$I_{heliostat}$	Heliostat field irreversibility (W)
$I_{receiver}$	Receiver irreversibility (W)
I_{pcycle}	Power cycle irreversibility (W)
Q^*_{total}	Total isolation (W)
Q^*_{inc}	Incident isolation (W)
$Q_{(inc,abs)}$	Heat absorbed by the receiver (W)
$Q_{(inc,loss)}$	Heat lost at the receiver (W)
ψ^*_{total}	Exergy input into the heliostat field (W)
ψ^*_{inc}	Exergy delivered to the receiver (W)
$\psi_{(inc,abs)}$	Exergy absorbed by the receiver (W)
HP-T	High pressure turbine
LP-T	Low pressure turbine
OFWH	Open feed water heater
CFWH	Close feed water heater
CO	Condenser
EV	Evaporator
S-H	Solar Heater

Introduction

Solar exergy is an important alternative exergy source used in many applications, especially in solar power systems which utilize the heat generated by collectors concentrating and absorbing the sun's exergy to drive heat engines/generators and produce electric power [1]. Most known types of the solar-thermal systems to produce electricity are trough/steam turbine, tower/steam turbine, and dish/heat engine systems. Out of all these technologies, tower/steam turbine looks like to be the best choice for high power production as it has the largest operating temperature range [2]. Tower/steam turbine or what so called Central Receiver System (CRS) is composed of the following main components: the heliostats, the receiver and the power block. The thermal storage and balance of plant components allow high temperatures which lead to high efficiency of the power conversion system [3]. However, the power generation efficiency of the CRS systems are found to be low which directly increase the capital cost of the electricity generation. To investigate the cause of the low generation efficiency in the power generation system an exergy analysis is required. The exergy analysis has proven to be a powerful tool in thermodynamic analyses of the system [4]. Therefore, a theoretical investigation based on the second law efficiency has been conducted for a CRS power plant.

---

*Corresponding author: Markus Haider, Professor, Vienna University of Technology, Institute for Exergy Systems and Thermodynamics, Getreidemarkt 9, 1060 Wien, Austria, E-mail: markus.haider@tuwien.ac.at

The Exergy is the maximum useful work that can be obtained from a system at a given state in a given environment; in other words, the most work you can get out of a system. In the last several decades exergy analysis has begun to be used for system optimization by analyzing the exergy destroyed in each component in a process. With the exergy analysis we can see where we should be focusing our efforts to improve system efficiency. The exergy analysis method is a useful tool for promoting the goal of more efficient exergy-resource use, as it enables the locations, types and true magnitudes of wastes and losses [5].

However, few papers have appeared on the exergy analysis and performance assessment of the solar thermal power plant. Bejan [6] showed that the amount of useful exergy (exergy) delivered by solar collector systems is affected by heat transfer irreversibility occurring between the sun and the collector, between the collector and the ambient air, and inside the collector. Singh [7] stated that the collector-receiver assembly is the part where the losses are maximum and the maximum exergy loss occurs in the solar collector field. Gupta and Kaushik [5] made an exergy analysis for the different components of a proposed conceptual direct steam generation (DSG) solar–thermal power plant and found that the maximum exergy loss is in the solar collector field while in other plant components is small. Chao [8] evaluated the exergy losses in each component and in a power tower solar plant and the results showed that the maximum exergy loss occurs in the receiver system.

The objective of this paper is to evaluate the exergy efficiencies and the irreversibilities of all components of a CRS power plant with an external central receiver and a supercritical Rankine cycle and pinpoint the causes and locations of the thermodynamic imperfection and the magnitude of the process irreversibilities in the system. Two parameters have been varied to see their effect on the exergy efficiency of the receiver and the overall system.

System Configuration

The schematic of the solar tower power plant is shown in Figure 1. The system consists of a heliostat field system, an external central receiver, a steam generator, two steam turbines with a reheat in between, a wet condenser, open and close feed water heaters and two pumps.

The target is to do an exergy analysis for each component of the power plant including the power cycle, the receiver and the heliostat to have more profound understanding of the performance of the CRS power plant. Also to compare all components and see which one has the biggest fractional exergy loss which will provide the guides of improving the performance and optimizing the operation.

Assumptions

The following assumptions are made in the analysis:

a.   The system runs of steady state with a constant solar insulation

b.   Capacity of the power plant is 30 MW

c.   The CRS power plant is using molten salt as HTF

d.   The temperature of molten salt entering the receiver 290°C and leaving it 565°C

e.   Kinetic and potential exergy are ignored

f.   The optical efficiency of the receiver is assumed to be 93%

g.   Conductive heat loss in the receiver is ignored

The Exergy Analysis

The Second Law Efficiency or what so called the exergy efficiency

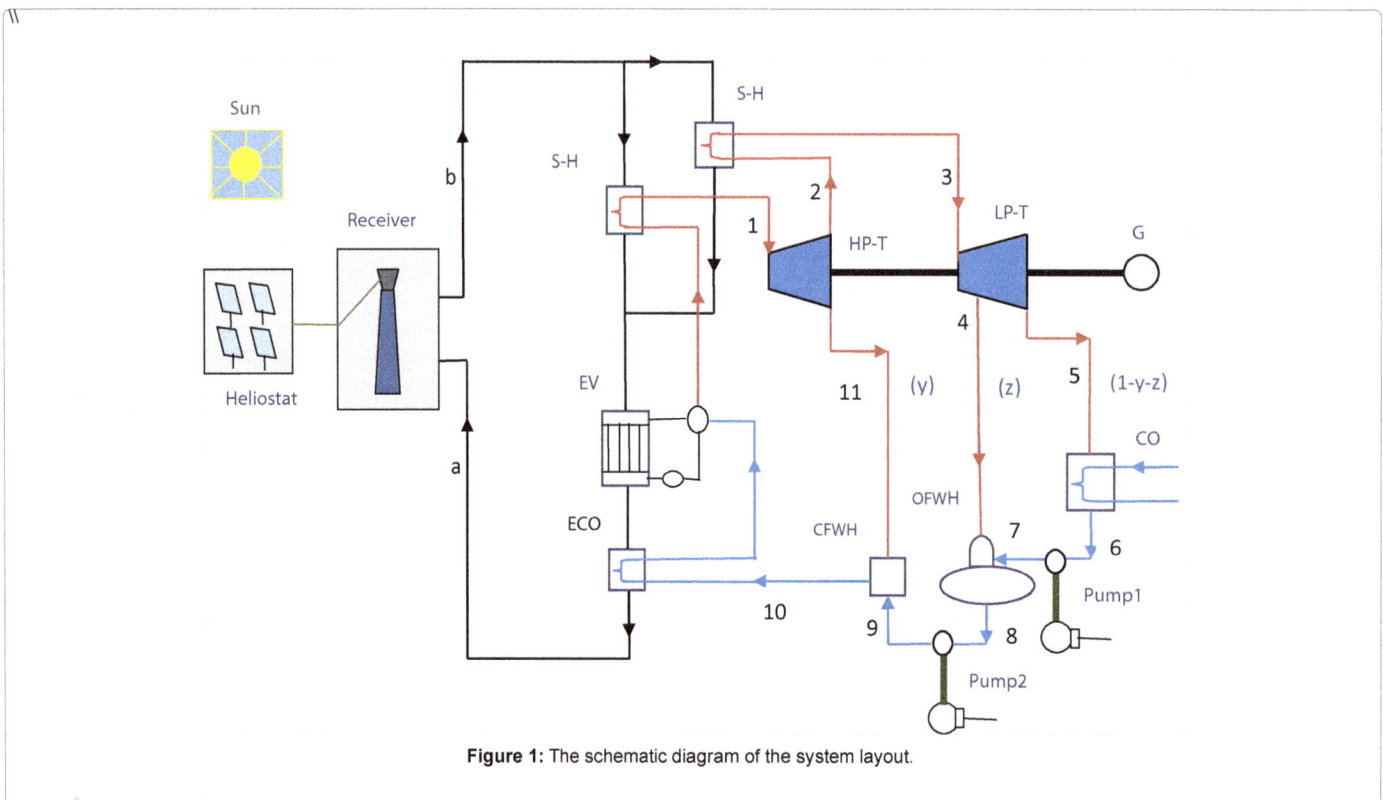

Figure 1: The schematic diagram of the system layout.

of the whole system can be defined as the ratio of net electricity output from the whole system to the exergy input associated with the solar irradiation on the heliostat surface [8]. The exergy input into the heliostat field (ψ^*_{total}) is basically the electricity output (W_{net}) plus the irreversibility (I_total) which is also called the exergy destroyed. The exergy destroyed represents exergy that could have been converted into work but was instead wasted. Therefore, the exergy efficiency of the overall system is given by:

$$\eta_{II} = \frac{W_{net}}{\psi^*_{total}} \tag{1}$$

$$\eta_{II} = \frac{W_{net}}{\left(W_{net} + I_{total}\right)} \tag{2}$$

Where the net output work (electricity) from the whole system is calculated as:

$$W_{net} = W_{hptur} + W_{lptur} - W_{pump1} - W_{pump2} \tag{3}$$

The total irreversibility (I_{total}) of the whole system is the summation of the irreversibility of the subsystems:

$$\mathbf{W}_{net} + \mathbf{I}_{pcycle} \tag{4}$$

Power cycle

To improve the exergy efficiency of the power cycle a regenerative Rankine cycle is used as the case of most existed solar tower power plants. The power cycle of our system as shown in Figure 1 consists of high and low-pressure turbine stages, a condenser, two pumps and feed water heater. Since a molten salt is being used as Heat Transfer Fluid, it is necessary to use an open and a closed feed water heaters to prevent the solidification of the molten salt in the steam generation subsystem.

The exergy efficiency of the power cycle subsystem can be calculated as:

$$\varsigma_{II} = \frac{W_{net}}{W_{net} + I_{pcycle}} \tag{5}$$

$$W_{net} = W_{turbine} - W_{pump} \tag{6}$$

$$W_{turbine} = W_{hptur} + W_{lptur} \tag{7}$$

$$W_{pump} = W_{pump1} + W_{pump2} \tag{8}$$

In order to calculate the power cycle output (W_{net}), it is necessary to apply an exergy balance in the regenerative Rankine cycle. As a result the power delivered by the turbines and consumed by the pumps can be evaluated as following:

$$W_{turbine} = m_{st}.\left[\left(h_1 - h_2\right) + \left(1-y\right)\left(h_3 - h_4\right) + \left(1-y-z\right)\left(h_4 - h_5\right)\right] \tag{9}$$

$$W_{pump} = m_{st}.\left[\left(1-y-z\right)\left(h_7 - h_6\right) + \left(1-y\right)\left(h_9 - h_8\right)\right] \tag{10}$$

y & z are the mass fractions and can be calculated by applying an exergy balance on the open and the closed feed water heaters. Therefore, the values of the mass fractions are:

$$y = \frac{h_{10} - h_9}{h_2 - h_9} \tag{11}$$

$$\tag{12}$$

$$z = \frac{\left(1-y\right)\left(h_8 - h_7\right)}{\left(h_4 - h_7\right)}$$

The irreversibility of the power cycle (I_{pcycle}) is the summation of the irreversibility of each component of the power cycle subsystem.

$$I_{pcycle} = I_{turbines} + I_{condenser} + I_{sgenerator} + I_{pump1} + I_{pump2} + I_{ofwh} + I_{cfwh} \tag{13}$$

The entropy generation in a system is the cause of the exergy destruction. The destruction of useful work because of the entropy generated knows also as irreversibility. The exergy destruction is a proportional of the entropy generated. The general equation for the exergy destruction can be expressed as:

$$I = T_o S_{gen} \tag{14}$$

S_{gen} is the entropy generated, for a steady state control volume, this leads us to

$$S_{gen} = \sum_{out} m_e s_e - \sum_{in} m_i s_i - \sum \frac{Q}{T_k} \tag{15}$$

where

m_i and m_e are the mass flow rate in and out respectively.

s_i and s_e are the entropy in and out respectively.

Q is the heat loss to the surrounding.

T_k is the temperature of the heat source.

As a result, the irreversibility of each component of the power cycle can be defined as:

$$I_{turbines} = m_{st} T_o \left[\left(s_2 - s_1\right) + \left(1-y\right)\left(s_4 - s_3\right) + \left(1-y-z\right)\left(s_5 - s_4\right)\right] \tag{16}$$

$$I_{sgenerator} = m_{st} T_o \left(\left(s_1 - s_{10}\right) + \left(1-y\right)\left(s_3 - s_2\right) - \left(\left(\left(h_1 - h_{10}\right) + \left(1-y\right)\left(h_3 - h_2\right)\right)/T_{rec,sur}\right)\right) \tag{17}$$

$$I_{pump} = I_{pump1} + I_{pump2} \tag{18}$$

$$I_{pump} = m_{st} T_o \left(1-y-z\right)\left(s_7 - s_6\right) + m_{st} T_o \left(1-y\right)\left(s_9 - s_8\right) \tag{19}$$

$$I_{ofwh} = m_{st} T_o \left(\left(1-y\right)s_8 - z s_4 - \left(1-y-z\right)s_7\right) \tag{20}$$

$$I_{cfwh} = m_{st} T_o \left(y s_2 + \left(1-y\right)\left(s_{10} - s_9\right)\right) \tag{21}$$

$$I_{condenser} = m_{st} T_o \left(1-y-z\right)\left(\left(s_6 - s_5\right) - \left(\frac{h_6 - h_5}{T_{wi}}\right)\right) \tag{22}$$

Receiver

In the solar receiver, the heat flux arriving at the aperture area is transferred into a heat flow directed to the HTF. The heat is used to raise the temperature of the HTF. On the way from the aperture to the fluid, some optical and thermal losses occur. Therefore, the incident isolation can be obtained by:

$$Q^*_{inc} = Q_{inc,abs} + Q_{inc,loss} \tag{23}$$

$$Q_{inc,abs} = m_{ms}(h_b - h_a) \tag{24}$$

h_a and h_b are the inlet and exist enthalpy of the molten slat at the receiver respectively.

Regarding the thermal losses, only the radiation and convective losses have been considered and the others are partially small and to be ignored.

$$Q_{inc,loss} = Q_{loss,radiation} + Q_{loss,convection} + Q_{loss,optical} \tag{25}$$

The radiation and convection heat loss is obtained from a CFD model for the same receiver geometry and same boundary conditions done by the author [9]. In the CFD model a Simulation tool ANSYS® FLUENT® was used to determine the thermal heat loss in both cavity and external (CRS) at wind speed varies from (2) to (10) m/s. Table 1 shows the values of the thermal heat loss at the receiver.

The optical efficiency is assumed to be 93%. Therefore, the $Q_{(loss,optical)}$ is equal to $0.07 Q^*_{inc}$

The exergy delivered to the receiver can be calculated by [8]:

$$\psi^*_{inc} = Q^*_{inc}\left(1 - \frac{T_o}{T^*}\right) \tag{26}$$

T^* is the apparent sun temperature as an exergy source and will considered 4500 K.

Similarly the exergy absorbed in the receiver can be given by:

$$\psi_{inc,abs} = Q_{inc,abs}\left(1 - \frac{T_o}{T_{rec,sur}}\right) \tag{27}$$

The exergy efficiency of the receiver is defined as

$$\eta_{II} = \frac{\psi_{inc,abs}}{\psi^*_{inc}} \tag{28}$$

The irreversibility in the receiver

$$I_{receiver} = \psi^*_{inc} - \psi_{inc,abs} \tag{29}$$

Heliostat field system

The incoming solar irradiance onto the receiver aperture area is concentrated by a large number of individually tracked heliostats. The flux density distribution on a defined aperture surface is the output of the heliostat field. This aperture surface can be used as an interface between the optical concentrator and the solar receiver.

The total isolation (Q^*total) is a proportional to the heliostat field aperture area and can be given by [8]:

$$Q^*_{total} = A_h.DNI \tag{30}$$

where

A_h is the heliostat field aperture area.

DNI is the direct normal irradiation.

$$Q^*_{total} = Q^*_{inc} + Q^*_{loss} \tag{31}$$

The total exergy (ψ^*_{total}) associated with the solar irradiation on the heliostat mirror surface (Q^*_{total}) can be expressed as [8]:

$$\psi^*_{total} = Q^*_{total}\left(1 - \frac{T_o}{T^*}\right) \tag{32}$$

$$\psi^*_{total} = \psi_{inc} + I_{heliostat} \tag{33}$$

The exergy efficiency of the heliostat will be as:

$$\eta_{II} = \frac{\psi^*_{inc}}{\psi^*_{total}} \tag{34}$$

The irreversibility in the heliostat is:

$$I_{heliostat} = \psi^*_{total} - \psi^*_{inc} \tag{35}$$

Fractional Exergy Loss

The ratio of the irreversibility of each component to the irreversibility of the whole system is the fractional exergy loss for that component. That illustrates the loss of useful exergy in each component which helps us to decide where we should be focusing our efforts to improve system efficiency.

The fractional exergy loss of each component can be calculated as:

$$Fractional\ Exegy\ Loss_{components} = \frac{I_{component}}{\sum I_{total}} \times 100 \tag{36}$$

Results

The exergy analysis has been carried out for a CRS power plant with Rankine cycle using one open and closed feed water heater. At a constant DNI the exergy efficiency differs from a component to another. The highest exergy efficiency is associated with the power cycle subsystem meanwhile the receiver subsystem has the lowest exergy analysis. It can be seen from the graph in the Figure 2 that at fixed DNI 800 W/m² and also fixed Outlet temperature of the HTF (molten salt) 565°C, the exergy efficiencies are 80%, 74% and 58% for the power cycle, the heliostat and the receiver respectively.

Figure 3 illustrates the effect of the DNI on the fractional exergy loss. The fractional loss at each component of the power cycle subsystem decreases slightly with the increase of the DNI. However, at the receiver and the heliostat field subsystems the case is opposite. The fractional loss at the receiver and the heliostat field increases with the increase the DNI and as a result the exergy efficiency in both subsystems decreases.

The total isolation (Q^*_{total}) at the heliostat field and the exergy (ψ^*_{total}) associated with it are both a proportional of the DNI as illustrated in equations (30) and (32). However, the second law efficiency of the heliostats ($\eta_{II,heliosat}$) is inversely proportional of the DNI and it gets lower because the irreversibility of the heliostat gets higher as shown in equation (34). Similarly when the DNI increase the exergy delivered to the receiver (ψ^*_{inc}) increase and also the irreversibility at the receiver increase ($I_{receiver}$) and this leads

DNI [W/m^2]	Wind speed [m/s]	Radiation loss [kW]	Convection loss [kW]
800	4	774	251

Table 1: Thermal loss at the receiver [9].

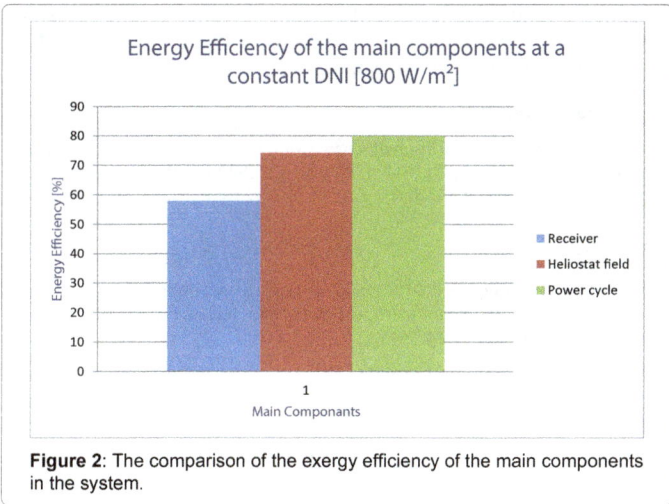

Figure 2: The comparison of the exergy efficiency of the main components in the system.

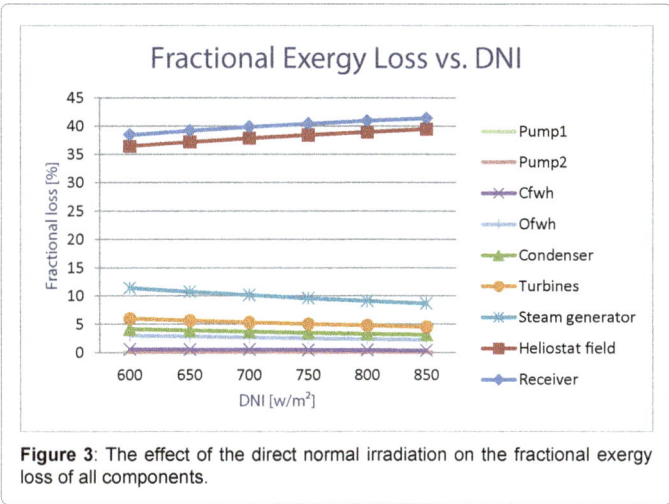

Figure 3: The effect of the direct normal irradiation on the fractional exergy loss of all components.

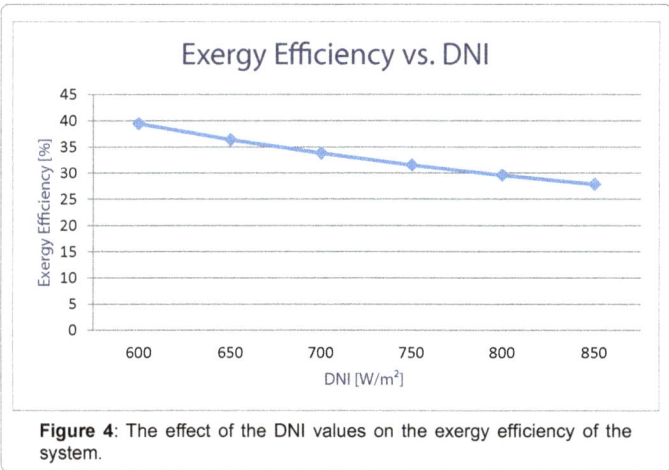

Figure 4: The effect of the DNI values on the exergy efficiency of the system.

to the decrease in the exergy efficiency of the receiver ($\eta_{II,receiver}$) as described in the equation (28).

Figure 4 shows the effect of the DNI on the exergy efficiency of the whole system. Since the irreversibility at the receiver and the heliostat field increase with the increase of the DNI, the total irreversibility (I_{total}) increases and as a result the exergy efficiency of the whole system

decreases.

The variation in the exergy efficiency of both the receiver and the overall system with the outlet temperature of the molten salt is shown in the Figure 5. It can be seen that the exergy efficiency at the receiver and the overall system both increase to a certain degree with the increase of the outlet temperature. The outlet temperature has more effect on the exergy efficiency of the receiver subsystem with comparison to the overall system. The increase of the outlet temperature of the molten salt from 450°C to 600°C will lead to an increase in exergy efficiency at the receiver to about 5% while the increase of the exergy efficiency at the overall system is only in the range of 1%.

The fractional exergy loss at the receiver and the heliostat subsystems varies with different values of the outlet temperature as shown in Figure 6. Regarding the receiver subsystem the fractional exergy loss decreases with the increase of the outlet temperature. However, the fractional exergy loss at the heliostat field subsystem increases and this lowers the exergy efficiency. As the outlet temperature of the molten salt gets higher the flow rate should be lowered in order to heat the molten salt to the required temperature

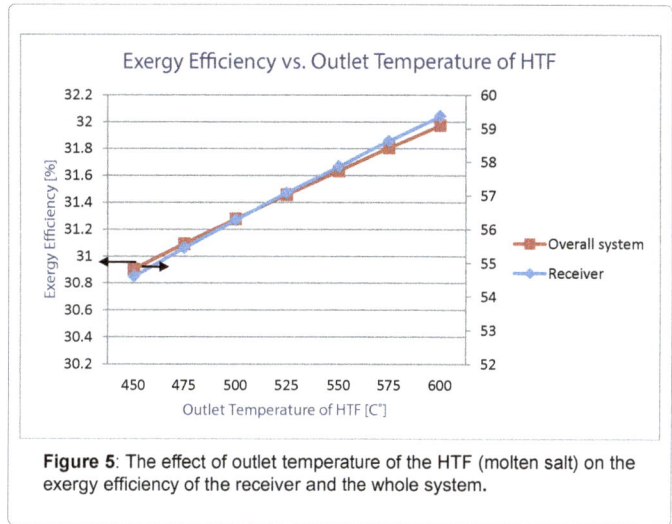

Figure 5: The effect of outlet temperature of the HTF (molten salt) on the exergy efficiency of the receiver and the whole system.

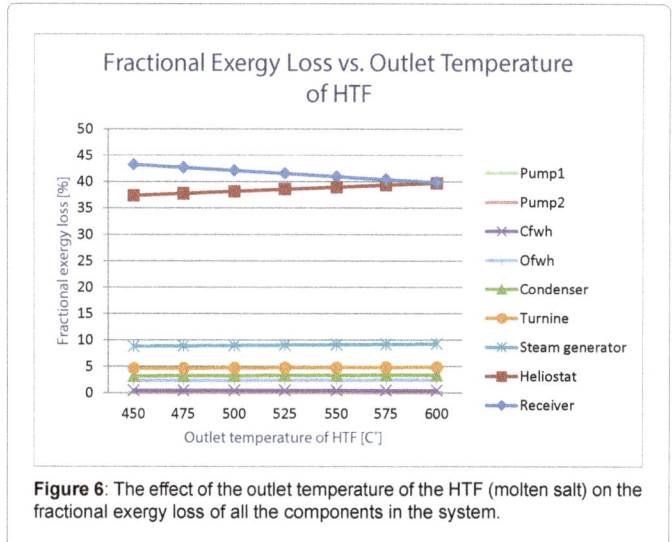

Figure 6: The effect of the outlet temperature of the HTF (molten salt) on the fractional exergy loss of all the components in the system.

and therefore, the absorbed heat at the receiver by the molten salt (Qinc, abs) is diminishing. As shown in equation (27) the exergy delivered to the receiver (ψinc,abs) is proportional of the absorbed heat by the receiver (Qinc, abs) Therefore, the exergy delivered to the receiver (ψinc,abs) also reduced and as a result the exergy efficiency of the receiver decreases as illustrated in equation (28). It can be also seen from the Figure 6 that the variation of the outlet temperature of the molten salt has no effect on the exergy efficiency of the power cycle subsystem components.

Conclusions

In this paper, a theoretical investigation based on the second law efficiency has been conducted for a CRS power plant with an external central receiver and a supercritical Rankine cycle to pinpoint the causes and locations of the thermodynamic imperfection and the magnitude of the process irreversibilities in the system. The following conclusions can be drawn from study:

The receiver subsystem has the highest fractional exergy loss and as a result it has the lowest exergy efficiency (58%) followed by the heliostat field subsystem (74%). While the power cycle subsystem having the lower fractional exergy loss and as result the highest exergy efficiency (80%).

The variation of the DNI affects the fractional exergy loss in all components. The increase of the DNI leads to a decrease of the fractional exergy loss in the power cycle subsystem components and an increase of the fractional exergy loss in the receiver and heliostat field subsystems. As result the increase of the DNI affects negatively the exergy efficiency of the overall system.

The variation of the outlet temperature of the HTF (molten salt) has no significant impact on the exergy efficiency of the receiver subsystem as well as the overall system. It is found that by increasing the outlet temperature of the molten salt from 450°C to 600°C will increase the exergy efficiency of the receiver subsystem to almost 5% and almost 1% to the exergy efficiency of the overall system.

Acknowledgements

The author is thankful for the institute for Exergy Systems and Thermodynamics at Vienna Technical University (TU). I am also grateful to Professor Haider for the help provided in guiding the study.

References

1. Qi LY, Ling HY, Feng WZ (2012) Exergy analysis of two phase change materials storage system for solar thermal power with finite-time thermodynamics. Renewable Energy 39: 447-454.

2. Ratlamwala TAH, Dincer I, Aydin M (2012) Energy and exergy analyses and optimization study of an integrated solar heliostat field system for hydrogen production. Int J Hydrogen Energy 37: 18704-18712.

3. IRENA (2012) Concentrating Solar Power. Cost analysis series. Renew Energy Technol.

4. Hepbasli A (2008) A key review on exergetic analysis and assessment of renewable energy resources for a sustainable future. Renew Sustain Energy Rev 12: 593-661.

5. Gupta MK, Kaushik SC (2010) Exergy analysis and investigation for various feed water heaters of direct steam generation solar–thermal power plant. Renew Energy 35: 1228–1235.

6. Bejan A, Kearney DW, Kreith F (1981) Second law analysis and synthesis of solar collector systems. Solar Energy Engg 103: 23-28.

7. Singh N, Kaushik SC, Misra RD (2000) Exergetic analysis of a solar thermal power system. Renew Energy 19: 135-143.

8. Chao X, Zhifeng W, Xin L, Feihu S (2011) Energy and exergy analysis of solar power tower plants. Appl Thermal Engg: 3904-3913.

9. Abdiwe R (2015) Investigations on heat loss in solar tower receivers with wind speed variation.

20

On-Line Maximum Power Point Tracking for Photovoltaic System Grid Connected Through DC-DC Boost Converter and Three Phase Inverter

MMA Mahfouz*

Electrical Power and Machines Department, Faculty of Engineering, Helwan University, Cairo, Egypt

Abstract

Photovoltaic solar electricity and solar thermal has the highest potential of all the renewable energies, since solar energy is a practically unlimited resource and available everywhere. These days photovoltaic energy has the potential to play an important role in the transition towards a sustainable energy supply system, to cover a significant share of the electricity needs, and is expected to be one of the key energy technologies of this century. This paper presents a grid-connected photovoltaic (PV) system with the functionality of on-line Maximum Power Point Tracking (MPPT). The integration topology is based on two-stage power conditioning modules, boost converter and three phase voltage source inverter. Therefore, the proposed MPPT algorithm is implemented to the boost converter to enable PV arrays to operate at maximum power point during variable climate and irradiation conditions. Fuzzy logic algorithm (MPPT) technique control is applied for on-line adapting the boost converter duty cycle. The Boost output is feeding the two level PWM inverter DC link. Adoringly, the Inverter modulation depth is tuned to have constant AC voltage and frequency for grid integration requirements. In addition, the three-phase output in phase with the voltage to optimize the PV power utilization .The results show that proposed on-line fuzzy algorithm performance has a fast response, stable and reliable to have MPP for the PV system during the different irradiation.

Keywords: Renewable energy; Photovoltaic; Maximum power point tracking (MPPT); Fuzzy logic; DC-DC converter; Three phase PWM inverter

Introduction

The growing energy demand coupled with the possibility of reduced supply of conventional fuels, along with growing concerns about environmental preservation, has driven research and development of alternative energy sources that are cleaner, renewable and produce little environmental impact [1]. Among the alternative sources, the electrical energy from PV is currently regarded as the natural energy source distributed over the earth and participates as a primary factor of all other processes of energy production on earth. Moreover, although the phenomena of reflection and absorption of sunlight by the atmosphere, it is estimated that solar energy incident on the surface of earth is of the order of ten thousand times greater than the world energy consumption [2]. In this context, the concept of distributed energy generation, became a real and present technical possibility, promotion various researches and standardizations in the world. Despite all the advantages presented by the generation of energy through the use of PV's, the efficiency of energy conversion is currently low and the initial cost for its implementation is still considered high, and thus it becomes necessary to use techniques to extract the maximum power from these panels, to achieve maximum efficiency in operation. It should be noted that there is only one point of maximum power (MPP), and this varies according to climatic conditions. The photovoltaic power characteristics is nonlinear, which vary with the level of solar irradiation and temperature, which make the extraction of maximum power a complex task, considering load variations. To overcome this problem, several methods for extracting the maximum power have been proposed in literature [3,4], and a careful comparison of these methods can result in important information for the design of these systems. There are a lot of techniques are used for MPPT such as Fixed Duty Cycle, Constant Voltage, Perturb and Observe (P&O) and Modified P&O, Incremental Conductance (IC) and Modified IC, Ripple Correlation and System Oscillation methods, which are briefly described in this section [5,6].

This paper, target is to have an on-line efficient optimization (MPPT) algorithm using fuzzy logic technique. In the studied power system PV, integration to the grid is done by using two stages; the PV array output is connected to DC-DC converter followed by two level voltage sources VSPWM inverter to the grid. Maximum power point tracking (MPPT) techniques is used to control the DC converter to extract maximum output power from under a given weather condition. The DC converter is on-line continuously controlled to have the maximum power point operation for the PV array in despite of the possible changes either on the climate condition and/or in the load impedance. To achieve this paper's target; the paper organized in the following sequences:

- Modeling of the PV–grid connected system for applying MPPT techniques using MATLAB/Simulink.

- Simulation of the PV–gird system during weather variation represented by an actual daily solar irradiation curve.

- The dynamic response of the algorithm is characterized by driving the PV system into its maximum power operation while ensuring that the injected current remains in phase with the grid voltage to achieve a unity power factor.

- Investigate the interactions between the PV system outputs and the grid.

***Corresponding author:** Mahfouz MMA, Faculty of Engineering, Electrical Power and Machines Department, Helwan University, Cairo, Egypt
E-mail: mohamed.mahfouz@yahoo.co.uk

Modeling of the PV-Grid Connected System

Power system under study single line diagram is illustrated in Figure 1. It consists of 22 kV interconnected network. The studied PV array is poly crystalline silicon each cell power maximum of 56 W and consists of 36 series cells per module with an open circuit 20.8 V, 16 series connected modules and 123 parallel with total maximum power rating of 100 kW is connected to the grid and to each other's. The PV arrays are connected to the grid through DC/DC boost converter and DC/AC three-phase two-level inverter. After inversion, the arrays are interfaced to the grid through a transformer 220 V/22 kV of rated 125 kVA. A local load of active power 50 kW is connected to the grid integration bus at PV point of common coupling.

PV array

The equivalent circuit model of a PV cell is needed to simulate its real behavior using the physics of p-n junctions; a cell can be modeled as a DC current source in parallel with diode that represents the most effective current escaping due to diffusion and charge recombination mechanisms. Two resistances, series and shunt are included to m represent the contact resistances and the internal PV cell resistance respectively. These two resistance values are usually obtained from measurements or applying the curve fitting technique based on the cell I-V characteristic [7]. The equivalent circuit model of a PV cell is needed to simulate its real behavior using the physics of p-n junctions, a cell can be modeled as a DC current source in parallel with diode that represents the most effective current escaping due to diffusion and charge recombination mechanisms. Two resistances, series and parallel, are included to represent the contact resistances and the internal PV cell resistance respectively. These two resistance values are usually obtained from measurements or applying the curve fitting technique based on the cell I-V characteristic [8]. PV cells are grouped in larger units called PV modules, which are further interconnected in a parallel-series configuration to form PV arrays or PV generators and the voltage and power of the PV module are depending on the number module No. of series and parallel cells. A grid-connected PV solar system is to transfer the maximum power obtained from the PV panel into the

grid independently of the climatic conditions. Therefore, the use of an appropriate electronic interface with maximum power point tracking (MPPT) capabilities is required. The output power of the PV cell has a nonlinear function with irradiance and temperature. Therefore, the power of the PV array changes continuously and consequently the PV operating point must change to maximize the energy produced by adjusting array voltage to its calculated maximum value at certain irradiation and temperature. Figure 2 shows the used power-voltage characteristics of the used PV array at different climate conditions and the target maximum power points for the proposed control system.

Boost chopper

A buck boost converter is a power converter with an output DC voltage greater or less than its input DC voltage depending on its duty cycle (d) as illustrated in Figure 3. By implementing the PWM technique on the buck boost converter, a stable output voltage from a non-stable input voltage can be obtained by changing the duty cycle (d) of the switched input pulse. The buck boost converter with a switching period (Ts), and on-period of (Ton) the relationship between output V_o and V_{in} voltage is governed by [9,10];

$$\frac{V_0}{V_i} = \frac{1}{1-d} \quad \text{where} \quad d = \frac{T_{on}}{T_s} \tag{1}$$

After determining the value for V_{MPP}, the DC/DC buck boost converter is switched to quickly force the array terminal voltage to its MPP. The fast switching action of the boost converter decouples the dynamics of the PV array due its changes in voltage or current from that of the DC link capacitor, which offers good performance under changing weather conditions. The other advantage of using the boost converter is to increase the array voltage to higher levels suitable for grid interconnection.

Three -phase voltage source PWM inverter

Among various modulation techniques, PWM is an attractive candidate due to efficient DC link voltage utilization, reducing commutation losses and total harmonic distortion THD. The proposed inverter is a three-phase, two-level DC/AC inverter using IGBTs due to their lower switching losses. Depending on the gating signals, the output terminal voltage is either equal to capacitor voltage or zero, and can be expressed by [11,12]:

$$V_{inv} = N * V_{cap} \tag{2}$$

where: V_{inv}, N, and V_{cap} are terminal voltage, gating signal and capacitor voltage, respectively. N takes the value of either one or zero. In the two-level inverter the switching of the upper and lower IGBT generates the output voltages with positive and negative levels (+V_{dc} and -V_{dc}). The inverter is generally connected to the network at the PCC through the equivalent inductance and resistance, which represent the impedance of the coupling transformer. The magnetizing inductance of the step-up transformer can also be taken into consideration through a mutual equivalent inductance. The aim of the control strategy is to regulate the current output from the inverter to follow a specified reference value. This can be achieved by transforming the three phase output currents of the inverter to the rotating reference frame (dq_0). The relation between the DC side voltage V_{dc} and the generated AC voltage V_{inv} can be described through the matrix in the dq-frame S_{av}, dq, as given by:

$$\begin{bmatrix} V_{inv,d} \\ V_{inv,q} \end{bmatrix} = S_{av,dq} V_{dc} \tag{3}$$

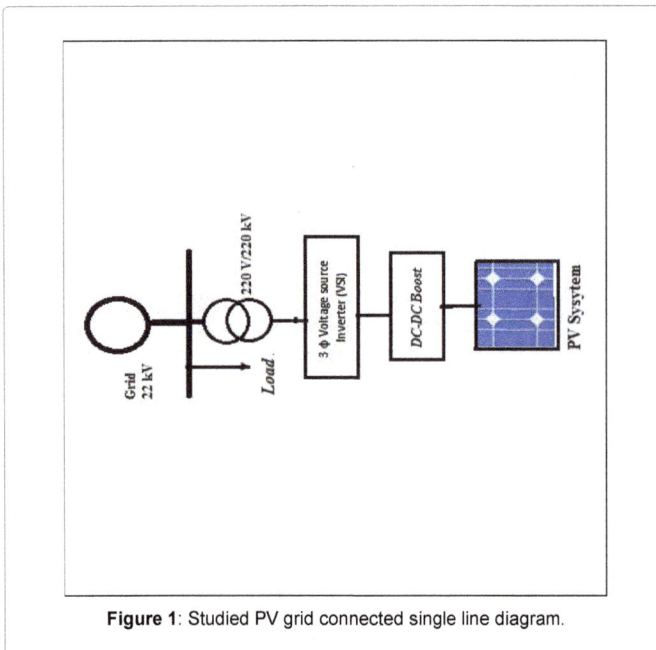

Figure 1: Studied PV grid connected single line diagram.

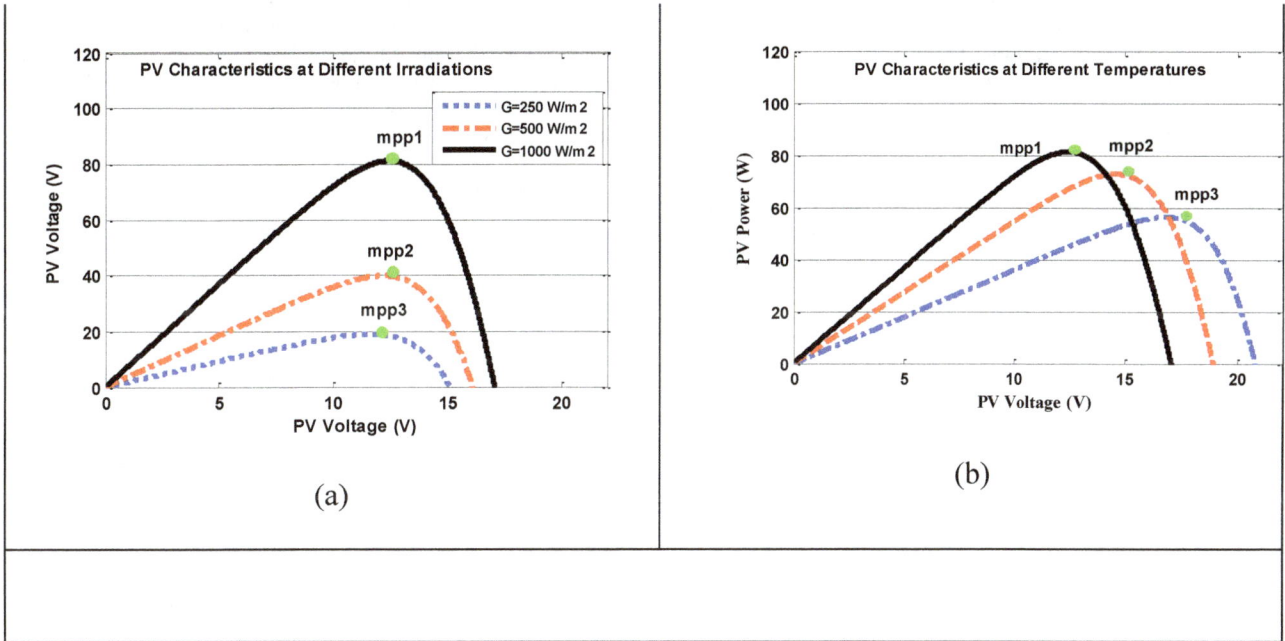

Figure 2: PV model characteristics; a) Different radiation and b) Different temperature.

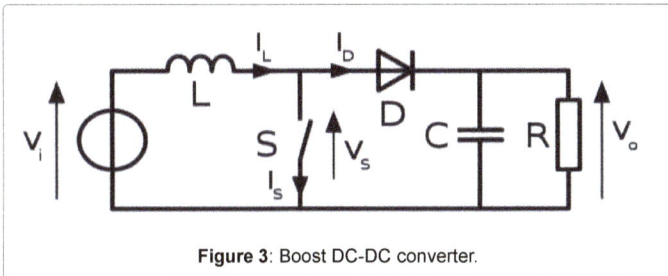

Figure 3: Boost DC-DC converter.

where

$$S_{av,dq} = 0.866 \, m_i \tag{4}$$

where; m_i is the inverter modulation index; and V_{dc} is equal to the Boost converter output (V_o).

Proposed Control Topology

This section describe two main parts the proposed global PV MPPT control scheme philosophy and secondly the proposed fuzzy logic algorithm to apply the MPPT concept. As shown in Figure 4, the block diagram of the global MPPT concept. A photovoltaic array is used to convert sunlight into DC current. The output of the array is connected to a boost DC converter that is used to perform MPPT functions and increase the array terminal voltage to a higher value so it can be interfaced to the grid at 22 kV. The DC converter controller is used to perform these two functions. A DC link capacitor is used after the DC converter and acts as a temporary power storage device to provide the voltage source inverter with a steady flow of power. The capacitor's voltage is regulated using a DC link controller that balances input and output powers of the capacitor. The voltage source inverter is controlled in the rotating dq frame is to inject a controllable three phase AC current into the grid. To achieve unity power factor operation, current is injected in phase with the grid voltage. A phase locked

loop (PLL) is used to lock on the grid frequency and provide a stable reference synchronization signal for the inverter control system, which works to minimize the error between the actual injected current and the reference current obtained from the DC link controller. A load is connected to the grid to simulate some of the loads that are connected to a distribution system network.

Proposed MPPT fuzzy algorithm

There are for Fuzzy logic MPPT algoritm has three sequential stages; fuzzification, rule based on the look-up table and finally the defuzzification. During fuzzification process, control variable are been converted from a numerical value to a linguistic representation [13]. In this paper, Mamadani fuzzy type is used as shown in Figure 5a with two inputs. These inputs are the error between the actual PV power and the maximum reference power cross pounding to the synchronized irradiation and the change of this error. The fuzzy controller output is the DC boost duty cycle. The fuzzy surface to

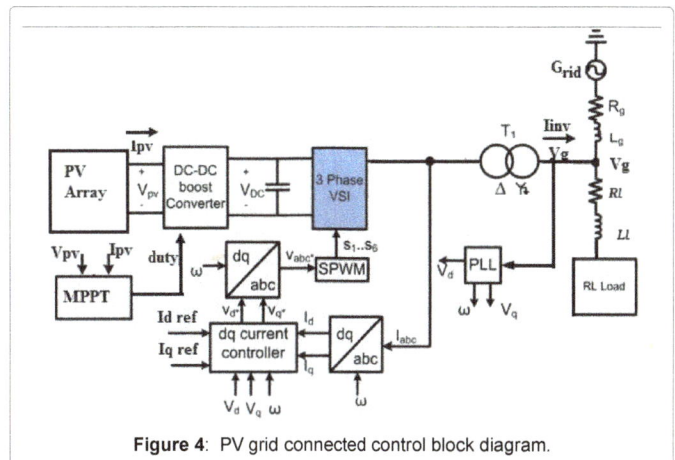

Figure 4: PV grid connected control block diagram.

show the ranges and the relation between the inputs and the output memberships is illustrated in Figure 5b. Twenty-one rules between the inputs to achieve the requested output target. This target is to minimize the error and the rate of error to zero to achieve the tracking for PV power equal to the MPP.

Simulation Results

Studied power system for PV grid connected system at a daily irradiation and temperature of 25˚C using MATLAB/Simulink program is simulated in this part to verify the dynamic and steady state response of the proposed fuzzy MPPT control system. In besides that monitoring of the DC and AC voltages and currents in the PV array and the associated power conditioning system. Figure 6a present the irradiation data which is typically get from the Arab Organization for Industrialization (AOI) Cairo irradiation records on summer 2011 at temperature of 25˚C to verify the performance at Cairo irradiation conditions. Figure 6b shows that the PV output after applying the proposed control to have MPPT is so closely and equal to the target calculated maximum power from the (Power /voltage) curve of the PV array at the different solar irradiation which already shown in Figure 2a. While Figure 6c illustrates the PV array dc voltage changes during the different conditions of radiation. The DC-DC Boost converter performance illustrate in Figure 7. The duty cycle behavior is shown in Figure 7a, it is changes to increase the Boost output voltage during the decreasing of the irradiation and its values ranged from 0.45 to 0.6. While Figure 7b represents the output V_{dc} with its reference values, it shows that the V_{dc} is follows the V_{dc} reference values that validate the proposed control technique. The second stage of voltage conditioning unit, three-phase voltage source inverter using pulse width modulation parameters are shown in Figure 8. In Figure 8a the modulation depth performances is illustrated while the output three phase voltage and current are shown in Figure 8b. It shows that the modulation depth is dynamically interacted with fast response to have a constant AC voltage output, which is essential for PV grid synchronizing. Also the three phase injected current is in phase with output voltage to have maximum utilization from the PV system. The three power components, load power, PV output and grid power

sharing at the point of common coupling at the grid Bus in presented at Figure 9. As shown, the load power is constant while the way of supplying this load is changed depending on the PV system output. When the irradiation is high; the PV power is greater than the load and the extra power is supplied to the grid. While if the irradiation decreased, and the PV output power is less than the load, in this case the load is complimentary supplied from both and the PV.

Conclusions

Modeling of grid connected photovoltaic system elements with two sequential stages of power electronic modules; DC-DC boost converter and three-phase two level voltage source PWM inverter was presented. Design of on-line control system to have Maximum power point tracking from the PV array using Fuzzy Logic Controller algorithms was introduced. The results show that the proposed on-line fuzzy logic controller of the MPP offers clearly short settling time, accurate tracking the reference voltage during all the weather conditions. In addition, the proposed MPPT control algorithm performance is stable, fast, reliable and adaptive to be implemented for PV micro grids and high voltage grid integration.

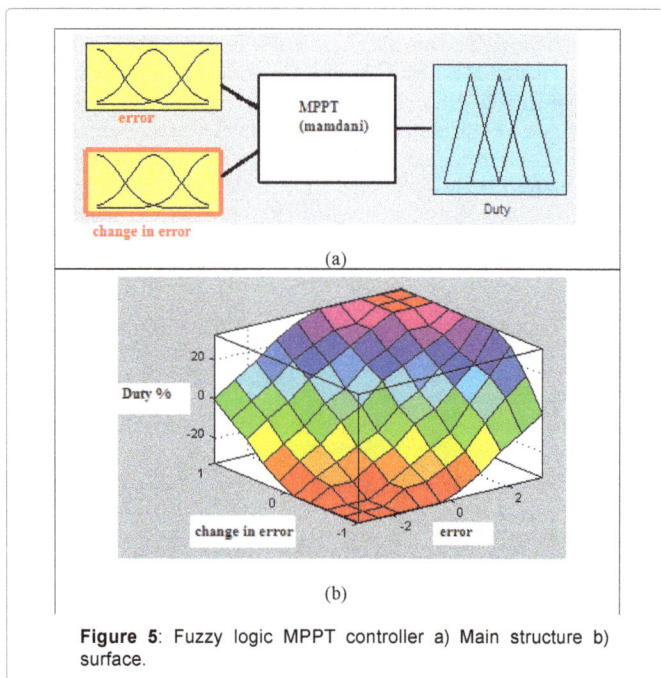

Figure 6: irradiation and PV array output; a) irradiation, b) PV power and C) PV voltage.

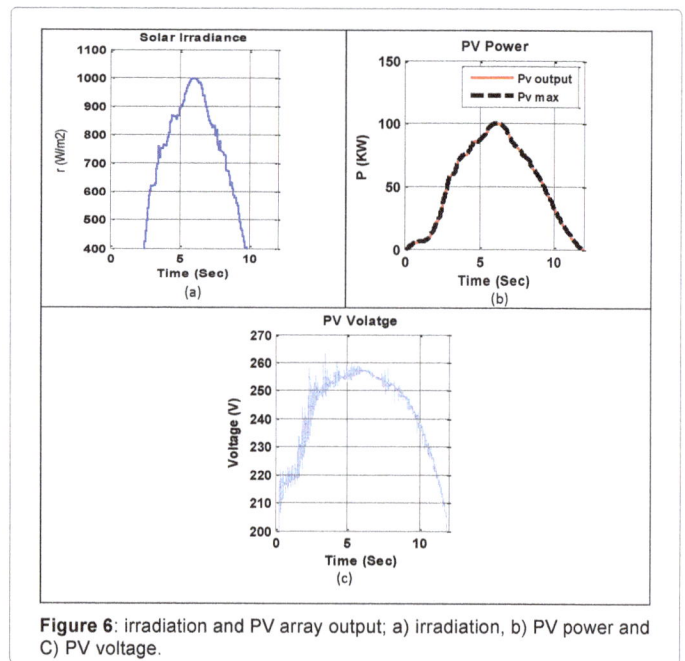

Figure 5: Fuzzy logic MPPT controller a) Main structure b) surface.

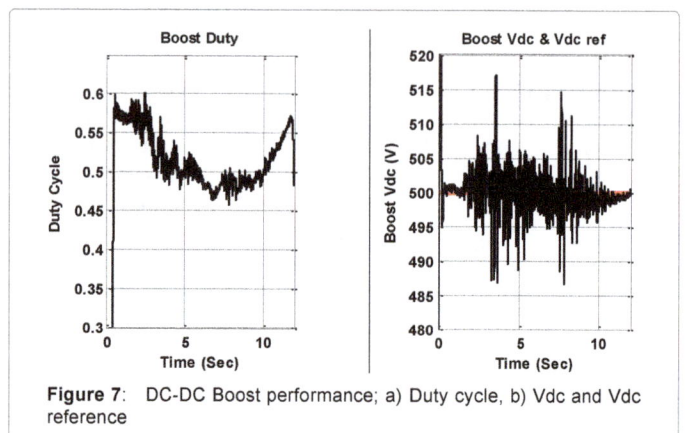

Figure 7: DC-DC Boost performance; a) Duty cycle, b) Vdc and Vdc reference

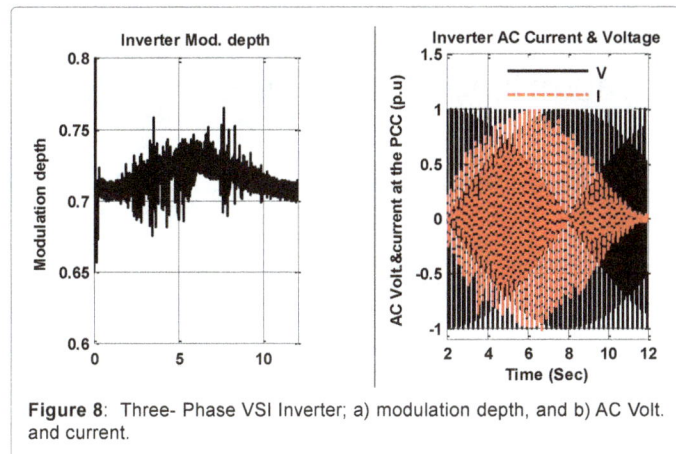

Figure 8: Three- Phase VSI Inverter; a) modulation depth, and b) AC Volt. and current.

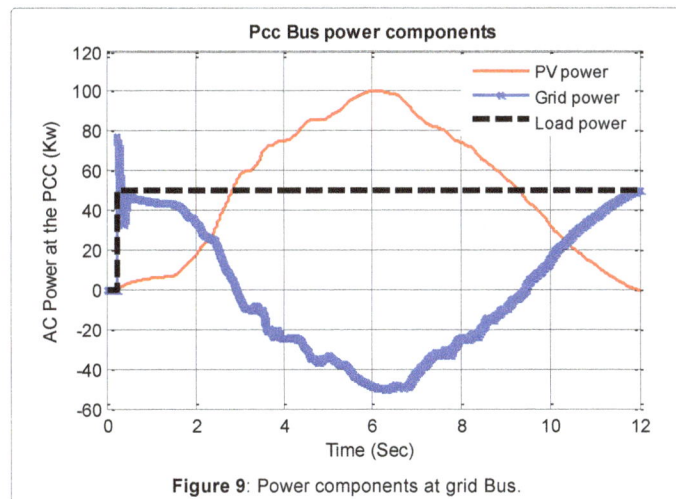

Figure 9: Power components at grid Bus.

References

1. Yazdani A, Dash PP (2009) A Control Methodology and Characterization of Dynamics for a Photovoltaic (PV) System Interfaced With a Distribution Network. Power Delivery IEEE Transactions 24: 1538-1551.

2. Wei LI, Gregoire, Belanger (2011) Control and Performance of a Modular Multilevel Converter System. Conference on Power Systems Halifax, CIGRÉ Canada.

3. Baroudi JA, Dinavahi V, Knight AM (2007) A Review of Power Converter Topologies for Wind Generators. El Sevier Renew Energy J 32: 2369–2385.

4. Khalifa AS, Saadany IFE (2011) Control of three phase grid connected Photovoltaic array with open loop maximum power point tracking. IEEE PES General Meeting, Detroit, MI, USA.

5. Selvaraj J, Rahim NA (2009) Multilevel Inverter For Grid-Connected PV System Employing Digital PI Controller. Industrial Electronics, IEEE Transactions 56: 149-158.

6. Libo W, Zhengming Z, Jianzheng L (2007) A Single-Stage three-Phase Grid-Connected Photovoltaic System with Modified MPPT Method And Reactive Power Compensation. IEEE Transactions on Energy Conversion 22: 881–886.

7. Khaehintung N, Pramotung K, Tuvirat B, Sirisuk P (2004) RISC-microcontroller built-in fuzzy logic controller of maximum power point tracking for solar-powered light-flasher applications. 30th Annual Conference of IEEE Industrial Electronics Society. IECON 3: 2673- 2678.

8. Patcharaprakiti N, Premrudeepreechacharn S (2012) Maximum power point tracking using adaptive fuzzy logic control for grid-connected photovoltaic system. IEEE Power Engineering Society Winter Meeting 1: 372- 377.

9. Mohamed Y, Saadany EFE (2008) Adaptive Decentralized Droop Controller to Preserve Power Sharing Stability of Paralleled Inverters in Distributed Generation Microgrids. IEEE Transactions on Power Electronics 23: 2806-2816.

10. Suul JA, Molinas M, Tore LN (2008) Tuning of Control Loops for Grid Connected Voltage Source Converters. 2nd IEEE International Conference on Power and Energy (PECon 08), Johor Baharu, Malaysia.

11. Mahfouz MMA, El-Sayed MAH (2013) Modeling and Control of Micro Grid Powered by Maximum Power PV Array and Fixed Speed Wind Energy Conversion System. International Conference on Renewable Energies and Power Quality, ICREPQ'13, Bilbao, Spain.

12. Mahfouz MMA (2014) Hybrid Micro Grid Maximum Power Condition Containing PV/Variable Speed Wind Generation. Engg Res J 37: 13-20.

13. The Arab Organization for Industrialization (AOI) Cairo irradiation records.

21

Physical, Spectroscopic and Thermal Characterization of Biofield treated Myristic acid

Mahendra Kumar Trivedi[1], Rama Mohan Tallapragada[1], Alice Branton[1], Dahryn Trivedi[1], Gopal Nayak[1], Rakesh K. Mishra[2] and Snehasis Jana[2]*

[1]*Trivedi Global Inc., 10624 S Eastern Avenue Suite A-969, Henderson, NV 89052, USA*
[2]*Trivedi Science Research Laboratory Pvt. Ltd., Hall-A, Chinar Mega Mall, Chinar Fortune City, Hoshangabad Rd., Bhopal- 462026, Madhya Pradesh, India*

Abstract

Myristic acid has been extensively used for fabrication of phase change materials for thermal energy storage applications. The objective of present research was to investigate the influence of biofield treatment on physical and thermal properties of myristic acid. The study was performed in two groups (control and treated). The control group remained as untreated, and biofield treatment was given to treated group. The control and treated myristic acid were characterized by X-ray diffraction (XRD), Differential scanning calorimetry (DSC), Thermogravimetric analysis (TGA), Fourier transform infrared (FT-IR) spectroscopy, and Laser particle size analyzer. XRD results revealed alteration in intensity of peaks as well as significant increase in crystallite size (27.07%) of treated myristic acid with respect to control. DSC study showed increase in melting temperature of treated myristic acid as compared to control. Nevertheless, significant change (10.16%) in latent heat of fusion (ΔH) was observed in treated myristic acid with respect to control. TGA analysis of treated myristic acid showed less weight loss (31.33%) as compared to control sample (60.49%). This may be due to increase in thermal stability of treated myristic acid in comparison with control. FT-IR results showed increase in frequency of $-CH_2$ and C=O stretching vibrations, probably associated with enhanced bond strength and force constant of the respective bonds. The particle size analyzer showed significant decrease in average particle size (d_{50} and d_{99}) of treated myristic acid with respect to control. Overall, the results showed significant alteration in physical, spectroscopic and thermal properties of myristic acid. The enhanced crystallite size, and thermal stability of treated myristic acid showed that treated myristic acid could be used as phase change material for thermal energy storage applications. .

Keywords: Myristic acid; Phase change materials; X-ray diffraction; Thermal stability; Fourier transform infrared spectroscopy; Particle size analysis

Abbreviation: XRD: X-ray diffraction; DSC: Differential scanning calorimetry; TGA: Thermogravimetric analysis; DTA: Differential thermal analysis; DTG: Derivative thermogravimetry; FT-IR: Fourier transform infrared; PCMs: Phase change materials

Introduction

Nowadays, the energy consumption and production has considered as an interesting topic and debatable among researchers. The demand for generation of newer energy sources are increasing steadily day by day. This calls for design and development of novel energy saving devices in order to reduce the consumption of energy [1]. The enormous increase in production of greenhouse gases in atmosphere and elevation in cost of fossil fuel have caused researchers to develop more efficient thermal energy storage devices. Thermal energy storage has grabbed significant attention worldwide for energy conservation from the available sources of heat [2,3]. Phase change materials (PCM) are known as substance with high latent heat of fusion, which are capable of storing and releasing large amount of energy whenever required. Phase change materials (PCMs) are smart devices that can be utilized for thermal energy storage. Moreover, form stable phase change PCMs are especially interesting due to high latent heat, shape stable and it can be directly used without encapsulation methods [4]. The materials used for fabricating the form stable solid-liquid PCMs are organic compounds such as paraffin [5], fatty acids [6], fatty alcohol [7,8], polyethylene glycol and their mixtures [9,10].

Fatty acids are commonly obtained from natural resources such as vegetable and animal oil products. These compounds have wide applicability in cosmetics, washing, environmental clean-up, encapsulation and drug delivery [11]. Additionally, fatty acids possess excellent properties such as high phase change enthalpy and tunable phase change nature [12]. Myristic acid has recently showed great potential as solid-liquid PCMs for thermal energy storage applications [13,14]. Hence, by considering the phase change property of myristic acid, authors decided to investigate the influence of biofield treatment on its physical, spectroscopic and thermal properties which can be further utilized for thermal storage applications.

Fritz, has first proposed the law of mass-energy interconversion and after that Einstein derived the well-known equation $E=mc^2$ for light and mass [15,16]. Though, conversion of mass into energy is fully validated, but the inverse of this relation, *i.e.* energy into mass is not yet verified scientifically. Additionally, it was stated that energy exist in various forms such as kinetic, potential, electrical, magnetic, nuclear etc. which have been generated from different sources. Similarly, neurons which are present in human brain have the ability to transmit the information in the form of electrical signals [17-20]. Thus, human beings have the ability to harness the energy from environment/Universe and can transmit into any object (living or non-living) around the Globe. The

---

*Corresponding author:Jana S, Trivedi Science Research Laboratory Pvt. Ltd., Hall-A, Chinar Mega Mall, Chinar Fortune City, Hoshangabad Rd., Bhopal- 462026, Madhya Pradesh, India, E-mail: publication@trivedisrl.com

object(s) always receive the energy and respond into a useful manner that is called biofield energy. This whole process is known as biofield treatment. Mr. Trivedi is known to transform the characteristics of various living and nonliving things. The biofield treatment has altered the physical and thermal properties in metals [21-24], improved the growth and production of agriculture crops [25-28] and significantly altered the phenotypic characteristics of various pathogenic microbes [29-31]. Additionally, biofield treatment has substantially altered the medicinal, growth and anatomical properties of ashwagandha [32].

By conceiving above mentioned excellent outcome from biofield treatment and phase change property of myristic acid, this work was undertaken to investigate the impact of biofield on physical, spectroscopic and thermal properties of myristic acid.

Materials and Methods

Myristic acid was procured from Sisco Research Laboratories (SRL), India. The sample was divided into two parts; one was kept as a control sample, while the other was subjected to Mr. Trivedi's biofield treatment and coded as treated sample (T). The treatment group (T) was in sealed pack and handed over to Mr. Trivedi for biofield treatment under laboratory condition. Mr. Trivedi provided the treatment through his energy transmission process to the treated group without touching the sample. The control and treated samples were characterized by XRD, DSC, TGA, FT-IR, and particle size analysis.

Characterization

X-ray diffraction (XRD) study

XRD analysis of control and treated myristic acid was carried out on Phillips, Holland PW 1710 X-ray diffractometer system, which had a copper anode with nickel filter. The radiation of wavelength used by the XRD system was 1.54056 Å. The data obtained from this XRD were in the form of a chart of 2θ vs. intensity and a detailed table containing peak intensity counts, d value (Å), peak width (θ°), relative intensity (%) etc. The crystallite size (G) was calculated by using formula:

$$G = k\lambda/(bCos\theta)$$

Here, λ is the wavelength of radiation used, b is full width half maximum (FWHM) of peaks and k is the equipment constant (=0.94). Percentage change in crystallite size was calculated using following formula:

$$\text{Percentage change in crystallite size} = [(G_t - G_c)/G_c] \times 100$$

Where, G_c and G_t are crystallite size of control and treated powder samples respectively.

Differential scanning calorimetry (DSC) study

DSC was used to investigate the melting temperature and latent heat of fusion (ΔH) of samples. The control and treated myristic acid samples were analyzed by using a Pyris-6 Perkin Elmer DSC on a heating rate of 10°C/min under air atmosphere and air was flushed at a flow rate of 5 mL/min.

Percentage change in latent heat of fusion was calculated using following equations:

$$\% \text{ change in Latent heat of fusion} = \frac{[\Delta H_{Treated} - \Delta H_{Control}]}{\Delta H_{Control}} \times 100$$

Where, $\Delta H_{Control}$ and $\Delta H_{Treated}$ are the latent heat of fusion of control and treated samples, respectively.

Thermogravimetric analysis-differential thermal analysis (TGA-DTA)

Thermal stability of control and treated Myristic acid were analyzed by using Mettler Toledo simultaneous TGA and Differential thermal analyzer (DTA). The samples were heated from room temperature to 400°C with a heating rate of 5°C/min under air atmosphere.

FT-IR spectroscopy

FT-IR spectra were recorded on Shimadzu's Fourier transform infrared spectrometer (Japan) with frequency range of 4000-500 cm^{-1}. The treated sample was divided in two parts T1 and T2 for FT-IR analysis.

Particle size analysis

The average particle size and particle size distribution were analyzed by using Sympetac Helos-BF laser particle size analyzer with a detection range of 0.1 micrometer to 875 μm. Average particle size d_{50} and d_{99} (size exhibited by 99% of powder particles) were computed from laser diffraction data table. The d_{50} and d_{99} values were calculated by the following formula:

$$\text{Percentage change in } d_{50} \text{ size} = 100 \times (d_{50} \text{ treated} - d_{50} \text{ control})/ d_{50} \text{ control}$$

$$\text{Percentage Change in } d_{99} \text{ size} = 100 \times (d_{99} \text{ treated} - d_{99} \text{ control})/ d_{99} \text{ control}$$

Results and Discussion

XRD study

XRD diffractogram of myristic acid (control and treated) are shown in Figure 1. XRD diffractogram of control sample showed intense crystalline peaks at 2θ equals to 18.58°, 18.93°, 20.18°, 20.79°, 21.51°, 23.92°, and 37.80°. However, the treated myristic acid showed alteration in intensity of the XRD peaks. The XRD data of the myristic acid was well supported by the literature [13]. The XRD peaks of treated sample were observed at 2θ equals to 18.62°, 18.93°, 20.74°, 21.51°, 23.80°, 23.97° and 37.65°. The result showed that control and treated myristic acid both has γ pattern α pattern after crystallization [33]. The crystallite size was calculated using Scherrer formula (crystallite size = kλ /b cos θ) and the result are presented in Figure 2. The crystallite size of control myristic acid was 61.58 nm and it was increased to 78.26 nm in treated sample. The result showed that crystallite size was increased significantly by 27.07% in treated myristic acid as compared to control. It was reported previously that increase in processing temperature significantly affects the crystallite size of the materials. The increase in temperature causes decrease in dislocation density and increase in number of unit cell which ultimately causes increase in crystallite size [34,35]. It is hypothesized that biofield may provide some thermal energy that possibly cause elevation in crystallite size of treated myristic acid with respect to control.

Additionally, recently it was showed that introduction of ultrasound to materials leads to substantial increase in crystallite size [36]. Hence, it is assumed that biofield treatment may provide electromagnetic waves similar like ultrasound to treated myristic acid atoms that led to increase in crystallite size with respect to control.

Thermal analysis

DSC study

DSC was conducted to investigate the melting nature and latent

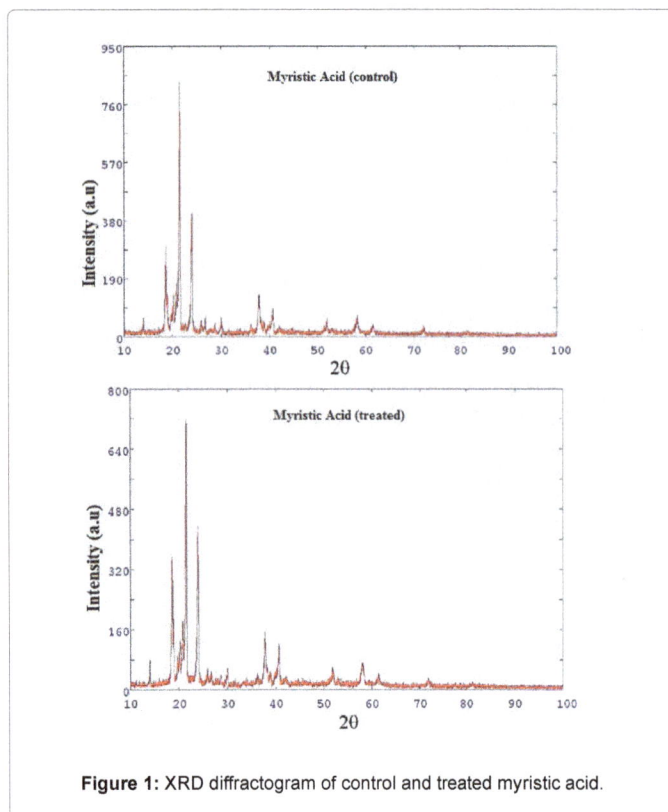

Figure 1: XRD diffractogram of control and treated myristic acid.

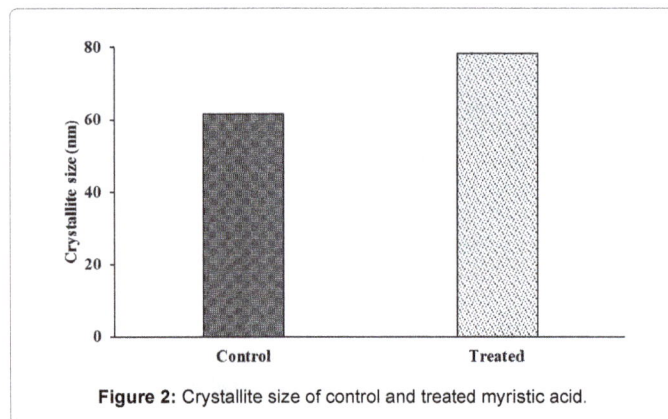

Figure 2: Crystallite size of control and treated myristic acid.

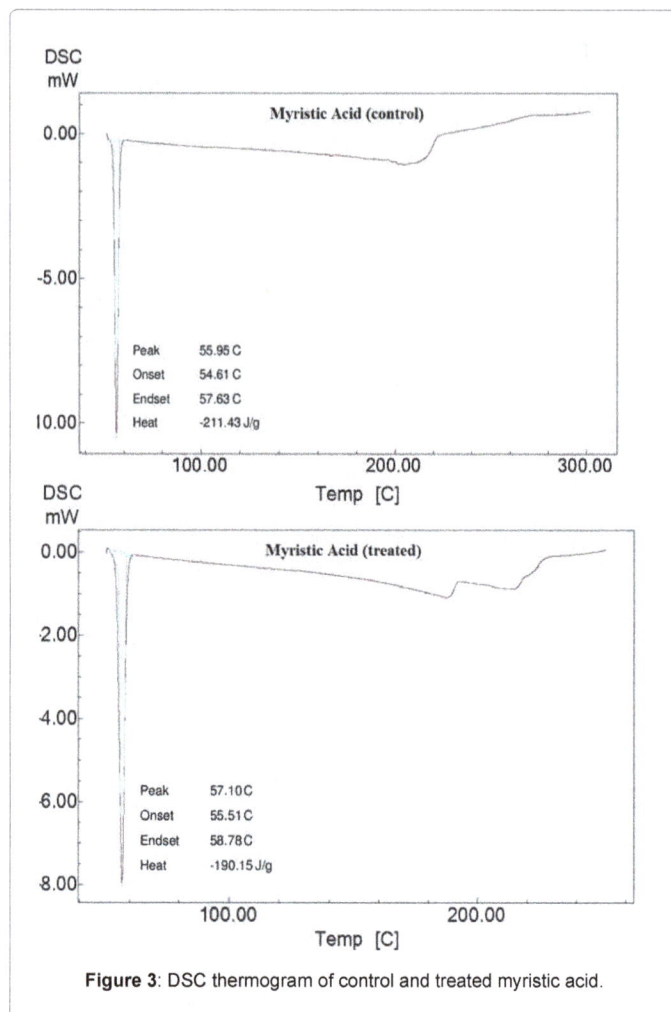

Figure 3: DSC thermogram of control and treated myristic acid.

heat of fusion of control and treated myristic acid. DSC thermogram of control myristic acid shows (Figure 3) an intense endothermic inflexion at 55.95°C; attributed to melting temperature of myristic acid. The melting temperature of myristic acid was well supported by the literature [37]. However, the treated myristic acid showed a melting endothermic peak at 57.10°C. It is assumed that increase in melting temperature may be due to interaction of treated myristic acid with biofield energy. It was previously reported that melting temperature of a material depends on its kinetic energy. Hence, it is assumed here that biofield may cause alteration in kinetic energy of treated myristic acid that led to increase in melting temperature. The latent heat of fusion (ΔH) was calculated from the DSC thermograms and results are presented in Table 1. The control myristic acid showed a ΔH of 211.43 J/g; however after biofield treatment it was changed to 190.15 J/g. The latent heat of fusion was decreased by 10.16% in treated sample as compared to control. Sari and Kaygusuz showed that myristic acid

has a suitable melting point of 49-51°C and a high latent heat of fusion [38]. They stated that myristic acid is good for thermal energy storage for domestic solar water systems [38]. Hence, in the present work the biofield treated myristic acid showed increase in melting temperature and appreciable latent heat of fusion which may be utilized for PCMs for thermal energy storage applications.

TGA Study

TGA was used to get further insights about the thermal stability of the control and treated myristic acid. TGA thermogram of control and treated myristic acid are shown in Figure 4. The thermogram of control myristic acid showed single step thermal degradation pattern. The onset temperature was noticed at around 196°C and the thermal degradation was stopped at around 237°C. During this step the control myristic acid lost 60.49% of its weight. However, the treated myristic acid also showed one step thermal degradation pattern. The sample started to degrade at around 182°C and the degradation terminated at around 237°C. The thermogram showed 31.33% of weight loss during this step. This showed that % weight loss of myristic acid was decreased after biofield treatment and this may be inferred as improvement in thermal stability of treated sample. Sharma et al., mentioned that good thermal stability, cheap and wide availability of fatty acids allows them to be used as PCMs [39]. Hence, it is presumed here that biofield treated myristic acid could be a good prospect as PCMs.

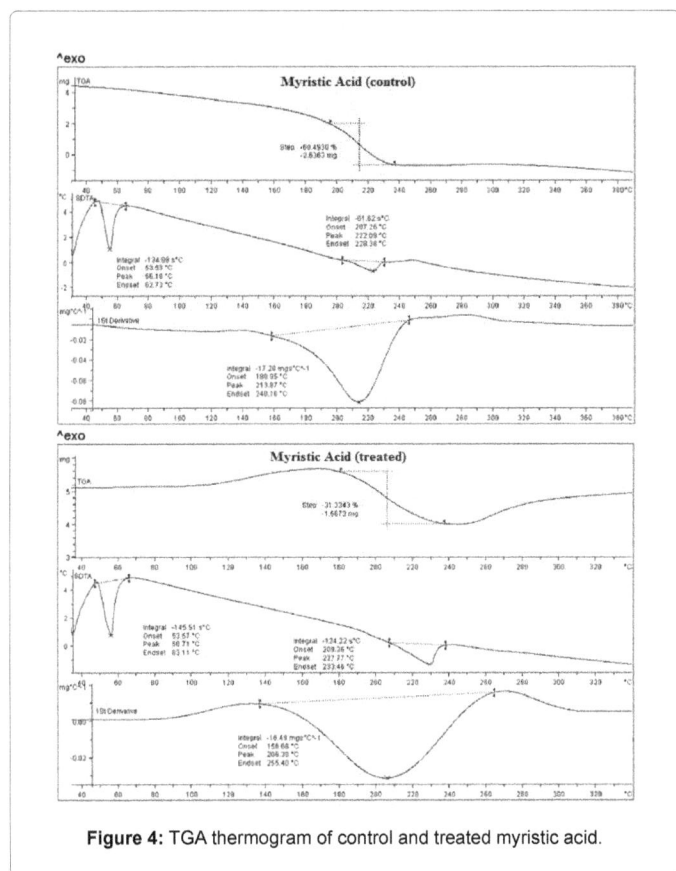

Figure 4: TGA thermogram of control and treated myristic acid.

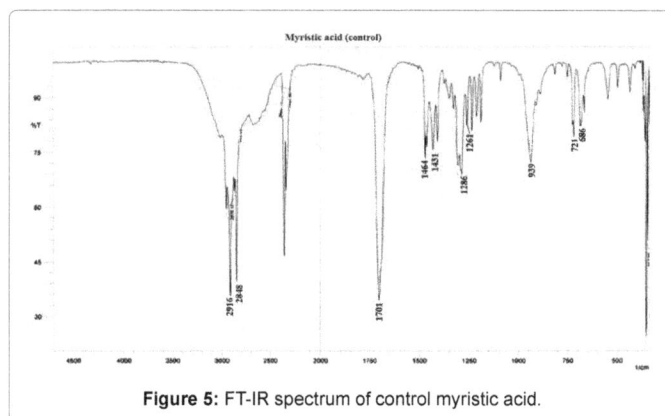

Figure 5: FT-IR spectrum of control myristic acid.

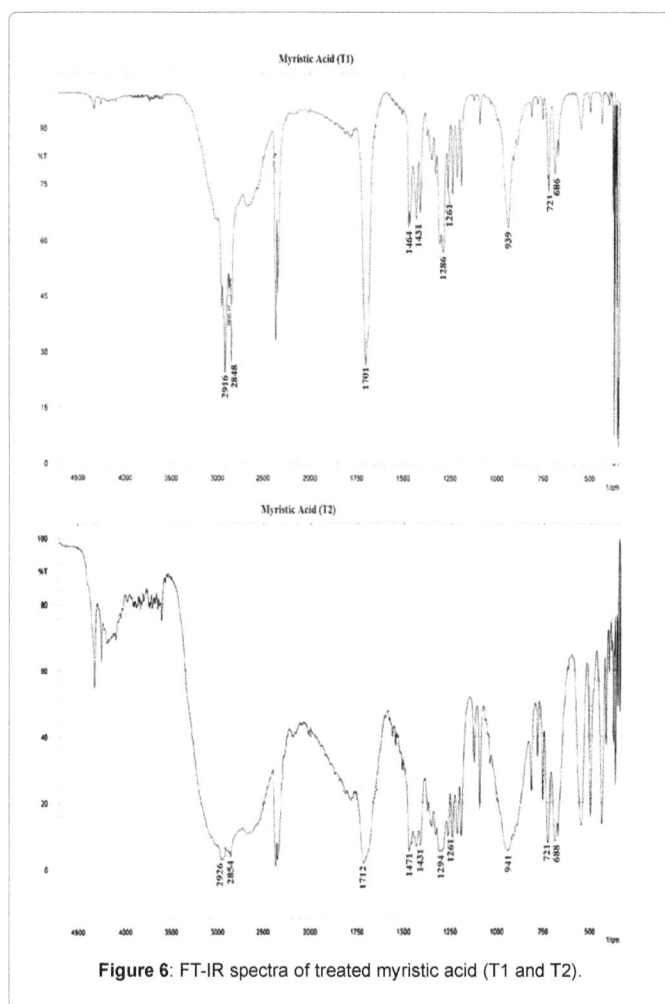

Figure 6: FT-IR spectra of treated myristic acid (T1 and T2).

The DTA thermogram of control myristic acid showed two endothermic peaks at 56.18°C and 222.09°C. The former peak attributed to melting of the myristic acid and second peak represented the thermal decomposition of the compound. However, the DTA thermogram of treated myristic acid showed two endothermic peaks; first one corresponded to melting endotherm (56.71°C) and second peak was due to thermal decomposition (227.77°C). Derivative thermogravimetry (DTG) of control myristic acid showed maximum thermal decomposition temperature (T_{max}) at 213.87°C; however it was decreased in treated myristic acid (206.38°C) (Table 1). Hence, the DTG result of treated myristic acid showed no improvement in T_{max} after biofield treatment.

FT-IR spectroscopy

FT-IR spectrum of control and treated myristic acid are presented in Figures 5 and 6, respectively. The control myristic acid showed band corresponding to C=O stretching vibration at 1701 cm⁻¹. Other peaks at 2916 and 2848 were due to symmetrical and asymmetrical stretching of –CH₂ functional group in fatty acid. The stretching peaks at 686, 721 and 939 cm⁻¹ were due to –OH swinging or rocking mode, which were characteristics of aliphatic chain of myristic acid [40]. The FT-IR peaks at 1431, and 1464 cm⁻¹ were due to –CH₂ bending vibration peaks. Other peaks were observed at 1286 and 1261 cm⁻¹ represents the C-H and C-C bending vibrations, respectively.

FT-IR spectrum of treated myristic acid T1 and T2 are presented in Figure 6. FT-IR spectrum of T1 showed typical absorption peaks of –CH₂ stretching vibrations at 2916 and 2848 cm⁻¹. Likewise, to control the peak at 1701, 1431 and1464 cm⁻¹ were due to C=O group stretching and –CH₂ bending vibrations. The absorption peaks at 1286 and 1261

cm⁻¹ corresponded to C-H and C-C bending vibrations, respectively. The FT-IR peaks at 686, 721, and 939 cm⁻¹ were due to –OH swinging or rocking vibrations. This showed no significant difference in FT-IR peaks of control and T1 myristic acid sample.

Whereas, the T2 sample showed substantial upward shifting of the FT-IR peaks of 2916→2926 cm⁻¹ (-CH₂ stretch), 2848→2854 cm⁻¹ (-CH₂ stretch), and 1701→1712 (C=O stretch). It was previously suggested that the frequency (ν) of vibrational peak depends on two factors i.e. force constant and reduced mass [41]. If mass is constant then frequency is directly proportional to force constant, hence, increase in frequency of any bond may cause possible enhancement in force constant of respective bond and *vice versa*. Hence, it is assumed that increase in frequency of $-CH_2$ and $C=O$ groups in T2 sample may increase the force constant and strength of these bonds. Moreover, the FT-IR of T2 also showed upward shifting of $-CH_2$ bending ($1464 \rightarrow 1471$ cm^{-1}) and C-C bending ($1286 \rightarrow 1294$ cm^{-1}) which may be due to some changes in chemical nature of the sample. Overall, the FT-IR results of treated myristic acid (T2) showed substantial alteration in bond strength and force constant after biofield treatment which may improve the stability of treated sample with respect to control.

Particle size analysis

Particle size of myristic acid (control and treated) was calculated from particle size distribution graph and presented in Figures 7 and 8. The average particle size (d_{50}) of control myristic acid was 63.19 μm and it was decreased to 21.67 μm in treated sample (Figure 7). Additionally, the d_{99} value (size exhibited by 99% of the particles) was decreased significantly in treated myristic acid 135.25 μm as compared to control myristic acid (679.95 μm). The d_{50} and d_{99} both were altered significantly by 65.7% and 80.1%, respectively in treated myristic acid

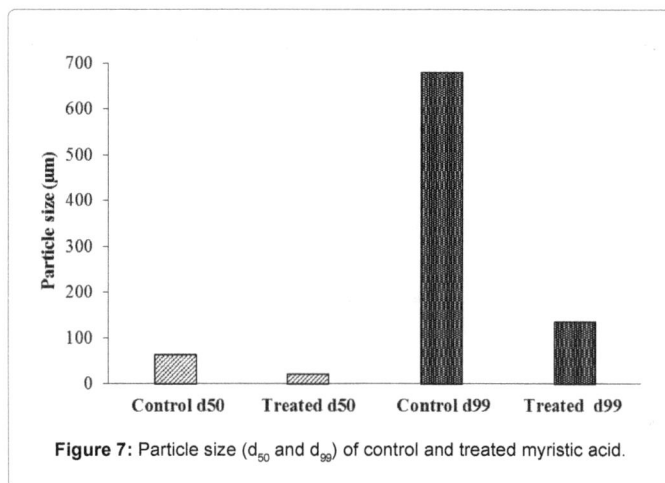

Figure 7: Particle size (d_{50} and d_{99}) of control and treated myristic acid.

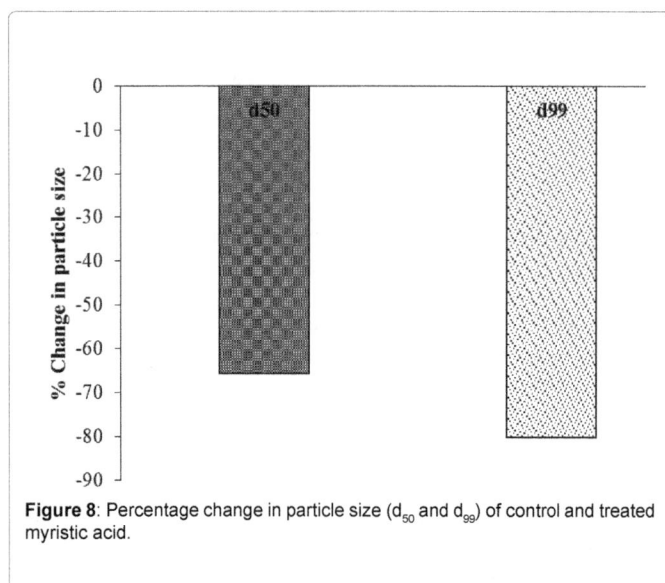

Figure 8: Percentage change in particle size (d_{50} and d_{99}) of control and treated myristic acid.

(Figure 8). The biofield treatment may cause fracture in the bigger size microparticles which led to reduction in particle size of treated myristic acid as compared to control.

Conclusion

The result showed significant impact of biofield treatment on physical, spectroscopic and thermal properties of myristic acid. XRD showed substantial increase in crystallite size of treated myristic acid with respect to control. DSC study on treated myristic acid showed increase in melting temperature with respect to control. A significant change in latent heat of fusion (10.16%) was observed in treated myristic acid as compared to control. TGA analysis revealed the lowering in weight loss of treated myristic acid as compared to control, which corroborated its high thermal stability. FT-IR spectroscopic study showed the alteration in force constant and bond strength of treated myristic acid with respect to control. Moreover, the biofield treated myristic acid showed decrease in particle size that may enhance the surface area as compared to control sample. Therefore, high melting temperature, thermal stability and appreciable latent heat of fusion of treated myristic acid may improve the phase change nature and it could be used for fabrication of thermal energy storage devices. Although, future studies such as thermal conductivity measurement can be further design to study the potential of biofield treated myristic acid for these applications.

Acknowledgement

The authors like to acknowledge the Trivedi Science, Trivedi Master Wellness and Trivedi Testimonials for their steady support during the work. This should be included in acknowledgment.

References

1. Feng L, Zheng J, Yang H, Guo Y, Li W, et al. (2011) Preparation and characterization of polyethyleneglycol/active carbon composites as shape-stabilized phase change materials. Sol Energ Mat Sol C 95: 644-650.

2. Chen C, Liu K, Wang H, Liu W, Zhang H (2013) Morphology and performances of electrospun polyethyleneglycol/poly (dl-lactide) phase change ultrafine fibers for thermal energy storage. Sol Energ Mat Sol C 117: 372-381.

3. Wang C, Feng L, Li W, Zheng J, Tian W, et al. (2012) Shape-stabilized phase change materials based on polyethyleneglycol /porous carbon composite: the influence of the pore structure of the carbon materials. Sol Energ Mat Sol C 105: 21-26.

4. Kenisarin MM, Kenisarina KM (2012) Form-stable phase change materials for thermal energy storage. Renew Sust Energ Rev 16: 1999-2040.

5. Zhang P, Hu Y, Song L, Ni J, Xing W, et al. (2010) Effect of expanded graphite on properties of high-density polyethylene/paraffin composite with in tumescent flame retardant as shape-stabilized phase change material. Sol Energ Mat Sol C 94: 360-365.

6. Fang G, Li H, Chen Z, Liu X (2011) Preparation and properties of palmitic acid/SiO2 composites with flame retardant as thermal energy storage materials. Sol Energ Mat Sol C 95: 1875-1881.

7. Zeng JL, Zhang J, Liu YY, Cao ZX, Zhang ZH, et al. (2008) Polyaniline/1- tetra decanol composites. J Therm Anal Calorim 91: 455-461.

8. Zeng JL, Liu YY, Cao ZX, Zhang J, Zhang ZH, et al. (2008) Thermal conductivity enhancement of MWNTs on the PANI/tetradecanol form-stable PCM. J Therm Anal Calorim 91: 443-446.

9. Karaman S, Karaipekli A, Sari A, Biçer A (2011) Polyethylene glycol (PEG)/diatomite composite as a novel form-stable phase change material for thermal energy storage. Sol Energ Mat Sol C 95: 1647-1653.

10. Karaipekli A, Sari A (2008) Capric–myristic acid/expanded perlite composite as form-stable phase change material for latent heat thermal energy storage. Renew Energ 33: 2599-2605.

11. Fameau AL, Arnould A, Saint-Jalmes A (2014) Responsive self-assemblies

based on fatty acids. Curr Opin Colloid Interface Sci 19: 471-479.

12. Yuan Y, Zhang N, Tao W, Cao X, He Y (2014) Fatty acids as phase change materials: A review. Renew Sust Energ Rev 29: 482-498.

13. Zeng J, Zhu F, Yu S, Xiao Z, Yan W, et al. (2013) Myristic acid/polyaniline composites as form stable phase change materials for thermal energy storage. Sol Energ Mat Sol C 114: 136-140.

14. Chen C, Liu X, Liu W, Ma M (2014) A comparative study of myristic acid/bentonite and myristic acid/Eudragit L100 form stable phase change materials for thermal energy storage. Sol Energ Mat Sol C 127: 14-20.

15. Hasenohrl F (1904) On the theory of radiation in moving bodies. Ann Phys (Berlin) 15: 344-370.

16. Einstein A (1905) Does the inertia of a body depend upon its energy-content? Ann Phys (Berlin) 18: 639-641.

17. Becker RO, Selden G (1985) The body electric: electromagnetism and the foundation of life, William Morrow and Company, New York City.

18. Barnes RB (1963) Thermography of the human body. Science 140: 870-877.

19. Born M (1971) The Born-Einstein Letters. (1stedn), Walker and Company, New York.

20. Cohen S, Popp FA (2003) Biophoton emission of the human body. Indian J Exp Biol 41: 440-445.

21. Trivedi MK, Patil S, Tallapragada RM (2013) Effect of biofield treatment on the physical and thermal characteristics of vanadium pentoxide powders. J Material Sci Eng S11: 001.

22. Trivedi MK, Patil S, Tallapragada RM (2013) Effect of biofield treatment on the physical and thermal characteristics of silicon, tin and lead powders. J Material Sci Eng 2: 125.

23. Trivedi MK, Patil S, Tallapragada RM (2014) Atomic, crystalline and powder characteristics of treated zirconia and silica powders. J Material Sci Eng 3: 144.

24. Trivedi MK, Patil S, Tallapragada RMR (2015) Effect of biofield treatment on the physical and thermal characteristics of aluminium powders. Ind Eng Manag 4: 151.

25. Shinde V, Sances F, Patil S, Spence A (2012) Impact of biofield treatment on growth and yield of lettuce and tomato. Aust J Basic Appl Sci 6: 100-105.

26. Sances F, Flora E, Patil S, Spence A, Shinde V (2013) Impact of biofield treatment on ginseng and organic blueberry yield. Agrivita J Agric Sci 35: 22-29.

27. Lenssen AW (2013) Biofield and fungicide seed treatment influences on soybean productivity, seed quality and weed community. Agricultural Journal 8: 138-143.

28. Patil SA, Nayak GB, Barve SS, Tembe RP, Khan RR (2012) Impact of biofield treatment on growth and anatomical characteristics of Pogostemon cablin (Benth.). Biotechnology 11: 154-162.

29. Trivedi MK, Patil S (2008) Impact of an external energy on Staphylococcus epidermis [ATCC –13518] in relation to antibiotic susceptibility and biochemical reactions – An experimental study. J Accord Integr Med 4: 230-235.

30. Trivedi MK, Patil S (2008) Impact of an external energy on Yersinia enterocolitica [ATCC –23715] in relation to antibiotic susceptibility and biochemical reactions: An experimental study. Internet J Alternative Med 6: 2.

31. Trivedi MK, Bhardwaj Y, Patil S, Shettigar H, Bulbule A (2009) Impact of an external energy on Enterococcus faecalis [ATCC – 51299] in relation to antibiotic susceptibility and biochemical reactions – An experimental study. J Accord Integr Med 5: 119-130.

32. Nayak G, Altekar N(2015) Effect of biofield treatment on plant growth and adaptation. J Environ Health Sci 1: 1-9.

33. Hua D, Zhang X, Zhan G, Hong Y, Su Y, et al. (2014) A high-pressure polar light microscopy to study the melt crystallization of myristic acid and ibuprofen in CO_2. J Supercrit Fluids 87: 22-27.

34. Gaber A, Abdel-Rahim MA, Abdel-Latief AY, Abdel-Salam MN (2014) Influence of calcination temperature on the structure and porosity of nanocrystalline SnO_2 synthesized by a conventional precipitation method. Int J Electrochem Sci 9: 81-95.

35. Raj KJA, Viswanathan B (2009) Effect of surface area, pore volume, particle size of P25 titania on the phase transformation of anatase to rutile. Indian J Chem 48A: 1378-1382.

36. Cherepanov PV, Melnyk I, Andreeva DV (2015) Effect of high intensity ultrasound on Al_3Ni_2, Al_3Ni crystallite size in binary AlNi (50 wt% of Ni) alloy. Ultrason Sonochem 23: 26-30.

37. Ince S, Seki Y, Ezan MA, Turgut A, Erek A (2015) Thermal properties of myristic acid/graphite nanoplates composite phase change materials. Renew Energ 75: 243-248.

38. Sari A, Kaygusuz K (2001) Thermal performance of myristic acid as a phase change material for energy storage application. Renew Energ 24: 303-317.

39. Sharma A, Tyagi VV, Chen CR, Buddhi D (2009) Review on thermal energy storage with phase change materials and applications. Renew Sust Energ Rev 13: 318-345.

40. Yang X, Yuan Y, Zhang N, Cao X, Liu C (2014) Preparation and properties of myristic–palmitic–stearic acid/expanded graphite composites as phase change materials for energy storage. Sol Energ 99: 259-266.

41. Pavia DL, Lampman GM, Kriz GS (2001) Introduction to spectroscopy. (3rd edn), Thomson learning, Singapore.

22

Experimental Measurements and Thermodynamic Modelling of Carbon Dioxide Capture with use of 2-(Ethylamino) Ethanol

Piotr Biernacki[1]*, **Sven Steinigeweg[1]**, **Wilfried Paul[1]** and **Axel Brehm[2]**

[1]*EUTEC Institute, University of Applied Sciences Emden/Leer, Emden, Germany*
[2]*Technische Chemie, Fk.V, Carl von Ossietzky Universität Oldenburg, Oldenburg, Germany*

Abstract

Carbon dioxide solubility was studied in 2.5 mass % and 5 mass % aqueous 2 (Ethylamino)ethanol (EAE; CAS 110-73-6) solution, an interesting secondary amine prepared mainly from renewable resources, at high loading rates, at three temperature ranges (293.00K, 313.15 K, 333.15 K), and in pressure range from 289 to 1011 kPa. In addition to that, heat capacity (cp) was measured of examined solution, allowing determination of temperature dependent coefficients of ideal gas heat capacity equation. Afterwards, the electrolyte Non Random Two-Liquid model's parameters were determined to represent the thermodynamic behaviour of the CO_2 – EAE – H_2O system with use of ASPEN® Plus V8.0 simulation software, indicating a good correlation between experimental and simulation results. As a consequence, model based calculation of the carbon dioxide capture with use of 2-(Ethylamino) ethanol is possible.

Keywords: Electrolyte NRTL model; 2-(Ethylamino) ethanol; Carbon dioxide capture; Biogas upgrading technology; Biomethane; Carbon capture and Storage

Introduction

Biogas produced through anaerobic digestion of organic wastes is already widely applicable process, which is often utilized in combined heat and power units (CHP). However, biogas can also be upgraded, which means removal of CO_2 to fulfil requirements as a vehicle fuel, or feeding it to the public natural gas grid. It is achieved with different techniques like pressure swing adsorption, water scrubbing or amine washing are applied to remove carbon dioxide, and allow maximal methane slippage [1].

The main goal of this research was preparation of the model, thus determination of binary interaction parameters, which can be applied for upgrading existing biogas power plants to biomethane plants. At such plants prior to biogas combustion at CHP unit H_2S is removed due to corrosion danger [1,2], therefore it was decided to concentrate only on CO_2 removal. Since amine scrubbing is the most technically and commercially mature method, which can be easily retrofitted to an existing plant [3], and according to Rochelle [4] in 2030 it probably will be the dominant method applied for coal-fired power plants, amine scrubbing was selected for this research. 2-(Ethylamino) ethanol (EAE) is a linear secondary amine which is linked to an ethyl group, was decided to be evaluated in this research. Unlike monoethanolamine (MEA), EAE has a small corrosion rate, even at higher concentrations. In addition, it requires less energy for regeneration, and the absorption rate is higher due to creation of moderate stability carbamate [5-8]. An additional advantage is that, produced from agriculture products or residues ethanol is used to produce ethylamine and ethylene oxide. Both those chemicals react to form EAE [9]. Moreover, methyldiethanolamine (MDEA) is often applied during amine washing to ensure H_2S removal [2], however its rate constant of second-order reaction is lower than for EAE [8], and existing biogas power plants already removed H_2S prior to combustion at CHP unit [1,2]. 2-(Ethylamino)ethanol has been already proved as an absorbent for CO_2 capture [9], and also as an activator in aqueous N,N-diethylethanolamine (DEEA) solutions [10]. However, there is still little experimental data on CO_2 capture with EAE at high loading rates (mol CO_2 mol EAE^{-1} > 1), and while the focus was rather on kinetics of reaction [8-12], no publication on thermodynamic modelling representing vapour-liquid equilibrium in the CO_2 –EAE –

H_2O system was found. Therefore in this research the Electrolyte Non Random Two-Liquid (eNRTL) model was applied for modelling the EAE's performance, since it has been widely applied for other amines like MEA [13], MEA and DEA [14], DGA [15], DGA and MDEA [16].

Summarizing, the main intention of this article is evaluation of the 2-(Ethylamino) ethanol for upgrading biogas, and preparation of the eNRTL model parameter, in order to promote model calculation of carbon capture with EAE. Moreover, those parameters are essential for the development of efficient industry upgrading installations.

Materials and Methods

Experimental

Materials and solutions

All chemicals used during this research were of analytical reagent grade, and employed without further purification. CO_2 was acquired from Linde˚ AG (purity 99,5 volume%), and 2-(Ethylamino)ethanol (EAE; CAS: 110-73-6) was acquired from Sigma-Aldrich˚ Co. LLC. (purity of ≥98 volume%). In order to ensure excellent water quality necessary for HPLC pump, Milli-Q water was used, due to its high degree of de-ionizing and purity. It was prepared by use of Milli-Q Biocel unit (®EMD Millipore Corporation).

Before each experiment is was crucial to guarantee that water is not containing CO_2, with the purpose of ensuring that solubility measurements are accurate. Therefore, prior to each experiment vacuum was applied to Duran˚ bottle, resistant to under- and over-pressure, filled with Milli-Q water. Afterwards aqueous alkanolamine

---

*Corresponding author: EUTEC Institute, University of Applied Sciences Emden/Leer Constantiaplatz 4, 26723 Emden, Germany, E-mail: piotr.biernacki@hs-emden-leer.de

solution was prepared gravimetrically. Subsequently prepared solution was purged with nitrogen, acquired from Linde' AG (purity 99,999 volume%), before the final stage of placing it in ultrasonic bath (Branson 2210) for one hour.

Apparatus: In order to measure CO_2 solubility in aqueous alkanolamine solutions an experimental apparatus was developed, based on modified approached of Cadours et al. [17] The unit consists of two reactors acquired from Parr' Instrument Company (4560 Pressure Reaction Apparatus; volume of 0.45 dm³; maximum working pressure of 20000 kPa; operating temperature from 263.15 K to 623.15 K) directly connected to each other with high pressure stainless steel capillary with double-sided conical bolt connection (A506HC; Hose Assembly 6FT T316), as presented on Figure 1.

The second reactor is equipped with a stirrer (A1120HC6 Parr' Magnetic Drives; Turbine Type Impeller) controlled by Parr' 4875 Power Controller. Gas bottles located in gas safety cabinet (Asecos', FWF 90) are connected to first reactor, also with use of Parr's' high pressure capillaries (A495HC, Hose Assembly 6FT Nylon). Both reactors were heated up with use of Lauda water bath (Ecoline Staredition 006), and the temperature inside each reactor was measured with use of Parr's' thermocouples (A472E2; Thermocouple 9-1/2, T316 stainless steel, Type J). Due to the measurement procedure (described in chapter 2.1.3) reactors were equipped with PR-33X pressure sensors, both acquired from Keller' Druckmesstechnik, but with different pressure ranges: Keller PR-33X 0-1000 kPa (accuracy ±0.1% of full scale) and Keller PR-33X 0-3000 kPa (accuracy ±0.1% of full scale). Both sensors accuracy is documented in 5 points test report prepared by firma Keller' Druckmesstechnik. In order to create a vacuum at both reactors, ILMVAC' P4Z vacuum pump was used. For pumping water or aqueous alkanolamine solutions into reactor, a HPLC pump (KNAUER' Smartline pump 100, 50 ml min⁻¹) was used. However, due to change in viscosity of the aqueous alkanolamine solutions, density of each solution was measured prior to pumping, with use of pyctometer corrected to three decimal places with thermometer (Assistant' 2572/325, volume of 25.003 cm³ in 293.15 K), and the pumped amount was controlled gravimetrically (Sartorius' BL1500S).

The data measured by sensors are collected in U12 LabJack' measurement and automation device, which is an interface between computers and the physical word. Afterwards collected data are sent via a Wi-Fi network at a PC workstation, where pressure and temperature of both reactors are recorded in a program, in ProfiLab' environment. The recording interval can be determined in a range of 1 to 10000 seconds.

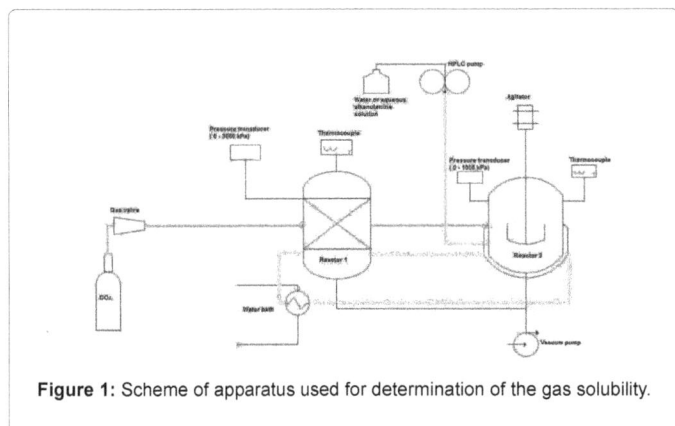

Figure 1: Scheme of apparatus used for determination of the gas solubility.

In addition to measuring the gas solubility, mixture's liquid heat capacity was measured with use of differential scanning calorimeter (Netzsch DSC 204 F1 Phoenix').

Measuring procedure: Initially the apparatus' functionality and accuracy was verified. In order to do so, solubility of CO_2 in water was measured at temperature of 292.95 K, pressure range of 500 up to 1200 kPa, and compared to the literature. The results are presented in section 3.1.1. The standard measuring procedure always starts with generating vacuum in both reactors, and simultaneously heating them up to a desired temperature. After reaching vacuum condition and constant temperature, reactors remained as such for 1 hour, to verify no pressure and temperature change, in order to confirm system's tightness. Afterwards CO_2 was introduced into the first reactor (Figure 1), and the second reactor was filled with 0.225 dm³ of water or amine solution. After obtaining desired temperature and steady pressure readings, CO_2 was introduced to the second reactor. Simultaneously the agitator was started. In the second reactor pressure increased (introducing CO_2) was observed, followed by pressure decrease (absorption process). The end of absorption process is indicated by a constant pressure in the second reactor, and the reaction's duration depends on the solvent and loading. However, to guarantee high accuracy of the results, each experiment lasted minimum one day, with agitator on during the whole measurement, despite equilibrium pressure was often obtained earlier. Each measurement was repeated, and also the correlation between points obtained was controlled.

In order to measure heat capacity with use of differential scanning calorimeter, for each measurement baseline profile (empty sample pan), a standard sample profile (24.9 mg sapphire standard), and a sample profile, as further described in [18,19], were determined. Additionally, measuring method was prepared for this application, where starting temperature was 293.15 K, followed by heating (heating rate 5 K min⁻¹) to 298.15 K, and then it is kept isothermally for 10 min, before the final heating to 355.15 K (heating rate 40 K min⁻¹), which is again followed by isothermal step for 10 min. Afterwards, cooling to 298.15 K was applied, allowing cp calculation during cooling step. Each measurement was prepared as triplicates.

The method is in line with Netzsch [19] recommendation, and the results were analysed using the Proteus' Analysis (version 6.1) data analysis program.

CO_2 **solubility calculation**: The solubility determination is based on approach presented by Park and Sandall [20]. However the calculation is modified, since Peng Robinson Equation of State (PREOS) [21] is used, available in ASPEN Plus™ V8.0 simulation software, rather than compressibility factors. As a consequence, number of CO_2 moles (n_{1CO2}) in the first reactor (Figure 1) just before feeding the gas to the second reactor (but after obtaining constant pressure and temperature in the first reactor) is calculated with use of PREOS. After introducing the gas to the second reactor, and obtaining constant pressure and temperature in the first reactor, n_{2CO2} is calculated with PREOS. Finally number of CO_2 moles (n_{1CO2}) introduced is calculated by subtracting n_{2CO2} from n_{1CO2}. The equilibrium pressure, obtained from the second reactor, is used for calculating the amount of remaining CO_2 (n_{eCO2}). Finally, number of moles absorbed is calculated by subtracting remaining CO_2 moles (n_{eCO2}) from introduced CO_2 moles (n_{1CO2}):

$$n_{CO_2}^{abs} = (n_{CO_2}^1 - n_{CO_2}^2) - n_{CO_2}^e = n_{CO_2}^i - n_{CO_2}^e \qquad (1)$$

Modelling of gas solubility

Physical solubility: Phase equilibria are described with use of fugacity coefficient in the following relations of vapour-liquid equilibrium [22]:

$$x_i \varphi_i^L = y_i \varphi_i^v \qquad (2)$$

Where fugacity coefficients are describing deviation from ideal gas behaviour are applied despite standard fugacity and activity coefficients. Moreover, fugacity coefficients can be calculated with use of cubic equations of state. However, because weak electrolytes are considered in this research, approach with Henry constant as standard fugacity was used instead [22]:

$$P.Y_1.\varphi_1 = H_{12}.X_1.\gamma_1^* \qquad (3)$$

Where system pressure (P), mole fraction in vapour phase (y_1) and in liquid phase (x_1), Henry's law constant of solute (1) in solvent (2) (H_{12}), fugacity coefficient in vapour phase (φ_1), and activity coefficient of solute in the solvent (γ_1^*) are included, as further described by Gmehling [22]. In this research fugacity coefficient was calculated with use of Redlich-Kwong EOS [22] , activity coefficient was determined with use of electrolyte Non Random Two Liquid model, and coefficients for Henry constant of 2-(Ethylamino) ethanol, were based on diglycolamine (DGA) from [23].

Chemical solubility: The chemical solubility, which is the chemical equilibrium for the aqueous phase chemical reactions between water, amines, acid gases (e.g. CO_2), together with physical solubility are representing the overall acid gases solubility in aqueous amines solutions. These equilibrium reactions were developed for 2-(Ethylamino) ethanol, and are based on reactions presented by Austgen et al. [14] and Zhange et al. [13]:

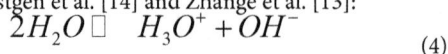

$$2H_2O \rightleftharpoons H_3O^+ + OH^- \qquad (4)$$

$$2H_2O + CO_2 \rightleftharpoons H_3O^+ + HCO_3^- \qquad (5)$$

$$HCO_3^- + H_2O \rightleftharpoons H_3O^+ + HCO_3 \qquad (6)$$

$$EAEH^+ + H_2O \rightleftharpoons EAE + H_3O^+ \qquad (7)$$

Reactions describe ionization of water (4), dissociation of carbon dioxide (5) and bicarbonate (6), and amine protonation (7). In addition to that, carbamate reversion to bicarbonate, firstly proposed by Caplow [24], is also included in the chemical solubility, which is only possible for primary and secondary amines [13,14]:

$$EAECOO^- + H_2O \rightleftharpoons EAE + HCO_3^- \qquad (8)$$

Equilibrium constants for reactions 4 – 8 are presented as temperature dependent via:

$$\ln(K) = C_1 + \frac{C_2}{T} + C_3.\ln(T) + C_4.T \qquad (9)$$

where the values for each reaction are presented in Table 1.

The Electrolyte-NRTL: Activity coefficient necessary for physical solubility calculation is acquired, when excess Gibbs energy is present [22]. The Electrolyte-NRTL (eNRTL), an excess Gibbs energy expression, presented by Chen and Evans [25], extended by Mock et al. [26] to mixed solvent electrolyte systems is implemented in ASPEN Plus® V8.0 engineering software as ELECNRTL [27] and used in this research. The proposed eNRTL model is a sum of two contributions. The first one is a long-range contribution, describing ion-ion interactions' outside the immediate neighbourhood of central ionic species. For this

Reaction	C_1	C_2	C_3	C_4	Source
4	132.899	-13445.9	-22.4773	0.0	[38]
5	216.049	-12431.70	-35.4819	0.0	[39]
6	1.6957	-8431.64	0.0	0.005037	[40]
7	231.465	-12092.10	-36.7816	0.0	[39]
8	8.8334	-5274.40	0.0	0.0	[32]

Table 1: Equilibrium constants for reactions [4-8].

contribution Chen and Evans [25] implemented Pitzer's reformulation of the Debye-Huckel formula [28], and the Born expression [29], which includes the difference between the dielectric constants of solvent mixture and water [22]. On the other side, the Non-Random Two Liquid theory developed by Renon and Prausnitz [30], based on the theory of like-ion repulsion and electroneutrality, represents the local contribution, resulting in [14,27]:

$$g^E = g^E_{Debye-Huckel} + g^E_{Born} + g^E_{NRTL} \qquad (10)$$

According to Austgen et al. [14] the adjustable parameters required by the eNRTL are only the NRTL binary energy interaction parameters, where, following the theory of like-ion repulsion and electroneutrality, three types of interaction can be determined,: molecule – molecule, molecule – ion pair (also ion pair – molecule), and ion pair – ion pair. However, as indicated by Chen and Evans [25] ion pair – ion pair parameters can be set to zero, because no significant impact on vapour-liquid equilibrium (VLE) is then caused. Moreover, because the experimental VLE data do not include in situ analysis of the VLE's composition, only the molecule-molecule binary interaction parameters were determined. Following the literature [14,25,26] all water – ion pair, and ion pair-water binary parameters were kept at default values (8 and -4). In addition, all other binary parameters (alkanolamine – ion pair; ion pair – alkanolamine; acid gas – ion pair; and ion pair – acid gas) were kept at values of 15 and -8. Besides that, binary interaction parameters for water and carbon dioxide (molecule – molecule interaction) are also already determined by Chen and Evans [25].

The non-randomness factor (α) for water – ion par and for molecule – molecule interactions was fixed at 0.2, as recommended by Chen and Evans [25], and as proposed by Mock et al. [26] it was kept at value of 0.1 for alkanolamine – ion pair and acid gas – ion pair.

The binary energy interaction parameters included in ASPEN Plus® V8.0 are adopted as a temperature dependent as given by Austgen et al. [14]:

$$\tau = a + \frac{b}{T} \qquad (11)$$

Values of a and b for alkanolamine – water and water – alkanolamine interactions were determined with use of Data Regression System (DRS). Following path proposed by Austgen et al. [14], for determination of the interaction parameters the Deming algorithm was used, and as an objective function maximum likelihood was selected. The binary energy interaction parameter τ and the nonrandomness parameter α are used for calculating Gibbs energy [22]:

$$G_{ij} = \exp(-\alpha_{ij}\tau_{ij}) \qquad (12)$$

Pure component properties: Most of the pure component parameters' for 2-(Ethylamino) ethanol were acquired from NIST Databank [31]. However, due to the limited number of data on EAE, it was decided to follow Austgen [32] concept, where the dielectric constants for pure diglycolamine (DGA) was set equal to dielectric constants for diethanolamine (DEA), due to missing data. Therefore coefficients for Henry's constants [23], the dielectric constants [32], equilibrium [14] and kinetic constants [16] were based on DGA [15]. Parameters for CO_2 and H_2O were acquired from ASPEN Plus' databanks (APV80.PURE27 and APV80.Binary).

Results and Discussion

Experimental results

Assessment of the apparatus precision: Aim of this article is to provide precise experimental results on CO_2 solubility in aqueous EAE solutions acquired with apparatus described in section 2.1.2. Therefore, in order to ensure correct functionality of the apparatus, and high accuracy of the experimental results, solubility of CO_2 in water was measured at temperature of 292.95 K, pressure range of 500 up to 1200 kPa, and compared to the literature found in Dortmund Data Bank® (DDBST GmbH): Silkenbäumer et al. [33], Crovetto [34], Landolt-Börnstein [35], and Addicks et al. [36] The results are presented as a Figure 2, indicating very good fit.

Solvent characteristics

Density: As explained in section 2.1.2., due to change in viscosity of the aqueous alkanolamine solutions, prior to each filling of the second reactor (Figure 1) with the solution, its' density was measured. The averaged density of 2,5 mass % solution was measured to be 0.9969 (± 0.1 mass%), and the averaged density of 5 mass % solution was measured to be 0.9959 (± 0.1mass%).

Mixture's liquid heat capacity: The binary NRTL interaction parameters are necessary prior to eNRTL model's application for activity coefficient calculations, which are then used for aqueous phase chemical equilibrium, phase equilibrium, enthalpy of absorption, liquid enthalpy and liquid heat capacity determination [14]. However, accurate prediction of mixture's liquid heat capacity is necessary for correct calculation of desorption step, necessary for complete assessment of industry upgrading installations. Therefore liquid heat capacity of pure EAE was measured and compared to the literature [37,38]. Together with aqueous solutions results are presented as a

Figure 3 (with ± 3% uncertainty). In addition to that, experimental mixtures' liquid heat capacity is also compared to the simulation, and is presented as a Figure 3. However, because used calorimeter was cooled with air, therefore precise measurement in the lower temperature range was not possible, as can be seen on the graph. The results were used for regressing CPIG Parameters given in Table 2, and used for calculating results on the Figure 3.

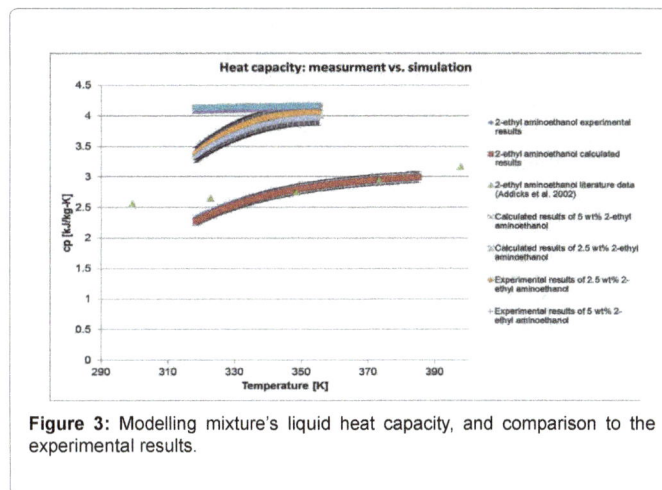

Figure 3: Modelling mixture's liquid heat capacity, and comparison to the experimental results.

Component	2-ethyl aminoethanol	Standard deviation
Temperature Unit	[C]	[-]
Property Unit	[kJ kmol $^{-1}$ K^{-1}]	
Coefficient 1	-1,58E+02	1.79E+03
Coefficient 2	9,98E+00	5.15E+00
Coefficient 3	-1,42E-01	5.21E+00
Coefficient 4	1,14E-03	4.18E-02
Coefficient 5	-4,98E-06	1.27E-04
Coefficient 6	9,49E-08	1.39E-07
Coefficient 7	-2,73E+02	-
Coefficient 8	7,27E+02	-
Coefficient 9	0,00E+00	-
Coefficient 10	0,00E+00	-
Coefficient 11	0,00E+00	-

Table 2: Temperature dependent coefficients of ideal gas heat capacity equation (*CPIG*) [14,15].

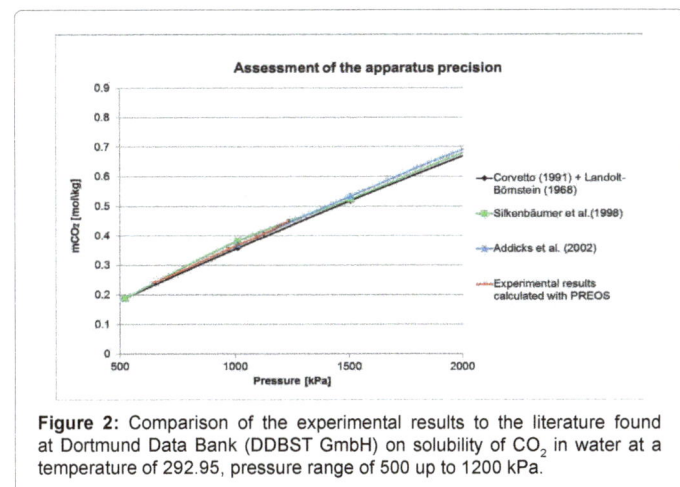

Figure 2: Comparison of the experimental results to the literature found at Dortmund Data Bank (DDBST GmbH) on solubility of CO_2 in water at a temperature of 292.95, pressure range of 500 up to 1200 kPa.

Results of solubility measurement

Experimental range was decided to be from 289 to 1011 kPa, at 293.00 K, 313.15 K, and 333.15 K. The solvent consisted of in 2.5 mass % and 5 mass % aqueous alkanolamine (Table 3). For each chosen temperature and concentration, presented data consists of five points, hence of five end pressures. Due to the measuring procedure, specifically filling of the first reactor with use of regular pressure regulator at the gas bottle, it was impossible to exactly repeat each measurement. Moreover, correction of the moles of carbon dioxide in the first reactor with use of the gas release valves was also attempted but did not deliver accurate results. As a consequence, for each chosen temperature and concentration minimum 10 measurements were conducted, and the five presented points were chosen based on standard deviation from the results obtained.

Thermodynamic Modelling

Carbamate stability parameters

Suda et al. [7] conducted NMR measurements for EAE, where

he indicated that EAE is forming moderate stability carbamate. An assessment of parameters used for the reduced power law expression, which were set equal to parameters used for DGA [15], was conducted, indicating a very good correlation with Suda et al.'s [7] research result, therefore it was decided not to modify used parameters.

NRTL's binary interaction parameters

Regressed values of the NRTL binary interaction parameters used for local contribution in eNRTL model are included in Table 4. Evaluation of the new values' applicability in representing the experimental results from Table 3 is reported as a Figure 4.

Summarizing, it can be stated that a good fit between model and experimental results was achieved, especially taking under consideration limited data, and pragmatic approach of fitting values from DGA for EAE.

Conclusion

To support model calculation of carbon capture with use of

Parameter	T	2.5 mass % EAE			5 mass % EAE		
		Loading	PCO2	Uncertainty	Loading	PCO2	Uncertainty
Unit	[K]	[mol CO_2 mol EAE^{-1}]	[kPA]	[kPa]	[mol CO_2 mol EAE^{-1}]	[kPa]	[kPa]
	293.00	1.4105	314	±1	1.1270	405	±1
	293.00	1.7591	541	±1	1.2222	514	±1
	293.00	1.9493	653	±1	1.2698	552	±1
	293.00	2.1078	71	±1	1.4841	721	±1
	293.00	2.5832	1011	±1	1.5873	840	±1
	313.15	1.1569	289	±1	1.0238	410	±1
	313.15	1.3154	416	±1	1.1905	668	±1
	313.15	1.4263	523	±1	1.2619	750	±1
	313.15	1.5214	561	±1	1.3730	856	±1
	313.15	1.5689	637	±1	1.3968	946	±1
	333.15	1.1569	444	±1	1.1032	713	±1
	333.15	1.2520	524	±1	1.1270	723	±1
	333.15	1.4263	653	±1	1.1905	790	±1
	333.15	1.6323	786	±1	1.2540	890	±1
	333.15	1.7116	852	±1	1.3492	1010	±1

Table 3: CO_2 solubility in 2-(Ethylamino)ethanol (EAE).

Parameter	Component i	Component j	Value	Standard deviation
A	H_2O	EAE	16.514	0.128
A	EAE	H_2O	-3.958	0.026
B	H_2O	EAE	-16.141	40.443
B	EAE	H_2O	-3.211	8.031

Table 4: NRTL's binary interaction parameters obtained with use of ASPEN Plus® V8.0 Data Regression System (DRS).

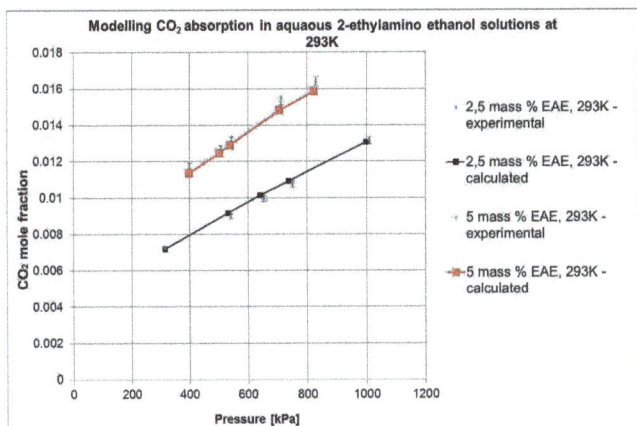

Figure 4a: Modelling CO_2 absorption in aqueous 2-(Ethylamino)ethanol solutions, and comparison to the experimental results.

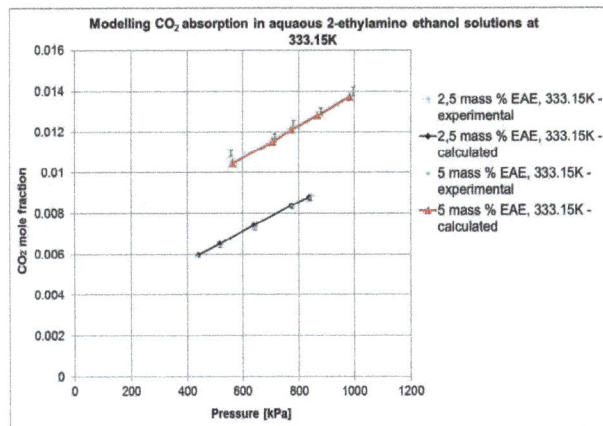

Figure 4c: Modelling CO_2 absorption in aqueous 2-(Ethylamino) ethanol solutions, and comparison to the experimental results.

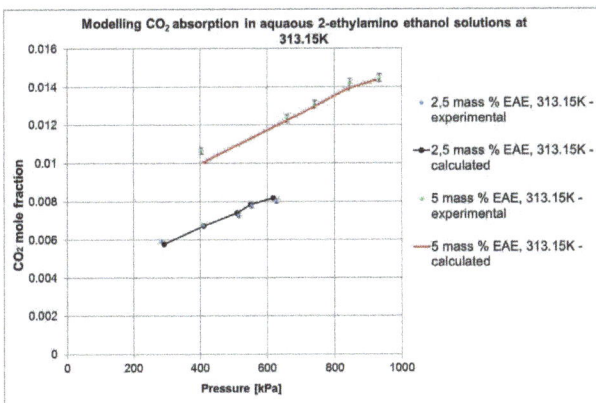

Figure 4b: Modelling CO_2 absorption in aqueous 2-(Ethylamino) ethanol solutions, and comparison to the experimental results.

2-(Ethylamino) ethanol, a promising alternative to diethanolamine (DEA) or monoethanolamine (MEA) was analysed, NRTL's binary interaction parameters necessary for eNRTL model, which can be used for modelling EAE's performance have been regressed. EAE's performance was modelled with use of eNRTL model indicating a good fit between experimental and simulation results (Figure 4). Model based optimization of the biogas power plants to biomethane power plants with use of 2-(Ethylamino) ethanol, for which the main raw material can be bio-ethanol, is now possible. The further research will focus on advanced economic and ecological analysis of the biogas upgrading with use of amine absorption processes.

Funding Sources

Funding for this study was provided by German Federal Ministry for Education and Research (BMBF): project FKZ 17N1710.

Acknowledgment

Support from Dipl. Ing. I. Stein and Dipl. Ing. S. Röefzaad in apparatus build up is greatly accredited. Dipl. Ing. M. Becker from Hochschule Emden/Leer is acknowledged for his time in development the data acquisition unit. Discussions with Prof. Dr. Axel Borchert, Prof. Dr. M. Schlaak, Dr. Frank Uhlenhut from EUTEC Institute, Hochschule Emden/Leer are highly valued.

References

1. Weiland P (2006) Biomass Digestion in Agriculture: A Successful Pathway for the Energy Production and Waste Treatment in Germany. Eng Life Sci 6: 302-309.

2. Abatzoglou N, Boivin S (2009) A review of biogas purification processes. Biofuels Bioprod Bioref 3: 42-47.

3. Kohl AL, Nielsen RB (1997) Gas purification. Houston TX: Gulf Professional Publishing.

4. Rochelle GT (2009) Amine scrubbing for CO2 capture. Science 325: 1652-1654.

5. Mimura T, Shimojo S, Suda T, Iijima M, Mituoka S (1995) Research and development on energy saving technology for flue gas carbon dioxide recovery and steam in power plant. Energy Convers Manage 36: 397-400.

6. Mimura T, Simayoshi H, Suda T, Iijima M, Mituoka S (1997) Development of energy saving technology for flue gas carbon dioxide recovery in power plant by chemical absorption method and steam system. Energy Convers Manage 38: 57-62.

7. Suda T, Iwaki T, Mimura T (1996) Facile Determination of Dissolved Species in CO2 – Amine – H2O System by NMR Spectroscopy. Chem Lett 25: 777-778.

8. Mimura T, Suda T, Iwaki T, Honda A, Kumazawa H (1998) Kinetics of Reaction Between Carbon Dioxide and Sterically Hindered Amines for Carbon Dioxide Recovery from Power Plant Flue Gases. Chem Eng Comm 170: 245-260.

9. Sutar PN, Jha A, Vaidya PD, Kenig EY (2012) Secondary amines for CO2 capture: A Kinetic investigation using N-ethylmonoethanolamine. Chem Eng J 207-208: 718-724.

10. Vaidya PD, Kenig EY (2007) Absorption of CO2 into aqueous blends of alkanolamines prepared from renewable resources. Chem Eng Sci 62: 7344-7350.

11. Bavbek O, Alper E (1999) Reaction Mechanism and Kinetics of Aqueous Solutions of Primary and Secondary Alkanolamines and Carbon Dioxide. Turk J Chem 23: 293-300.

12. Li J, Henni A, Tontiwachwuthikul P (2007) Reaction Kinetics of CO2 in Aqueous Ethylenediamine, Ethyl Ethanolamine, and Diethyl Monoethanolamine Solutions in the Temperature Range of 298-313 K, Using the Stopped-Flow Technique. Ing Eng Chem Res 46: 4426-4434.

13. Zhang Y, Que H, Chen C (2011) Thermodynamic modelling for CO2 absorption in aqueous MEA solution with electroyle NRTL model. Fluid Phase Equilibria 311: 67-75.

14. Austgen DM, Rochelle GT, Peng X, Chen C (1989) Model of Vapor-Liquid Equilibria for Aqueous Acid Gas-Alkanolamine Systems Using the Electrolyte-NRTL Equation. Ind Eng Chem Res 28: 1060-1073.

15. Aspen Technology Inc (2008) Aspen Plus: Rate-Based Model of the CO2 Capture Process by DGA using Aspen Plus (Version Number: V8.0).

16. Pacheco MA, Kaganoi S, Rochelle GT (2000) CO2 Absorption into Aqueous Mixture of Diglycolamine and Methyldiethanolamine. Chem Eng Sci 55: 5125-5140.

17. Cadours R, Roquet D, Perdu G (2007) Competitive Absorption – Desorption of Acid Gas into Water-DEA Solutions. Ind Eng Chem Res 46: 233-241.

18. Chiu LF, Liu HF, Li MH (1999) Heat capacity of Alkanolamines by Differential Scanning Calorimetry. J Chem Eng Data 44: 631-636.

19. NETZSCH GmbH & Co. Holding KG (2007) Artifacts in data curves Service Trouble shooting – getting the best from your DSC,TG,STA. International Customer Service Training: 26-27.

20. Park MK, Sandall OC (2001) Solubility of Carbon Dioxide and Nitrous Oxide in 50 mass % Methyldiethanolamine. J Chem Eng Data 46: 166-168.

21. Peng DY, Robinson DB (1976) A New Two-Constant Equation of State. Ind Eng Chem Fundam 15: 59-64.

22. Gmehling J, Kolbe B, Kleiber M, Rarey J (2012) Chemical Thermodynamics for Process Simulation, Wiley.

23. Martin JL, Otto FD, Mather AE (1978) Solubility of Hydrogen Sulfide and Carbon Dioxide in a Diglycolamine Solution. J Chem Eng Data 23: 163-164.

24. Caplow M (1968) Kinetics of Carbamate Formation and Breakdown. J Am Chem Soc 90: 6795-6803.

25. Chen CC, Evans LB (1986) A Local Composition Model for the Excess Gibbs Energy of Aqueous Electrolyte Systems. AIChE J 32: 444-454.

26. Mock B, Evans LB, Chen CC (1986) Thermodynamic Representation of Phase Equilibria of Mixed-solvent Electrolyte Systems. AIChE J 32: 1655-1664.

27. Aspen Technology Inc (2012) Aspen Physical Property System: Physical Property Models (Version Number: V8.0).

28. Pitzer KS (1980) Electrolytes. From dilute solutions to fused salts. J Am Chem Soc 102: 2902-2906.

29. Robinson RA, Stokes RH (1970) Electrolyte solutions. (2nd edn.) Butterworth and Co, London, UK.

30. Renon H, Prausnitz JM (1968) Local Compositions in Thermodynamic Excess Functions for Liquid Mixtures. AIChE J 41: 135-144.

31. National Institute of Standards and Technology (NIST) (2014) Thermo Data Engine. Thermodynamics Research Center.

32. Austgen DM (1989) A Model of Vapour – Liquid Equilibria for Acid Gas – Alkanolamine – Water Systems [dissertation]. Austin (TX): University of Texas at Austin.

33. Silkenbäumer D, Rumpf B, Lichtenthaler RN (1998) Solubility of Carbon Dioxide in Aqueous Solutions of 2-Amino-2-methyl-1-propanol and N-Methyldiethanolamine and Their Mixtures in Temperature Range from 313 to 353 K and Pressure up to 2.7 MPa. Ind Eng Chem Res 37: 3133-3141.

34. Crovetto R (1991) Evaluation of Solubility Data of the System CO2 – H2O from 273 K to the Critical Point of Water. J Phys Chem Ref Data 20: 575-589.

35. D"Ans J, Bartels J, Bruggencate Pt, Eucken A, Joos G, et al. (1968) Landolt-Börnstein. Zahlenwerte und Funktionen aus Physik, Chemie, Astronomie, Geophysik und Technik. Heidelberg: Springer: 6.

36. Addicks J, Owren GA, Fredheim AO, Tangvik K (2002) Solubility of Carbon Dioxide and Methane in Aqueous Methyldiethanolamine Solutions. J Chem Eng Data 47: 855-860.

37. Maham Y, Hepler LG, Mather AE, Hakin AW, Marriott RA (1997) Molar heat capacities of alkanolamines from 299.1 to 397.8 K. J Chem Soc Faraday Trans 93: 1747-1750.

38. Maurer G (1980) On the Solubility of Volatile Weak Electrolytes in Aqueous Solutions. In: Newman SA. Thermodynamics of Aqueous Systems with Industrial Applications. (133rd edn.) ACS Symposium Series, American Chemical Society, Wasington DC, USA.

39. Edwards TJ, Newman J, Prausnitz JM (1978) Thermodynamics of Aqueous Solutions Containing Volatile Weak Electrolytes. AIChJ 24: 966-976.

40. Dingman, JC, Jackson JL, Moore TF, Branson JA (1983) Equilibrium Data For The H2S-CO2-Diglycolamine Agent -Water System. San Franciso.

23

Production of Biodesiel from Animal Tallow via Enzymatic Transesterification using the Enzyme Catalyst Ns88001 with Methanol in a Solvent-Free System

Kumar S, Ghaly AE * and Brooks MS

Department of Process Engineering and Applied Science, Dalhousie University, Halifax, Nova Scotia Canada

Abstract

The effectiveness of enzymatic transesterification of animal fat using the experimental enzyme catalyst NS88001 with no solvent was studied. The effects of oil:alcohol molar ratio (1:1, 1:2, 1:3, 1:4 and 1:5), reaction temperature (35, 40, 45 and 50°C) and reaction time (4, 8, 12 and 16 h) on the biodiesel yield were evaluated. The highest biodiesel yield was obtained at the 1:4 molar ratios. No reactions were observed with the 1:1 and 1:2 (oil:alcohol) molar ratios and increasing the oil:alcohol molar ratio above 1:4 decreased the biodiesel conversion yield. The rate of conversion of fatty acid esters increased with increases in the reaction time. The reaction proceeded slowly at the beginning and then increased rapidly due to the initial mixing and dispersion of alcohol into the oil substrate and activation of enzyme. After dispersion of alcohol, the enzyme rapidly interacted with fatty acids esters giving a maximum conversion yield. Increasing the reaction time from 4 to 16 h increased the conversion yield of biodiesel by 114.95-65.59%. The interactions between enzyme polymer surface and substrate appears to be dependent on reaction temperature due to hydrogen bonding and ionic interactions which play an important role in maintaining the thermostability of lipase in the system. The optimum reaction temperature for the experimental enzyme catalyst (NS88001) in the solvent free system was 45°C. Increasing the reaction temperature from 40 to 45°C increased the biodiesel conversion yield while higher temperatures above 45°C denatured the specific structure of enzymes and resulted in decreased methyl esters formation. The activity of experimental enzyme catalyst NS88001 in the presence of methanol without solvent at the optimum conditions (a reaction temperature of 45°C, an oil:alcohol molar ratio of 1:4 and a reaction time of 16 h) remind relatively constant for 10 cycles and then decreased gradually reaching zero after 50 cycles.

Keywords: Animal tallow; Transesterification; Enzyme NS88001; Biodiesel; Temperature; Time; Oil:alcohol molar ratio; Solvent-free system

Introduction

The high demand for fossil fuels, their limited and insecure supply and their environmental impact prompted the search for alternative renewable fuel sources such as biodiesel from biomass materials [1-3]. Biodiesel is a renewable energy source that can be used as a fuel in compression-ignition engines instead of diesel [4,5]. It is sulfur free, non-toxic and biodegradable [6]. These characteristics make it more greener and eco-friendly than diesel [7-11].

Biodiesel can be produced from many raw materials including plant oils (jatropha, canola, coconut, cottonseed, groundnut, karanj, olive, palm, peanut, rapeseed, safflower, soybean and sunflower), animal fats (beef tallow, chicken fat, lamb fat, pig lard, yellow grease, waste cooking oil and the greasy by-product from omega-3 fatty acid production) and algae biomass [4,8,11-14]. The main component of fats and oils are triacylglycerols (triglycerides) which are made of different types of fatty acids with one glycerol (glycerine) being the backbone. The types of fatty acids present in the triglycerides determine the fatty acids profile. Fatty acid profiles from plants and animal sources are different and each fatty acid has its own chemical and physical properties which can be a major factor influencing the properties of biodiesel.

Transesterification is a chemical process used to convert oils and fats to biodiesel. A short chain alcohol is used with the feedstock to convert it to methyl esters and glycerin. The process is achieved with one of three catalysts: acid, alkali or enzyme. With an acid catalyst, the proton is donated to the carbonyl group which makes it more reactive while with a base catalyst, the proton is removed from alcohol which makes the reactants more reactive [15]. Base catalysts are widely in use by the biodiesel industry. However, both acid and alkali methods require more energy and a downstream processing step for removing the by-product glycerin. An enzymatic catalyst cleaves the backbone of the glycerol which makes the reactants more reactive, thereby giving the product without the need for a costly downstream processing step.

Objectives

The aim of this study was to investigate the effectiveness of the enzymatic transesterification process for the production of biodiesel from animal tallow using methanol in a solvent free system. The specific objectives were: (a) to study the effectiveness of experimental enzyme catalyst (NS88001) with methanol in a solvent-free system, (b) to study the effects of alcohol feedstock ratio (1:1, 1:2, 1:3, 1:4 and 1:5), reaction temperature (35, 40, 45 and 50°C) and reaction time (4, 8, 12 and 16 h) on the biodiesel yield and (c) to evaluate the usability of the enzyme.

Materials and Methods

Animal tallow

The animal rendering waste used in the study was obtained as beef tallow from the Company S.F Rendering, Centreville, Nova Scotia. Samples (10 Kg) were collected and stored at - 20°C in the Biotechnology Laboratory of Dalhousie University, Halifax, Nova Scotia. The sample material was yellowish in colour.

Chemicals and enzymes

The immobilized Lipase was an experimental enzyme catalyst (NS88001) obtained from Novozyme (Franklinton, North Carolina, USA). The chemicals used in the study included: methanol, tertrahydrofuran, N, O-Bis (Trimethylsilyl)-trifluroacetamide (BSTFA)

*Corresponding author: Ghaly A, Professor, Department of Process Engineering and Applied Science, Faculty of Engineering, Dalhousie University, Halifax, Nova Scotia, Canada, E-mail: abdel.ghaly@dal.ca

and hilditch reagent. They were purchased from Sigma Aldrich (St. Louis, Missouri, USA). The FAME standards, which included methyl myristate, methyl pentadecanote, methyl cis-11-eicosenoate, methyl all-cis-5,8,11,14,17- eicosapentaenoate (EPA), methyl erucate, methyl all-cis-7,10,13,16,19-docosapentaenoate (DPA) and methyl all-cis-4,7,10,13,16,19-docosahexenoate (DHA), were purchased from Sigma Aldrich (St. Louis, Missouri, USA). The other FAME standard, which included methyl palmitate, methyl palmitoleate, methyl stearate, methyl oleate, methyl linoleate and methyl linolenate, were purchased from Alltech Associates, Inc. (Deerfield, Illinois, USA). The FAME standard methyl-stearidonate was purchased from Cayman Chemical (Ann Arbor, Michigan, USA).

Experimental procedure

Purification of crude animal tallow: The animal tallow was first heated to 110°C with constant stirring at 50 rpm in a round bottom flask for one hour. During the process of melting the fats, the top layer consisting of bubbles and impurities was discarded regularly. The extracted crude oil from animal tallow filtered four times using vacuum filtration with ultra-filter paper (Whatman No.40, Fisher Scientific, Toronto, Ontario, Canada). The oil percentage was calculated as follows

$$Percent\ Oil = \left(\frac{Weight\ of\ oil}{Weight\ of\ Total\ fat} \right) \times 100 \qquad (1)$$

Enzymatic transesterification: The enzymatic transesterification of biodiesel was carried out in order to extract fatty acid methyl esters from the animal fats by the experimental enzyme catalyst NS88001 according to the procedure described in Figure 1. Five oil:alcohol molar

ratios (1:1, 1:2, 1:3, 1:4 or 1:5), four reaction temperatures (35, 40, 45 or 50°C) and four reaction times (4, 8, 12 or 16 hours) were investigated. No solvent was used.

The homogenized oil (2.3 ml corresponding or 2 g of fat) was placed into a 50 ml conical flask and heated on a hot plate (PC-620, Corning, New York, New York, USA). The experimental enzyme catalyst NS88001 (25% of the oil weight or 0.5 g) was added to the flask. The appropriate amount of alcohol (Methanol) was added based on the oil:alcohol molar ratio (1:1, 1:2, 1:3, 1:4 or 1:5). The solution was mixed using a reciprocal shaking bath (2850 Series, Fisher Scientific, Toronto, Ontario, Canada) at 200 rpm. The desired temperature (35, 40, 45 or 50°C) was selected. After the desired reaction time was completed (4, 8, 12 or 16 hours), the enzyme was filtered by vacuum filtration as recommended by Nelson et al. [19] Samples (100 μl) were taken from the mixture and analyzed using a gas chromatography system (Hewlett Packard 5890 series II, Agilent, Mississauga, Ontario, Canada). The same procedure was repeated with all oil:alcohol molar ratios, reaction temperatures and reaction times.

Determination of Biodiesel Yield: The preparation steps for the gas chromatography analysis of the biodiesel samples are shown in Figure 2. A 100 μL aliquot was taken from the transesterification process at selected time intervals (4, 8, 12 and 16 hours) and flushed with nitrogen gas in a reciprocating water bath (280 series, Fisher Scientific, Toronto, Ontario, Canada) at 45°C. A 10 mg portion of the residue was dissolved in 100 μL of tertrahydrofuran and 200 μL of BSTFA. Then, the mixture was heated in a microprocessor-controlled water bath (280 series, Fisher Scientific, Toronto, Ontario, Canada) at 90-95°C for 15 minutes.

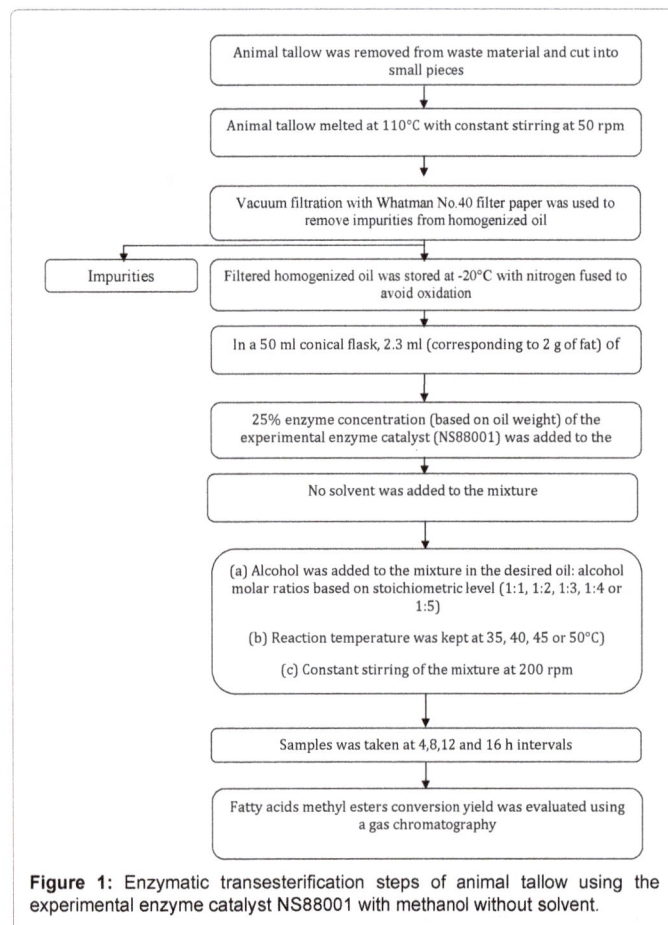

Figure 1: Enzymatic transesterification steps of animal tallow using the experimental enzyme catalyst NS88001 with methanol without solvent.

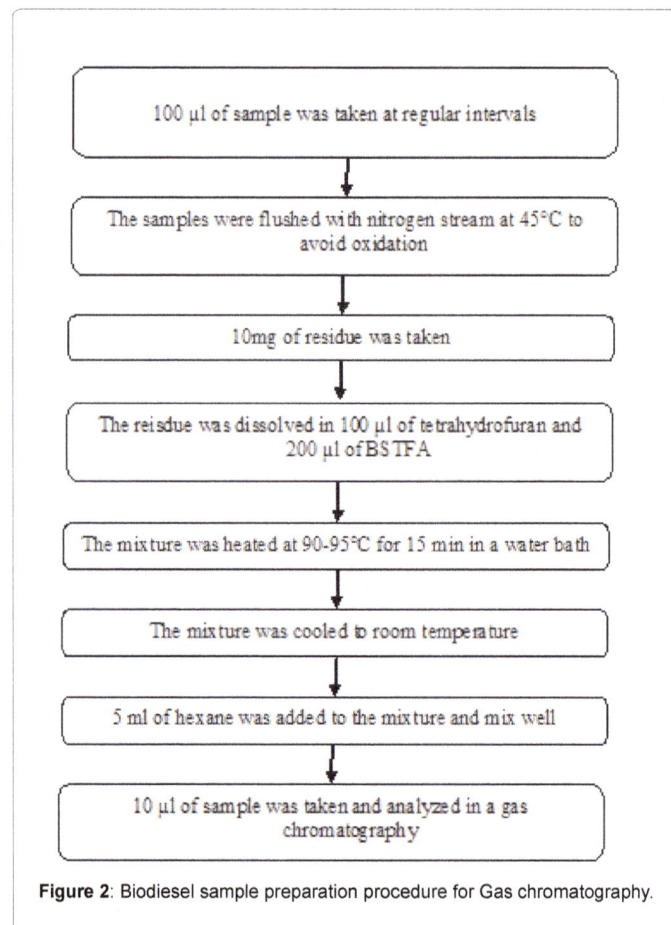

Figure 2: Biodiesel sample preparation procedure for Gas chromatography.

The sample was then cooled to room temperature for few minutes after which 5 mL of hexane was added. An aliquot of 1.5 mL mixture was transferred to the GC crimp vials and capped tightly for further analysis using GC.

An aliquot of 10 μL of the mixture was separated by fatty acid class (methyl ester, MAG, DAG and TAG) based on the carbon atom by a gas chromatography system, coupled with flame ionization detector (FID) (HP5890 Series II, Agilent Technologies, Mississauga, Ontario, Canada). An AT-FAME capillary column, 30 m in length, 0.32 mm of internal diameter and 0.25 μm film thicknesses, (Alltech Associates, Inc., Deerfield, Illinois, USA) was used for analyses. The column is a highly polar and stable bonded polyethylene glycol phase. The separated samples were injected directly into the column with the initial oven temperature of 60°C, followed by a flow rate of 20°C/min. A final temperature of 280°C was held for 10 minutes. The detection system was equipped with a flame ionization detector (FID) operating at 275°C with helium as a carrier gas at a flow rate of 0.6 mL/min. The total run time was 40 minutes.

$$\text{Conversion yield}\,(\text{wt \%}) = \frac{Peak\,area\,A \times 100}{\sum (Peak\,area\,A + Peak\,area\,B + \ldots + Peak\,area\,N)} \quad (2)$$

Statistical analyses: Statistical analyses were performed on the results using Minitab Statistics Software (Ver 16.2.2, Minitab Inc., State College, Pennsylvania, USA). Both analysis of variance (ANOVA) and Tukey's grouping were carried out.

Results

Characterization of animal tallow

Table 1 shows the composition of the animal tallow used in this study. The filtration process removed about 7.5 % of the total weight of tallow as impurities present in the animal fats. The homogenized oil was characterized by gas chromatography to identify and quantify the fatty acid composition of the tallow. Five fatty acids were identified in the animal tallow: oleic acids (44%), palmitic acids (28%), stearic acids (26%), linoleic acids (1%), and myristic acids (1%).

Enzymatic transesterification

Enzymatic transesterification by the experimental enzyme catalyst (NS88001) was carried out to investigate the effects of reaction time (4, 8, 12 and 16 h), oil:alcohol molar ratios (1:1, 1:2, 1:3, 1:4 and 1:5) and reaction temperature (35, 40, 45 and 50°C) on biodiesel yield in a solvent free system. The results are shown in Table 2.

Table 3 shows the analysis of variance performed on the biodiesel yield data. The effects of oil:alcohol molar ratio, reaction time and reaction temperature were highly significant at the 0.001 level. All

Parameters	Value
Impurities (Kg)	0.375
Oil (%)	92.5
Impurities (%)	7.5
Fatty acids (wt%)	
Oleic acid	44
Palmitic acid	28
Stearic acid	26
Linoleic acid	1
Myristic acid	1

Tallow Sample Size =5 kg

Table 1: Composition of animal tallow.

interactions (two, three and four way interactions) between the parameters were also highly significant at the 0.001 level.

The results obtained from Tukey's Grouping (Table 4) indicated that the three levels of oil:alcohol molar ratio (1:3, 1:4 and 1:5) were significantly different from one another at the 0.05 level. Two levels of alcohol: oil molar ratios (1:1 and 1:2) did not produce any biodiesel. The highest mean biodiesel yield of 80.42% was obtained with the 1:4 oil:alcohol molar ratio. The four reaction times (4, 8, 12 and 16 h) were significantly different from one another at the 0.05 level. The highest mean biodiesel yield of 49.00% was achieved with 16 hour reaction time. The three reaction temperatures (40, 45 and 50°C) were significantly different from each other at the 0.05 level. The highest mean biodiesel yield of 48.56% was obtained at the reaction temperature 45°C.

Effect of Oil:alcohol molar ratio

Figure 3 shows the effect of oil:alcohol molar ratio on the biodiesel yield using the experimental enzyme catalyst (NS88001) at different reaction temperatures and reaction times in a solvent free system. Generally, there was an increase in the biodiesel yield with increases in the oil:alcohol molar ratios from above 1:2 to 1:4 followed by decrease in the biodiesel yield with a further increase in the oil : alcohol ratios from 1:4 to 1:5 for all reaction times (4, 8, 12 and 16 h) and reaction temperatures (40, 45 and 50°C). No reaction was observed at 1:1 and 1:2 oil:alcohol molar ratios.

The biodiesel yield at the 4 h reaction time increased from 72.80 to 77.86% (6.95%), from 74.16 to 80.1% (8.00%) and from 38.4 to 58.4% (52.08%) with increases in the oil:alcohol molar ratios from 1:2 to 1:4 for the reaction temperatures of 40, 45 and 50°C, respectively. A further increase in the oil:alcohol from 1:4 to 1:5 decreased the biodiesel yield from 77.86 to 47.69% (-38.74%), from 80.1 to 59.74% (-25.41%) and from 58.4 to 32.6% (-44.17%) for the reaction temperatures of 40, 45 and 50°C, respectively. Similar trends were observed with the 8, 12 and 16 h reaction times at all reaction temperatures.

Effect of reaction time

Figure 4 shows the effect of reaction time on the biodiesel yield using the experimental enzyme catalyst (NS88001) at different reaction temperatures, oil:alcohol molar ratios and reaction times in the solvent free system. Generally, there was an initial rapid increase in the biodiesel conversion yield with increases in the reaction time during the first 4 hours followed by a slow gradual increase thereafter (between 4 and 16 h) for all reaction temperatures (40, 45 and 50°C) and the oil:alcohol molar ratios of 1:3, 1:4 and 1:5.

The biodiesel conversion yield at the 40°C reaction temperature and 4 h reaction time reached 72.80%, 77.86% and 47.69% for the oil:alcohol molar ratios of 1:3, 1:4 and 1:5, respectively. No reaction was observed at the 1:1 and 1:2 oil:alcohol molar ratio at the 40°C reaction temperature. Further increases in reaction time from 4 h to 16 h increased the biodiesel yield from 72.80 to 83.05% (14.95%), from 77.86 to 87.09% (11.85%) and from 47.69 to 78.97% (65.59%) the oil:alcohol molar ratios of 1:3, 1:4 and 1:5, respectively. Similar trends were observed at the 45 and 50°C reaction temperatures for the oil:alcohol molar ratios of 1:3, 1:4 and 1:5.

Effect of reaction temperature

Figure 5 shows the effect of reaction temperature on the biodiesel conversion yield using the experimental enzyme catalyst (NS88001) at different reaction times, reaction temperatures and oil:alcohol molar ratios in the solvent free system. There was an increase in biodiesel yield

Time Time (h)	Oil:alcohol Molar Ratio	Reaction Temperature (°C)		
		40	45	50
4	1:1	Not extractable	Not extractable	Not extractable
	1:2	Not extractable	Not extractable	Not extractable
	1:3	72.80 ± 1.46	74.16 ± 1.48	38.40 ± 1.77
	1:4	77.86 ± 1.56	80.10 ± 1.60	58.40 ± 1.17
	1:5	47.69 ± 0.95	59.74 ± 1.19	32.60 ± 0.65
8	1:1	Not extractable	Not extractable	Not extractable
	1:2	Not extractable	Not extractable	Not extractable
	1:3	77.58 ± 1.55	80.22 ± 1.60	46.60 ± 0.93
	1:4	82.46 ± 1.65	84.85 ± 1.70	67.90 ± 1.36
	1:5	75.05 ± 1.50	76.23 ± 1.52	40.10 ± 0.80
12	1:1	Not extractable	Not extractable	Not extractable
	1:2	Not extractable	Not extractable	Not extractable
	1:3	77.92 ± 1.56	82.10 ± 1.64	57.90 ± 1.16
	1:4	83.40 ± 1.67	87.67 ± 1.75	77.08 ± 1.54
	1:5	78.61 ± 1.57	78.97 ± 1.58	53.51 ± 1.07
16	1:1	Not extractable	Not extractable	Not extractable
	1:2	Not extractable	Not extractable	Not extractable
	1:3	83.05 ± 1.66	93.16 ± 1.86	62.81 ± 1.26
	1:4	87.09 ± 1.74	94.04 ± 1.88	84.19 ± 1.68
	1:5	78.97 ± 1.58	80.76 ± 1.62	70.98 ± 1.42

Table 2: Biodiesel yield (wt%) from animal tallow using 0.5 grams of experimental enzyme catalyst (NS88001) with methanol without hexane at different reaction times, oil : alcohol molar ratios and reaction temperatures.

Source	DF	SS	MS	F	P
Total	179	248498.3			
Model					
MR	4	227313.9	56828.5	57224.62	0.001
RTI	3	3989.0	1329.7	1338.95	0.001
RTE	2	6750.7	3375.3	3398.87	0.001
MR*RTI	12	3355.6	279.6	281.58	0.001
MR*RTE	8	5533.5	691.7	696.51	0.001
RTI*RTE	6	500.8	83.5	84.06	0.001
MR*RTI*RTE	24	935.5	39.0	39.25	0.001
Error	120	119.2	1.0		

DF: Degree of freedom; SS: Sum of square; MS: Mean of square; MR: Molar Ratios; RTI: Reaction Time; RTE: Reaction Temperature
R^2=99.95%

Table 3: ANOVA of biodiesel yield.

Factors	Level	N	Mean Yield (%)	Tukey Grouping
Oil:alcohol Molar Ratios	1:1	36	0.00	A
	1:2	36	0.00	A
	1:3	36	70.55	B
	1:4	36	80.42	C
	1:5	36	64.36	D
Reaction Time (h)	4	45	36.11	A
	8	45	42.06	B
	12	45	45.09	C
	16	45	49.00	D
Reaction Temperature (°C)	40	60	46.12	A
	45	60	48.56	B
	50	60	34.52	C

Groups with the same letter are not significantly different from one another at the 0.05 level.

Table 4: Tukey's grouping of the biodiesel yield.

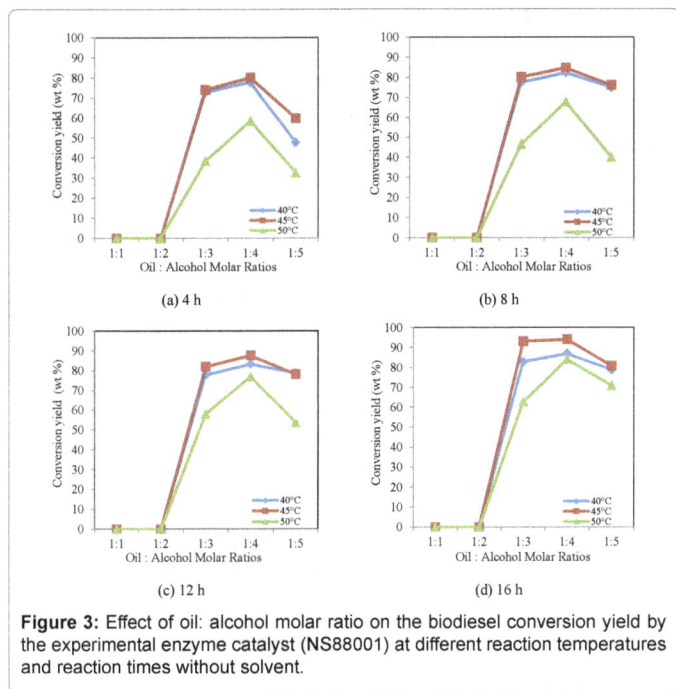

Figure 3: Effect of oil: alcohol molar ratio on the biodiesel conversion yield by the experimental enzyme catalyst (NS88001) at different reaction temperatures and reaction times without solvent.

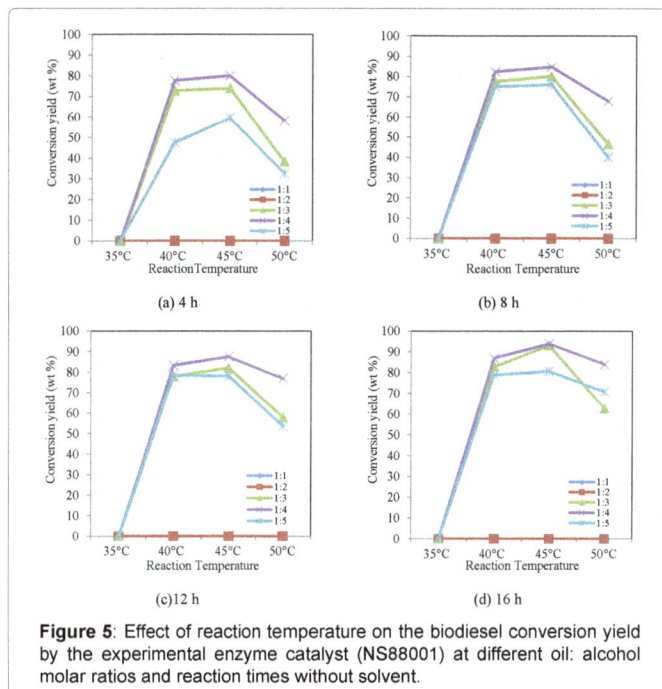

Figure 4: Effect of reaction time on the biodiesel conversion yield by the experimental enzyme catalyst (NS88001) at different reaction temperatures and oil: alcohol molar ratios without solvent (R= oil: alcohol molar ratios).

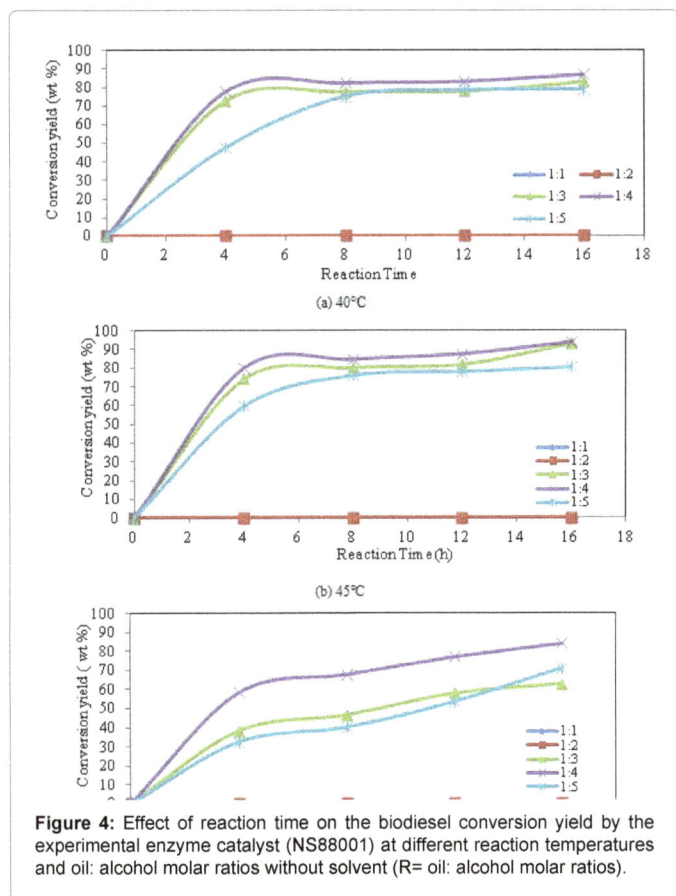

Figure 5: Effect of reaction temperature on the biodiesel conversion yield by the experimental enzyme catalyst (NS88001) at different oil: alcohol molar ratios and reaction times without solvent.

(4, 8, 12 and 16 hrs) and oil:alcohol molar ratios (1:1, 1:2, 1:3, 1:4 and 1:5).

The biodiesel yield at the 4 h reaction time increased from 72.80 to 74.16% (1.86%), from 77.86 to 80.10% (2.87%), from 47.69 to 59.74% (25.26%) for the oil:alcohol molar ratios of 1:3, 1:4 and 1:5, respectively. No reactions were observed with the 1:1 and 1:2 oil:alcohol molar ratios at the 4 h reaction time. A further increase in the reaction temperature from 45 to 50°C decreased the biodiesel yield from 74.16 to 38.40% (-48.22%), from 80.10 to 58.40% (-27.09%), from 59.74 to 32.60% (-45.76%) for the oil:alcohol molar ratios of 1:3, 1:4 and 1:5, respectively. Similar trends were observed with the 8, 12 and 16 h reaction times for the oil:alcohol molar ratios of 1:3, 1:4 and 1:5. No reactions were observed with the 1:1 oil:alcohol molar ratio for the 8, 12 and 16 h reaction times.

Discussion

Extraction profiles of the raw material

After melting and homogenizing the animal tallow, the impurities (7.5%) were removed by filtration. The fatty acids analysis indicated that the homogenized oil contained high percentages of oleic acid (44%), palmitic acid (28%) and stearic acid (26%) as well as lower percentages of myristic acid (1%) and linoleic acid (1%). A high concentration of oleic acid improves the characteristics of biodiesel resulting in a high cetane index and combustion temperature [11]. Biodiesel produced from feedstocks containing a high level of oleic acid showed similar characteristics to these of conventional diesel [5,11]. Therefore, the biodiesel produced from oil extracted from animal tallow is expected to have good characteristics as a biofuel.

The extracted oil from animal tallow can be transformed to biodiesel by chemical or enzymatic transesterification. Watanabe et al. [16], Dorado et al. [17] and Kulkarni and Dalai [18] reported that oxidized oil can inhibit the chemical transesterification process and increase the oxidation of methyl esters. Kulkarni and Dalai [18] stated that an increase in the oxidation of methyl esters might increase the cetane number which tends to delay the ignition time in the engine. However,

when the reaction temperature was increased from 40 to 45°C followed by a decrease in the biodiesel conversion yield when the reaction temperature was further increase from 45 to 50°C for all reaction times

Nelson et al. [19] and Watanabe et al. [16] reported that oxidation in crude tallow or oil containing high free fatty acids is a common problem and no negative effects of the oxidized oil substrate on the enzymatic transesterification process was observed. Kulkarni and Dalai [18] found that oxidized oil did not inhibit the formation of methyl esters from the methanolysis process by *Candida antarctica* lipase. Watanabe et al. [16] stated that in the enzymatic process, the oxidized substrate becomes a non-recognition site for the enzyme to bind and the process continues with the substrates which are not oxidized. The authors stated that using oxidized oil might reduce the biodiesel stability. Nelson et al. [19] reported that the stability of biodiesel can be increased by blending the biodiesel with conventional diesel especially in cold environment. In this study, enzymatic transesterification was carried out and no oxidation stability test was performed on crude tallow or oil nor was antioxidants used.

Effect of oil:alcohol molar ratios

Reports in the literature suggested that the theoretical stoichiometric oil:alcohol molar ratio of 1:3 is needed to complete the reaction in the following continuous steps: (a) conversion of triglycerides to diglycerides, (b) conversion of diglycerides to monoglycerides and (c) conversion of monoglycerides to methyl esters and glycerol [12,20,21]. In this study, increasing in the oil:alcohol molar ratio from 1:1 to 1:4 at the 4 h reaction time with no solvent increased the biodiesel yield by the experimental enzyme catalyst NS88001 by 6.95, 8.00 and 52.08% and when the oil:alcohol molar ratio was further increased from 1:4 to 1:5, the biodiesel yield was decreased by 38.74, 25.41 and 44.10% at the reaction temperatures of 40, 45 and 50°C, respectively. Similar trends were seen with all reaction times.

Kumari et al. [22] reported that the biodiesel yield increased when the oil:alcohol molar ratio was increased up to 1:4 and then decreased when the oil:alcohol molar ratio was further increased to 1:5. Chen et al. [23] reported that increasing the oil:alcohol molar ratio from 1:1 to 1:4 promoted the methanolysis reaction with waste cooking oil, but the formation of methyl esters decreased when the oil:alcohol molar ratios was increased from 1:4 to 1:5 due to an excess of methanol in the system. The decreases in the formation of methyl esters observed in these studies were similar to that observed in the present study.

The decrease in the conversion yield of methyl esters from oil at higher oil: alcohol molar ratios might be due to the presence of insoluble methanol in the reaction system which may have deactivated the experimental enzyme catalyst (NS88001). Tamalampudi et al. [24] suggested that the presence of soluble methanol would cause the active site on the surface of the lipase to be locked resulting in less access of enzyme to the surface of oil substrate. Dizge and Keskinler [25] reported that the use of excessive amount of methanol might deactivate the lipase in the reaction. Nelson et al. [19] and Bernardes et al. [26] stated that it is likely that once the maximum level of esters is formed, a further increase in number of moles of alcohol decreases the formation of methyl esters in the reaction due to enzyme inactivation. Chen et al. [23] reported that the excess methanol distorted the essential water layer needed to stabilize the structure of the enzyme. Chen and Wu [27] and Samukawa et al. [28] stated that short chain alcohols such as methanol are responsible for deactivation and inhibition of immobilized lipase. Salis et al. [29] and Al-zuhair et al. [30] reported that deactivation of enzyme occurred by the insoluble alcohol present in the reaction due to its tendency to be absorbed by the surface support matrix. They also indicated that exceeding the stoichiometric oil:alcohol molar ratio of 1: 3 ensures proper rate of reaction and leads to higher biodiesel yield but the excess amount of the alcohol might decrease the activity and

distort the spatial confirmation of lipase structure and cause the lipase to deactivate.

In this study, the highest biodiesel conversion yield of 94.04% was achieved using the experimental enzyme catalyst 88001 at the 45°C reaction temperature with the 1:4 oil:alcohol molar ratio and the 16 h reaction time.

Effect of reaction time

When the reaction time was increased from 4 to 16 h at the 40°C reaction temperature, the increases in biodiesel conversion yield by the experimental enzyme catalyst NS88001 in the solvent free system were 14.95, 11.88 and 65.89% for the oil:alcohol molar ratios of 1:3, 1:4 and 1:5, respectively. No reactions were observed with the 1:1 and 1:2 oil:alcohol molar ratios. Several researchers observed similar trends from crude tallow, waste cooking oil and vegetable oil.

Nelson et al. [19] reported a maximum biodiesel yield of 83.8 % after 16 h with the 1:3 molar ratio using 25% concentration of the enzyme *Candida antarctica* (SP 435). Chen et al. [23] achieved a maximum biodiesel yield of 85.12% after 30 h with the 1:4 oil:alcohol molar ratio using 30% concentration of the immobilized enzyme *Rhizopus oryzae* and waste cooking oil as a substrate. Modi et al. [31] reported that a maximum biodiesel conversion yield of 93.4% was achieved after 8 h with the 1:4 oil:alcohol molar ratio using the enzyme *Candida antarctica* (Novozyme 435) with vegetable oil. A high biodiesel yield (94.04%) was obtained with the 16 h reaction time in this study, which indicates that NS88001 is non-regiospecific.

At the initial phase of the reaction, the enzymes, oil and alcohol appeared to be static and the reaction started when the stirring speed reached 200 rpm which increased the mass transfer between the substrate and enzyme catalyst. Formation of esters increased with increases in reaction time from 1 to 4 h. Several authors [20,32-35] reported that the rate of conversion of fatty acid esters increased with increases in reaction time and the reaction proceeds rapidly due to the initial mixing and dispersion of alcohol into the oil substrate and the activation of enzyme. After the alcohol is dispersed, it rapidly interacts with fatty acids giving a maximum conversion yield. Kose et al. [36] and Li et al. [37] reported that the initial reaction in a solvent-free system might take a longer period to activate the enzyme in the system. However, a further increase in the reaction time may decrease conversion yield due to the backward reaction of transesterification [23].

Effect of reaction temperature

In this study, when the reaction temperature was increased from 40 to 45°C at the 4 h reaction time for the experimental enzyme catalyst 88001 with no solvent, the increases in biodiesel yield were 1.86, 2.87 and 26.64 % and when the reaction temperature was further increased from 45 to 50°C, the biodiesel yield decreased by 48.22, 27.09 and 45.76% for the 1:3, 1:4 and 1:5 oil:alcohol molar ratios, respectively. No reactions were observed with the 1:1 and 1:2 oil:alcohol molar ratios at the 4 h reaction time. Similar tends were observed with other reaction times. Several researchers observed similar trends from crude tallow, waste cooking oil, canola oil and soybean oil.

Chen et al. [23] reported that the biodiesel yield increased (reaching a maximum of 87%) when the reaction temperature was increased from 30 to 40°C and then decreased when the reaction temperature was further increased from 40 to 70°C during conversion of waste cooking oil to methyl esters using Lipozyme RM IM. Dizge and Keskinler [25] reported that the biodiesel yield increased (reaching a maximum of 85.8%) when the reaction temperature was increased from 30 to 40°C and then decreased when the reaction temperature was further increased from 40 to 70°C when converting canola oil to methyl esters using Lipozyme TL. Rodrigues et al. [38] reported that a maximum biodiesel yield of 53% was achieved at 35°C which then decreased with increases in reaction temperature above 35°C during conversion of soybean oil to methyl esters using Novozyme 435. Nie et al. [39] reported that a maximum biodiesel yield of 90% was obtained at 40°C and increasing the reaction temperature above 40°C decreased the biodiesel yield. In this study, the highest conversion yield (94.04) was obtained at 45°C which was higher than those reported in the literature.

Increasing the reaction temperature from 40 to 45°C reduced the viscosity of the oil and enhances the mass transfer between substrate and enzyme catalyst. Due to this effect, an increase in conversion yield of biodiesel was obtained. However, when the reaction temperature was further increased from 45 to 50°C the biodiesel yield decreased. A higher temperature may denature the specific structure of enzymes resulting in a decreased methyl esters formation. Denaturation of enzyme support matrix may also promote the enzyme leakage from the outer layer of the support matrix. However, the optimum reaction temperature is dependent on other parameters such as oil:alcohol molar ratio, enzyme activity, stability and type of system used. Reetz et al. [40], Kumari et al. [22] and Antczak et al. [14] reported that interactions between enzyme polymer surface and substrate appears to be dependent on reaction temperature due to hydrogen bonding and ionic interactions which play important roles in maintaining the thermostability of lipase in the system. Kose et al. [36] reported that increasing the reaction temperature over 50°C in a solvent free-system decreased the biodiesel yield of methyl esters due to inhibition of enzyme activity by higher temperature. Nie et al. [39] reported that higher temperature can give faster reaction but exceeding the optimum temperature may lead the enzyme denaturing.

Enzyme usability

In this study, the activity of experimental enzyme catalyst NS88001 in the presence of methanol without solvent at the optimum conditions (a reaction temperature of 45°C, an oil:alcohol molar ratio of 1:4 and a reaction time of 16 h) remind relatively constant for 10 cycles and then decreased gradually reaching zero after 50 cycles. Ghamgui et al. [41], Xu et al. [42], and Bernardes et al. [26] obtained similar results from immobilized Lipozyme *Thermomyces lanuginosus*, immobilized Lipozyme *Rhizomucor miehei* and immobilized *Rhizopus oryzae*. Several researchers [41,43,44] stated that repeated use of enzyme in the reaction without removing glycerol from the system might inhibit the interaction between the substrate and lipase.

Conclusions

The effectiveness of enzymatic transesterification of animal fat using the experimental enzyme catalyst NS88001 with no solvent was studied. The effects of oil:alcohol molar ratio (1:1, 1:2, 1:3, 1:4 and 1:5), reaction temperature (35, 40, 45 and 50°C) and reaction time (4, 8, 12 and 16 h) on the biodiesel yield were evaluated. The effects of oil:alcohol molar ratio, reaction time and reaction temperature on the biodiesel yield were highly significant at the 0.001 level. There were also significant interactions among the parameters at the 0.001 level. The highest biodiesel yield was obtained at the 1:4 molar ratio. No reactions were observed with the 1:1 and 1:2 oil:alcohol ratios and increasing the oil:alcohol molar ratio above 1:4 decreased the biodiesel conversion yield. The rate of conversion of fatty acid esters increased with increases in reaction time. The reaction proceeds slowly at the beginning and

then increased rapidly due to the initial mixing and dispersion of alcohol into the oil substrate and activation of enzyme. After dispersion of alcohol, the enzyme rapidly interacted with fatty acids esters giving a maximum conversion yield. Increasing the reaction time from 4 to 16 h increased the conversion yield of biodiesel by 14.95-65.59%. The interactions between enzyme polymer surface and substrate appears to be dependent on reaction temperature due to hydrogen bonding and ionic interactions which play an important role in maintaining the thermostability of lipase in the system. The optimum reaction temperature for the experimental enzyme catalyst (NS88001) in the solvent free system was 45°C. Increasing the reaction temperature from 40 to 45°C increased the biodiesel conversion yield while higher temperatures above 45° C denatured the specific structure of enzymes and resulted in decreased methyl esters formation. The activity of experimental enzyme catalyst NS88001 in the presence of methanol without solvent at the optimum conditions (a reaction temperature of 45°C, an oil:alcohol molar ratio of 1:4 and a reaction time of 16 h) remind relatively constant for 10 cycles and then decreased gradually reaching zero after 50 cycles.

Acknowledgement

The research was supported by the Natural Sciences and Engineering Research Council (NSERC) of Canada.

References

1. Saka S, Kusidana D (2001) Biodiesel fuel from rapeseed oil as prepared in supercritical methanol. Fuel 80: 225-231.

2. Xie W, Li H (2006) Alumina-supported potassium iodide as a heterogenous catalyst for biodiesel production frok soybean oil. J Mol Catal A: Chem 225: 1-9.

3. Kumar RR, Rao HP, Muthu A (2014) Lipid extraction methods from microalgae: A comprehensive review. Frontiers Energy Res 2: 61.

4. Demirbas A (2003) Biodiesel fuels from vegetable oils via catalytic and non-catalytic supercritical alcohol transesterifications and other methods, a survey. Energy Conver Manage 44: 2093-2109.

5. Knothe G (2005) Dependence of biodiesel fuel properties on the structure of fatty acid alkyl esters. Fuel Process Technol 86: 1059-1070.

6. Venkataraman C, Negi G, Sardar SB, Rastogi R (2002) Size distributions of polycyclic aromatic hydrocarbons in aerosol emissions from biofuel combustion. J Aerosol Sci 33: 503-518.

7. Bondioli P, Gasparoli A, Lanzani A, Fedeli E, Veronese S, et al (1995) Storage stability of biodiesel. J Am Oil Chem Soc 72: 699-702.

8. Akoh CC, Chang S, Lee G, Shaw J (2007) Enzymatic approach to biodiesel production. J Agric Food Chem 55: 8995-9005.

9. Basha S A, Gopal KR, Jebaraj S (2009) A review on biodiesel production, combustion, emissions and performance. Renew Sustain Energy Rev 13:1628-1634.

10. Shafiee S, Topal E (2009) When will fossil fuel reserves be diminished?. Energy Policy 37: 181-189.

11. Robles-Medina A, Gonzalez-Moreno PA, Esteban-Cerdán L, Molina-Grima E (2009) Biocatalysis: Towards ever greener biodiesel production. Biotechnol Adv 27: 398-408.

12. Marchetti JM, Miguel VU, Errazu AF (2008) Techno-economic study of different alternatives for biodiesel production. Fuel Process Technol 89: 740-748.

13. Ranganathan SV, Narasimhan SL, Muthukumar K (2008) An overview of enzymatic production of biodiesel. Bioresour Technol 99: 3975-3981.

14. Antczak MS, Kubiak A, Antczak T, Bielecki S (2009) Enzymatic biodiesel synthesis-key factors affecting efficiency of the process. Renew Energy 34:1185-1194.

15. Schuchardt U L F, Sercheli R, Vargas RM (1998) Transesterification of vegetable Oils: A review. J B Chem Soc 9: 199-210.

16. Watanabe Y, Shimada Y, Sugihara A, Tominaga Y (2002) Conversion of

degummed soybean oil to biodiesel fuel with immobilized Candida antarctica lipase. J Mol Catal B: Enz 17: 151-155.

17. Dorado MP, Ballesteros E, Mittelbach M, Lopez FJ (2004) Kinetics parameters affecting the alkali-catalyzed transesterification process of used olive oil. Energy Fuels 18: 1457-1462.

18. Kulkarni M G, Dalai AK (2006) Waste cooking oil-An economic source for biodiesel: A-review. Ind Eng Chem Res 45: 2901-2913.

19. Nelson LA, Foglia TA, Marmer WN (1996) Lipase-catalyzed production of biodiesel. J Am Oil Chem Soc 73: 1991-1994.

20. Freedman B, Pryde EH, Mounts TL (1984) Variables affecting the yields of fatty esters from transesterified vegetable oils. J Am Oil Chem Society 61: 1638-1643.

21. Noureddini H, Zhu D (1997) Kinetics of transesterification of soybean oil. J Am Oil Chem Soc 74: 1457-1463.

22. Kumari A, Mahapatra P, Garlapati VK, Banerjee R (2009) Enzymatic transesterification of Jatropha oil. Biotechnol. Biofuels 2: 1-7.

23. Chen G, Ying M, Li W (2006) Enzymatic conversion of waste cooking oils into alternative fuel-biodiesel. Appl Biochem Biotechnol 129: 911-921.

24. Tamalampudi S, Talukder RM, Hamad S, Numatab T, Kondo A (2008) Enzymatic production of biodiesel from Jatropha oil: A comparative study of immobilized-whole cell and commercial lipases as a biocatalyst. Biochem Eng J 39: 185-189.

25. Dizge N, Keskinler B (2008) Enzymatic production of biodiesel from canola oil using immobilized lipase. Biomass Bioenergy, 32: 1274-1278.

26. Bernardes OL, Bevilaqua JV, Leal MCM, Freire DMG, Langone MAP (2007) Biodiesel fuel production by the transesterification reaction of soybean oil using immobilized lipase. Appl Biochem Biotechnol 137: 105-114.

27. Chen JW, Wu WT (2003) Regeneration of immobilized Candida antarctica lipase for transesterification. J Biosci Bioeng 95: 466-469.

28. Samukawa T, Kaieda M, Matsumoto T, Ban K, Kondo A (2000) Pretreatment of immobilized Candida antarctica lipase for biodiesel fuel production from plant oil. J Biosci Bioeng 90: 180-183.

29. Salis A, Pinna M, Monduzzi M, Solinas V (2005) Biodiesel production from triolein and short chain alcohols through biocatalysis. J Biotechnol 119: 291-299.

30. Al-Zuhair S, Ling FW, Jun LM (2007) Proposed kinetic mechanism of the production of biodiesel from palm oil using lipase. Process Biochem 42: 951-960.

31. Modi MK, Reddy JRC, Rao BVS, Prasad RBN (2006) Lipase-mediated

transformation of vegetable oils into biodiesel using propan-2-ol as acyl acceptor. Biotechnol Lett 28: 637-640.

32. Ma FLD Clements, M A Hanna (1998) Biodiesel fuel from animal fat. Ancillary studies on transesterification of beef tallow. Ind Eng Chem Res 37: 3768-3771.

33. Leung DYC, Guo Y (2006) Transesterification of neat and used frying oil: optimization for biodiesel production. Fuel Process Technol 87: 883-890.

34. Peter F, Zarcula C, Kiss C (2007) Enhancement of lipases enantioselectivity by entrapment in hydrophobic sol-gel materials: Influence of silane precursors and immobilization parameters. J Biotechnol 131: S109.

35. Eevera T, Rajendran K, Saradha S (2009) Biodiesel production process optimization and chracterization to access the suitability of the product for varied environmental conditions. Renew Energ 34: 762-765.

36. Kose O, Tuter M, Aksoy HA (2002) Immobilized Candida antarctica lipase-catalyzed alcoholysis of cotton seed oil in a solvent-free medium. Bioresour Technol 83: 125-129.

37. Li L, Du W, Liu D, Wang L, Li Z (2006) Lipase catalyzed transesterification of rapeseed oils for biodiesel production with a novel organic solvent as the reaction medium. J Mol Catal B: Enz, 43: 58-62.

38. Rodrigues RC, Volpato G, Wada K, Ayub MAZ (2008) Enzymatic synthesis of biodiesel from transesterification reactions of vegetable oils and short chain alcohols. J Am Oil Chem Soc 85: 925-930.

39. Nie K., Xie F, Wong F, Tan T (2006) Lipase catalyzed methanolysis to produce biodiesel: Optimization of the biodiesel production. J Mol Catal B: Enz. 43: 142-147.

40. Reetz MT (2002) Lipases as practical biocatalysts. Curr Opi Chem Biol 6: 145-150.

41. Ghamgui HM, Karra-Chaabouni, Gargouri Y (2004) 1-Butyl oleate synthesis by immobilized lipase from Rhizopus oryzae: a comparative study between n-hexane and solvent-free system. Enz Microbiol Technol 35: 355-363.

42. Xu Y, Nordblad M, Nielsen PM, Brask J, Woodley JM (2011) In situ visualization and effect of glycerol in lipase-catalyzed ethanolysis of rapessed oil. J Mol Catal B: Enzym 72: 213-219.

43. Dossat VD, Combles, Marty A (1999) Continuous enzymatic transesterification of high oleic sunflower oil in a packed bed reactor: influence of the glycerol production. Enz Microb Technol 25: 194-200.

44. Soumanou MM, Bornscheuer UT (2003) Improvement in lipase-catalyzed synthesis of fatty acid methyl esters from sunflower oil. Enz Microbiol Technol 33: 97-103.

24

Performance and Emission Characteristics of Annona-Ethanol Blend Fuelled with Diesel Engine

Senthil R* and Silambarasan R

Department of Mechanical Engineering, University College of Engineering Villupuram, Tamilnadu, India

Abstract

In this present work aims at evaluate the performance and emission characteristics of a diesel engine fueled with annona-ethanol blend as a fuel. A single cylinder water-cooled four stroke diesel engine was used. The ethanol is blended with Annona Methyl Ester (AME) in the proportions of 60-40, 55-45, 50-50, 45-55. The performance and emission characteristics of annona-ethanol blends are evaluated by operating the engine at different load conditions. The performance parameters such as Brake Specific Fuel Consumption (BSFC), Brake Thermal Efficiency (BTE) and Exhaust Gas Temperature (EGT) were evaluated. Further, the exhaust emissions such as oxides of nitrogen (NOx) unburned hydrocarbon (HC), carbon monoxide (CO) and smoke were measure. It is found that annona-ethanol blend (A-E-50-50) showed slight increase in brake thermal efficiency with the reduction of exhaust gas temperature. Further, it is found that the slight reduction in NOx emission and smoke emission. It is also found that reduction in HC and CO emission was achieved. Hence, it is concluded that A-E 50-50 can be used as alternate fuel for DI diesel engine without any major modification.

Keywords: Annona methyl ester; Ethanol; Performance; Emission; Diesel engine

Introduction

Diesel engines are commonly used as prime movers in the transportation, industrial and agricultural sectors because of their high brake thermal efficiency and reliability. The increasing industrialization and motorization of the world has led to a steep rise in the demand of petroleum based fuels. Petroleum based fuels are obtained from limited reserves. These finite reserves are highly concentrated in certain regions of the world. Therefore, those countries not having these resources are facing energy/foreign exchange crisis, mainly due to the import of crude petroleum. Hence, it is necessary to look for alternative fuels which can be produced from resources available locally within the country such as alcohol, biodiesel, vegetable oils etc. Ethanol is also an attractive alternative fuel because it is a renewable bio-based resource and it is oxygenated, thereby providing the potential to reduce particulate emissions in compression ignition engines.

Renewable fuels like biodiesel and ethanol are carbon neutral fuels, which will remove carbon dioxide from the atmosphere while they grow and emits the same amount CO_2 while combustion.

Studies on the use of ethanol in diesel engines have been continuing since the 1970s. The initial investigation was focused on reduction of the smoke and particle levels in the exhaust. Ethanol addition to diesel fuel results in different physical-chemical changes in diesel fuel properties, particularly reductions in cetane number, viscosity and heating value. Therefore, different techniques involving alcohol-diesel dual fuel operation have been developed to make diesel engine technology compatible with the properties of ethanol based fuels.

The vegetable oil, animal fats, used frying oil, waste cooking oil and edible oils such as soybean, sunflower, canola, palm and non-edible oils such as *Jatropha curcas*, *Pongamia pinnata*, *Madhuca indica*, *Ficus elastica*, *Nicotina tabacum,* and *Calophyllu inophyllumm* can be used as an alternate fuel for diesel [1]. The performance, emission and combustion of DI diesel engine using rapeseed oil and its blends of 5%, 20%, 70% and standard fuel. He has reported that the biodiesel produces lower smoke emission and higher brake, specific fuel consumption compare to the diesel fuel [2]. The effects of biodiesel

types, biodiesel fraction and physical properties on combustion and performance characteristics of a CI engine were studied. They have conducted on experiments on 4 cylinders 4 stroke DI and turbo charged diesel engine using biodiesel blends of waste oil and rapeseed oil and corn oil with normal diesel [3]. In this study performance and emission characteristics of a DI diesel engine using blends of diesel fuel with vegetable oils. They have conducted the experimental study on 4 strokes DI, Ricardo/cussons using various bio diesels such as cotton seed oil, soyabean oil, sunflower oil, rapeseed oil, palm oil, corn oil and olive kernel oil and their corresponding methyl ester at the blended ratio of 10/90 and 20/80. These biodiesel produces lower emission and improved performance [4]. The vegetable oils and their methyl esters (raw sun flower, raw cotton seed oil, raw soybean oil and their methyl esters, refined corn oil, distilled opium poppy oil and refined rapeseed) performance and emission of a four strokes, direct injection diesel engine was study [5]. They have conducted the experiments using Soybean oil, peanut oil, corn oil, sunflower oil, rapeseed oil, palm oil, palm kernel oil, and waste fried oil (vegetable oil basis). They have found that diesel engine fueled with vegetable oil methyl ester could potentially produce the same engine power as one fueled with diesel fuel, but with a reduction in the exhaust gas temperature (EGT), smoke and total hydrocarbon (THC) emissions, with a slight increase in nitrogen oxides (NO_x) emissions [6]. The bio diesel was production from Mahua (*Madhuca indica*) oil through esterification followed by transesterification. The result shows that 4% H_2SO_4, 0.33% v/v alcohol/oil ratio, 1 hr reaction time and 65°C temperature are the optimum conditions for esterification [7]. The suitability of transesterified mahua oil as a fuel in C.I. engine was evaluated. The conducted experiments

---

*Corresponding author: Senthil R, Department of Mechanical Engineering, University College of Engineering Villupuram, Tamilnadu, India
E-mail: drrs1970@gmail.com

7B.H.P single cylinder four stroke and vertical, water cooled Kirloskar diesel engine at rated speed of 1500 rpm [8]. The mechanism of a dual process adopted for the production of biodiesel from Karanja oil containing FFA up to 20%. The conventional alkali-catalyzed route of biodiesel production does not work out effectively with high FFA feedstock such as Karanja oil [9]. Biodiesel from karanja oil (pongamia pinnata), properties and effect of biodiesel on engine the performances and emissions were measured. They conducted experiment in a single cylinder water cooled, naturally aspired, 4-strokes DI diesel engine. They have found that B100 reduced CO and smoke emissions by 50% and 45% respectively, while 15% increase in the NO_x emissions was experimented with the same fuel [10]. The experiments on single cylinder 4 strokes DI diesel engine using Annona methyl ester and its blends with diesel. They have reported that AME shows at 20% blends showed better performance and lower exhaust emissions. They have also found that CO, HC and Smoke emission was reduced and slight increase of NO_x for the the various proportions of Annona methyl ester [11]. The performance and emission characteristics of the different ethanol-jatropa and ethanol-pongamia blends were compared with that of diesel and the perfect blend is estimated by considering all the parameters pongamia-ethanol (50-50) is considered as a better fuel when compared with fuel blends [12]. The performance and emission tests were carried on Compression Ignition Engine using blends (B20, B40, B60, B80 and B100) of Jatropa Methyl Esters (JME) and diesel. Also 5% of Ethanol was injected into the intake manifold by port injection method with the assistance of a mechanical fuel injection pump. The ethanol injection assisted in getting an improved combustion process in diesel and jatropa blends [13].

Studies conducted at different injection pressures (200, 250, 300 and 350 bar) on different loading conditions showed that the higher injection pressure reduces CO and smoke emissions with respect to diesel fuel [14]. In this study E100 (100% ethanol fuel) can improve full load engine performance around whole engine speed range in a high compression ratio engine, compared to that of a base compression ratio engine operated on a premium gasoline [15].

Reduction of Particulate Matter and NO_x emissions with no serious fuel consumption penalty is achievable when the diesel-ethanol blends are used with a combination of the modern combustion control methods [16]. Ethanol is of particular interest because it is a fuel produced from all biomass including cereals, rice, and corn, and potatoes etc., crops widely produced in many places in the world [17]. A blended ethanol-based fuel is used as a substitute to conventional petroleum fuels in existing engines without modification, comprising water, a gaseous hydrocarbon (such as acetylene or propane), a binding component (such as benzene), and a lubricating oil [18].

An alcohol fuel lubricating additive mixture for use as a fuel in internal combustion engines as well as methods of preparing such mixtures [19].

The inhibitors provide a reasonable method for preventing corrosion when hydrated ethanol fuel is used [20].

This study was to increase the efficiency of rapeseed oil recovery by pressure shockwaves and to assess the changes related to energetically utilization of the seedcake obtained. Mass balances and several design parameters (along with their manifestations on the seedcake) were analyzed to allow further optimization of the technology. It was found that the use of pressure shockwaves, in combination with the mechanical expeller, may increase oil yields up to the theoretical 100% maximum, or alternatively reduce expeller energy requirements while maintaining the same oil yield. Decreased amounts of oil in the seedcake correlate with reduced amounts of volatile matter, which means lower quantities of hazardous fumes generated during direct combustion. In addition, higher levels of seedcake disintegration accelerated the biogas production [21]. Kinetic data regarding the intensity of maceration and subsequent pretreatment with pressure shockwaves (50 MPa to 60 MPa) are described in detail and evaluated statistically. Mass balances as well as the study on liquid environment are reported, allowing further process optimization according to financial aspects. It was verified on a laboratory scale by Soxhlet apparatus that oil extraction over 94% may be reached. Achieving such a high level of disintegration opens wide options for application of hydrolysis in order to break apart the remaining lignocellulose cell walls and access the last oil remaining in the vacuoles [22].

The operating principle consists of gasification of deshelled oil seeds mash using small amounts of gas. These were subsequently subjected to the pressure waves generated externally by underwater high-voltage discharges, which are then followed by expansion of the water plasma. It was observed that gasification using small amounts of gas may enhance the destruction effects of the edges of the pressure waves, resulting in deeper lignocellulose disintegration. Breakage of the cell walls increased the level of oil extraction from oil-rich vacuoles up to 93% and also accelerated the subsequent anaerobic fermentation of the presscake residue [23].

Ethanol as Fuel

General

Ethanol is ethyl alcohol (C_2H_5OH) is nowadays used as an alternative fuel for diesel. Unlike diesel, ethanol is a form of renewable energy that can be produced from agricultural feed stocks.

such as sugar cane, potato, manioc and corn.

Physical properties of ethanol

Ethanol is a volatile, colorless liquid that has a strong characteristic odor. It burns with a smokeless blue flame that is not always visible in normal light.

The physical properties of ethanol stem primarily from the presence of its hydroxyl group and the shortness of its carbon chain. Ethanol's hydroxyl group is able to participate in hydrogen bonding, rendering it more viscous and less volatile than less polar organic compounds of similar molecular weight. The properties of ethanol are compared with neat diesel fuel and Gasoline as shown in Table 1.

Combustion of ethanol

During combustion ethanol reacts with oxygen to produce carbon dioxide, water, and heat:

$$C_2H_5OH + 3O_2 \rightarrow 2CO_2 + 3H_2O + Heat$$

After doubling the combustion reaction because two molecules of ethanol are produced for each glucose molecule, and adding all three reactions together, there are equal numbers of each type of molecule on each side of the equation, and the net reaction for the overall production and consumption of ethanol is just light into heat.

There is a reduction in calorific value by adding the ethanol along with diesel fuel is compensated by varying the supply of fuel (the fuel supply system was modified during the lower & higher load operation) during the part and full load engine operation. At full load operation, the combustion chamber temperature is high and minimum fuel will be supplied. At low load operations, the fuel supply will be maximum due to biodiesel have low calorific value and more oxygen content.

The heat of the combustion of ethanol is used to drive the piston in the engine by expanding heated gases.

Selection of biodiesel (Annona Methyl Ester) for ethanol

Annona squamosa is a member of the family of Custard apple trees called Annonaceae and a species of the genus *Annona* known mostly for its edible fruits Annona. It is commonly found in India and Cultivated in Thailand and originates from the West Indies and South America. *Annona squamosa* produces fruits that are usually called sugar apple or custard apple in English, sitafal in Marathi, sharifa in Hindi and sitaphalam in Tamil, in India and corossolier and cailleux, pommiercannelle in French. It is mainly grown in gardens for its fruits and ornamental value. It is considered as beneficial for cardiac disease, diabetes hyperthyroidism and cancer. The root is considered as a drastic purgative. The properties of AME compare with neat diesel fuel as shown in Table 2.

Ethanol blended with Annona oil

Ethanol and Annona oil are blended in various proportions and the fuel stability is studied. The fuel properties are studied and compared with diesel and shown in Table 3.

Ethanol is blended with Annona methyl ester in the proportions of 60-40, 55-45, 50-50, 45-55 and the blends are kept in observation for a week the blends are observed for any separation or precipitate formation (Figure 1).

Experimental Setup

A single cylinder, water cooled, four stroke direct injection compression ignition engine with a displacement volume of 661 cc, compression ratio of 17.5:1, developing 5.2 kW at 1500 rpm was used for the present study as shown in fig. Initially the engine was allowed to run with diesel at a constant speed of 1500 rpm for nearly 30 minutes to attain the steady state conditions at the lowest possible load. During the investigation, the temperature of lubricating oil and temperature of the engine cooling water were held constant to eliminate their influence

Properties	Diesel	Gasoline	Ethanol
Boiling point (°C)	188 - 343	27 – 225	78
Auto ignition temperature (°C)	210	300	420
Stoichiometric A/F ratio	14.6	14.5	9
Lower heating value (MJ/kg)	43.2	44.0	26.9
Rich Flammability Limit	7.6	6	19
Lean Flammability Limit	1.4	1	4.3
Density gm/cc	0.84	0.72	0.789

Table 1: Comparison of Ethanol properties with Diesel and Gasoline.

Properties	Diesel	AME
Cetane no	48	52
Specific gravity	0.83	0.862
Viscosity @ 40°C	3.9	5.18
Calorific value (KJ/Kg)	43000	41000
Density (Kg/m³)	830	880.2
Flash point (°C)	56	76
Fire point (°C)	64	92
Oxygen Content (wt %)	-	10.8

Table 2: Fuel properties of AME and Diesel.

on their results. The speed of the engine was stabilized with injected fuel to attain the temperature of lubricating oil as 65°C. Then the following observations were made twice for concordance. The exhaust gas analyzer and smoke meter was switched on quite early so that all its systems will get stabilized before the commencement of equipment and the following observations were documented (Figure 2).

Time for 50 cc of fuel consumption(s).

- Exhaust gas temperature (°C).

- Measurement of smoke using AVL smoke meter and heated Vacuum NO_x Analyser is used.

- Measurement of CO, CO_2, HC and O_2 using CRPTYON gas analyzer.

- Combustion parameters were analyzed using pressure transducer and combustion analyzer.

Testing Procedure

Experiments were carried out at steady state for different engine loads at constant speed of 1500. The engine was allowed to run for few minutes until the exhaust gas temperature, the cooling water temperature, the lubricating oil temperature, as well as the emission have attained steady-state values and data's were recorded subsequently. All the gas concentrations were continuously measured for 10 min and the average results presented. The experiment uncertainties are shown in Table 4.

The steady-state test was repeated thrice. Since the error value is too low compared with the actual values thereby error values not included in the actual result and also the equipment are often calibrated and it was kept in error free condition.

For each load condition the engine was run for five minutes and the data were collected during the last two minute of operation. The readings are tabulated and the various performance characteristics such as brake power, total fuel consumption, specific fuel consumption, brake mean effective pressure, brake thermal efficiency are calculated and various graphs are plotted.

Results and Discussion

Performance analysis of Annona - ethanol blends

Brake Specific Fuel Consumption (BSFC): The brake specific fuel consumption of various proportions of Annona Methyl Ester-Ethanol blends and diesel is shown in Figure 3. It is observed that BSFC of diesel is minimum compared with other proportions of Annona Methyl Ester-Ethanol blends at all loads. Among various proportions of Annona Methyl Ester-Ethanol blends, Annona-Ethanol (A50-E50) showed better SFC than conventional diesel fuel. BSFC is 0.297 kg/kW-hr for Annona-Ethanol (A50-E50) and 0.287 kg/kW-hr for diesel fuel at maximum load. This is due to complete combustion and also excess oxygen, high specific gravity, high viscosity and lower calorific value of biodiesel when compared to diesel. This is also due to the lower calorific value of annona methyl ester-ethanol compared with that of neat diesel fuel.

Brake thermal Efficiency (BTE): The brake thermal efficiency of various proportions of Annona Methyl Ester-Ethanol blends and diesel is shown in Figure 4. It is observed that BTE of AME and its blends are slightly lower than that of diesel fuel. The maximum BTE of diesel fuel is 30% and that of Annona-Ethanol (A50-E50) is 31.21%. Among various proportions of Annona Methyl Ester-Ethanol blends, Annona-

	DIESEL	A-E-50-50	A-E-40-60	A-E-30-70	A-E-80-20
ETHANOL	-	50%	60%	70%	80%
ANNONA METHYL ESTER	-	50%	40%	30%	20%
INFERENCE	-	STABLE	STABLE	STABLE	STABLE
EXPERIMENTAL VISCOSITY(at 38°C) cSt	3.85	3.475	3.02	2.56	2.11
THEORETICAL CETANE NO.	45	30	25.6	21.2	16.8
THEORETICAL CALORIFIC VALUE (MJ/kg)	42.5	34.72	33.71	32.71	31.71

Table 3: Stability of Ethanol and Annona Methyl Ester.

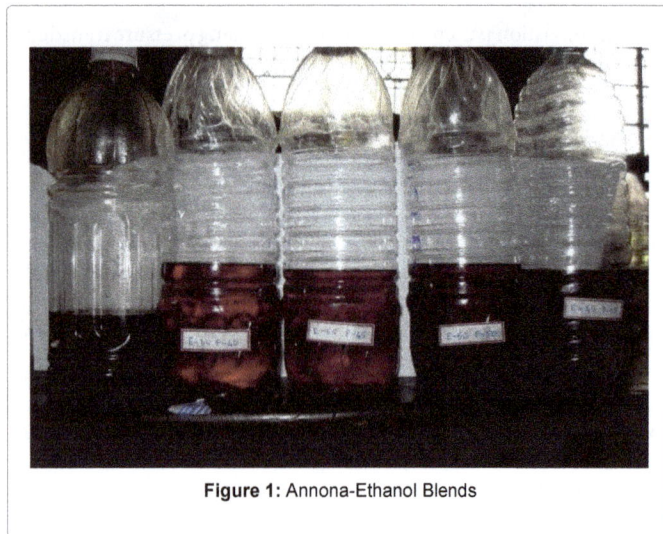

Figure 1: Annona-Ethanol Blends

Parameters	Systematic Errors (±)
Speed	1 ± rpm
Load	± 0.1 N
Time	± 0.1 s
Brake power	± 0.15 kW
Temperature	± 1°
Pressure	± 1 bar
NOX	± 10 PPM
CO	± 0.03%
CO2	± 0.03%
HC	± 12 PPM
Smoke	± 1 HSU

Table 4: Experiment Uncertainties.

its blends than diesel under various load conditions. Among various proportions of Annona Methyl Ester-Ethanol blends, Annona-Ethanol (A50-E50) showed lower EGT than conventional diesel fuel. EGT for Annona-Ethanol (A50-E50) is 189°C and 235°C for diesel at maximum load. This is due to the improved combustion provided by the annona-ethanol blends lower heating value, higher density and increased viscosity which leads to poor atomization and fuel vaporization reducing reduction of exhaust gas temperature.

Emission characteristics of Annona - ethanol

Oxides of nitrogen: The variation of oxides of Nitrogen of various proportions of Annona Methyl Ester-Ethanol blends and diesel is shown in Figure 6. The formation of NO_x in the cylinder is affected by oxygen concentration, combustion flame temperature and residence time in the high temperature zone. It is observed that NO_x emission of Annona-Ethanol (A50-E50) is minimum when compared with other biodiesels blends-ethanol and diesel at all loads. Among various proportions of Annona Methyl Ester-Ethanol blends, Annona-Ethanol (A50-E50) showed minimum NO_x emissions than other blends. It is observed that NO_x emission of 5.8% higher than conventional diesel fuel under full load condition. Obviously, with biodiesel the combustion temperature as well as the oxygen contents could be higher which leads to the higher NOx emissions. However, the higher oxygen contents of ethanol could also enhance NOx emissions. For annona-ethanol blends the cooling effect of ethanol associated with its lower calorific value and higher latent heat of evaporation could reduce the combustion temperature and hence reduce the NO_x emissions. Further annona-ethanol blends, the cooling effect of ethanol seems to be dominating effect leading to the overall reduction of NO_x emission.

Carbon monoxide emission (CO): CO is one of the intermediate compounds formed during the intermediate combustion stage of hydrocarbon fuels. CO formation depends on air fuel equivalence

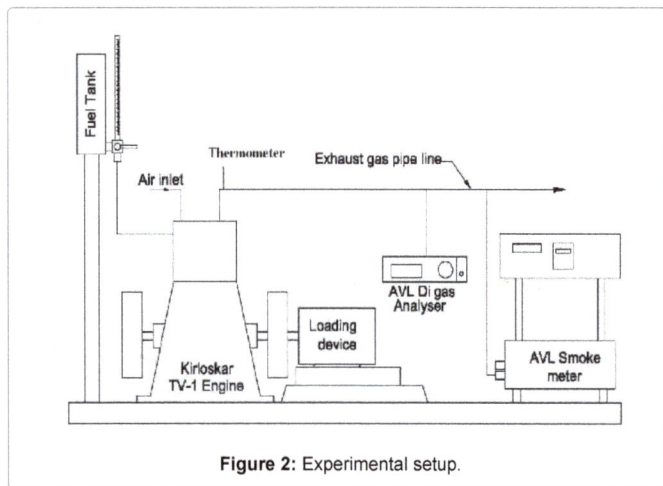

Figure 2: Experimental setup.

Ethanol (A50-E50) showed better BTE than conventional diesel fuel. This is due to The increase of BTE is due to the improvement of the combustion process on account of increased oxygen content on the annona-ethanol blends. From the fig, the faster combustion process of the annona-ethanol blend in all modes could be a contribution of the increase in BTE.

Exhaust gas temperature (EGT): The exhaust gas temperature of various proportions of Annona Methyl Ester-Ethanol blends and diesel is shown in Figure 5. It is observed that TFC is higher for all AME and

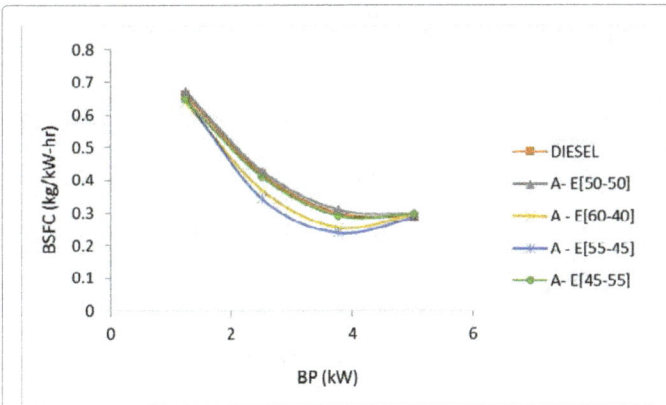

Figure 3: Brake Power (BP) vs Brake Specific Fuel Consumption (BSFC).

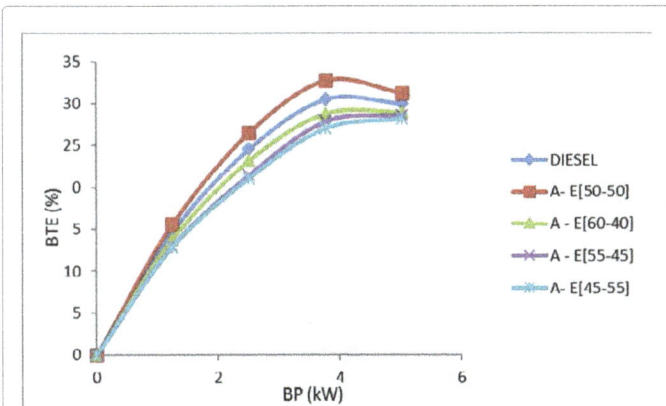

Figure 4: Brake Power (BP) vs Brake thermal Efficiency (BTE).

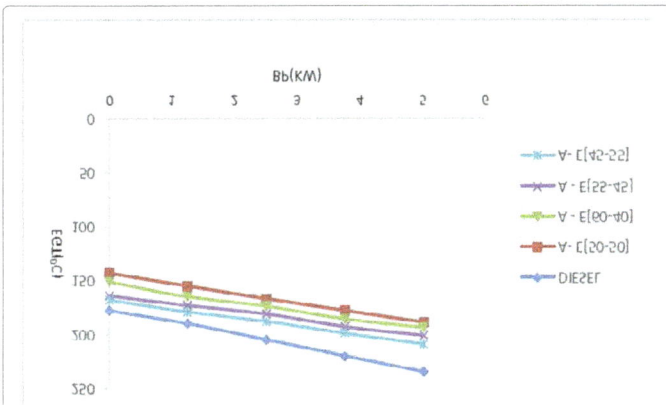

Figure 5: Brake Power (BP) vs Exhaust Gas Temperature (EGT).

in turn helps in reduction of CO. For annona-ethanol blend higher oxygen content maybe the major factor leading to the reduction of CO emission. Further, the cooling effect of ethanol can increase the in cylinder gas temperature, leading to more fuel in combusted and hence reduce CO emission.

Hydro Carbon emission (HC): The hydrocarbon emission of various proportions of Annona Methyl Ester-Ethanol blends and diesel is shown in Figures 8 and 9. Among all various proportions of Annona Methyl Ester-Ethanol blends, Annona-Ethanol (A50-E50) has lower

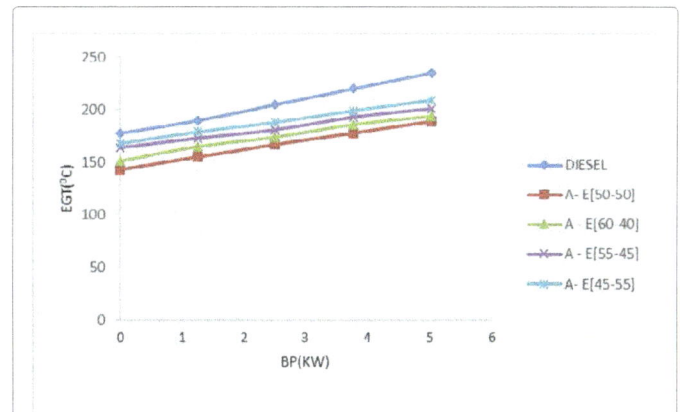

Figure 6: Brake Power (BP) vs Oxides of Nitrogen (NO$_x$).

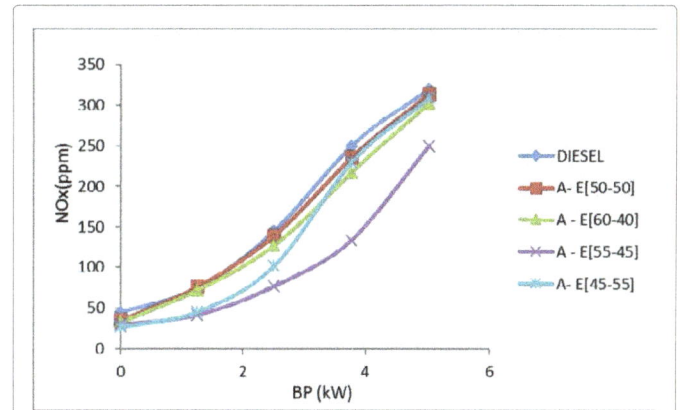

Figure 7: Brake Power (BP) vs Carbon Monoxide (CO).

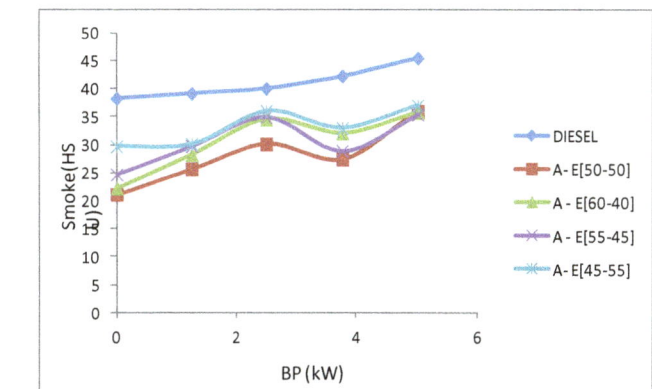

Figure 8: Brake Power (BP) vs Hydro Carbon Emission (HC).

ratio, fuel type, design of combustion chamber, start of injection timing, injection pressure and speed. The hydrocarbon emission of various proportions of Annona Methyl Ester-Ethanol blends and diesel is shown in Figure 7. It shows that among various proportions of Annona Methyl Ester-Ethanol blends; Annona-Ethanol (A50-E50) has lower CO emission than that of diesel at all loads. It is observed that, CO of Annona-Ethanol (A50-E50) is 0.13 % and is 0.16% for diesel at maximum load. This is due to more oxygen molecules present in the biodiesels -ethanol blends, leads to complete combustion which

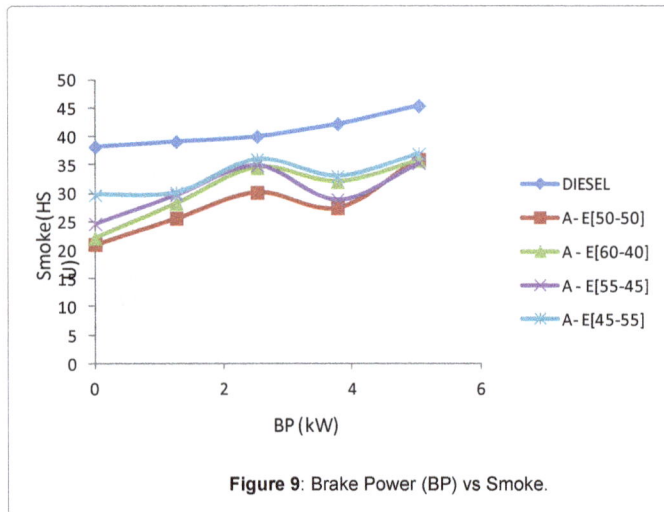

Figure 9: Brake Power (BP) vs Smoke.

HC emission than that of diesel at all loads. It is observed that HC of Annona-Ethanol (A50-E50) is 1348 ppm and 1918 ppm for diesel at maximum load. A 38% reduction of HC emission in the case Annona-Ethanol (A50-E50) as compared to diesel indicated better combustion of AME. The hydro carbon content at biodiesels-ethanol blends, which leads to better combustion when compared with diesel. HC emission for different blends of biodiesels-ethanol blends is high and Annona-Ethanol (A50-E50) shows better reduction of HC. It is due to the increase in oxygen content and reduce in viscosity and density of the blended fuel, leading to improved spray and atomization, better combustion and hence lower HC emission.

Smoke: The smoke of various proportions of Annona Methyl Ester-Ethanol blends and diesel is shown in Fig.9. It is found that smoke emission for biodiesel blends are lower than that of diesel. Among the various proportions of Annona Methyl Ester-Ethanol blends, Annona-Ethanol (A50-E50) showed much lower smoke emission. It is observed that smoke of Annona-Ethanol (A50-E50) is 36 HSU and 45.5 HSU for conventional diesel fuel. This is due to the inbuilt oxygen presence in the biodiesel- ethanol which helps in better and nearly complete combustion. It is also due to the dilution of aromatics, which are soot producers. The ethanol reduces the soot precursors due to the production of OH radicals by the ethanol. Finally it is found that the reduction of smoke could be attributed to the improved premixed combustion mode.

Conclusion

The performance and emission characteristics of the different ethanol- Annona blends are compared with that of neat diesel and the perfect blend is estimated using the results obtained. Based on the experimental results, the conclusion can be summarized as follows

1. Compared with neat diesel fuel the brake thermal efficiency slightly increases with annona-ethanol (50-50), while there is no significant difference with other proportional blends.

2. Compared with neat diesel fuel, annona-ethanol (50-50) gives slightly low HC and CO emission in all test conditions while other proportions of annona-ethanol blends have slight increase of HC and CO emission at low and high loads.

3. The annona-ethanol (50-50) have lower NO_x and smoke emissions compared with neat diesel fuel, while there is no significant difference among the biodiesel-ethanol blends at medium and high loads

4. By considering all the parameters Annona -ethanol (50-50) is considered as a better fuel compared to other fuel blends.

References

1. Sharma YC, Singh B, Upadhyay SN (2008) Advancements in development and characterization of bio-diesel: A review. Fuel 87: 2355-2373.

2. Buyukkaya E (2010) Effects of bio-diesel on a DI diesel engine performance, emission and combustion characteristics. fuel 89: 3099-3105.

3. Tesfa B, Mishra R, Zhang C, Gu F, Ball AD (2013) Combustion and performance characteristics of CI (Compression ignition) engine running with bio-diesel. Energy 51: 101-115.

4. Rakopoulos CD, Antonopoulos KA, Rakopoulos DC (2006) Comparative performance and emission study of a direct injection diesel engine using blends of diesel fuel with vegetable oil or bio-diesel of various origins. Energy conserv manag 47: 3272-3287.

5. Altin R, Cetinkaya S, Yucesu HS (2001) The potential of using vegetable oil fuel as fuel for diesel engines. Energy conserv manag 42: 529-538.

6. Lin BF, Huang JH, Huang DY (2009) Experimental study of the effects of vegetable oil methyl ester on DI diesel engine performance characteristics and pollutant emissions. Fuel 88: 1779-1785.

7. Padhi SK, Singh RK (2010) Optimization of esterification and trans-esterification of mahua (madhuca indica) oil for production of bio-diesel. J Chem Pharm Res 2: 599-608.

8. Nandi S (2013) Performance of C.I engine by using Bio-diesel-Mahua oil. Amr J Engg Res 2: 22-47.

9. Naik M, Meher LC, Das LM (2008) Production of bio-diesel from high free fatty acid karanja (pongamia pinnata) oil. Biomass Bioenergy 32: 354-357.

10. Nabi MN, Rahman MM, Akhter MS (2009) Biodiesel from cottonseed oil and its effect on engine performance and exhaust emissions. Appl Therm Engg 29: 2265-2270.

11. Senthil R, Silambarasan R (2014) Annona: A new biodiesel for diesel engine: A comparative experimental investigation. J Energy inst

12. Senthil R, Silambarasan R (2014) Effect of ethanol blend addition on performance and emission of diesel engine operated with jatropha & pongamia methyl esters. J Sci Industrial Res 73: 453-455.

13. Kannan D, Nabi MN, Hustad JE (2009) Influence of Ethanol Blend Addition on Compression Ignition Engine Performance and Emissions Operated with Diesel and Jatropha Methyl Ester. Norwegian Univ Sci Technol SAE: 1808.

14. Kumar C, Athawe M, Agahav YV, Babu MKG, Das LM (2007) Effects of Ethanol Addition on Performance, Emission and Combustion of DI Diesel Engine Running at Different Injection Pressures. SAE-2007-01-0626.

15. Taniguchi S, Yoshida K, Tsukasaki Y (2007) Feasibility Study of Ethanol Applications to A Direct Injection Gasoline Engine. SAE-2007-01-2037.

16. Mohammadi A, Ishiyama T, Kakuta T, Kee SS (2005) Fuel Injection Strategy for Clean Diesel Engine Using Ethanol Blended Diesel Fuel. SAE-2005-01-1725.

17. Moriya S, Yaginuma F, Watanabe H, Kodama D, Kato M, et al. (1999) Utilization of Ethanol and Gas Oil Blended Fuels for Diesel Engine (Addition of Decanol and Isoamyl ether). SAE-1999-01-2518

18. Neves AM (1982) Blended ethanol fuel" United States Patent-4333739.

19. Smith EJ (1986) Lubricating and additive mixtures for alcohol fuels and their method of preparation. United States Patent-4595395.

20. Walker MS, Chance RL (1983) Corrosion of Metals and the Effectiveness of Inhibitors in Ethanol Fuels. SAE-8318.

21. Maroušek J (2013) Use of continuous pressure shockwaves apparatus in rapeseed oil processing. Clean Technol Environ Policy 15: 721-725.

22. Maroušek J, Itoh S, Higa O, Kondo Y, Ueno M, et al (2013) Pressure Shockwaves to Enhance Oil Extraction from Jatropha Curcas L. Biotechnol Biotechnological Equip 27.

23. Maroušek J (2015) Novel technique to enhance the disintegration effect of the pressure waves on oilseeds. Ind Crops Prod 53: 1-5.

25

The Environmental Challenges of Biomass Utilisation for Combined Heat and Power Generation in a Paper Mill in Tanzania

Sisty Basil Massawe[1]*, AO Olorunnisola[2] and A. Adenikinju[3]

[1]*Pan African University, Institute of Life and Earth Sciences (Including Health and Agriculture), University of Ibadan, Nigeria*
[2]*Department of Agricultural and Environmental Engineering, University of Ibadan, Nigeria*
[3]*Department of Economics, University of Ibadan, Nigeria*

Abstract

Biomass-driven, combined heat and power (CHP) also known as co-generation plants are said to provide reliable, efficient, clean power and heat worldwide. However, it is known that the use of biomass for energy applications may lead to land use competition, environmental degradation and food in-security. This study was therefore carried out at a Paper Mill and the seven surrounding villages with the aim of assessing the environmental challenge of wood biomass utilisation for CHP generation.

Data were collected by interviewing technical staff at the paper mill, Sao Hill Plantation, Government officials from Ministry of energy and other energy regulatory bodies. A questionnaire was used to collect data from seven villages surrounding the paper mill while a check list was used to collect information on environmental management aspect within the paper mill departments. Descriptive Statistics was used in assessing environmental challenge of biomass use at the Paper Mill while a chi square was used also to establish the relationship and association between variables.

Findings revealed that there were negative impacts on air quality, land use and water. The chi square test revealed that there was no significant difference (x^2=0.253 and p > 0.05) in having environmental problems and distance from Paper Mill. It was also observed that arable land which was needed to grow trees was becoming scarce affecting the sustainable supply of raw materials.

Keywords: Wood biomass resources; Cogeneration of electricity; Pulp and paper mill; Forestry; Environmental management

Introduction

Biomass is a versatile raw material that can be used for production of heat, power, transport fuels, and bio-products. When generated and used on a sustainable basis, it is a carbon-neutral carrier that can make a large contribution to reducing greenhouse gas emissions. Currently, biomass accounts for about 10% of the total primary energy consumption in the world [1]. Despite the fact that traditional biomass in the form of wood fuel still remains a major source of bio energy; liquid biofuel and processed biomass production have shown rapid growth during the last decade [2].

Several studies including Dasappa et al., Smeets et al., Smeets et al. and Marrison et al. [3-6] have highlighted the potential for bio-energy production on the African continent. In Tanzania for example several studies [7-9], have been conducted on the use of sisal, charcoal, animal sludge and bagasse as raw materials for energy use and generation of electricity, But at the country level, the use of wood biomass residue have not received high attention in the context of specific assessments, associated environmental impacts as well as awareness on electricity generated despite the fact that wood biomass is currently contributing more than 11 MW of electricity to the national grid [10].

The Tanzanian energy demand is estimated at 22 million tonnes of oil equivalent (TOE) per annum or 0.7 TOE per capita. According to MEM-2013 [11], the quantitative distributions of the different energy sources to the energy balance were biomass fuels 90%, Petroleum 8%, electricity 1.2% and others less than 1% (including coal and renewable energy sources). These percentages show low per capita consumption of commercial energy (petroleum, coal and electricity) and relatively high dependence on biomass fuels in Tanzania. According to MFA-2011 [12], only 14% of the population had access to electricity (approximately 2% of rural population where 80% of country's population live and 37% of urban population) despite the fact that the country had very huge potential of renewable energy sources especially wood biomass.

Tanzanian industries using wood or agricultural feedstock (e.g., sugar, tannin, and sisal) have been generating their own power from waste biomass materials. It is estimated that about 58 MW of such generation is taking place [11]. According to Gwang'ombe [9], the estimated co-generation potential in Tanzania was more than 315 GWh per year. This was 10.5% of the national electricity generation. Songela [7] asserts that the energy generation potential from excess bagasse in sugar mills was about 99 GWh per year which was 3.5% of the national electricity generation; Private sector has been leading in utilizing biomass to generate heat and power.

The Paper Mill Combined Heat and Power Capacity

The paper mill has two product lines; Line 1 for manufacturing 30,000 tons per annum of industrial packaging grades, Line 2 for manufacturing 30,000 tons per annum of graphic paper grades; newsprint, mechanical printing and wood free printing paper grades.

*Corresponding author: Sisty Basil Massawe, Pan African University, Institute of Life and Earth Sciences (Including Health and Agriculture), University of Ibadan, Nigeria, E-mail: stiba7@gmail.com

The production lines are integrated with a Chemical Pulp Mill (Kraft) with a designed capacity to produce 150 tons per day of unbleached chemical pulp, 80 tons per day of mechanical pulp; Chemical Recovery Plant for handling 320 tons per day dry black liquor solids and supplying 640 m³/day white Liquor to the Kraft Pulp Mill.

Process heat and part of the electrical energy requirement are met through a captive co-generation plant comprising one (1) 10.5 MWe Extraction-Back Pressure Turbine, one (1) 60 Tonnes Per Hour (TPH), 45 bar of pressure, temperature of 450°C, coal/Wood biomass fired Steam Boiler and one (1) 40 TPH, 45 bar, 450°C Chemical Recovery Boiler firing the dissolved organics from the Kraft Mill Spent Chemicals. The total electrical energy demand at optimum operating levels is at 25 MW, out of which, approximately 9.0 MW is met from the co-generation plant and the balance 16 MW drawn from the grid; TANESCO [13].

The objective of this study was to examine the environmental impacts of biomass utilisation for heat and power generation at a Paper Mill and also to identify available environmental management programmes.

Paper Mill Raw Material Requirements

Information collected from Paper Mills on their current wood raw material are as follows:

Sufficiency of Wood Raw Material Requirements for Paper Mill Medium and Long Term Requirements

Data collected from the paper mill indicated that, with first level upgrades on Paper Machine No.1, It had increased the installed rated capacity of Paper Machine No.1 of 30,000 FTPA, to a new level capacity of 54,545.50 FTPA from Year 2010/2011.

Increase in the mill capacity means increase demand in the raw materials for both power plant and paper making. Paper Mill current projections on raw materials stands at.

Materials and Methods

This study was carried out at a paper mill and in the seven surrounding villages in southern highlands of Tanzania. The following sample determination formula based on Kothari [14] was used to generate a sample size to be used in this study.

$$n = \frac{z^2 pq}{d^2} \qquad (1)$$

Where:

n =sample size in the study area when population > 10 000.

z = Standard normal deviation, set at 1.96 (2.0 approximate) corresponding to the 95% confidence interval level.

p = Proportion of the target population (50% if population is not known).

q = 1.0 – p (1-50) (1-0.5) = 0.5

d = degree of accuracy desired, (set at the 95% equivalent to 0.05)

Based on the above formula, the sample size for this study was supposed to be 384 respondents, but due to number of households which were at a distance of less than 30km from the paper mill 28% of the cases were selected for this study. Therefore, 106 respondents were selected to participate in the study, based on the fact that a sample of 30 respondents, according to Bailey [15] irrespective of the population

size is the bare minimum for a study in which statistical analysis is to be done while, Kumar [16], observes that a sample size of between 80 and 120 respondents is suitable for rigorous statistical analysis.

Purposive sampling of the seven surrounding villages was done based on accessibility and proximity to the Mufindi Paper Mill site [17] as well as the wood plantations within a radius of 30 kilometres. Systematic sampling technique was used to select the required 106 households and from each, a household head or spouse to the household head was enumerated. A survey of the seven villages was conducted to determine the geographical location of the village as well as household distribution; therefore data was collected from every 5th household in each of the seven villages.

Primary data were collected using structured questionnaire containing both open and closed-ended questions on biomass utilization from the selected villages. Key informant interview was used to collect data from government officials and other stakeholders; these included Ministry of Energy and Minerals (MEM), National Environmental Management Council (NEMC), Rufiji Water Basin Authority - Iringa, Rural Energy Agency (REA), Tanzania forest services (TFS), Energy and Water Regulatory Authority (EWURA),Tanzania Traditional Energy Development and Environment Organization *(TaTEDO)* and Tanzania Renewable energy Association (TAREA) and a checklist was used to collect data during Focus group discussion from various departments at the Paper mill.

Quantitative data were analysed using Statistical Package for Social Sciences (SPSS), while chi-square test was used to establish the relationship between socio-demographic characteristics of the respondents and their awareness of the cogeneration activities at the paper mill as well as environmental impacts.

Results and Discussion

Size of land owned by the villagers

The size of land owned by respondents varied from one village to another and from one household to another, the study indicated that 70% of the respondents owned ≤ 10 hectares of land, 19.8% of the respondents owned 11-20 hectares while the remaining 9.4% of the respondents owned ≥ 21 hectares (Table 2). Land ownership was one of the crucial factors as the bigger the land the household possesses the more the income derived from agricultural activities and tree plantations. However, the presence of larger tree plantations and increase in tree product prices had led to not only increased land prices but, also land scarcity and land use related-conflicts [18] had argued that Sub-Saharan Africa, including Tanzania, would witness an a 8% increase in the total land use for wood fuel cultivation, offset by fall, incomes decline, and their ability to access food depreciate roughly 3.4% decrease in forested land and a 4.5% reduction in pastureland. In a rural area, like the study area, having the larger percentage of people owning less than 10 hectares of land is a typical sign that most of land is now under wood cultivation by larger companies.

The findings also showed that due to increasing lack of sufficient land, the available natural forest had been encroached upon in opening new farms. Also, there had been frequently burning of existing larger plantations and this was associated with the increasing scarcity in land ownership by the villagers. The same argument had been canvassed by Narain et al. [19] who found that households with less land tend to perceive conservation programmes as a limitation to their subsistence needs and therefore are likely to have negative attitude toward conservation. Masozera [20], Reardon and Vostii [21] furthermore,

S. N	Area	Unit	Conversion formula	Total wood	Remarks
1	Paper production	54,545.50 FTPA	5.5 m³ tonne of FTPA	300,000 m³	
2	Biomass for power generation and power boiler	292,000 tonnes/a (800t/d˙365 days)	575 kg/m³	167,900 m³	About 44.6 (35.6+9.0) MW will be generated.
	Total wood requirement			467,900 m³	

Table 1: Paper mill wood requirement.

argued that households with less land tend to be poor in off-farm capital and therefore cannot afford to continue sustainable agriculture.

Size of land used for tree plantation

About 84% of the respondents used ≤ 10 ha for tree planting, 11.3% used 11-20 ha, while 2.8% used ≥ 21 hectares for tree planting. There was a noticeable shift from growing of food crops to cash crops, especially tree planting. The shift was motivated by the increasing prices of wood products especially timber and the huge market for electricity poles and wood fuel at the Paper Mill. At the time of this study, there were already cases of food price increases. Despite the fact that many respondents believed that this might have been caused by increased demand from the number of people who were working at the mill, another reason could be due to the decline in the number of farmers who were involved in growing food crops. These results correlate the findings of ABN-2007 [22] who reported that in Zambia, farmers were persuaded by agribusinesses to grow cotton instead of maize only to see market prices.

Wood waste utilization

Fuel wood used at the Paper Mill for electrical power generation was in form of wood waste. The researcher wanted to know what the respondents did with the wood waste after they had harvested their trees. The findings indicated that 71.7% of the respondents had no idea of what they would do with such wood waste after harvesting. This was perhaps because 78.3% of the respondents had not yet harvested their trees. However 13.2% of the respondentsleft their wood waste on the farm after harvesting, 12.3% used as firewood, while 2.8% were burning it on the field as a means of land clearing for the next planting season. The researcher also observed that even Paper Mills left most of the waste at the field after harvesting trees (Figure 1). When asked why they were leaving the tree branches and roots while they could be used as fuel at the cogeneration plant the harvesting manager said:

"The branches and roots are the smallest parts and for now we don't have any mechanism to transfer them to the mill. Also we have a lot of raw materials in forms of wood chips from other supplies and from our sister company".

Environmental problems resulting from wood biomass use at paper mill

Data collected from the field indicated that 78.3% of the respondents believed that there were environmental impacts associated with Mufindi paper mill, while 21.7% believe that there were no environmental impacts. From such finding it is clear that the majority of

the respondents believed that the mill operation caused environmental problems.

The mill had different levels of impact environmental impacts across the villages (Figure 2), of all the respondents, 31.1% mentioned air pollution in form of smoke, bad smell and ashes from the Paper Mills, 25.5% mentioned bad smell only, while 20.8% reported that there had not been any significant environmental impact. Less than 10% mentioned smoke from the power Plant, dust pollution especially that which was caused by moving cars carrying tree logs from harvesting sites to the paper mill.

These findings corroborate those by WWF-2006 which argued that plantations and biomass use have negatively impact on biodiversity, water resources, soil quality, and air pollution. An environmental impact assessment done at Mufindi Paper Mill by Nzalalila et al. [23] also indicated that, the likely key environmental issues relating to mill operations included generation of solid, liquid, heat, and gaseous wastes which, if not properly disposed could lead to environmental pollution. The same source also asserted that solid wastes can result in abnormal piling of debris and emission of noxious and malodorous gases and that sometimes fire may result, dust might lead to breathing and lung problems.

Again the type of environmental problems mentioned differed depending on the distance of the village from the Paper Mill. Air pollution by ashes from the power generation plant was recorded only at the distances greater than 17 km. Dust had environmental impact at 6-17 km or more (Figure 3).

This is because most of these villages are close to the main road heading to the paper mill. Hence, there was higher vehicle traffic especially during the transportation of both raw materials from the forest to the paper mill, and the paper products to Dar es Salaam.

When the distance from the paper mill and the associated environmental problem was statistically tested, however there was no significant association between the two variables (Table 3). Therefore the null hypothesis was accepted. This means that despite the change in distance from the paper mill, most villages experienced the same type of environmental problems. This could be due to the height of the smoke chimney of the paper mill.

Effects of environmental pollution on human health, physical facilities and biodiversity

About 34% of the respondents opined that the pollution caused by the paper mill led to frequent coughing, 14.2% reported being diagnosed with chest diseases as the result of inhaling the polluted air from the paper mill while 6.6% reported food contamination by ashes coming from the paper mill. About 4% reported death of fishes as the result of discharge of untreated effluent from the Paper mill to the nearby river. About 32.2% of the respondents were not aware of any environmental impact resulting from operation of the paper mill. These respondents believed that further scientific studies should be carried out to identify the likelihood of any impact as some of the impacts might take years to identify. The remaining 6.6% believe that the environmental pollution problems had led to iron sheet rusting, flue and coughing (Table 4).

Availability of environmental management programmes

From the findings, about 82.2% of the respondents reported that there were environmental management programmes and activities being undertaken. Also this study indicated that there were several types of environmental programmes mostly aimed at mitigating the

Where	Upto 2012		2015-2020		2025 And Beyond	
Paper Machine I/li	300'000 m³	45,000-54,545.50	495000 m³	90,000 FTPA	900,000 m³	180,000 FTPA
Biomass Powered Power Plant	167'900 m³	44.6MW	167900 m³	44.6 MW	167,000 m³	44.6 MW
Total	467-500,000 m³		662,900 m³		1,157,900 m³	

Table 2: Short and long term paper mill wood requirements.

Figure 1: Paper Mill harvesting site and typical type of waste left at the site.

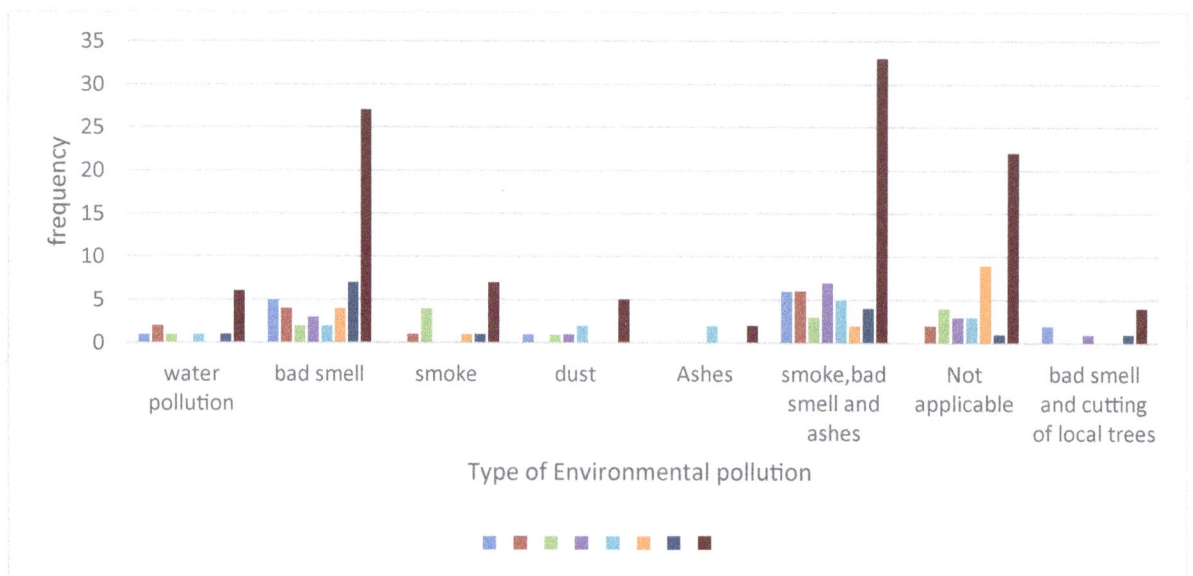

Figure 2: Environmental impacts per each village.

impact of climate change and controlling unsustainable use of natural resources. About 17.9% of these environmental management activities were in the form of fire burning control. 17% on tree planting activities and 20.8% on tree planting, water sources management and bush burning control. About 17.9% of the respondents were not aware of any environmental management programme, activity or campaign at the study area (Figure 4).

Effectiveness of the available environmental programmes

Findings showed that the available environmental management programmes had been effective at different levels. About 28.3% of the respondents believed that the available environmental management programmes had led to an increase in tree planting activities, 22%, believed that awareness towards environment management had led to decrease in forest fires. The decrease in forest burning activities was

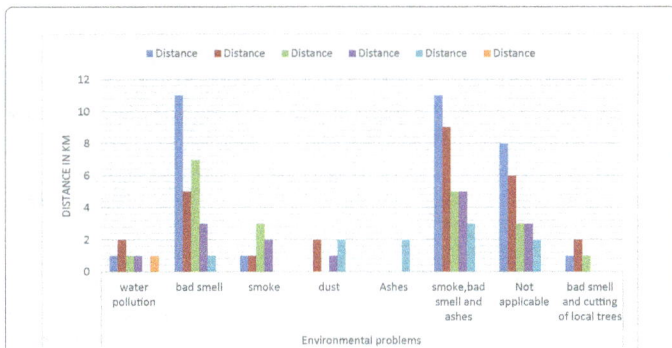

Figure 3: Effect of distance from the paper mill on environmental impact

Size of Land	Frequency (N)	Percent (%)
≤ 10 hectares	75	70.8
11- 20hectres	21	19.8
≥ 21 Hectares	10	9.4
Total	106	100

Table 3: Size of land owned.

Distance from Mufindi paper mill	Environmental problems				Chi-square value (x^2)	P-value
	Yes	%	No	%		
< 6 km	25	75.8%	8	24.2%		
6-17 km	37	80.4%	9	19.4%	0.253	0.881
> 17 km	21	77.8%	6	22.2%		

Table 4: Chi-square test of association between having environmental problems and distance from Mufindi Paper Mill.

associated with the increased environmental management programmes as confirmed by 19% of all respondents (Figure 5). Despite these achievements there is still much to be done on improving the quality of the environment as well as solving the land use conflicts.

Conclusion

The focus of this paper was on the environmental challenge of wood biomass utilisation for energy cogeneration in one of the paper mill in Tanzania. The sustainability of wood biomass cogeneration will mostly depend on the awareness of the people; this is because majority of villagers where this study was conducted were not aware of electricity generation at the mill. this lack of awareness in a way affected the raw material supply to the mill due to the fact that people are mostly planting trees for other uses than wood fuel such as for timber which takes up to 15 years before harvesting while if they were to plant for fuel purposes it would take them up to only 5 years and also increasing their income. The study also found that there are environmental problems being caused by Paper Mill, with direct impacts on the air quality, land use and water .Although some cases might need technical evaluation, numerous complaints from various stakeholders signifies the extent of the problem. Despite the presence of environmental programmes at the study area, their effectiveness is also a matter of concern. In villages like Kitasengwa where fire prevention education has been preached every day and despite the presence of Sao hill plantation division office, fire cases have been occurring repeatedly. Also due to the fact that most programs at village have been championed by the villagers themselves the financial and technical operation has always been a problem.

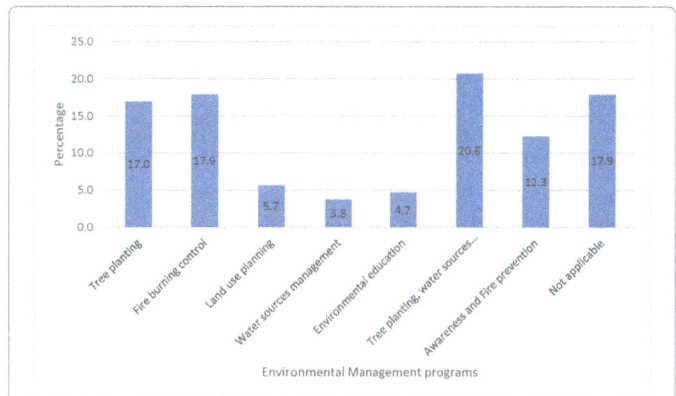

Figure 4: Environmental management programs.

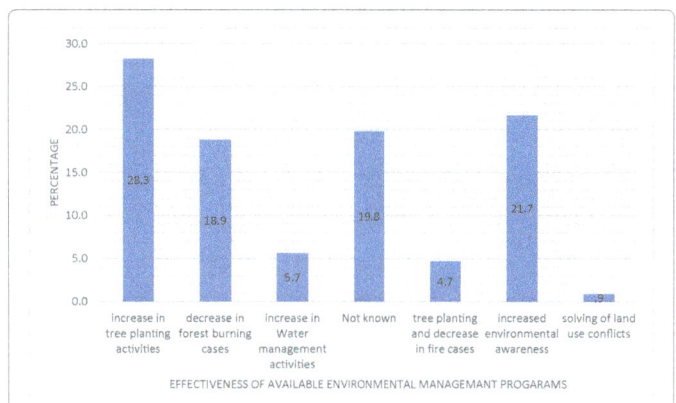

Figure 5: Effectiveness of available environmental management programs.

Coughing of residents near the industry	36	34.0
Chest diseases	15	14.2
Dying of fish	4	3.8
Crop diseases 'burning'	3	2.8
Food contamination	7	6.6
Not known	34	32.1
Iron sheet rust, flue and coughing	4	3.8
burning of crops, dust, coughing	3	2.8
Total	**106**	**100.0**

Table 5: Effects of the environmental problems at the study area.

References

1. REN21 (2012) Renewables 2012: Global Status Report.

2. United Nations Environment Programme (UNEP) (2009) Towards sustainable production and use of resources: Assessing Biofuels.

3. Dasappa S (2010) Potential of biomass energy for electricity generation in sub-Saharan Africa. Energy Sustain Dev 15: 203-213.

4. Smeets E, Faaij A (2007) Bioenergy potentials from forestry in 2050 - An assessment of the drivers that determine the potentials. Climatic Change 8: 353-390.

5. Smeets E, Faaij A, Lewandowski I (2004) A quick scan of global bioenergy potentials to 2050. An analysis of the regional availability of biomass resources for export in relation to the underlying factors. Report NWS-E-2004-109, Utrecht University, Netherlands.

6. Marrison I, Larson ED (1996) A preliminary estimate of the biomass energy

production potential in Africa in 2025. Considering projected land needs for food production. Biomass and Bioenergy 10: 337-351.

7. Songela AF (2010) A Capacity Building for Renewable Energy SMEs in Africa (CABURESA).

8. United Republic of Tanzania (2013) Scaling up renewable energy programme (Srep). Ministry of Energy and minerals.

9. Gwang'ombe FRD (2004) Renewable Energy Technologies in Tanzania. Biomass Based Cogeneration.

10. Jeppe B (2011) Biomass 2020 Opportunities, Challenges and Solutions.

11. Ministry of Energy and Minerals (2013) Power System Master Plan 2012 update.

12. MFA (2011) Private forestry and carbon trading project. Ministry for Foreign Affairs of Finland.

13. Clean development mechanism project design document form.

14. Kothari CR (2004) Research methodology. Methods and techniques.

15. Bailey BK (1994) Methods of Social Research.

16. Kumar R (2005) Research Methodology: A step by step guide for beginners.

17. Sandwell (1964) Mufindi Pulp Project site study. Tanganyika Government, Ministry of agriculture forest and wild life.

18. Thomas H, Tyner W, Birur D (2008) Biofuels for all? Understanding the Global Impacts of Multinational Mandates. GTAP.

19. Narain U, Gupta S, Veld K (2008) Poverty and the environment: Exploring the relationship between household incomes, private assets, and natural assets. Land Economics 84: 148-167.

20. Masozera KM (2002) Socio-economic impact analysis of the conservation of the Nyungwe forest reserve, Rwanda.

21. Reardon T, Vostii S (1995) Links between rural poverty and the environment in developing countries: Asset categories and investment poverty. World Development 23: 1495-1506.

22. African Biodiversity Network (2007) Agrofuels in Africa – The Impacts on Land, Food and Forests. Case Studies from Benin, Tanzania, Uganda and Zambia.

23. Nzalalila EV, Musokwa JWA, Haule AM (2012) Environmental audit report for Mufindi Paper Mills (MPM), National environmental management council.

26

Modified Fractionation Process via Organic Solvents for Wheat Straw and Ground Nut Shells

Prashant Katiyar[1],*, Shailendra Kumar Srivastava[2] and Vinod Kumar Tyagi[3]

[1]Sam Higgin Bottom Institute of Agriculture, Technology and Sciences, Allahabad, India
[2]Department of Biochemistry and Biochemical Engineering, Jacob School of Biotechnology and Bioengineering, Sam Higgin Bottom Institute of Agriculture, Technology and Sciences (Deemed University), Allahabad,India
[3]Department of Oil and Paint Technology, HBTI, Kanpur, India

Abstract

Modified organic solvent fractionation process involves the degradation of lignin, which is the main barrier of lignocelluloses biomass and the other two elements are cellulose and hemicelluloses. The treatment of organic solvent is given at a different concentration ratio, i.e. varies according to 10 ml mixture of ethyl alcohol and water, acetone and water is made for ground nut shells and wheat straw residues at elevated temperature and pressure conditions. Resulting in the decomposition of lignin and hemicelluloses hydrolysis and remaining solid residues mainly contains cellulose, which is further undergoing for enzymatic or microbial fermentation. The main aim of the study is to produce biodiesel in an efficient way by following the treatment of organic solvents. The present approach depends totally on the type of organic solvent and effect of process condition is highlighted that is useful for fractionation of lignocelluloses biomass. Statistical analysis is done to show the significant and non-significant values as well as standard deviation and standard errors are discussed here in this paper.

Keywords: Biomass pretreatment; Lignin; Lignocellulose biomass; Organic solvents

Introduction

The main objective of bio-refineries is to modify the fractionation process of biomass by utilizing the chemical functionalities process to coproduce bio-fuels as well as its byproducts [1].

Objective of Study

1.   To develop a modified method fractionation process using organic solvent and water combination, i.e. is capable to break the main barrier Lignin of lignocelluloses biomass.

2.   Determination of required optimization condition for organic solvent- water mixture.

3.   Selection of better residue, i.e. Wheat Straw or Groundnut shell for biodiesel production.

Lignocellulose biomass consists of three polymers: cellulose, hemicelluloses and lignin. These polymers are associated with each other to form a hetero-matrix in different degrees and found in varied relative composition depending on the type of resources [2-4] and the relative abundance of cellulose, hemicelluloses and lignin are, inter alia, key factors in determining the optimum energy conversion route for each type of lignocellulosic biomass [5].

The main aim of pretreatment process is to break the fibrous structure and separate the main barrier, i.e. lignin via following the physically/chemically methods, therefore, it allows the easy accessibility of cellulose fraction for saccharification and hydrolysis of hemicelluloses [6].

Traditionally, pretreatment technologies are optimized for sugar production and produce a solid residue containing lignin, un-hydrolyzed sugar polymers, minerals, added process chemicals and fermentable inhibitors like organic acids but the hydrolyzed product obtained from sugar and lignin fraction only. In addition, novel pretreatment technologies overcome traditional conventional technology because its prime targets are to achieve full fractionation of biomass including lignin and remaining solid mass has been utilized for the generation of power and heat and biochemical products as well.

Currently, pretreatment technologies are utilizing the combination of organic solvents and water or combination of organic solvents and catalyst (synthetic or natural catalyst) [7] for degrading the lignocelluloses biomass.

The big advantage of pretreatment technology: combination of organic solvents and water or combination of organic solvents and synthetic or natural catalyst is that the full biomass conversion into bio-fuels occurs with an efficient rate. Organic solvent methodology depends upon on process parameters (Physical and chemical parameters), type of feedstock and type of organic solvent used. Another Similar technology has been developed recently, Using tetra-hydro-furan (THF) as a co-solvent to aid in the breakdown of raw biomass feed stocks to produce valuable primary and secondary fuel precursors at high yields at moderate temperatures. Those fuel precursors can then be converted into ethanol, chemicals or drop-in fuels. Drop-in fuels have similar properties to conventional gasoline, jet and diesel fuels and can be used without significant changes are made in vehicles [8].

Here, the organic solvent methodology is performed on residues: Wheat straw, Groundnut shells.

Methodology

This experiment is performed in a lab scale; organic solvent

---

*Corresponding author: Katiyar P, Department of Biochemistry and Biochemical Engineering, Sam Higgin Bottom Institute of Agriculture, Technology and Sciences, Allahabad, India, E-mail: katprashant27@gmail.com

treatment is given to non edible biomass of Wheat straw and Ground nut shell. The procedure to perform the experiment is as follows: first of all the biomass is milled up to the approximate size of >0.7 mm. Subsequently, a mixture of organic solvent-biomass-water in a suspended form was made for 100 mg biomass approx. According to the requirement of 10 ml quantity of organic solvent and water (5% organic solvent and 5% water, 6% organic solvent and 4% water, 7% organic solvent and 3% water) of different concentration ratio were made (Figures 1 and 2).

A 10 ml mixture of organic solvent and water were made to treat the biomass under different conditions of temperature and pressure. Treated biomass undergo to the next step of biochemical analysis via (Spectro-photometric method is used for the analysis of extracts, Hemi cellulose content Cellulose content and Lignin content) [9] of both the filtrate and solid mass remain over the filter. The filtrate is heated at a temperature range of 150°C-210°C for 1 hour and then the suspension is settled in reaction time of around 60 mins and kept it for cooling.

Filtration of resulting slurry is performed and remaining solid residues were washed with an identical organic solvent-water mixture and then dried at 60°C. Both the Filtrate and washed suspension were undergoing for Spectro-photometric analysis. This analysis checks the level of oligomeric sugar and lignin content present after and before giving the treatment of organic solvent and water mixture.

Figure 1: Physical appearance of untreated biomass

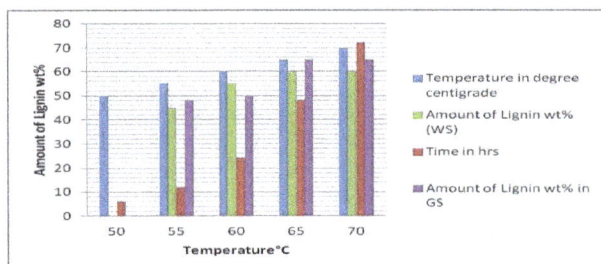

The above graph (1) is plotted between the Temperature (°C) vs. Amount of Lignin (weight %) which shows the amount of Lignin present in Groundnut Shells (GS) and Wheat Straw sample at a temperature range of 55°C-70°C. No treatment of organic solvents. (Reference graph) is given.
Figure 2: Wheat Straw and Ground nut shell before the treatment of organic solvent and water (Standard data)

Results

Hemicelluloses hydrolysis

Hemicelluloses are branched, heterogeneous polymers of pentoses (xylose, arabinose), hexoses (mannose, glucose, galactose) and acetylated sugars. The second most abundant polymer (20-50% of Lignocellulose Biomass) and it can be easily hydrolyzed thermochemically sensitive [10,11] due to presence of branches with short lateral chains [4,12]. In the present study, hydrolysis of hemicellulose take place via using organic solvent Ethyl alcohol and the Acetone and water combination. Hydrolysis of hemicelluloses occurs faster at lower concentration of organic solvent and water combination, i.e. 5:5 as compared to higher concentration of organic solvent and water combination i.e. 7:3. As depicted in Figures 3 and 4 Xylan hydrolysis was increasing. During the treatment of organic solvent, un-branched homo-polymers i.e. cellulose was hardly unaffected.

Lignin decomposition

The separation of lignin is a complex process known as delignification [13]. This process involves the degradation of lignin macromolecules into smaller fragments [14]. Almost all pretreatment methods can be used to fractionate the lignocelluloses biomass and the term "delignification" is associated with treatment that uses solvents. Here, organic solvents: Ethyl alcohol and water, Acetone and water combinations are used in a different concentration ratio of 5:5, 6:4 and 7:3. At these concentration all the sample were kept in Hot air oven at elevated temperature (55°C, 60°C and 65°C for 12 hrs, 24 hrs and 48 hrs) conditions required for the full fractionation of Lignocelluloses biomass i.e. lignin. (Figure 2 is easily compares with reference figure) This depicts that how much lignin remaining in the sample (GS) Groundnut shells and (WS) Wheat straw after the treatment with combination of organic solvents and water.

Alternative solvent used

Acetone, ethyl alcohol, acetyl chloride, hexane, dioxane or other organic solvents are used for the purpose of lignin decomposition. Lignin decomposition is indicated via mass loss and amount of lignin precipitated during the organosolvent experiment. Based upon the total mass loss, influence of solvent type on hemicelluloses hydrolysis is limited, but makes an easily accessible of other sugar, i.e. cellulose remaining in solid residues over the filter paper and the exposed sugar is utilized for further enzymatic sacchrification or microbial fermentation.

Discussion

Researchers are tried to develop the cost effective pretreatments technology using an organic solvent with water in combination. This technology is known as co-solvent technology. Recently another similar pretreatment technology develops by researchers of University of California, River side i.e. co-solvent-enhanced lignocellulosic fractionation (CELF) [15]. This technology definitely reduces the amount of enzymes required to breakdown the raw material which forms biofuels. This development could mean reducing enzyme costs from about $1 per gallon of ethanol to about 10 percents or less as well as more effective in terms of yield of ethanol twice as compared to dilute acid pretreatment technology.

Potential error

In the present study, co solvent technology is effective for bioconversion into bio-fuels but its real problem is that if organic solvents are not properly

WS 5:5 Ethyl Alcohol: Water

GS 5:5 Ethyl Alcohol: Water

GS 5:5 Acetone: Water

WS 5:5 Acetone: Water

WS 6:4 Ethyl alcohol: Water

WS 6:4 Acetone: Water

GS 6:4 Acetone: Water

GS 6:4 Ethyl alcohol: Water

WS 7:3 Ethyl alcohol: Water

WS 7:3 Acetone: Water

GS 7:3 Acetone: Water

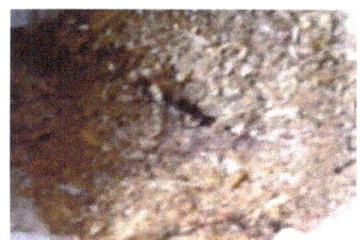

GS 7:3 Ethyl alcohol: Water

Figure 3: Physical appearance of Wheat straw (WS) and Groundnut shell (GS) residues after the treatment of organic solvent at different solvent concentration mixture.

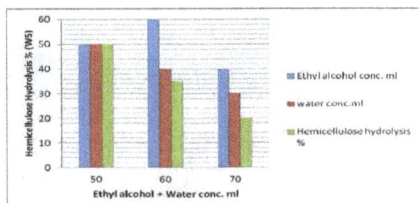

The most frequent hydrolysis are occurring at 5:5 (ethyl alcohol: water) as compared to higher concentration of 7:3 (ethyl alcohol: water).

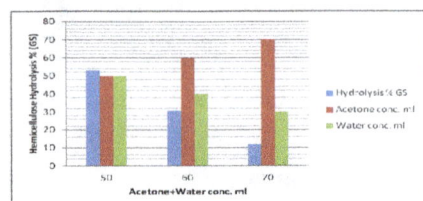

The most frequent hydrolysis were occurring at 5:5 (acetone: water) as compared to higher concentration of 7:3 (acetone: water).

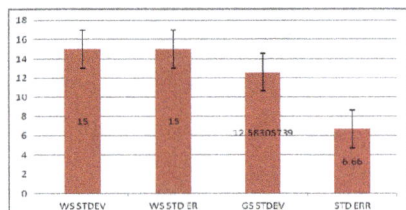

Graph shows standard deviation and standard error values according to the above graph plotted between the amount of lignin decomposed (wt%) vs Acetone +water conc.(ml) in wheat straw (WS) and Ground nut shells (GS).

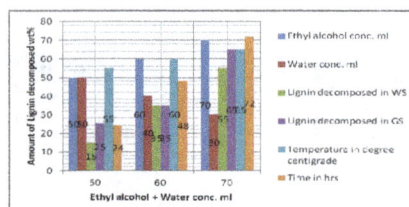

Representation of decomposition of Lignin in Wheat Straw (WS) and Groundnut Shells (GS) expressed in wt% by giving the treatment of organic solvent (Ethyl alcohol) and (Water) concentration expressed in ml at different Temperature (°C) for 24 hrs, 48 and 72 hrs.

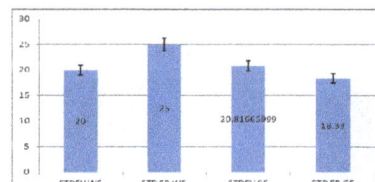

Standard Deviation and Standard Errors between the amount of lignin decomposed (wt%) vs. Ethyl alcohol + water conc.(ml) in wheat straw (WS) and Ground nut shells (GS).

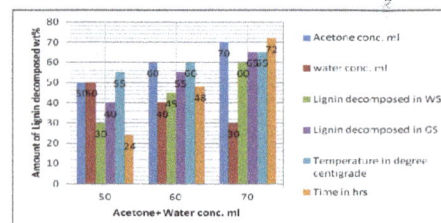

Plot of decomposition of Lignin in Wheat Straw (WS) and Groundnut Shells (GS) expressed in wt% by giving the treatment of organic solvent (Acetone) and (Water) concentration expressed in ml at different Temperature (°C) for 24 hrs, 48 hrs and 72 hrs.

Figure 4: According to figure 2 graphs is plotted between % hydrolysis of Hemicellulose vs. different used organic solvent.

eliminated by evaporation, some unwanted reactions occurring such as inhibitors production during fermentation which inhibits the actual yield of bio-fuels. To avoid this problem, researchers are utilizing solvents such as Tetra hydro furan (THF) having more volatility.

Conclusion

Based on the result drawn so far, the following important points are drawn out:

- Modification of organo solvent and water combination is an interesting approach which is useful for enhancing the delignification of biomass and increased hemicelluloses hydrolysis. Another approach acid or base used as a catalyst to increase the more hydrolysis of hemicelluloses. This approach becomes the subject of further study.

- Selection of the best residues for an efficient production of Bio-energy: this is decided on the basis of amount of lining remaining and reduced moisture content after the treatment of organic solvent. In present study clearly indicates the ground nut shells sample is better than the wheat straw sample.

- Additions of catalyst or organic solvents are used in different a physical condition which inhibit the formation of inhibitors and decreases the processing cost.

- Here the required condition is defined, i.e. At a different concentration ratio is 5:5, 6:4 and 7:3 of Ethyl alcohol + water and acetone + water at different temperature 55°C, 60°C, 65°C and ≥70°C for 24 hrs, 48 hrs and 72 hrs. Out of obtaining data the required optimum condition for maximum delignification of biomass is occurring at 65°C for 72 hrs.

- Acetone is more effective than ethyl alcohol in terms of degree of delignification at same physical condition.

The reason behind this color change is that the re-condensation or decomposition of lignin occurs from the lignin - carbohydrate complex. In addition, above pictures clearly indicate that the structure of residues has been dramatically changed as well as it becomes very harder as compared to fresh sample. In addition, the present study clearly indicates that acetone is more effective organic solvent as compared to ethyl alcohol at a concentration of 7%: 3% (% of acetone or alcohol: water) at 65°C for 72 hrs i.e. required condition to dissolve the lining completely.

Acknowledgement

I am thankful to my advisor Dr. Shailendra Kumar Shrivastava, SHIATS, Allahabad and Co-advisor Dr. Vinod Kumar Tyagi, Professor, Department of Oil and Paint Technology, HBTI, Kanpur for giving me a valuable suggestion and show me a right direction to accomplish this work. I am so much thankful to Dr. Alok Milton Lall (Head of the Department of Biochemistry and Biochemical Engineering), SHIATS (Jacob school of Biotechnology

Sample	Concentration (mg/ml) of		Color	Temperature (°C)	Time (hrs)
Wheat straw (WS) 100 mg	5% 5%	Ethyl alcohol: 5% water and Acetone: 5% water per 100 mg of WS	No color changes occur	55°C	24 hrs
Groundnut shells (GS) 100 mg	5% 5%	Ethyl alcohol: 5% water and Acetone: 5% water per 100 mg of GS	No color changes	55°C	24 hrs
Groundnut shells (GS) 100 mg	6% 6%	Ethyl alcohol: 4% water and Acetone: 4%water per 100 mg of GS	Dark Brown & Hard residue	60°C	48 hrs
Wheat straw (WS) 100 mg	6%	Ethyl alcohol: 4% water and 6% acetone: 4% water per 100 mg of WS	Darkish brown, thickness& Hardness increases	60°C	48 hrs
Wheat straw (WS) 100 mg		7% ethylalcohol:3% water and 7% acetone: 3% water per 100mg of WS	Darkish Yellowish Color & hardness Increases	65°C	72 hrs
Groundnut shells (GS) 100 mg	7%	Ethyl alcohol: 3% water and 7% acetone: 3%water per 100 mg of GS	Light brown, Hardness& Thickness increases	65°C	72 hrs

Table 1: Shows the physical changes occur in Wheat straw and Ground nut shells sample under different physicochemical conditions. A biochemical change is observed after the treatment with organic solvents such as color of residue changes from light yellow to dark brown.Statistical analysis: All the experiment was conducted in a triplicate manner in a laboratory. Using the Annova software, statistical analysis can be carried out to find out the experimental significant values at P<0.05 level.
Statistical analysis: Organic solvent: Solvent Conc. ratio of (Acetone: Water)

Acetone: water conc.(ml)	Lignin reduction (wt%) in (WS*)	Temperature °C	Time interval (hrs)
5:5	30%	55°C	24 hrs
6:4	45%	60°C	48 hrs
7:3	60%	65°C	72 hrs

X^2 cal=10>X^2tab (5%) 5.991 (S) *WS stands: Wheat Straw sample This table clearly indicates the significant values (S) of lignin reduction in WS sample after the treatment with acetone: water at different temperature range of 55-65°C for 24-72 hrs.
Table 2: Lignin reductions (wt %) in WS after the treatment with acetone: water mixture.

Acetone: water conc. (ml)	Lignin reduction (wt%) in (GS*)	Temperature °C	Time interval (hrs)
5:5	40%	55°C	24 hrs
6:4	55%	60°C	48 hrs
7:3	65%	65°C	72 hrs

X^2 cal=5.932<X^2tab (5%) 5.991 (NS) *GS stands: Groundnut shells sample This table clearly indicates the non significant values (NS) of lignin reduction in GS sample after the treatment with acetone: water at different temperature range of 55-65°C for 24-72 hrs.
Table 3: Lignin reductions (wt%) in GS after the treatment with Acetone: water mixture.

Ethyl alcohol: water conc. (ml)	Lignin reduction (wt%) in (WS*)	Temperature °C	Time interval (hrs)
5:5	15%	55°C	24 hrs
6:4	35%	60°C	48 hrs
7:3	55%	65°C	72 hrs

X^2 cal=22.85>X^2tab (5%) 5.991 (S) *WS stands: Wheat straw sample This table clearly indicates the significant values (S) of lignin reduction in WS sample after the treatment with ethyl alcohol: water at different temperature range of 55-65°C for 24-72 hrs.
Table 4: Lignin reductions (wt%) in WS after the treatment with Ethyl alcohol: water mixture.

Ethyl alcohol: water conc. (ml)	Lignin reduction (wt%) in (GS*)	Temperature °C	Time interval (hrs)
5:5	25%	55°C	24 hrs
6:4	35%	60°C	48 hrs
7:3	65%	65°C	72 hrs

X^2 cal=20.79>X^2tab (5%) 5.991 (S)   *GS stands: Groundnut shells sample This table clearly indicates the significant values (NS) of lignin reduction in GS sample after the treatment with ethyl alcohol: water at different temperature range of 55-65°C for 24-72 hrs.(S) = Significant values (NS) = Non significant values

Table 5: Lignin reductions (wt %) in WS after the treatment with Ethyl alcohol: water mixture

and Bioengineering), Allahabad who provided me a good facility in a laboratory which makes this work possible. In addition, my colleague Deepshika Kushwaha and lab attended Mr. Hemant and Mr. Sanjay gives me a lot of support during a practical work in a lab. No funding agencies are involved to support and grant this project. (Tables 1-5)

References

1. Kamm B, Gruber P, Kamm M (2005) Biorefineries- Industrial Processes and Products. Wiley-VCH, USA 964.

2. Carere CR, Sparling R, Cicek N, Levin DB (2008) Third generation biofuels via direct cellulose fermentation. Int J Mol Sci 9: 1342–1360.

3. Chandra RP, Bura R, Mabee WE, Berlin A, Pan X, et al. (2007) Substrate pretreatment: the key to effective enzymatic hydrolysis of lignocellulosics?. Adv Biochem Eng Biotechnol 108: 67–93.

4. Fengel D, Wegener G (1984) Wood Chemistry, Ultrastructure Reactions. Wiley, New York, USA.

5. Mckendry P (2002) Energy production from biomass (part 1): overview of biomass. Bioresource Technol. 83: 37–46.

6. Moiser N, Wyman C, Dale B, Elander R, Lee Y, et al. (2005) Features of promising technologies for pretreatment of Lignocellulosic biomass. Bioresource Technology 96, 673-86.

7. Huijgen WJJ, Theodosiadias K, Bakker RR, Reith JH (2007) Pretreatment of Lignocellulosic Biomass for production of Fermentable sugar and high quality Lignin by a Modified organosolv process. 15th European Biomass conference and Exhibition, Berlin, Germany.

8. Charles E Wyman (2014) Enhancing Biofuel Yields from Biomass with Novel New Method. University of California, Riverside's Bourns College of Engineering.

9. SK Thimmaiah (2004) Standard method of biochemical analysis. (1st edn.), Kalyani publisher, Ludhiana, Punjab, India .

10. Hendricks AT, Zeeman G (2009) Pretreatments to enhance the digestibility of lignocellulosic biomass. Bioresour Technol 100: 10–18.

11. Levan SL, Ross RJ, Winandy JE (1990) Effects of fire retardant chemical bending properties of wood at elevated temperatures. Research paper FPF-RP-498 Madison, WI: USDA, forest service. 24.

12. Ruiz HA, Ruzene DS, Silva DP, da Silva, FFM Vicente AA, et al. (2011) Development and characterization of an environmentally friendly process sequence (autohydrolysis and organosolv) for wheat straw delignification. Appl Biochem Biotech 164: 629-641.

13. Saha BC (2003) Hemicellulose bioconversion. J Ind Microbiol Biotechnol 30: 279–91.

14. Sun N, Rahman M, QinY, Maxim ML, Rodriguez H, et al. (2009) Complete dissolution and partial delignification of wood in the ionic liquid 1-ethyl-3-methylimidazolium acetate. Green Chem.11: 646-655.

15. Charles EW, Thanh YN, Charles MC, Rajeev K (2014) "Co-solvent Pretreatment Reduces Costly Enzyme Requirements for High Sugar and Ethanol Yields from Lignocellulosic Biomass". J Chem Sus Chem 8: (10) 1016.

27

Enhancing Photoelectric Conversion Efficiency of Solar Panel by Water Cooling

M Mohamed Musthafa*

School of Mechanical Engineering, SASTRA University, Thanjavur-613401, Tamilnadu, India

Abstract

Photovoltaic solar cell generates electricity by receiving solar irradiance. The electrical efficiency of photovoltaic (PV) cell is adversely affected by the significant increase of cell operating temperature during absorption of solar radiation. This undesirable effect can be partially avoided by fixing a water absorption sponge on the back side of the photovoltaic panel and maintain wet condition by circulation of drop by drop water through sponge. The objective of the present work is to reduce the temperature of the solar cell in order to increase its electrical conversion efficiency. Experiments were performed with and without water cooling. A linear trend between the efficiency and temperature was found. Without cooling, the temperature of the panel was high and solar cells achieved an efficiency of 8–9%. However, when the panel was operated under water cooling condition, the temperature dropped maximally by 4°C leading to an increase in efficiency of solar cells by 12%.

Keywords: Photovoltaic cell; Solar panel cooling; Photo-electric conversion efficiency; Water absorption sponge

Introduction

As the world is facing the problem of energy deficit, global warming and detoriation of environment and energy sources, there is need for an alternative energy resource for power generation other than use of fossil fuels, water and wind. Fossil fuel get depleted in next few decades, hydro power plants are depends on annual rainfall and wind power is also depends on climate changes. Solar energy is one of the comparable candidate for alternate energy source. Solar energy is a very inexhaustible source of energy. The power from the sun intercepted by the earth is approximately $1.8×10^{11}$MW which is larger than the present consumption rate on the earth of all commercial energy sources. Thus solar energy could be supply all the present and future energy needs of the world on a continuing basis. This makes it one of the most promising of the unconventional energy sources [1-4].

A solar cell is a device that directly converts the energy from sunlight in to electrical energy through the process of photovoltaics. The first solar cell was built around 1883 by Charles Fritts, who used junctions formed by coating selenium (a semiconductor) with an extremely thin layer of gold. In 2009, a thin film cell sandwiched between two layers of glass was made.

A typical PV module has an ideal conversion efficiency in the range of 15%. The remaining energy is converted into heat and this heat increases the operating temperature of PV system which affects the electrical power production of PV modules and this can also cause the structural damage of PV modules leads to shorting its life span and lowering conversion efficiency. The output power of PV module drops due to rise in temperature, if heat is not removed [5]. The temperature of the solar cell generally reach to the 80°C or more when the solar cell is a silicon series solar cell.

The various literatures reveal that cell temperature has a remarkable effect on its efficiency. The temperature increase of 1K corresponds to the reduction of the photoelectric conversion efficiency by 0.2%-0.5% [6]. Various studies have been conducted in order to improve the PV conversion efficiency, among these cooling provides a good solution for the low efficiency problem. Both water and air are suitable as the cooling fluid to cool the PV module in order to avoid the drop of electrical efficiency [7-12].

Performance of a solar-photovoltaic (PV) system not only depends on its basic electrical characteristics; maximum power, tolerance rated value %, maximum power voltage, maximum power current, open-circuit voltage, short-circuit current, maximum system voltage, but also is negatively influenced by several obstacles such as ambient temperature, relative humidity, dust storms and suspension in air, shading, global solar radiation intensity, spectrum and angle of irradiance [13,14].

There are several reasons which motivate the development of the PV/T system. One of the main reasons is that PV/T system can provide higher efficiency than individual PV and thermal collector system. With increased the efficiency, the payback period of the system can also be shorten [15]. Many efforts have been made to find an efficient cooling technology by analyzing the performance of solar cells using different technologies and various cooling liquids. The technique used in this study is the cooling of solar panel back side using water as the coolant. The main focus of this work is on comparison of the electrical conversion efficiency of the PV panel with and without cooling at optimum flow rate.

Materials and Methods

A commercial polycrystalline solar panel with an area of 36×27 cm² was tested. PV panel specifications are listed in Table 1. The experimental setup is consists of 12W power rating solar panel, 12V battery, volt meter, ammeter, solar lamp and cooling system. The photographic view of experimental set up is shown in Figure 1. The cooling system consists of 5 litre capacity water cane, hose with flow regulating knob, water absorbing sponge and drain pipe for collecting

*Corresponding author: Musthafa MM, School of Mechanical Engineering, Sastra University, Thanjavur-613401, Tamilnadu, India,
E-mail: mdm_712003@yahoo.co.in

Peak power	12W
Type	Poly-crystalline
Open circuit voltage	21.3V
Maximum power voltage	17.5V
Maximum power current	0.68A
Operating temperature	-40°C to 80°C
Number of cells	36
Dimensions	32×27 cm

Table 1: Solar panel specification.

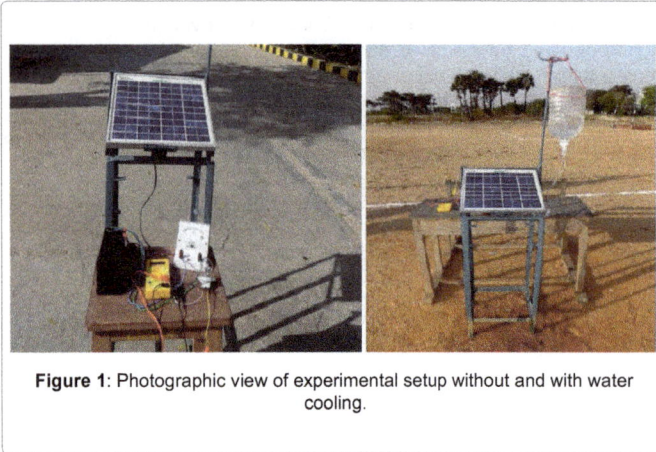

Figure 1: Photographic view of experimental setup without and with water cooling.

the water. The solar panel is placed on 3 feet mild steel stand with a tilt angle of 45°. The solar panel is connected to the positive and negative terminals of the battery through the voltmeter and ammeter. Battery is discharged with bulb load. Schematic diagram of output characteristics test system of solar panels is shown in Figure 2 Voltmeter and ammeter were used in range of 0-50 V and 1-10A respectively. 8 W dc bulb was used as the load. A solarimeter was used to measure the real-time solar radiation intensity (W/m²). Temperatures of the solar panel, ambience and the water in the tank was monitored with digital thermometer.

The water is supplied from the five litre capacity water cane to the sponge which is fixed on the back side of the solar panel through the hose. The flow rate of water is controlled by knob in the hose pipe line. The setup is placed towards south in the direct sunlight and the readings of ammeter and voltmeter are noted in hour by hour and the panel temperature was also noted using digital thermometer. Readings were recorded for every one hour on 3rd May 2014 from 8.00 am to 18.00 pm without water cooling. The same procedure was repeated from 4-8th May 2014 with water cooling by varying the flow rate from one litre/hr up to 3 litre/hr in step of 0.5 litre/hr. Weather conditions on those days are more or less same.

Results and Discussion

In order to find electrical conversion efficiency of the solar panel, the following parameters were measured, such as the output power in terms of voltmeter and ammeter reading, the panel surface temperature and real time solar radiation intensity(W/m²). In addition that ambience temperature, the inlet and outlet temperature of water flow and water flow rate were measured and recorded.

The photoelectric conversion efficiency is calculated as:

$$\eta = \frac{P_{max}}{AI}$$

where $\eta_=$ the photoelectric conversion efficiency (%), P_{max}(W) is the

maximum power generated from the PV panel, A(m²) is the surface area of the panel, and I(W/m²) is the solar irradiance incident on the panel. The maximum power generated is estimated by voltmeter and ammeter readings.

The theoretical cell electrical efficiency (ηe) and this parameter are functioned of the cell temperature [16].

$$\eta e = \eta o [1 - \beta (Tc - To)] \tag{1}$$

$$\eta e = \int VI\ dt Ac \int G(t) dt \tag{2}$$

The electrical efficiency of the PV module can be described as following equation:

$$\eta o = Vmp\ ImpG \tag{3}$$

The thermal efficiency can be computed with the following equation [17]:

$$\eta th = m \cdot Cp \int (Tout - Tin)\ dt Ac \int G(t) dt \tag{4}$$

The total efficiency of the hybrid PV/T system is:

$$\eta total = \eta th + \eta e = m \cdot Cp \int (Tout - Tin) dt + \int VI\ dt Ac \int G(t) dt \tag{5}$$

The electrical and thermal efficiencies are presented in Equation (2) and (4). It can be seen that the solar irradiation is a function of time and those parameters which are affected by the solar irradiation, such as inlet and outlet temperatures, PV voltage and PV current, are also functions of time. That is the reason to integrate the equation with time.

Figure 3 represents peak output efficiency of solar panel against mass flow rate of cooling water. As seen from Figure 3, two litres per hour mass flow rate of cooling water gave better performance of solar panel. It might be the water absorption capacity of the sponge. This describes that beyond two litres of water flow per hour is not stay in the sponge results in decrease in peak efficiency of the panel. It concludes that 2 litres per hour is an optimal flow rate of water for conducting the test.

Figures 4-6 shows the comparing results between the solar panel without cooling and two litres per hour flow of cooling water. The average air temperature, the radiation intensity, the maximal and average wind speeds are 39.6°C, 1070 W/m², 4.32 m/s and 0.61 m/s, respectively. The daily net radiation is 24.9 MJ from 8:00 to 19:30 hours

Figure 4 represent comparison on temperature of solar panel between cooling and without cooling. From the result, it is observed that the temperature of the solar panel with water-cooling reduces maximally by 4°C and averagely by 1.7°C at two litres per hour flow rate of water compared with ordinary one.

Figure 5 show that comparison of power output per hour of solar panel between cooling and without cooling. As seen from the Figure 5, the output power of solar panel first increases and then decreases. The highest values of power output appears in the time range between 12:00 to 13:00. The output power of the solar panel with cooling increases maximumly by 6.4% and averagely by 4.3% compared with ordinary one.

Figure 6 shows comparisons on output efficiency per hour of solar panel between cooling and without cooling. From the experimental result it is found that the efficiency of solar panel with cooling increases maximally by 2.69% and averagely by 0.39% compared with ordinary one. The maximum efficiency of 11.84% was achieved with water cooling of the panel and corresponding maximum efficiency ordinary

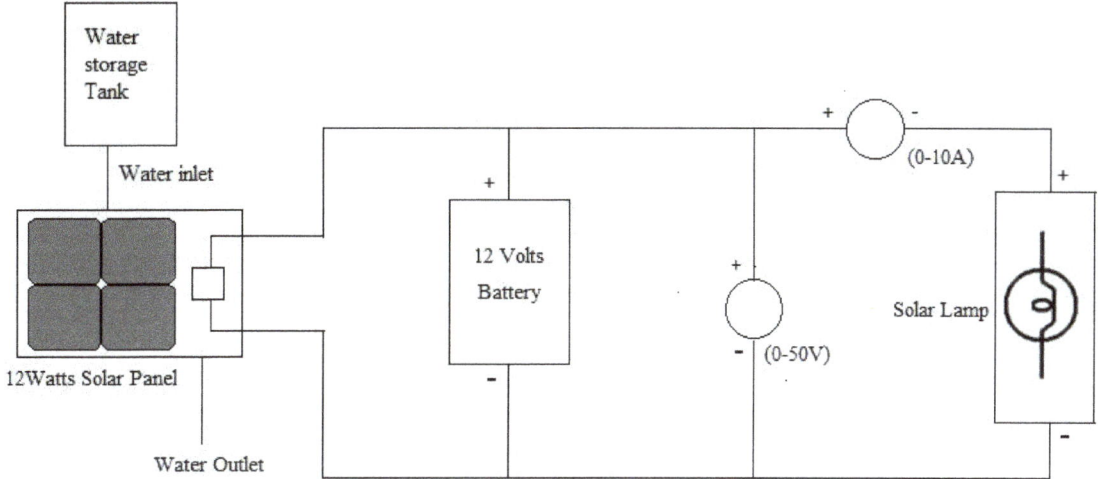

Figure 2: Output characteristics test system of solar panels with water cooling.

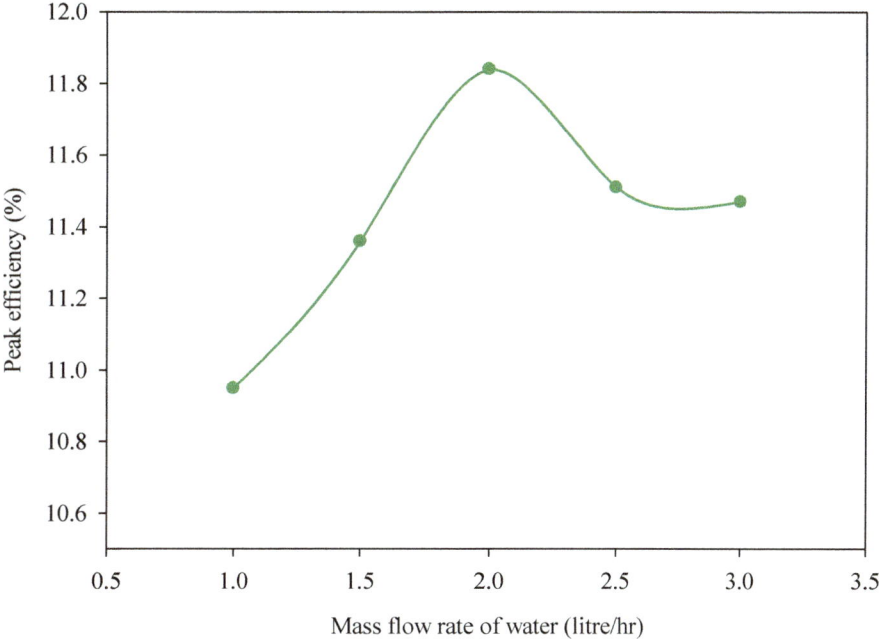

Figure 3: PeaTk efficiency of solar panel against mass flow rate of water.

panel is 9.15%.

Conclusions

A novel sponge arrangement at back side of solar panel for cooling is

proved better results. The results indicate that under cooling condition, the temperature can be reduced to effectively increase the photoelectric conversion efficiency of solar panel.

Compared with the ordinary solar panel, the water cooling

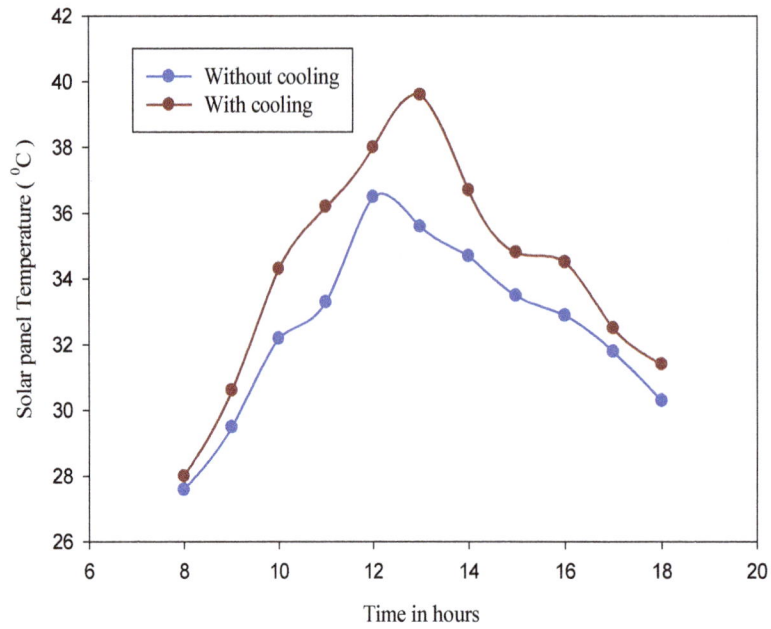

Figure 4: Comparisons on solar panel temperature between cooling and without cooling.

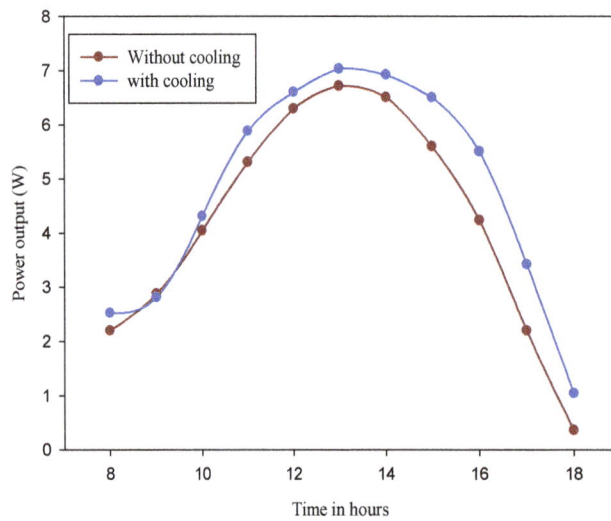

Figure 5: Comparisons on power output per hour of solar panel between cooling and without cooling.

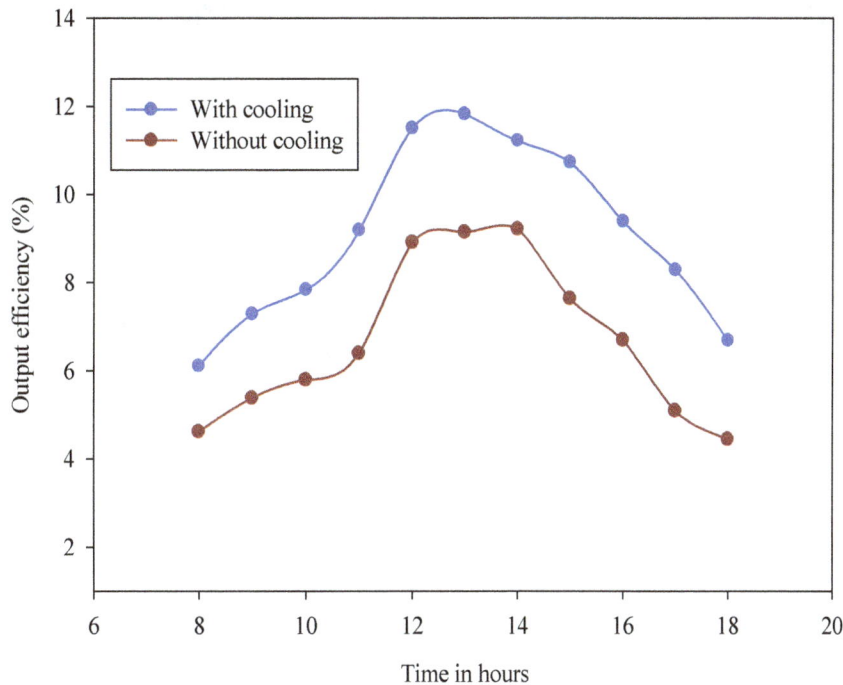

Figure 6: Comparisons on output efficiency per hour of solar panel between cooling and without cooling.

arrangement reduces cell temperature maximally by 4⁰C, the output power increases maximally by 6.4%, and increase in output efficiency by 2.6%.

The very low cost of water absorption sponge may be used as component for solar panel cooling for enhancing photoelectric conversion efficiency.

Simple attachment and life of the sponge is also six month.

Replacement of the sponge is easy and quick.

References

1. Zhu L, Wang YP, Fang ZL, Sun Y, Huang QW (2010) An effective heat dissipation method for densely packed solar cells under high concentrations. Solar Energy Mat Solar Cells 94: 133.

2. Sayran A, Abdulgafar, Omar S, Kamil M, Yousif (2014) Improving The Efficiency Of Polycrystalline Solar Panel Via Water Immersion Method. Int J Innovative Res Sci Engg and Technol 3: 83-89.

3. Teo HG, Lee PS, Hawlader MNA (2012) An active cooling system for photovoltaic modules. Appl Energy 90: 309–315.

4. Tang X, Quan Z, Zhao Y (2010) Experimental Investigation of Solar Panel Cooling by a Novel Micro Heat Pipe Array. Energy Power Engg 2: 171-174.

5. Mehrotra S, Rawat P, Debbarma M, Sudhaka K (2014) Performance of A Solar Panel With Water Immersion Cooling Technique. Int J Sci Environ Technol 3: 1161 –1162.

6. Weng ZJ, Yang HH (2010) Primary Analysis on Cooling Technology of Solar Cells under Concentrated Illumination. Energy Technol 29: 507-517.

7. Dinesh S, Borkar, Sunil, Prayagi V, Gotmare J (2011) Performance Evaluation of Photovoltaic Solar Panel Using Thermoelectric Cooling. Int J Engg Res 3: 536-539.

8. Gardas BB, Tendolkar MV (2012) Design of Cooling System for Photovoltaic Panel for Increasing its Electrical Efficiency. Int J Mechanical Prod Engg 1: 63-67.

9. Rodriguez M, Horley D, Gonzalez-Hernandez PP, Vorobiev J, Gorley PN (2005) Photovoltaic solar cells performance at elevated temperatures. Solar Energy 78: 243–250.

10. Royne A, Dey CJ, Mills DR (2005) Cooling of photovoltaic cells under concentrated illumination: a critical review. Solar Energy Mat Solar Cells 86: 451–453.

11. Zhu L, Robert F, Boehm, Wang Y, Halford C, et al. (2011) Water immersion cooling of PV cells in a high concentration system. Solar Energy Mat Solar Cells 95: 538-535.

12. Chapin DM, Fuller CS, Pearson GL (2006) Hybrid photovoltaic and thermal solar-collector Designed for Natural circulation of water. Appl Energy 83: 199-210.

13. Joshi AS, Tiwari A, Tiwari GN, Dincer I, Reddy BV (2009) Performance evaluation of a hybrid photovoltaic thermal (PV/T) (glass-to-glass) system. Int J Thermal Sci 48:154.

14. Dubey S, Sandhu GS, Tiwari GN (2009) Analytical expression for electrical efficiency of PV/T hybrid air collector. Appl Energy 8: 697-705.

15. Chow TT (2010) A review on photovoltaic/thermal hybrid solar technology. Appl Energy 87: 365-9.

16. Rott N (1990) Note on the history of the Reynolds number. Annual Review of Fluid Mechanics 22: 1–11.

17. Gnielinski V (1996) New equations for heat and mass transfer in turbulent pipe and channel flow. Int Chem Engg 16: 359–368.

28

Thermodynamic Analysis of a Two Stage Vapour Compression Refrigeration System Utilizing the Waste Heat of the Intercooler for Water Heating

Aftab Anjum*, Mohit Gupta, Naushad A Ansari and Mishra RS

Department of Mechanical Engineering, Delhi Technologial University, Shahbad Daulatpur, Main Bawana Road, Delhi, India

Abstract

This article focuses on the various aspect of use of waste heat from intercooler of a multi-stage vapor compression refrigeration system using ammonia as a refrigerant. Energy and exergy analysis of the multi- stage refrigeration system having an intercooler is performed. The COP of such a system is found to be increased by 4 to 5%, and calculated to be approximately 3.24. Heat recovery through the intercooler proved to be beneficial as the COP of the system is improved along with heat recovery of 20 kJ/s. The analysis can represent a physical system where water can be heated through the heat extracted from the intercooler.

Keywords: Two-stage vapour compression system; Intercooler; Heat recovery; Coefficient of performance; Waste heat recovery

List of Symbols/Abbreviations:

GWP: Global Warming Potential;
ODP: Ozone Depletion Potential;
COP: Coefficient of Performance;
MMVCS: Modified Multistage Vapour Compression System;
EDR: Exergy Destruction Ratio;
HP: High Pressure;
LP: Low Pressure;
$\dot{Q}$: Rate of Heat Transfer (kW);
$\dot{W}$: Work Rate (kW);
X : Exergy Destruction Rate (kW);
$\dot{X}$: Exergy Rate (kW);
$\dot{m}_r$: Rate of Thermal Exergy Flow Rate (kW);
$\dot{m}_r$: Mass Flow Rate of Refrigerant (kg/s);
T: Temperature (K);
S: Specific Entropy (kJ/kg-K);
H: Specific Enthalpy (kJ/kg);
V: Velocity of Fluid (m/s);
Greek symbols
H: Efficiency;
E: Effectiveness;
X: Specific Exergy;
Δ: Efficiency defect;
Subscripts
E: Evaporator;
Comp: Compressor;
C: Condenser;
T: Throttle valve;
J: j^th component of the system;
R: Region to be cooled or refrigerant;
I: Inlet to the control region;
E: Outlet to the control region;
R: Refrigerator;
O: Ambient state;
Rev: Reversible;
Sub: Sub-cooling;
Su: Superheat;
Vcr: Vapour compression refrigeration system

Introduction

Heat rejected from refrigeration air conditioning plants is of low grade quality. Due to the high costs related with the recovery of such heat and the availability of alternate means for meeting low grade heat requirements, low grade waste heat is generally rejected to the atmosphere.

Kaushik et al. [1] presented an investigation of the feasibility of the heat recovery from the condenser of a simple vapour compression refrigeration system through a Canopus heat exchanger which acts as an auxiliary condenser between the compressor and condenser components. Results were compared for different working fluids and found that heat recovery factor of the order of 2.0 and 40% of condenser heat can be removed through the canopus heat exchanger.

Zubair and Yakub [2] investigated and compared the results with the existing practices in the industry. Furthermore, the theoretical results of a two-stage refrigeration system performance were also compared with experimental values for the refrigerants mentioned above.

Nikolaidis et al. [3] observed the performance of two-stage compression refrigeration having flash-chamber and water-intercooler using refrigerants R22 has been demonstrated by the thermodynamic analysis. Any reduction in irreversibility rate of the condenser gives approximately 2.40 times greater reduction in the irreversibility rate for the whole plant, because the changes in the temperature in the condenser and the evaporator contributed significantly to the overall plant performance.

Rahman et al. [4] presented the performance of the recently developed integrated space condition system. A conventional split type

*Corresponding author: Aftab Anjum, Department of Mechanical Engineering, Delhi Technologial University, Shahbad Daulatpur, Main Bawana Road, Delhi 110042, India, E-mail: aftabanjum915@gmail.com

air conditioner is modified to reclaim the superheated portion of the heat leaving the compressor to be utilized for the space conditioning purposes. The end result was expected to be faster cooling and prolonged compressor life.

Cabello et al. [5] proposed the main operating variables on the energetic characteristics of the vapour compression plant. They concluded that the refrigerant mass flow rate is largely dependent on the suction specific volume and therefore on the suction conditions.

Apera et al. [6] performed of a vapour compression refrigeration plant using as working fluids R22 and R417a. The problem related to the replacement of the fully HCFC and of the partially HCFC have been only partially solved. Refrigerant fluid experimentally tested as a substitute for R22 is the non-azeotropic mixture R417a.

Ouadha et al. [7] calculated components exergetic loses by operating at constant evaporating temperature of -30°C and condensation temperatures of 30, 40, 50 and 60°C with two natural substitutes of HCFC22, namely propane (R290) and ammonia (R717) as working fluids. They found that the optimum interstage pressure for a two-stage refrigeration system is very close to the saturation pressure.

Arora and Kaushik [8] proposed a detailed exergy analysis of an actual vapour compression refrigeration (VCR) cycle. A computational model was developed for computing coefficient of performance (COP), exergy destruction, exergetic efficiency and efficiency defects for R502, R404A, and R507A. The results indicate that R507A is a better substitute to R502 than R404A. The efficiency defect in condenser is highest and lowest in liquid vapour heat exchanger for the refrigerants considered [9-12].

Santiago et al. [13] found that two stage cycle in case of high temperature difference between heat sink and heat source in order to overcome the high pressure ratios that deteriorate compressor volumetric and isentropic efficiency. Results showed that optimum intermediate pressure is close to the arithmetic mean in case of R-404a but there is a significant difference in case of ammonia.

Mishra et al. [14] worked on the replacement of R22 with several environment friendly refrigerants. They used hydrocarbon (HCF) refrigerants like R134a, R410a, R407C and M20. Out of the above stated refrigerants R407C proved to be a potential HFC refrigerant which can replace R22 with minimum investment and efforts. It proved to be a non-ozone depleting refrigerant giving high system efficiency [15-17].

Mishra [18] described a thermodynamic modeling of a vapour compression refrigeration system using R134a in primary circuit and Al_2O_3-water based nanofluids in the secondary circuit. Simulation showed that for the same geometric characteristics of the system performance increases from 17% to 20% by application of nanofluid as a secondary fluid in VCS [19-22].

Xiaoui She et al. [23] proposed a new sub-cooling method for vapor compression refrigeration system depending on expansion power recovery. Liquid refrigerant is sub-cooled by using evaporative cooler. This makes a hybrid refrigerant system. Analysis is to done by using different refrigerants and results shows that hybrid vapor compression refrigeration have more (C.O.P) than conventional vapor compression refrigeration system.

N Upadhya [24] presented a concept of effect of sub-cooling on performance of refrigeration system. In this a diffuser is used after condenser which converts kinetic energy in to the pressure energy of refrigerant, it results in reduction of power consumption and reduction

of condenser size. After studying of all above techniques, concludes that it will be helpful for future research.

M Yang et al. [25] Performed enhancement and exergy destruction analyses were conducted numerically for vapor-compression refrigeration systems using R22, R134a, R410A, and R717. The effects of cooling water in a subcooler, refrigerant pressure drop among heat exchangers, and superheating in an evaporator were also considered and compared with evaporation and condensation temperature. Optimal degrees of subcooling obtained according to the second law of thermodynamics were consistently higher than those obtained according to the first law of thermodynamics.

Going through the historical background of various research articles, the gap in the literature review is found that the thermodynamic analysis of a two stage vapor compression refrigeration system with intercooler and flash chamber both in the cycle using ammonia as a refrigerant yet to be performed and also to evaluate the amount of heat from the intercooler which can be recovered for various applications.

Thermodynamic modelling of a modified two stage vapour compression refrigeration system

The present study focuses on the waste heat recovery from a two-stage VCR system from intercooler. The purpose of incorporating heat recovery is to further decrease the temperature of refrigerant (ammonia) coming out of the intercooler exit thereby reducing the net work done on the system and hence increasing the overall coefficient of performance of the system. A computational model was developed for carrying out the analysis of the system using Engineering Equation Solver software EES [22].

System description: The present work is on this modified multistage vapour compression refrigeration (MMVCS) system. The compressed refrigerant is passed through a water intercooler where the temperature is reduced down to the state 3' is directed into the flash chamber (process 3-4), where saturated liquid and saturated vapour refrigerant at intermediate pressure is separated out, and then it passes through high stage compressor (process 4-5) adiabatically, followed by condenser where it condenses (process 5-6) at constant pressure as a result of the removal of heat of condenser Q_{cond} and undergoes adiabatic expansion through high stage expansion valve, accompanied by a drop in pressure at constant enthalpy (process 6-7), it passes through again to flash chamber at intermediate pressure (process 7-8). The liquid refrigerant coming from the flash chamber undergoes adiabatic expansion through the low stage expansion valve, accompanied by a drop in pressure at constant enthalpy (process 8-9), low pressure vapour from the evaporator is compressed by the compressor and the cycle is repeated. In order to recover heat from the intercooler, an external fluid (water) is used to remove heat as shown in (Figures 1 and 2).

Thermodynamic modelling

Energy analysis:

The first law of thermodynamics or energy balance for the steady flow process of an open system is given by:

$$\dot{Q}_i + W_i + \sum \dot{m}_i \left(h_i + \frac{v_i^2}{2} + gz_i \right) = \dot{Q}_e + W_e + \sum m_e \left(h_e + \frac{v_e^2}{2} + gz_e \right) \quad (1)$$

Evaporator:

Heat extracted in the evaporator:

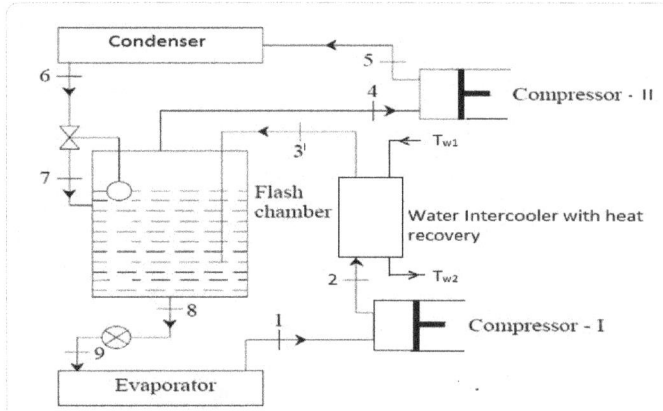

Figure 1: Two stage VCRS with waste heat recovery from intercooler using water as heat carrying media.

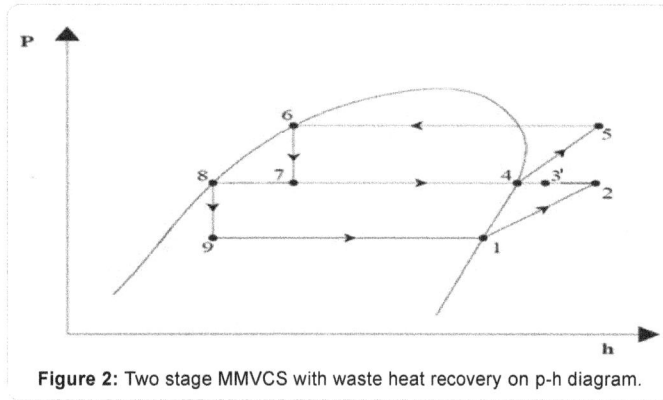

Figure 2: Two stage MMVCS with waste heat recovery on p-h diagram.

$$\dot{Q}_{9-1} = \dot{Q}_e = \dot{m}_{LP}(h_1 - h_9) \tag{2}$$

Where m_{lp} = mass flow rate of refrigerant in low stage compressor

$$\dot{m}_{LP} = TR\left[\frac{3.5167}{h_1 - h_9}\right] \tag{3}$$

Low stage compressor:

Ideal work input to compressor:

$$\dot{W}_{LP} = W_1 = \dot{m}_{LP}\left(h_2 - h_1\right) \tag{4}$$

$$\eta_{comp,lp} = \frac{W_{isentropic}}{W_{actual}} = \left(\frac{h_3 - h_1}{h_2 - h_1}\right) \tag{5}$$

$$W_{LP} = \dot{m}_{LP}\left(h_2 - h_1\right) \tag{6}$$

$$W_{HP} = \dot{m}_{HP}\left(h_5 - h_4\right) \tag{7}$$

$$W_{net} = W_{LP} + W_{HP} \tag{8}$$

$$COP = \frac{Q_e}{W_{net}} \tag{9}$$

$$W_{HP,1} = \dot{m}_{HP,1}\left(h_5 - h_4\right) \tag{10}$$

$$W_{net,1} = W_{LP} + W_{HP,1} \tag{11}$$

$$COP_{new} = \frac{Q_e}{W_{net,1}} \tag{12}$$

Condenser":

Heat rejected by the condenser to the condenser to the surrounding is given by:

$$Q_c = Q_e + W_{net} \tag{13}$$

$$Q_{wic} = \dot{m}w * C_w * \left(T_{w2} - T_{w1}\right) \tag{14}$$

The effectiveness ε_{wic} of the water intercooler with heat recovery can be written as

$$T_3 = T_{3,1} + \varepsilon_{wic} * \left(T_3 - T_{w1}\right) \tag{15}$$

Exergy analysis

Evaporator:

$$\dot{E}D_E = \dot{X}_9 + \dot{Q}_E\left(1 - \frac{T_o}{T_e}\right) - \dot{X}_1 = \dot{m}_{LP}\left[(h_9 - h_1) - T_o(s_9 - s_1)\right] + \dot{Q}_E\left(1 - \frac{T_o}{T_e}\right) \tag{16}$$

Compressor-1:

$$\dot{E}D_{comp1} = \dot{X}_1 + \dot{W}_{comp1} - \dot{X}_3 = \dot{m}_{LP}\left[T_o\left(s_1 - s_2\right)\right] \tag{17}$$

Condenser:

$$\dot{E}D_c = \dot{X}_5 - \dot{Q}_c\left(1 - \frac{T_o}{T_c}\right) - \dot{X}_6 = \dot{m}_{HP,1}\left[(h_5 - h_6) - T_o(s_5 - s_6)\right] - \dot{Q}_c\left(1 - \frac{T_o}{T_c}\right) \tag{18}$$

Water Intercooler with heat recovery:

$$\dot{E}D_{wic} = \left(\dot{X}_3 + \dot{X}_{w1} - \dot{X}_{31} - \dot{X}_{w2}\right) = \dot{m}_{LP}\left[(h_3 - h_{31}) - T_o(s_3 - s_{31})\right] + \dot{m}_w\left[(h_{w1} - h_{w2}) - T_o(s_{w1} - s_{w2})\right] \tag{19}$$

Flash chamber:

$$\dot{E}D_{flchm} = \left(\dot{X}_{3,1} + \dot{X}_7 - \dot{X}_8 - \dot{X}_4\right) = \dot{m}_{LP}\left[(h_{3,1} - h_8) - T_o(s_{3,1} - s_8)\right] + \dot{m}_{HP,1}\left[(h_7 - h_4) - T_o(s_7 - s_4)\right] \tag{20}$$

Throttle valve-1:

$$\dot{E}D_{TV1} = \dot{X}_6 - \dot{X}_7 = \dot{m}_{hp,1}\left[(h_6 - h_7) - T_o(s_6 - s_7)\right] \tag{21}$$

Throttle valve-2:

$$\dot{E}D_{TV2} = \dot{X}_8 - \dot{X}_9 = \dot{m}_{LP}\left[(h_8 - h_9) - T_o(s_8 - s_9)\right] \tag{22}$$

Compressor-2:

$$\dot{E}D_{comp2} = \dot{X}_4 + \dot{W}_{comp2} - \dot{X}_5 = \dot{m}_{HP,1}\left[T_o\left(s_4 - s_5\right)\right] \tag{23}$$

Total exergy destruction

It is the sum of exergy destruction in different components of the system.

$$\dot{E}D_{Total} = \dot{E}D_E + \dot{E}D_{comp1} + \dot{E}D_{comp2} + \dot{E}D_{wic} + \dot{E}D_c + \dot{E}D_{TV1} + \dot{E}D_{TV2} + \dot{E}D_{flchm} \tag{24}$$

Exergetic efficiency

$$\eta_{exergetic} = \frac{\dot{Q}_E\left|\left(1 - \frac{T_o}{T_r}\right)\right|}{\dot{W}_{comp}} = \frac{COP_{ver}}{COP_{rev}} \tag{25}$$

Exergy destruction ratio (EDR)

$$EDR = \frac{\dot{E}D_{total}}{\dot{X}_E^Q} \tag{26}$$

$$EDR = \frac{1}{\eta_{exergetic}} - 1 \qquad (27)$$

Efficiency defect (δ_j):

$$\delta_j = \frac{\dot{ED}_j}{\dot{w}_{comp}} \qquad (28)$$

Selection criteria of the refrigerant, input parameters and their values used for analysis

Ammonia has better heat transfer properties than most of chemical refrigerants and therefore it allows for the use of equipment with a smaller heat transfer area, it has highest refrigerating capacity per pound of any refrigerant, specific volume of ammonia is high, and the compressor displacement required per ton of refrigeration is quite small because wet state ammonia is compressed when an intercooler is used along with flash chamber.

A mathematical computational model is developed for performing the energy and exergy analysis of the Integrated Refrigeration System as shown in (Figure 1) using EES software [22].

(a)   Refrigerant: Ammonia

(b)   Effectiveness of the water intercooler ε_{wic} : 0.85

(c)   Condenser temperature T_c : 40°C-52°C

(d)   Evaporator temperature T_E : -40°C

(e)   Isentropic efficiency of compressor: (ηcomp,lp)=0.8

(f)   Ambient state temperature (To): 298 K

(g)   Water inlet temperature flowing through the condenser (Tw1):290 K

(h)   Ambient atmospheric pressure (Patm): 101.325 Kpa

(i)   Tons of refrigeration (TR): 10

Mass flow rate of cooling fluid (water) is varied from 0.18 to 0.85, while discussing its effect on the system performance.

Results and Discussion

An extensive exergy analysis has been performed and the results obtained are arranged in the tabular form. Also the comparison between various parameters has been done e.g. after varying evaporator temperature, ambient state temperature, and mass flow rate of water has been represented in (Figures 3-9).

The effect of evaporator temperature and condenser temperature on COP, heat recovery by water intercooler, exergetic efficiency, exergy destruction ratio, efficiency defect, rise in temperature and mass flow rate with various heat recovery parameters are presented in (Figures 3-9), it reveals that:

• With the increase in condenser temperature the total exergy destruction increases at a fixed evaporator temperature.

• Exergetic efficiency decreases with the increase in condenser temperature. Exergetic efficiency has a maximum value of 0.6852 at T_c=313 K, T_e=223 K.

• Evaporator is the most efficient component in the system whereas

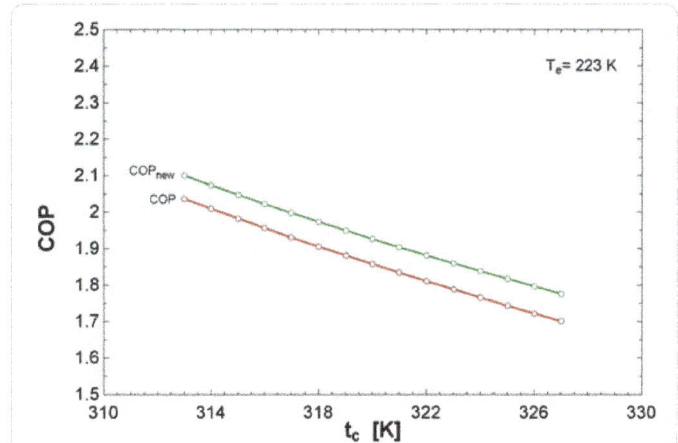

Figure 3: Showing variation of COPs with condenser temperature (t_c) at T_e=223 K.

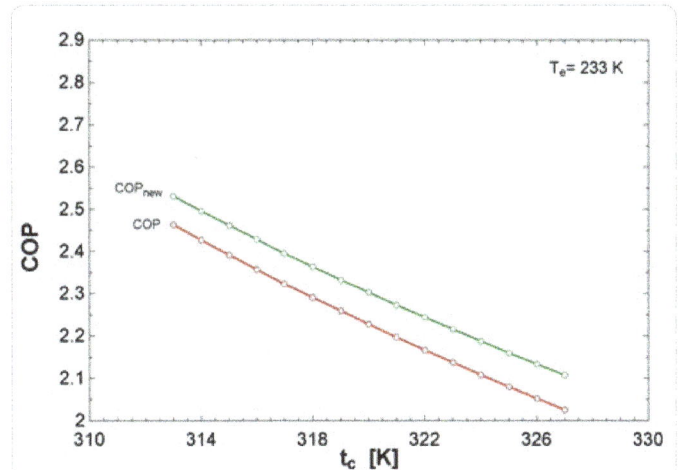

Figure 4: Showing variation of COPs with condenser temperature (t_c) at T_e=233 K.

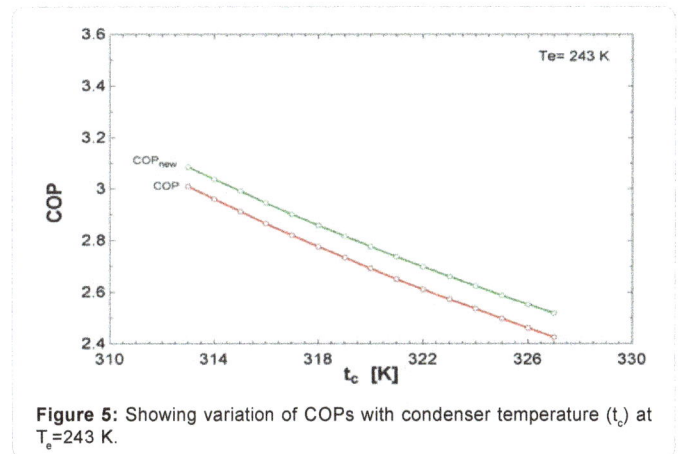

Figure 5: Showing variation of COPs with condenser temperature (t_c) at T_e=243 K.

throttling valve (corresponding to the HP side) is the worst component followed by the throttling vale of LP side, condenser, compressor, water intercooler and flash chamber.

• With the increase in ε_{wic} there is an increase in the COP_{new} for a given evaporator and condenser temperature. The maximum value of COP_{new} obtained is approx. 3.24 at T_C = 303 K, T_e = 303

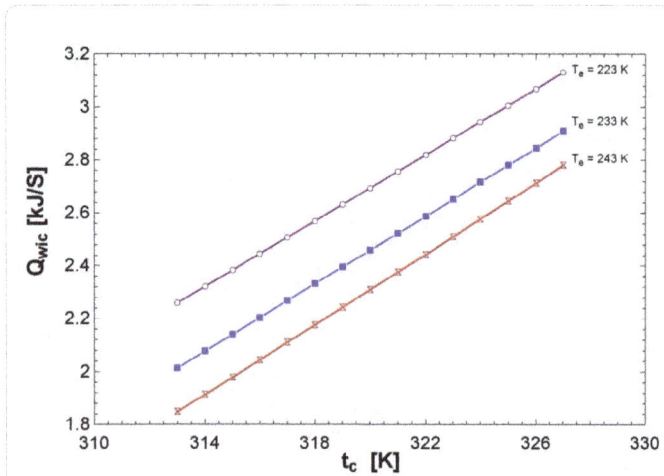

Figure 6: Showing variation of Q_{wic} with condenser temperature (t_c) at different evaporator temperatures.

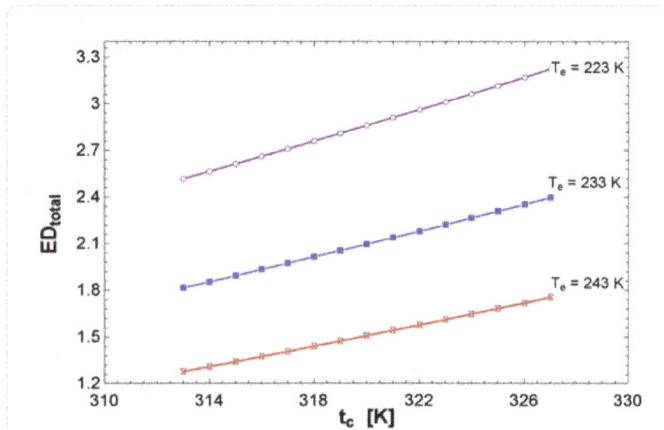

Figure 7: Showing variation of ED_{total} with condenser temperature (t_c) at different evaporator temperatures.

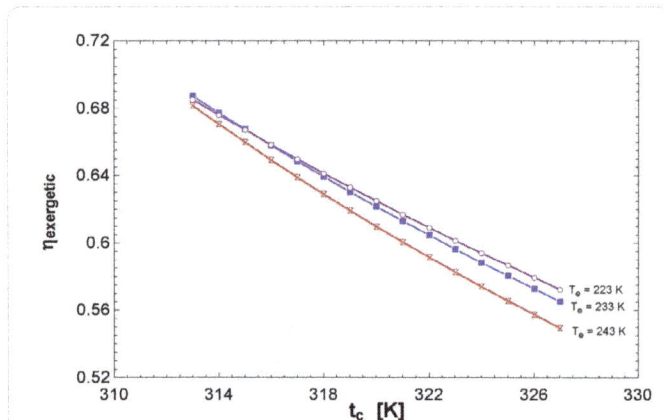

Figure 8: Showing variation of $\eta_{exergetic}$ with condenser temperature (t_c) at different evaporator temperatures.

K and ε_{wic} =0.95.

- Q_{wic} decreases with the increase in T_{w1}. Q_{wic} is maximum at T_{w1} = 280 K, T_c= 313 K, T_e = 223 K with a value of 2.83 kJ/s.

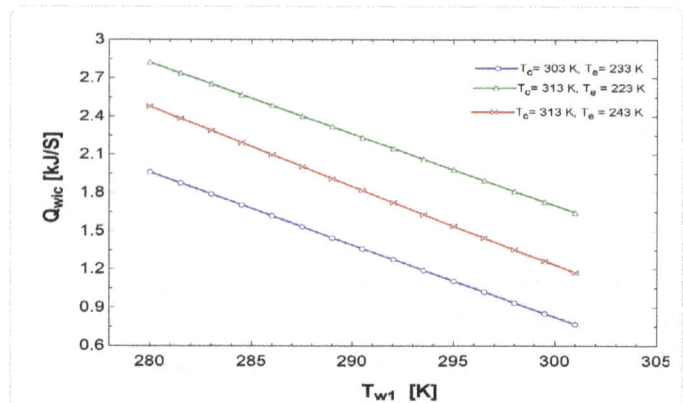

Figure 9: Showing variation of Q_{wic} with inlet water temperature (T_{w1}) at different evaporator & condenser temperatures.

- With the increase in $\eta_{comp,lp}$, $\eta_{exergetic}$ decreases at a given evaporator and condenser temperature.

- T_{w2} increases with increase in TR from 10 TR to 100 TR. At TR =100 heat recovery of 20.15 Kj/s is achieved with T_{w2}=337.2 K. 100TR.

Model validation

Thermodynamic modeling of similar system done by Ouadha et al. predicted components exergetic loses by operating at constant evaporating temperature of -30°C and condensation temperatures of 30, 40°C, with two natural substitutes of HCFC22, namely propane (R290) and ammonia (R717) as working fluids, which can be validated for the same as calculated in the present study to be constant evaporating temperature of -30°C and condensation temperatures of 40°C respectively. The error may be due to approximation but in agreement to be accepted. Similarly, optimal degrees of sub-cooling obtained according to the second law of thermodynamics, Yang et al. enhancement and exergy destruction analyses were conducted numerically for vapor-compression refrigeration systems using R22, R134a, R410A, and R717. The effects of cooling water in a sub-cooler, refrigerant pressure drop among heat exchangers, and superheating in an evaporator were also considered and compared with evaporation and condensation temperature. In the present study the same has been computed as -30°C to -50°C evaporator temperature and 30°C, 40°C condenser temperature for ammonia and with intercooler COP is improved. The error may be due to approximation and could be agreed upon.

Conclusion

During extensive energy and exergy analysis with refrigerant discussed in results in a modified two stage vapour compression cycle, following conclusions are summarized below:

The COP of the system is improved by 4-5% with incorporation of heat recovery through intercooler. It is observed that COP decreases with increase in condenser temperature at a fixed evaporator temperature. Also, as the evaporator temperature increases COP increases and it is maximum (3.087) at evaporator temperature of 243 K and condenser temperature of 313 K. With the increase in condenser temperature the total exergy destruction increases at a fixed evaporator temperature.

This particular result is significant from the point of view of the application of hot water coming from the intercooler. At lower tonnage of the system, this hot water is useful for household works in cold climate condition. At high tonnage (100TR) of the system, it could be useful for process industries also as difference of the outlet and inlet temperature obtained is approximately 50°C.

Hence, it can be accomplished that the available waste heat of the intercooler of a two stage modified vapor compression system could be utilized and also results in overall increase of COP of the system. Hence, heat recovery through a water intercooler of a two stage vapor compression refrigeration system is found to be feasible and can be maximized by selecting optimum water flow rate, inlet water temperature, suitable operating conditions, and working fluid.

References

1. Arora RC (2006) Multi Stage Vapour Compression System, Refrigeration and Air Conditioning.

2. Arora CP (2006) Multi-pressure System: Refrigeration and air Conditioning.

3. Prasad M (2013) Vapour Compression Systems: Refrigeration and Air Conditioning.

4. Kotas TJ (1995) The Exergy Method of Thermal Plant Analysis. 3 Butterworths, London. pp: 57-98.

5. Dincer I, Cengel YA (2001) Energy, Entropy and Exergy Concepts and their Roles in Thermal Engineering. Entropy 3: 116-149.

6. Zubair SM, Yaqub M, Khan SH (1996) Second-Law-Based Thermodynamic Analysis of Two-Stage and Mechanical Sub-Cooling Refrigeration Cycle. Int J Refrig 19: 506-516.

7. Nikolaidis C, Probert D (1998) Exergy-Method Analysis of a Two-Stage Vapour Compression Refrigeration- Plants Performance. Journal of Applied Energy Department 60: 241-256.

8. Rahman MM, Rahman HY (2011) Performance of Newly Developed Integrated Space Conditioning and Domestic Water Heating Device. Journal of Energy and Environment 3: 23-27.

9. Torrela E, Cabello R, Sanchez D (2003) Second-Law Analysis of Two-Stage Vapours Compression Refrigeration Plants. Int J Exergy 7: 641-653.

10. Aprea C, Renno C (2004) Experimental Comparision of R-22 with R417A, Performance in a Vapour Compression Refrigeration Plant Subjected to a Cold Store. Energy Conversion and Management 45: 1807-1819.

11. Ouadha A, En-Nacer M, Imine O (2005) Exergy Analysis of Two- Stage Refrigeration Cycle Using Two Natural Substitutes of HCFC2. Int J Exergy 2: 14-30.

12. Arora A, Kaushik SC (2008) Theoretical Analysis of Vapour Compression Refrigeration System with R502, R404A, and R507A. International Journal of Refrigeration 3: 998-1005.

13. Cabello R, Torrela E, Sanchez D (2010) Comparative Evaluation of the Intermediate Systems Employed in Two Stage Refrigeration Cycles Driven by Compound Compressors. Int J Energy 35: 1274-1280.

14. Ballester SM, Macia JG, Corbera JM (2011) Optimum Performance of External Intercooling Two- Stage Cycle with Real Compressor Curves. Journal of Institute of Energy Engineering Valencia 32: 1282-1292.

15. Mishra RS, Jain V, Kachhwaha SS (2011) Comparative Performance Study of Vapour Compression Refrigeration System with R22/R134A/R410A/R407C/M20. International Journal of Energy and Environment 2: 297-310.

16. Mukesk K, Agrawal A, Matani G (2012) Refrigeration System Using Different Refrigerants. International Journal of Engineering and Innovative Technology 2: 227-275.

17. Soni J, Gupta RC (2013) Performance Analysis of Vapour Compression Refrigeration System with R404A, R407C, and R410A. Int J Mech Eng 2: 49-165.

18. Mishra RS, Sahni V, Chopra K (2014) Methods for Improving First and Second Law Efficiencies of Vapour Compression Refrigeration System Using Flash Intercooler With Eco-friendly Refrigerant. International Journal of Advance Research and Innovation 2: 50-64.

19. Mishra RS (2014) Methods for Improving Thermodynamic Performance of Vapour Compression Refrigeration System Using Twelve Eco-friendly Refrigerants in Primary Circuit and Nanofluid (Water-Nano Particle Based) in Secondary Circuit. International Journal of Engineering Technology and Advanced Research 4: 878-890.

20. Kumar GR (2014) Performance Analysis of Household Refrigerator with Alternate Refrigerants. International Journal of Innovative Research in Science, Engineering and Technology 3: 19-53.

21. Jerald L, Senthikumaran AD (2014) Investigation on the Performance of Vapour Compression system Retrofitted with Zeotropic Refrigerant R404A. Am J Env Sci 10: 35-43.

22. http://fchart.com/ees

23. Yonggao XS, Zhang X (2014) A Proposed Subcooling Method for Vapor Compression Refrigeration Cycle Based on Expansion Power Recovery. International Journal of Refrigeration 43: 50-61.

24. Upadhyay N (2014) Analytical Study of Vapor Compression Refrigeration System Using Diffuser and Subcooling. Journal of Mechanical and Civil Engineering 11: 92-97.

25. Yang M, Yeh RH (2015) Performance and Exergy Destruction Analyses of Optimal Subcooling for Vapor Compression Refrigeration System. International Journal Of Heat and Mass Transfer 87: 1-10.

PERMISSIONS

LIST OF CONTRIBUTORS

Jaakko Saastamoinen
VTT Technical Research Centre of Finland, Jyväskylä, Finland

Nabil N Atta, Amro A El-Baz, Noha Said and Mahmoud M Abdel Daiem
Department of Environmental Engineering, Zagazig University, Egypt

George Passas and Charles W Dunnill
Energy Safety Research Institute (ESRI), Swansea University, College of Engineering, Bay Campus, United Kingdom

Zhenglong Jiang, Yunfei Zhang and Kangning Xu
School of Marine Sciences, China University of Geosciences, Beijing 100083, China

Yajun Li
School of Energy Resources, China University of Geosciences, Beijing 100083, China

Renato Cataluña Veses, Zeban Shah, Pedro Motifumi Kuamoto, Elina B. Caramão, Maria Elisabete Machado
Instituto de Química, Federal University of Rio Grande do Sul, Brazil

Rosângela da Silva
Pontifical Catholic University of Rio Grande do Sul, RS, Brazil

Renato Cataluña Veses, Zeban Shah, Pedro Motifumi Kuamoto, Elina B. Caramão, Maria Elisabete Machado
Instituto de Química, Federal University of Rio Grande do Sul, Brazil

Rosângela da Silva
Pontifical Catholic University of Rio Grande do Sul, RS, Brazil

Darren Greetham, Nattha Pensupa and Gregory A. Tucker
Department of Bioenergy, University of Nottingham, UK

Jwan J. Abdullah
Department of Bioenergy, University of Nottingham, UK

Department of Environment, Salahaddin University-Erbil, Iraq

Chenyu Du
Department of Bioenergy, University of Nottingham, UK
Department of Environment, University of Huddersfield, UK

MKh Rumi, MA Zufarov, EP Mansurova and NA Kulagina
Institute of Material Sciences SPA, Physics – Sun, Academy of Sciences Republic of Uzbekistan, Tashkent, Uzbekistan

Rui-na Liu and Yong-feng Li
School of Forestry, Northeast Forestry University, Harbin, China

Ning Li, Jianhui Zhao and Nan-qi Ren
School of Municipal and Environmental Engineering, Harbin Institute of Technology, Harbin, China

Umish Srivastva
Indian Oil Corporation Limited, RandD Centre, Faridabad, Haryana, India

RK Malhotra
MREI, Faridabad, Haryana, India

SC Kaushik
Indian Institute of Technology Delhi, New Delhi, India

Nassereldeen Ahmed Kabbashi, Nurudeen Ishola Mohammed, Md Zahangir Alam and Mohammed Elwathig S Mirghani
Bioenvironmental Engineering Research Centre (BERC), Department of Biotechnology Engineering, Faculty of Engineering, International Islamic University Malaysia

Mugwang'a FK
Department of Physics, Pawni University, Kenya

Karimi PK, Njoroge WK and Omayio O
Department of Physics, Kenyatta University, Kenya

Temilola T Olugasa and Oluwafemi A Oyesile
Department of Mechanical Engineering, University of Ibadan, Ibadan, Nigeria

Georgina Izquierdo Montalvo and Alfonso Aragón Aguilar
Instituto de Investigaciones Eléctricas, Reforma 113, Col. Palmira, Cuernavaca, Morelos, CP, Mexico

F. Rafael Gómez Mendoza
Paseo Cuauhnáhuac 8532, Col. Progreso Jiutepec, Morelos, Mexico

Magaly Flores Armienta
Comisión Federal de Electricidad, GPG. Morelia, Michoacán, Mexico

Eduard Oró and Jaume Salom
Catalonia Institute for Energy Research, IREC, Spain

Alvaro Vergara
University of Freiburg, Friedrichstr 39, 79098, Freiburg, Germany
Pontificia Universidad Católica de Chile, Santiago, Chile

Sen Zhang
Science Island Branch of Graduate School, University of Science and Technology of China, China
China National Tobacco Quality Supervision and Test Center, Zhengzhou, China

Ping-huai Liu, Jiang-wei Wu and Qing Wang
College of Materials and Chemical Engineering, Hainan University, China

Arash Farnoosh and Fendric Lantz
IFP Énergies Nouvelles, IFP School, 228-232 Avenue Napoléon Bonaparte, F-92852 Rueil-Malmaison, France

Petit P, Aillerie M and Charles JP
Lorraine University, LMOPS-EA 4423, 57070 Metz, France

Nguyen TV
Lorraine University, LMOPS-EA 4423, 57070 Metz, France
Quang Ninh University of industry, Quang Ninh, Vietnam

Le QT
Quang Ninh University of industry, Quang Ninh, Vietnam

DI Ramadan Ali Abdiwe and Markus Haider
Vienna University of Technology, Institute for Exergy Systems and Thermodynamics, Austria

MMA Mahfouz
Electrical Power and Machines Department, Faculty of Engineering, Helwan University, Cairo, Egypt

Mahendra Kumar Trivedi, Rama Mohan Tallapragada, Alice Branton, Dahryn Trivedi, Gopal Nayak
Trivedi Global Inc., 10624 S Eastern Avenue Suite A-969, Henderson, NV 89052, USA

Rakesh K. Mishra and Snehasis Jana
Trivedi Science Research Laboratory Pvt. Ltd., Hall-A, Chinar Mega Mall, Chinar Fortune City, Hoshangabad Rd., Bhopal- 462026, Madhya Pradesh, India

Piotr Biernacki, Sven Steinigeweg and Wilfried Paul
EUTEC Institute, University of Applied Sciences Emden/Leer, Emden, Germany

Axel Brehm
Technische Chemie, Fk.V, Carl von Ossietzky Universität Oldenburg, Oldenburg, Germany

Kumar S, Ghaly AE and Brooks MS
Department of Process Engineering and Applied Science, Dalhousie University, Halifax, Nova Scotia Canada

Senthil R and Silambarasan R
Department of Mechanical Engineering, University College of Engineering Villupuram, Tamilnadu, India

Sisty Basil Massawe
Pan African University, Institute of Life and Earth Sciences (Including Health and Agriculture), University of Ibadan, Nigeria

AO Olorunnisola
Department of Agricultural and Environmental Engineering, University of Ibadan, Nigeria

A. Adenikinju
Department of Economics, University of Ibadan, Nigeria

Prashant Katiyar
Sam Higgin Bottom Institute of Agriculture, Technology and Sciences, Allahabad, India

Shailendra Kumar Srivastava
Department of Biochemistry and Biochemical Engineering, Jacob School of Biotechnology and Bioengineering, Sam Higgin Bottom Institute of Agriculture, Technology and Sciences (Deemed University), Allahabad,India

Vinod Kumar Tyagi
Department of Oil and Paint Technology, HBTI, Kanpur, India

M Mohamed Musthafa
School of Mechanical Engineering, SASTRA University, Thanjavur-613401, Tamilnadu, India

Aftab Anjum, Mohit Gupta, Naushad A Ansari and Mishra RS
Department of Mechanical Engineering, Delhi Technologial University, Shahbad Daulatpur, Main Bawana Road, Delhi, India

Index